The Lucy Mission

Harold F. Levison • Simone Marchi • Keith S. Noll
Editors

The Lucy Mission

Previously published in *Space Science Reviews, Topical Collection: "The Lucy Mission"*

Springer

Editors
Harold F. Levison
Southwest Research Institute
Boulder, CO, USA

Simone Marchi
Southwest Research Institute
Boulder, CO, USA

Keith S. Noll
NASA Goddard Spaceflight Center
Greenbelt, MD, USA

ISBN 978-94-024-2372-3

Cover illustration: A rendering of the Lucy spacecraft with a schematic view of the two clouds of the Jupiter's Trojan asteroids, which are target of the mission.
Credit: Image: Spencer/Marchi/Levison.
Spacecraft model: NASA's Goddard Space Flight Center Conceptual Image Lab.
Background image: JWST image 2022-041.

This Springer imprint is published by the registered company Springer Nature B.V.
The registered company address is: Van Godewijckstraat 30, 3311 GX Dordrecht, The Netherlands

If disposing of this product, please recycle the paper.

Contents

Space Science Reviews (2025) 221:119
https://doi.org/10.1007/s11214-025-01252-9

The Lucy Mission: First to the Jupiter Trojan Asteroids – Editorial

Keith S. Noll[1] · Harold F. Levison[2] · Simone Marchi[2]

Received: 14 October 2025 / Accepted: 14 October 2025 / Published online: 27 November 2025

NASA's Lucy spacecraft, launched in October 2021, is carrying out an ambitious mission designed to study the Jupiter Trojan asteroids — one of the largest dynamically distinct populations of objects in the Solar System that has not yet been explored by a spacecraft. The Trojans, which share Jupiter's orbit, are clustered near the stable L4 and L5 Lagrange points, 60 degrees ahead of or behind Jupiter. They have a distinct compositional mix compared to asteroids in the Main Belt, with a prevalence of dark, red objects. Our current understanding of the early history of the Solar System predicts that the Trojans were formed further from the Sun and were scattered inward from the protoplanetary disk during the era of giant planet migration. The Lucy mission was designed to test this hypothesis by encountering five different Trojan systems that consist of at least eight distinct bodies over the course of its 12-year flight.

In the time since these articles were written, the Lucy spacecraft has completed two flybys of Main Belt asteroids. In November 2023, Lucy encountered the small Main Belt asteroid (152830) Dinkinesh and discovered a unique contact-binary satellite. On 20 April 2025, Lucy had a close approach flyby of the asteroid (52246) Donaldjohanson which was found to be an elongated and bilobed body, similar in architecture to other observed small Solar System bodies. Donaldjohanson, a member of the Erigone family, is the first carbonaceous asteroid that is a collisional family member to be encountered by a spacecraft. Lucy is now cruising on its way to its first Trojan encounter in August 2027.

Lucy is the thirteenth mission selected under the aegis of NASA's Discovery program. Lucy is led by Principal Investigator Harold Levison at Southwest Research Institute in Boulder in partnership with multiple institutions. The spacecraft was built and integrated by Lockheed Martin Space in Waterton, Colorado. Instruments were built and supplied by Johns Hopkins Applied Physics Laboratory in Laurel, Maryland, the Goddard Space Flight Center in Greenbelt, Maryland, and Arizona State University in Tempe, Arizona. Navigation is led by KinetX Aerospace. The Lucy science team includes participants from numerous universities and research institutes.

A high-level overview of the scientific motivation and implementation for the Lucy mission is detailed in the article by Levison et al. drawing on and updating earlier descriptions of the overall project. In brief, this article describes why the Trojans are scientifically

✉ K.S. Noll
keith.s.noll@nasa.gov

1 NASA Goddard Space Flight Center, Greenbelt, MD, USA

2 Southwest Research Institute, Boulder, CO, USA

compelling generally and how the specific mission targets will accomplish Lucy's scientific goals. Fulfilling science goals requires mission objectives which are broken down into a series of requirements that specify the technical capabilities required for mission success and creating a framework for detailed observation sequences that must be constructed for each flyby encounter.

Altogether, the 15 articles in this collection capture the state of knowledge near the beginning of Lucy's journey. It is intended to serve as a baseline from which it will be possible to measure the impact of the mission. Four main themes comprise this collection: the current state of knowledge of the Trojan asteroids, mission design and operations, Lucy's instrumentation, and outreach.

1 The Trojan Asteroid Population

The origin of the Jupiter Trojan asteroids is at the center of the scientific goals of the Lucy mission. Rather than having formed in situ, it is now thought that that these bodies were captured into these orbits early in Solar System history at the end of a period of giant planet dynamical instability. The properties of Trojan objects themselves are most consistent with an origin the primordial Kuiper belt. Bottke and coauthors describe the current understanding of the Trojans in detail and outline how the Lucy mission will advance our understanding of planet formation and early Solar System evolution.

The surface geology of the Trojan asteroids is a key component of planned Lucy science and is described in the article by Marchi et al. Panchromatic, color, and spectral imaging will combine to carry out a wide range of investigations. Crater counts and morphology will be used to constrain the ages of terrains. Surface topography, mass wasting, boulders, pits and other geological features will inform a variety of investigations constraining composition, and internal structure. Comparisons between targets and with other spacecraft-imaged bodies will be employed to establish relationships between small body populations and the variety of processes that sculpt small bodies in the Solar System.

The bulk composition of Trojans is expected to reflect that of the protoplanetary disk at the time and location of their formation. The handful of Trojans binary systems whose orbits have been measured have revealed consistently low bulk densities of around 1000 kg m^{-3}. Those estimates, together with the presence of a spin barrier between 4 and 4.8 h rotation period, suggest that low densities are a general property of the population — a key link to low density populations found in the outer Solar System. The photometric properties of Trojans are diagnostic of both composition and macroscopic surface roughness. The spectral slopes and the phase functions appear to have a bimodal distribution, reinforcing the notion of two major compositional sub-populations among the Trojans. These important aspects of the Trojan population are described in the article by Mottola and colleagues.

An important part of planning the Lucy mission was to ensure encounters with objects that span the range of compositional types found in the Trojans. Planned investigations of surface composition will take advantage of the unique opportunity to examine compositionally distinct units on the surface and access to spectral ranges not accessible from the ground. As a background to this investigation, the article by Emery et al. describes the current state of knowledge of the Trojans from Earth-based spectroscopic investigations.

The Lucy mission will encounter five Jupiter Trojans during its mission with three of the five already known to be multiple systems. These include a near-equal-mass binary, a small and widely separated satellite, and one intermediate-size satellite system. The article by Noll et al. reviews the current state of knowledge of Trojan asteroid satellites in the context of similar satellite systems in other small body populations. The prospects for the detection of additional satellites as well as other near-body phenomena are considered. The scientific utility of satellites makes their observation with Lucy an important scientific priority for the mission.

2 The Lucy Mission: Design, Instruments, and Management

Cox and colleagues describe the design of the Lucy spacecraft, its heritage, and novel design components. Of particular relevance is the description of the uniquely large solar arrays, one of which failed to fully deploy after launch. Fortunately, after intensive modeling and laboratory testing, it was possible to further deploy the array to an almost-fully-open configuration that has proved to be adequate to meet the mission's needs.

The article by Olkin and others describes how mission operations – the trajectory, instrumentation, and flyby geometries, are designed to meet mission requirements. Over the course of the 12-year mission, the spacecraft will conduct five flybys of Trojan targets containing at least eight distinct bodies. In addition, two Main Belt asteroids provide opportunities to test mission operations, as well as serendipitous science. The complex combination of spacecraft maneuvers, target tracking, range-finding, and detailed science instrument operation sequences together comprise Lucy's mission operations.

Getting Lucy to the launchpad required more than technical acumen as described in the article by Douglas-Bradshaw and colleagues. Of the many challenges that arise in the building of a spacecraft, there is always the threat of "unknown unknowns". The Lucy mission was planned with a relatively short timeline in phase C/D – the development and integration phases leading up to launch. On top of this known challenge was the completely unanticipated interruption caused by the COVID-19 pandemic. Extraordinary mission management creativity was required to reach the narrow launch period in October 2021. The success of Lucy to this point is a testament to the unflinching dedication of the Lucy team.

3 Instrumentation

The Lucy spacecraft carries a complement of three scientific instruments, complemented by two spacecraft systems that together are used to meet the mission requirements. Lucy's instruments all have high heritage from previous missions and experienced team members with familiarity with the details of instrument operations. Each of the instruments is described in a separate article.

TTCam is a wide-angle camera that is an integral part of the onboard terminal tracking system. Its primary function is to solve for range during approach and to keep instruments pointed at the target during the flyby. However, its wide field of view adds a critical global imaging capability during the close flyby when the L'LORRI camera is able to observe only a small portion of the surface at a time. Bell and coauthors describe this in their article.

The L'LORRI instrument is Lucy's workhorse high-angular-resolution camera that will obtain the most detailed imagery of the targets as well as being used for optical navigation. Its design, operations, and calibration are described in the article led by Weaver.

L'Ralph is Lucy's spectrographic instrument and consists of two separate modules. MVIC is a visible and near-infrared imager fitted with five broad-band color channels and one panchromatic channel. LEISA is L'Ralph's infrared imaging spectrograph. Both are described in the article by Reuter and colleagues. Together MVIC and LEISA are designed to produce spectroscopic information that will be used to constrain the composition of Trojan surfaces.

L'TES is an infrared Michelson interferometer operating from 7–50 μm. At the distance of the Trojans, the primary mode of operation will be to determine spectral energy distributions from which temperature of the surface within the instrument aperture can be derived. L'TES is one of a long line of similar instruments that have flown — it is described in the article headed by Christensen.

Finally, radio science will be carried out by making use of Lucy's high gain antenna system. Doppler tracking during flybys will be used to determine the mass of the Trojan targets as described in the article by Levison et al.

4 Outreach

The Lucy spacecraft relies on three Earth Gravity Assist maneuvers over the course of the mission to modify the spacecraft trajectory. The first of these occurred on October 16th, 2022, providing one of the first opportunities to carry out instrument calibrations including imaging of the Earth and Moon. As part of the calibration Lucy obtained scientifically valuable thermal images of the DART mission impact into the Didymos system, from a unique geometry, 20 days before the Earth flyby. Spencer and coauthors describe the planning of these unique observations and their use as a channel to introduce the Lucy mission to the public.

Finally, there is the L'SPACE program (Lucy Student Pipeline and Competency Enabler), a student technical STEM workforce development program associated with the Lucy mission. L'SPACE recruits students from U.S. colleges and universities and has enrolled more than ten thousand since its inception in 2018. This program is described in detail in the article by Boonstra and colleagues.

5 Contents

OVERVIEW

1. Levison, HF, Marchi, S, Noll, KS et al (2025) NASA'S Lucy mission to Trojan asteroids: unraveling the history of the outer solar system. Space Sci Rev 221:63. https://doi.org/10.1007/s11214-025-01173-7

THE TROJAN ASTEROIDS

2. Bottke, WF, Marschall, R, Nesvorný, D & Vokrouhlický, D (2023) Origin and evolution of Jupiter's Trojan asteroids. Space Sci Rev 219:83. https://doi.org/10.1007/s11214-023-01031-4S
3. Marchi, S, Bell, JF, Bierhaus, B & Spencer, J (2023) Surface geology of Jupiter's Trojan asteroids. Space Sci Rev 219:44. https://doi.org/10.1007/s11214-023-00985-9
4. Emery, JP, Binzel, RP, Britt, DT et al (2024) Surface compositions of Trojan Asteroids. Space Sci Rev 220:28. https://doi.org/10.1007/s11214-024-01060-7
5. Noll, KS, Brown, ME, Buie, MW et al (2023) Trojan Asteroid satellites, rings, and activity. Space Sci Rev 219:59. https://doi.org/10.1007/s11214-023-01001-w
6. Mottola, S, Britt, DT, Brown, ME et al (2024) Shapes, rotations, photometric and internal properties of Jupiter Trojans. Space Sci Rev 220:17. https://doi.org/10.1007/s11214-024-01052-7

THE MISSION: DESIGN, OPERATIONS, AND MANAGEMENT

7. Olkin, C, Vincent, M, Adam, C et al (2024) Mission design and concept of operations for the Lucy mission. Space Sci Rev 220:47. https://doi.org/10.1007/s11214-024-01082-1
8. Cox, M, Cuddy, C, Schneider, J, et al (2025) The Lucy spacecraft. Space Sci Rev 221:94. https://doi.org/10.1007/s11214-025-01190-6
9. Douglas-Bradshaw, D, Bartels, A, Andrews, J et al (2025) Managing the Lucy mission. Space Sci Rev 221:91. https://doi.org/10.1007/s11214-025-01208-z

INSTRUMENTS

10. Weaver, HA, Wilson, JP, Conard, SJ et al (2023) The Lucy Long Range Reconnaissance Imager (L'LORRI). Space Sci Rev 219:82. https://doi.org/10.1007/s11214-023-01028-z
11. Reuter, DC, Simon, AA, Lunsford, A et al (2023) L'Ralph: a visible/infrared spectral imager for the Lucy mission to the Trojans. Space Sci Rev 219:69. https://doi.org/10.1007/s11214-023-01009-2
12. Bell, JF, Zhao, Y, Cisneros, E et al (2023) The Terminal Tracking Camera system on the NASA Lucy Trojan Asteroid discovery mission. Space Sci Rev 219:86. https://doi.org/10.1007/s11214-023-01030-5
13. Christensen, PR, Hamilton, VE, Mehall, GL et al (2024) The Lucy Thermal Emission Spectrometer (L'TES) instrument. Space Sci Rev 220:1. https://doi.org/10.1007/s11214-023-01029-y

OUTREACH

14. Spencer, JR, Bell, JF, Christensen, PR et al (2024) The First Lucy Earth flyby (EGA1). Space Sci Rev 220:3. https://doi.org/10.1007/s11214-023-01034-1
15. Boonstra, SK, Kretke, K, Garcia, D et al (2025) Technical STEM workforce development re-imagined: the Lucy Student Pipeline and Competency Enabler (L'SPACE). Space Sci Rev 221:28. https://doi.org/10.1007/s11214-024-01123-9

Declarations

Competing Interests The authors declare no competing interests.

Space Science Reviews (2025) 221:63
https://doi.org/10.1007/s11214-025-01173-7

NASA'S *Lucy* Mission to Trojan Asteroids: Unraveling the History of the Outer Solar System

Harold F. Levison[1] · Simone Marchi[1] · Keith S. Noll[2] · John R. Spencer[3] · Catherine B. Olkin[4] · Thomas S. Statler[5]

Received: 9 August 2023 / Accepted: 15 May 2025 / Published online: 3 July 2025

Abstract

Lucy is a NASA Discovery-class mission to send a highly capable and robust spacecraft to investigate primitive bodies near both the L_4 and L_5 Lagrange points with Jupiter; the Jupiter Trojan asteroids. This heretofore unexplored population of planetesimals from the outer planetary system holds vital clues to deciphering the history of the Solar System. Due to an unusual and fortuitous orbital configuration, *Lucy* will perform a comprehensive investigation that visits eight Trojans, including all the recognized taxonomic classes, a collisional family member and a near equal-mass binary. It will visit objects with diameters ranging from roughly 1 to 100 km. In particular, *Lucy* will perform flybys of (3548) Eurybates and its satellite Queta (L_4, C-type), (15094) Polymele and its currently unnamed satellite (L_4, P-type), (11351) Leucus (L_4, D-type), (21900) Orus (L_4, D-type), and the (617) Patroclus-Menoetius binary (L_5, P-types). This diverse array of targets will supply invaluable constraints on the formation and early dynamical evolution of the giant planets. In addition, *Lucy* will visit two main-belt asteroids, (152830) Dinkinesh and (52246) Donaldjohanson, in order to practice its encounters. *Lucy*'s payload suite consists of a color camera and infrared imaging spectrometer, a high resolution panchromatic imager, and a thermal infrared spectrometer. Additionally, two spacecraft subsystems will also contribute to the science investigations: the terminal tracking cameras will supplement imaging during closest approach and the telecommunication subsystem will be used to measure the mass of the Trojans. *Lucy* launched on October 16, 2021 and will have encounters with the Trojans from August 2027 until March 2033.

✉ H.F. Levison
hal.levison@swri.org

1 Department of Space Studies, Southwest Research Institute, 1050 Walnut St., Suite 300, Boulder, CO 80302, USA

2 NASA Goddard Spaceflight Center, Greenbelt, MD, USA

3 Southwest Research Institute, 1301 Walnut St., Suite 400, Boulder, CO 80302, USA

4 Muon Space, Mountain View, CA, USA

5 NASA Headquarters, Washington, DC, USA

1 Introduction

Jupiter Trojan asteroids have yet to be studied close up, making them one of the last accessible stable small body reservoirs in the Solar System that has not been visited by a spacecraft. Owing to the critical role they play in understanding the formation and evolution of the Solar System, Trojans have been a high priority for space missions for over a decade. This is evidenced by calls for their reconnaissance in the 2013 Planetary Decadal Survey, *Vision and Voyages*, that identified a Trojan survey as one of the highest priority planetary missions. The *Lucy* mission fulfills this goal.

Our mission is called *Lucy* in recognition of the influence that the primitive *Australopithecus afarensis* hominin fossil named 'Lucy' has had in advancing understanding of the history of our species. The *Lucy* name embodies the aspiration that this mission will similarly advance our understanding of the formation and evolution of our Solar System.

The Trojan Asteroids were long thought to be a population that formed near Jupiter's orbital distance representing the composition of the nebula at that distance (Barucci et al. 2002; Yoder 1979; Kary and Lissauer 1995; Marzari and Scholl 1998; Fleming and Hamilton 2000). Thus it is surprising that detailed Earth-based observations show the Trojans to be an inhomogeneous population (these differences are explored below). This unexpected diversity may be understood in the context of a class of models that has been developed within the last 20 years or so, which suggest that the objects currently found in the Trojan swarms were originally formed far beyond their current home (at $\sim 15-30$ AU), and were transported to their current locations by early orbital evolution of the giant planet orbits (Tsiganis et al. 2005, see Nesvorný 2018 for a review). According to this hypothesis, any objects originally trapped in the L_4 and L_5 Lagrange points were lost during the early phase of giant planet migration and were subsequently restocked with objects scattered in from the outer protoplanetary disk (Morbidelli et al. 2005, Nesvorný et al. 2013; Fig. 1). These ideas are explored in more detail in Bottke et al. (2023) in this collection. These models suggest that the observed diversity of Trojans is the result of the fact that they originated over a large range of heliocentric distances with varying physical and compositional conditions. Understanding the diversity of Trojans, by interrogating as many observables as possible, will allow us to determine whether these ideas are true, and if true allow us to constrain the orbital evolution of the giant planets. If *Lucy* proves these ideas incorrect, it will provide us with vital clues to develop new hypotheses. In any case, *Lucy* will allow us to disentangle multiple and complex factors of origin and subsequent evolution of the Solar System.

Much of what follows is based on the text in Levison et al. (2021a,b). In Sect. 2, we discuss the diversity of the Trojans in more detail. In Sect. 3, we describe *Lucy*'s remarkable trajectory. We present our targets in Sect. 4, and *Lucy*'s science goals and our formal requirements in Sect. 5. We conclude in Sect. 6.

2 Diversity of Trojan Asteroids

Small bodies are classified by the shape of their visible and near-infrared spectra (Bus et al. 2002; DeMeo et al. 2009). This classification scheme divides the majority of bodies in the main asteroids belt into two main groupings. The S-complex, which are generally thought to be stony with low volatile and carbon content, are most common in the inner asteroid belt. C-complex, which are expected to be richer in volatiles and carbon, and more common in the middle and outer main belt.

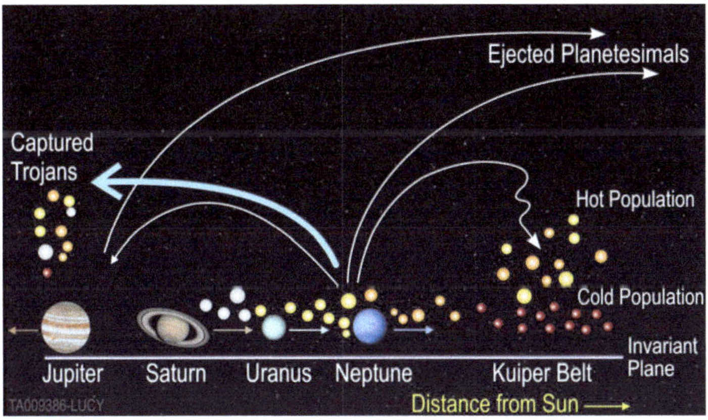

Fig. 1 A schematic of the effects of planet migration on the small body reservoirs in the outer Solar System (i.e., the Trojans and Kuiper belt). Objects initially in a disk are gravitationally scattered by the migrating planets. Most are ejected from the Solar System, but a small fraction are captured in the Trojan swarms and Kuiper belt. Any compositional gradients in the original population (represented by arbitrary colors) are mixed in the Trojans. The degree of mixing is model dependent

Trojans, which lie outside of the main belt, are members of the C-complex. In particular, they have predominately D- and P-type spectra, although there are a small number of C-types, see Fig. 2. C-type asteroids are characterized by featureless visible-near infrared spectra with a neutral or "gray" slope. P- and D- types are similarly featureless, but have progressively redder spectral slopes. There is a general trend seen in the Solar System as a whole that small bodies are found to be redder with increasing heliocentric distance. This trend can be attributed to the interaction of solar UV radiation with organic compounds and volatile materials, which are not stable close to the Sun (Wong and Brown 2016). It is consistent with the idea that the redder Trojans formed at greater heliocentric distances and were embedded in the swarms during the migration of the planets. DeMeo and Carry (2014) estimate that less than 10% of large Trojans are C-types, while the rest are approximately evenly divided between D-types and P-types (Fig. 2).

The range of spectral diversity can also be seen in the Trojans' visible-near-infrared colors, which fall into two broad groupings (Szabó et al. 2007; Roig et al. 2008; Emery et al. 2011; Grav et al. 2012), (Figs. 3 and 4). This color bimodality is a property that is qualitatively shared with Centaurs and smaller, dynamically excited trans-Neptunian objects (TNOs) (Wong and Brown 2017; Nesvorný et al. 2020a) suggestive of possible genetic links.

Another way of visualizing the diversity of the Trojan asteroids can be seen in Fig. 5, which shows the distribution of visible and 3.4 μm (W1) albedos of a large number of Trojans as determined by WISE spacecraft (Grav et al. 2012). Trojans are very dark, the brightest has a visible albedo of about 0.13, which is about the same as aged asphalt. However, Trojans can be as dark as 3.5%, which is darker than newly poured asphalt or black acrylic paint. This factor of 4 range in albedo implies that the surface composition of the objects are likely very different from one another.

As described above, the Trojan diversity is likely due to the fact that they formed at different heliocentric distances, and were delivered to the current location close to the orbit of Jupiter as the outer Solar System evolved. So, Trojans likely harbor objects that formed throughout the outer Solar System. If true, a detailed study of these objects, similar to the

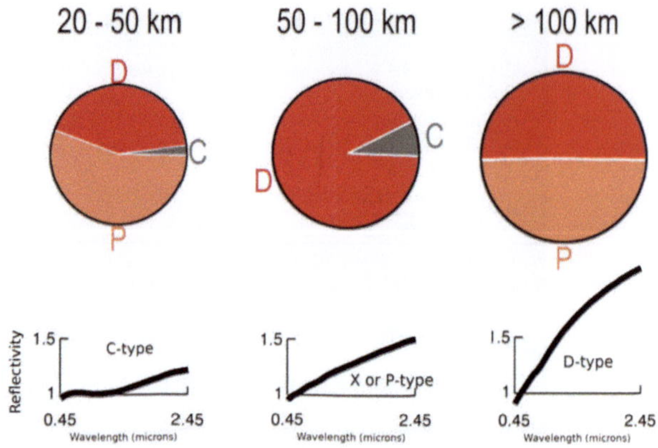

Fig. 2 Top: The fraction of different spectral types as a function of size in the Trojan swarms DeMeo and Carry (2014). Bottom: The idealized reflectance spectra of primitive asteroids according to the Bus–DeMeo classification scheme (DeMeo et al. 2009). This taxonomy is based on the spectra of asteroids from 4500 Å to 2.45 microns. C-type asteroids are gray in color. D-type are very red, while P-types are moderately red. Reproduced from Levison et al. (2021b)

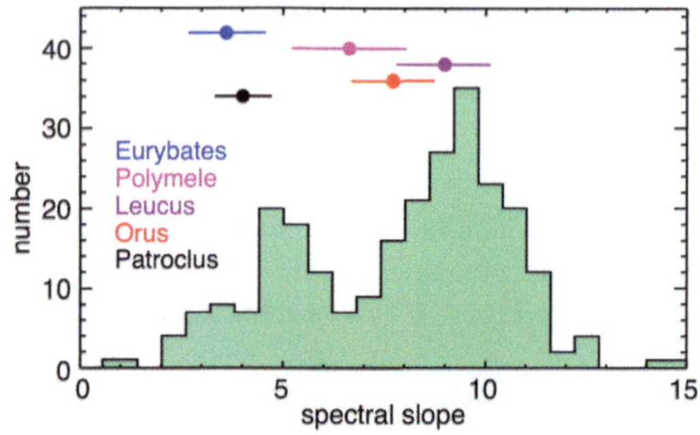

Fig. 3 The visible color slopes of Trojans are bimodal, as shown in this histogram (Wong et al. 2014; Schemel and Brown 2021). The *Lucy* targets, which will be discussed in detail in Sect. 4, fall as indicated, sampling this color range. The black symbol is the combined, unresolved color of the Patroclus–Menoetius binary

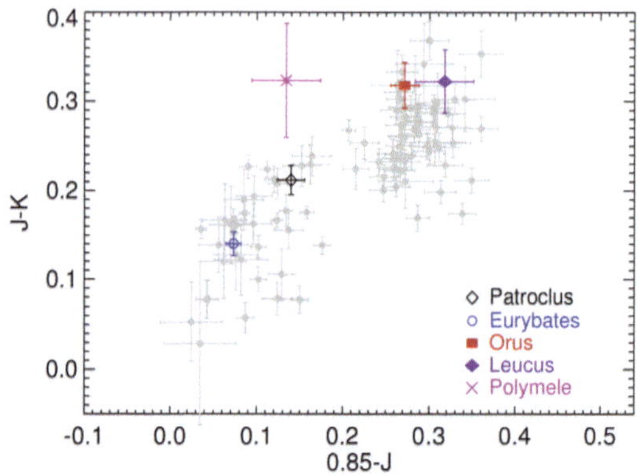

Fig. 4 An infrared color-color plot of Trojan asteroids with the *Lucy* targets identified (Emery et al. 2011). Reproduced from Levison et al. (2021a)

one that *Lucy* will perform, will provide vital clues about their formation locations, which, in turn, will constrain the dynamical process that delivered them to 5 AU. Such a study could, in fact, show that these ideas are false and that Trojans formed near their current location. Only time, and perhaps *Lucy*, will tell. In any case, Trojans present us with a unique opportunity to constrain planet formation and evolution models. It is only by sampling their diversity, however, that their true scientific potential can be realized.

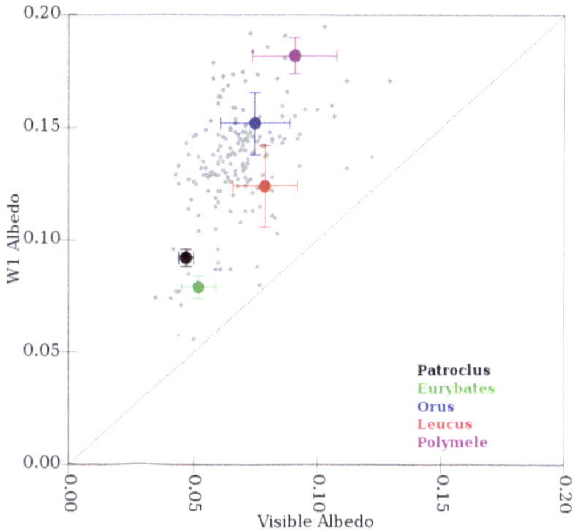

Fig. 5 The distribution of Trojan albedos in two wavelength bands: visible and 3.4 microns (W1). The small symbols represent all Trojans with well-measured values (Grav et al. 2012), while the large symbols represent the *Lucy* targets. Note that our targets cover the same range as the population as a whole. They will be described in more detail in Sect. 4

3 The Trajectory of *Lucy*

Through a series of favorable circumstances, we have constructed a comprehensive tour for *Lucy* that covers the richness of the Trojan diversity, see Sect. 4 for more detail. In particular, *Lucy* will perform a series of targeted close flybys of eight known Trojans and two main belt asteroids (Fig. 6). *Lucy*'s 11.6 year journey began with launch on October 16, 2021. It launched at 5:34 AM local time, which was at the very beginning of its 21-day launch period (Fig. 7). It launched on an Atlas V 401 launch vehicle (LV) from Cape Canaveral directly onto an Earth escape trajectory. Initially, the LV inserted the spacecraft into a heliocentric orbit with a period of $\sim$1 year and an aphelion slightly above 1.16 AU.

Lucy uses two Earth gravity assists (EGAs) and a Deep Space Maneuver (DSM) to reach Jupiter's L_4 Trojan swarm. On its way to the Trojans the spacecraft intercepts two main-belt asteroids, which are used as in-flight demonstrations of the spacecraft's capability and to reduce mission risk prior to the Trojan encounters. The first was (152830) Dinkinesh, which took place on November 1, 2023. (152830) Dinkinesh is a S-type asteroid (Bolin et al. 2023; de León et al. 2023) with an diameter of only $\sim$ 1.0 km (Mottola et al. 2023a). It was named after an Ethiopian nickname for the *Lucy* fossil, ድንቅነሽ, which means 'you are marvelous.' The second encounter is with (52246) Donaldjohanson (named in honor of the discoverer of the *Lucy* fossil) and occurs on April 20, 2025. Although this encounter is intended as a rehearsal for the main Trojan events, Donaldjohanson is also interesting as a scientific target because of its membership in the $\sim 130 + 30$ Myr-old Erigone collisional family (Nesvorný et al. 2006; Vokrouhlický et al. 2006; Bottke et al. 2015b).

After encountering Donaldjohanson, *Lucy* will perform another DSM that will align the trajectory of the spacecraft so that it can execute two L_4 Trojan encounters that are only 34 days apart. The first will be with (3548) Eurybates and its satellite Queta (Noll et al. 2020c). This encounter will occur on August 12, 2027. Next *Lucy* will flyby (15094) Polymele on September 15, 2027. Stellar occultations have recently shown that Polymele has a relatively large satellite in its equatorial plane (Buie et al. 2022). After the Polymele encounter, *Lucy* will execute a third DSM. *Lucy* will finish its exploration of the L_4 swarm by encountering (11351) Leucus and (21900) Orus on April 18, 2028 and November 11, 2028, respectively. Between Leucus and Orus, *Lucy* will perform its last DSM. This not only enables the Orus encounter, but sets the altitude of the third EGA, which is discussed below.

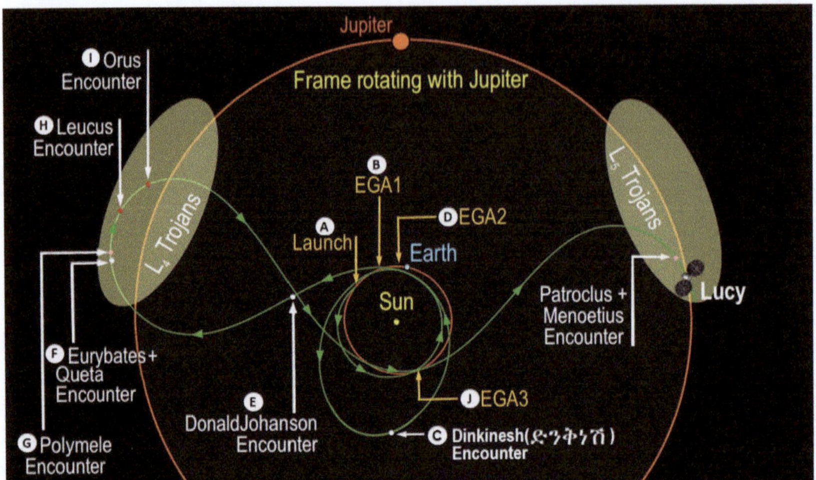

A	10/16/21	Launch 5:34 am
B	10/16/22	EGA 1 Altitude = 350 km
C	11/1/23	Dinkinesh (S-type) Encounter v = 4.4 km/s, ϕ = 114° $d_{C/A}$ = 450 km
D	12/13/24	EGA 2 Altitude = 344 km
E	4/20/25	DonaldJohanson (C-type) Encounter v = 13.4 km/s, ϕ = 15° $d_{C/A}$ = 1000 km
F	8/12/27	Eurybates + Queta (C-type) Encounter v = 5.8 km/s, ϕ = 81° $d_{C/A}$ = 1000 km
G	9/15/27	Polymele & Satellite (P-type) Encounter v = 6.0 km/s, ϕ = 82° $d_{C/A}$ = 434 km
H	4/18/28	Leucus (D-type) Encounter v = 5.9 km/s, ϕ = 104° $d_{C/A}$ = 1000 km
I	11/11/28	Orus (D-type) Encounter v = 7.1 km/s, ϕ = 126° $d_{C/A}$ = 1000 km
J	12/27/30	EGA 3 Altitude = 626 km
K	3/3/33	Patroclus + Menoetius (P-type) Encounter v = 8.8 km/s, ϕ = 63° $d_{C/A}$ = 1620 km (P)

Fig. 6 The trajectory of *Lucy* in a frame that rotates with Jupiter as it orbits the Sun. *Lucy*'s orbit is shown in green, while the orbits of Jupiter and Earth are shown in orange. Since this is not an inertial frame, the trajectory of *Lucy* does not appear Keplerian. The approximate region of space that the Trojans occupy are shown in brown. Major mission events are labeled alphabetically in chronological order. For each encounter, the relative velocity of the spacecraft with respect to the target, v_{rel}, the approach phase angle, ϕ, and the close approach distance, $d_{C/A}$, are listed. The (P) indicates that $d_{C/A}$ is for Patroclus

Lucy then takes advantage of a fortunate geometric characteristic of our planetary system – an object on an orbit with a perihelion near Earth's orbit and an aphelion near Jupiter's has an orbital period of roughly half that of Jupiter. This means that it is possible for a spacecraft to visit both the L_4 and L_5 swarms on consecutive orbits. Following the last L_4 encounter with Orus, *Lucy* returns to Earth for its third and final EGA that sets up an encounter with the final Trojan targets, (617) Patroclus and its binary companion, (617) I Menoetius, on March 2, 2033. The encounter date can be adjusted using the final EGA altitude to arrive in the L_5 cluster so that the relative position of Patroclus and Menoetius optimizes the viewing geometry. At the end of mission, *Lucy* will be parked on the orbit that it used for its entire exploration of the Trojans — one crossing both Jupiter's and Earth's orbit.

Lucy will encounter its Trojan targets when it is near aphelion (see Fig. 6), and so its velocity with respect to the Sun will be much slower than that of the asteroids. The Trojans will basically pass *Lucy* as they orbit. As a result, *Lucy* will approach its targets at a high phase angle, pass through low phase angles during the encounter (we will target the line between the asteroid and the Sun), and depart at high phase angles once again. Most of the encounters (Eurybates, Polymele, and Patroclus) will occur before *Lucy* reaches aphelion.

Fig. 7 Photos leading up to the *Lucy* launch on October 16, 2021

Thus, it will be moving away from the Sun and approach phase angles will be less than 90° — between 63° and 82°. These targets will be seen in a gibbous configuration during approach. *Lucy* will be moving toward the Sun during the Leucus and Orus encounters, and thus will approach these targets from the dark side. The resulting approach phase angles are 104° and 126°, and the asteroids will appear as a crescent on approach. The impact that this geometry will have on the concept of operations of the encounters can be found in Olkin et al. (2024) in this collection.

In all, *Lucy* will study eight Trojans and two main belt asteroids. Not only will its innovative trajectory allow it to visit more asteroids than any other spacecraft in history, but it was designed to cover the known diversity of the Trojan swarms. Such a study is necessary to take full advantage of the Trojans to constraining models of the formation and evolution of the Solar System.

4 *Lucy's* Trojan Targets

For reasons discussed in Sect. 2, the overarching theme of the *Lucy* mission is that in order to understand a diverse, never-before-visited population, it is necessary to have a broad sample of objects that span as many axes of physical properties as possible. The Trojans that *Lucy* will encounter with the trajectory described above fulfill this diversity requirement as measured by a range of target properties. The large red dots in Fig. 5 represent the visible and W1 albedos of our targets. As can be seen in the figure, the distribution of our targets covers the distribution of the Trojans as a whole (gray symbols). Similarly, the *Lucy* targets also well represent the distribution seen in Trojan colors (Figs. 3, 4). Two of the targets are in the redder of the two color groups (Orus and Leucus), while two are clearly in the bluer group (Eurybates and Patroclus). Polymele, which is the smallest of the targets (see Table 1), is likely in the bluer group, but there is some uncertainty.

In addition to convering the necessary diversity, the *Lucy* targets are interesting for their own sake. We now discuss each of these in turn.

Table 1 *Lucy* Target Properties: Size and Light curve

Object	shape (km)	T_{rot} (hr)	Amplitude (mag)	Density (gm/cm^3)	Reference
(3548) Eurybates	$[77.5\times71.3\times61.8]^a$	8.7027283 ± 0.0000029	0.20	0.8	[1,2]
(3548) I Queta	$1.2(4)^b$				[3]
(15094) Polymele	$[27.0\times24.4\times10.4]^c$	11.5(1)	0.09		[4]
(15094) I S/2022	$>5^d$	-	-		[5]
(11351) Leucus	$[60.8\times39.2\times27.8]^a$	445.683(7)	0.61		[6,7]
(21900) Orus	$[70.7\times63.0\times51.4]^a$	13.486190 ± 0.000017			[1,2]
(617) Patroclus	$[127\times117\times98]^c$	102.78432(15)	0.07	0.9	[8,9]
(617) I Menoetius	$[117\times108\times90]^c$	102.78432(15)			
(152830) Dinkinesh	1.01 ± 0.18^e	52.66(15)	0.39		[10]
(52246) Donaldjohanson	3.9^f	251(1)	1.7		[4,11,12]

Uncertainty in last digit(s), when known, is shown in parentheses; [a]The maximum orthogonal dimension of a convex shape fit to lighcurves; [b]relative brightness to Eurybates assuming same albedo; [c]Fit to occultations., [d]Longest occultation chord, [e]Based on an assumed albedo of 0.3, [f]Based WISE thermal radiometry.

[1] Mottola et al. (2016), [2] Mottola et al. (2023b), [3] Noll et al. (2020c), [4] Mottola et al. (2023) in prep, [5] Buie et al. (2022), [6] Mottola et al. (2020), [7] Buie et al. (2018), [8] Buie et al. (2015), [9] Grundy et al. (2018), [10] Mottola et al. (2023a) [11] Masiero et al. (2012), [12] Ferrais et al. (2021)

Table 2 *Lucy* Target Properties: Orbital

Object	a^a (AU)	e^a	i^{ab} (°)	q^a (AU)	Encounter (JD-245000)	$R_{enc}{}^d$ (AU)
heliocentric						
(3548) Eurybates	5.20	0.089	8.06	4.73	11630	5.67
(15094) Polymele	5.17	0.095	12.99	4.68	11664	5.69
(11351) Leucus	5.29	0.064	11.56	4.95	11880	5.65
(21900) Orus	5.13	0.037	8.47	4.94	12087	5.31
(617) Patroclus	5.22	0.139	22.05	4.49	13660	5.38
(152830) Dinkinesh	2.19	0.112	2.10	1.94	9975	2.26
(52246) Donaldjohanson	2.38	0.187	4.42	1.94	10786	2.07
satellite/binary[c]	(km)					
(3548) I Queta	2321	0.098	132.75		11630	5.67
(15094) I S/2022	204^f	0^g	0^g		11664	5.69
(617) I Menoetius	688.5	0^e	164.11		13660	5.38

[a]Osculating orbital elements, [b]Equatorial J2000, [c]Binary/satellite orbit, *a* given in km, [d]Heliocentric distance at encounter, [e]fixed zero eccentricity solution (Grundy et al. 2018), [f]Levison et al. (2023, in prep) [g]Assumed to be tidally damped.

(3548) Eurybates (Encounter date 8/12/27): Eurybates is a C-type Trojan in the L_4 swarm with a volumetric effective diameter and WISE-determined visible albedo of 70 km and 5.2%, respectively. Its rotational period is 8.7 hours. Using a comprehensive set of photometric data, Mottola et al. (2023b) derive a convex shape model for Eurybates with axis

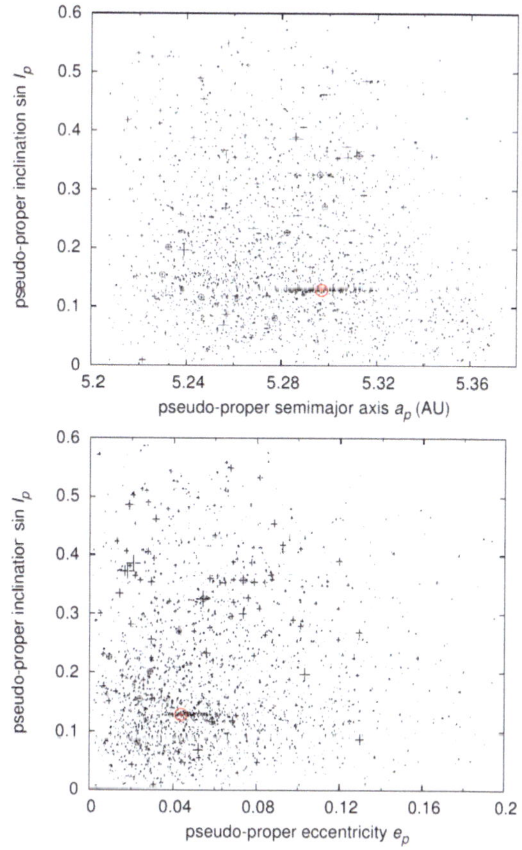

Fig. 8 The distribution of L_4 Trojans in a space defined by the pseudo-proper semi-major axis (a_P), pseudo-proper eccentricity (e_p) and pseudo-proper inclination (I_p). These pseudo-proper elements are defined in Brož and Rozehnal (2011). The sizes of the crosses represent the relative diameters of bodies. The asteroid (3548) Eurybates, around which a significant cluster is visible, is shown in red. This figure is reproduced from Brož and Rozehnal (2011)

ratios of $A/B = 1.08$ and $B/C = 1.16$. (The axes dimensions are $A > B > C$, and C is coincident with the spin axis.) The uncertainties in A/B are $\pm\sim 10\%$, and for B/C are $\pm\sim 20\%$.

Eurybates is particularly interesting because it is the largest member of the largest catastrophic collisional family in the Trojan swarms (Brož and Rozehnal 2011). This is illustrated in Fig. 8, which shows the distribution of proper orbital elements for objects in the L_4. Fragments from a collisional breakup are ejected at relative velocities that are small compared to their orbital velocities about the Sun. As a result, all the fragments have similar heliocentric orbits. Collisional families are therefore seen as concentrations in these distributions. As the figure shows, there is such a structure associated with Eurybates (shown in red).

Marschall et al. (2022) estimates that Eurybates family is the result of a collision between a $\sim$100 km target and $\sim$35 km impactor. This implies that Eurybates, itself, only contains $\sim$34% of the total mass of the family. The fact that this fraction is small indicates that impact was powerful enough to catastrophically disrupt the parent body. *Lucy* will be the first spacecraft to study such an object up close. By accomplishing this *Lucy* will supply vital information about collisions and the planet formation process.

A small satellite of Eurybates, now named Queta, was detected in Hubble Space Telescope images obtained on two dates in September 2018 (Noll et al. 2020c). A third detection in January 2020 confirmed its existence. Additional observations, including non-detections, have been used to determine Queta's orbit, which has a period of 83 days, a semi-major axis of 2200 km, and an eccentricity of 0.06. The orbit pole is well constrained. The satellite has a diameter of roughly 1.2 km assuming that its albedo is the same as that of Eurybates (Brown et al. 2021). Importantly, the existence of a satellite makes it possible to determine the system mass and to estimate the density of Eurybates prior to the *Lucy* encounter. The current best orbit, combined with shape provided in Table 1, gives a density of $\sim 0.8\ \mathrm{g/cm^3}$. A well-determined orbit will also make it possible to plan targeted observations of Queta at the time of the *Lucy* encounter.

Perhaps it is not surprising that Eurybates has been determined to have a small satellite. Both small satellites and rings can form as a body re-accretes after a catastrophic collision (Durda et al. 2004). It is surprising, however, that Queta's eccentricity is so small because collision models predict large eccentricities. The cause of the small eccentricity is a mystery.

Given its status as a member of this family, it is somewhat surprising that Eurybates is a C-type because objects of this type are rare in the Trojan swarms. Less than 10% of Trojans fall into this class (DeMeo et al. 2009), and many of these are members of the Eurybates family. It is unlikely that the one object that was involved in such a rare collision is independently a member of rare spectral class. This suggests a connection between Eurybates' gray color and the collision. Two possible explanations come to mind. First, P- and D-types could be so porous that they can absorb an major impact without generating a lot of ejected fragments. Second, whatever substance makes P- and D-types red could be only skin-deep, and their interiors could be gray. When such an object is broken open due to a collision, the gray material will dominate the fragments. If the second hypothesis is correct, then young craters on the surface of P- and D-type asteroids should be gray in color — a situation that *Lucy* will be able to see.

(15094) Polymele (Encounter date 9/15/27): Polymele is in the L_4 swarm. Very little was known about this object before NASA selected the *Lucy* mission for flight because it is faint and difficult to observe from the Earth. It has a volumetric effective diameter of 19 km and a WISE-determined visible albedo of 0.09 (Grav et al. 2012). The spectrum of Polymele has yet to be classified, but it has a Sloan Spectral Slope of $2.6 \times 10^{-5}\ \text{Å}^{-1}$ that is consistent with it being a P-type asteroid (Roig et al. 2008).

Because of its small size, the *Lucy* team has invested significant resources into observing Polymele because we needed to ensure that we could satisfy our Level 1 science requirements (see Sect. 5). This includes intensive astronmetric and photometric campaigns, as well as five successful occultation expeditions (see Mottola et al. 2024, this collection). These data have allowed us to fill in many of the details about this object. Assuming that Polymele is ellipsoidal, its axes have been found to be $A = 27.0 \pm 2.0$ km, $B = 24.4 \pm 1.6$ km, and $C = 10.4 \pm 1.6$ km. With a $B/A = 0.91$ and $C/A = 0.40$, Polymele is almost a highly-flattened oblate spheroid and so is reminiscent of the cold classical Kuiper belt object, Arrokoth (Spencer et al. 2020). Recent planetesimal formation models predict highly-flattened oblate objects are the natural outcome of the collapse of a rotating cloud of meter-sized objects (Nesvorný et al. 2021). This suggests that, like Arrokoth, Polymele is a remnant of the first generation of macroscopic bodies in the Solar System. See Bottke et al. (2023) in this collection for a more detailed discussion.

A satellite of Polymele was serendipitously discovered as Polymele occulted a star on March 27, 2022. The discovery was confirmed during a stellar occultation on February 4, 2023. The occultation data suggest that the satellite, (15094) I S/2022, is roughly 5 km is diameter, although it is clearly not spherical. The *Lucy* team has given (15094) I S/2022 the nickname 'Shaun'. The projected separation between Polymele and Shaun is roughly 200 km during both events. A satellite this close to Polymele should have tidally evolved onto a circular orbit that lies in the primary's equatorial plane on a timescale of $\lesssim 400$ Myr (Nesvorný et al. 2020b, also see Noll et al. 2023 in this collection).

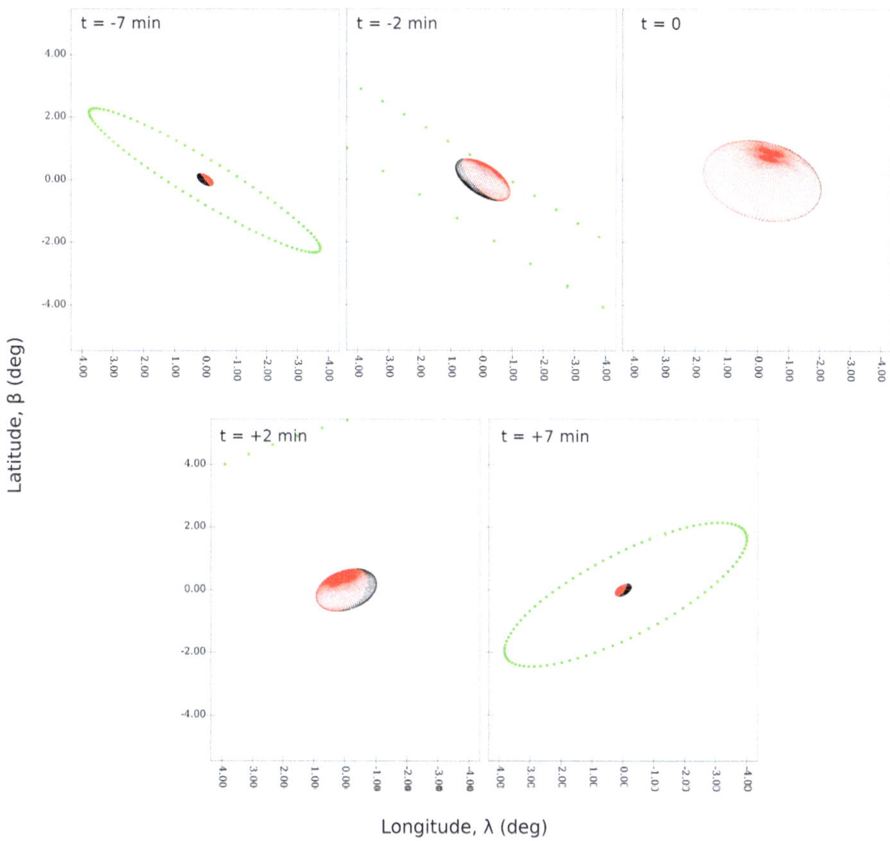

Fig. 9 The view of Polymele as seen from the spacecraft during its encounter. The time relative to the instant of close approach is listed in each panel. The sunlit region of Polymele is shown in red, while the shadowed region is in black. Polymele is assumed to be a ellipsoid with principal axis of [27.0×24.4×10.4] km. Polymele's north pole can be seen as the cyan dot in the left-most panel. The possible locations of Shaun are shown in green, assuming an orbital semi-major axis of 204.4 km

By combining the projected shapes of Polymele observed during the occulations with the position of Shaun with respect to Polymele, and assuming that Shaun is on a zero-inclination and zero-eccentricity orbit, we have found that the semi-major axis of its orbit is 204.4 ± 2.6 km. We do not have enough data to fit a period, but assuming that Polymele's density is between 0.7 and 1.3 gm/cm^3, we find that Shaun's orbital period is between 11 and 18 days.

The analysis employed above allowed us also to determine the position of the rotational pole for Polymele, which is the same as the pole of Shaun's orbit for reasons described above. The combination of the pole position and Polymele's flattened, nearly-oblate shapes creates a challenge for *Lucy* (Fig. 9). For the other encounters, *Lucy* will see the largest fraction of the target's lit surface on the asymptote with the smallest phase angle. Since the surface brightness of a target decreases with increasing phase angle, *Lucy* will see the majority of the lit surface when the surface is at its brightest. As Fig. 9 shows, this will not be true for Polymele — most of its surface will been seen at phase angles larger than 90°

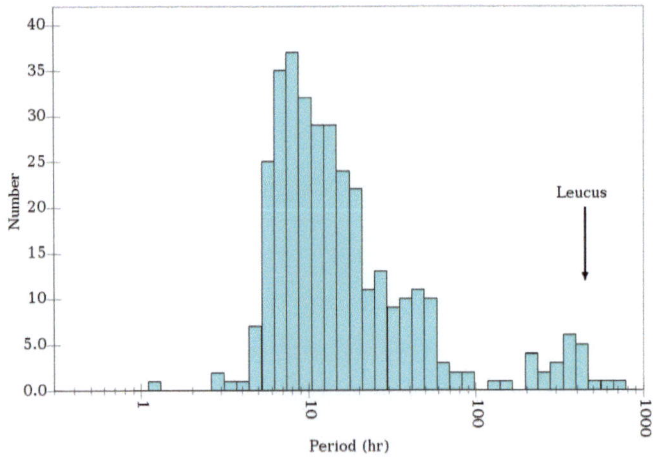

Fig. 10 The observed rotational period distribution of Trojan asteroids in the Asteroid Light Curve Database (Warner et al. 2021). The location of Leucus, which has a rotational period of 446 hours, is shown

when the surface brightness will be faint. This needs to be accounted for in *Lucy*'s science planning.

(11351) Leucus (Encounter date 4/18/28): Leucus is a Trojan in the L_4 swarm with a volumetric effective diameter of 40 km and a WISE-determined visible albedo of 8% (Grav et al. 2012). The spectrum of Leucus has yet to be classified, but it has a 3.4 μm albedo of 12% (Grav et al. 2012) and a Sloan Spectral Slope of 9.7×10^{-5} Å^{-1}, which is consistent with it being a D-type asteroid (Roig et al. 2008). The *Lucy* team has successfully determine limb profiles for Leucus during five stellar occultations. These results combined with photometric measurements (Mottola et al. 2020), allow us to fit an ellipsoidal shape for Leucus: $60.8 \times 39.1 \times 27.8$ km. All these values are full axes diameters.

Leucus has rotational period of 446 hours (Buie et al. 2018), making it one of the longest rotational periods in the Solar System. One surprising aspect of the Trojan population in general is that the distribution of their rotation periods is bimodal with roughly 8% of them having anomalously long rotational periods (Szabó et al. 2017; Ryan et al. 2017, see Fig. 10). A natural explanation for these slow rotators is that they are members of binary systems that de-spun due to tides, and so *Lucy* might discover a satellite when it arrives. However, Nesvorný et al. (2020b) point out that such a binary system might not have survived the to the current epoch.

(21900) Orus (Encounter date 11/11/28): Orus is an object in the L_4 swarm with a volumetric effective diameter and WISE-determined visible albedo of 61 km and 0.075 (Grav et al. 2012), respectively. Its rotational period is 13.5 hours. The spectrum of Orus has yet to be classified, but its 3.4 μm albedo and near-inferred (NIR) color (Emery et al. 2011) are consistent with it being a D-type asteroid. The *Lucy* team has successfully observed the limb of Orus during three stellar occultations. These results combined with photometric measurements, allow us to fit an ellipsoidal shape to Orus with full axes $[70.7\times63.0\times51.4]$ km Mottola et al. (2023b).

Figures 3 and 4 show that the colors of Trojan asteroids at both visible and near-infrared wavelengths, as determined from telescopic observations, fall into two distinct color groups: a "red" group analogous to the asteroidal D-type (and comparable to "gray" Centaurs) and a "less-red" group analogous to the asteroidal P and C-types (Szabó et al. 2007; Roig et al.

2008; Emery et al. 2011; Grav et al. 2012; see Emery et al. 2024 in this collection). In this regard, Orus and Eurybates are a very interesting pair for understanding the diversity of Trojans. These two objects have similar sizes (volumetric effective diameters of 70 and 61 km, respectively, cf. Table 1), orbital eccentricities (0.09 and 0.04, respectively), inclinations (8.1° and 8.5°, respectively, cf. Table 2), but very different colors ($0.85\mu - \mathrm{J} = 0.07$ and 0.27, respectively, cf. Table 1). The similar sizes and orbits guarantee that these objects have experienced similar solar radiation and collisional environments — at least since their formation and/or delivery to their current orbits. The color differences mean that they represent very different physical/chemical classes. These two objects, therefore, allow us to compare these two Trojan types, which represent the extremes of the population, without the complications associated with evolutionary and environmental differences. As a result, they make a excellent case study for *Lucy*.

(617) Patroclus (Encounter date 3/2/33): The Patroclus system is a near-equal mass binary in the L_5 swarm consisting of two P-type asteroids. The visible albedo of the pair is estimated to be 0.05. This is a doubly-synchronous binary, where the rotation period of each of the asteroids is the same as their mutual orbital period of 104 hours, with a semi-major axis of the orbit is 680 km (Grundy et al. 2018).

Shapes of the binary members have been determined through a combination of stellar occultations and photometric lightcurves (Buie et al. 2015). In particular, the occultation observations of the 2013 event returned a total of 12 chords across the two bodies. These data allowed ellipsoids to be fit to the bodies. The dimensions of Patrolcus were estimated to be 127 × 117 × 98 km, and Menoetius were 117 × 108 × 90 km.

The Patroclus–Menoetius binary is remarkable because of its similarity to the numerous binaries observed in the transneptunian object (TNO) population. The Cold Classical TNOs, a subset of TNOs with low eccentricity, low inclination orbits that are not in resonance and are not suffering gravitational encounters with Neptune, have a particularly high abundance of near-equal mass binaries (Noll et al. 2008, 2020a, also see Noll et al. 2023 in this collection). The high abundance of binaries in this population including weakly bound wide systems suggests that the Cold Classicals are a nearly undisturbed remnant of the protoplanetary disk. The New Horizons images of the Cold Classical TNO Arrokoth (Stern et al. 2019), itself a contact binary, strongly supports this point of view. Models of planetesimal formation predict the formation of near-equal mass binary systems as a natural outcome of rotation of the collapsing cloud of meter-sized objects (Nesvorný et al. 2010).

The existence of a potentially primordial binary in the Trojans is a remarkable opportunity to study directly what may be an example of the most common building blocks of planetary systems in general. Moreover, the survival of the PM binary, formed in a massive protoplanetary disk and then scattered to the Jupiter L_5 constrains the timing of the scattering event (Nesvorný et al. 2018).

5 *Lucy's* Science Requirements

The *Lucy* mission has 17 Level 1 science requirements (*R*.1 – *R*.17; Table 3) that specify the Trojan targets of the mission and the science data to be collected. In addition to target diversity, which is discussed above, the science objectives break down into four areas:

1. *Interior Structure and Bulk Properties*: *Lucy* will determine the volume and mass of each of its target. The goal is to determine the densities of these bodies.

Table 3 *Lucy* Level 1 Requirements

R.	Descriptor	Requirement	Section
1	Target Selection	Targets: Patroclus, Meneotius, Eurybates, Leucus, Polymele, and Orus	Sect. 2
2	Shape and Geology	Pan images: full rotation spaced by 1/25 to 1/13 of a rotation	Sect. 5.1
3	Shape and Geology	Pan images: series of phase angles separated by 15°– 25°	Sect. 5.1
4	Elevation Models	Pan images: area $\geq$ 100 km^2; resolution $\leq$ 200 m, two stereo emission angles	Sect. 5.2
5	Landform Degradation	Pan images: area $\geq$ 500 km^2; equator to 60° lat.; resolution $\leq$ 100 m	Sect. 5.2
6	Cratering	Pan images: area $\geq$ 700 km^2; resolve craters d > 7 km	Sect. 5.2
7	Cratering	Pan images: area $\geq$ 10 km^2; resolve craters d > 70 m	Sect. 5.2
8	Satellites	Search: satellites d $\geq$ 2 km, $p_v > 0.04$ within R_{S_Z}	Sect. 5.4
9	Compositional Units	Color images: full rotation spaced by 1/6 to 1/3 of a rotation	Sect. 5.3
10	Exposed Materials	Color images: area $\geq$ 700 km^2; resolution $\leq$ 1.5 km	Sect. 5.3
11	Exposed Materials	Color images: area $\geq$ 150 km^2; resolution $\leq$ 600 m	Sect. 5.3
12	Spectral Range	Spec: spectral range 1.0 – 3.8 µm	Sect. 5.3
13	Spectral Resolution	Spec: detect features with spectral depth $\geq$ 4% and width of $\geq$ 70 nm	Sect. 5.3
14	Compositional Variations	Spec: full rotation spaced by 1/6 to 1/3 of a rotation	Sect. 5.3
15	Compositional Variations	Spec: resolution (r) and areal coverage (A) satisfy $r \leq 2(A/1470.6)^{0.473}$ km	Sect. 5.3
16	Mass Determination	Targets: mass accuracy $\leq 25\%$, for $\rho \geq 1$ g/cm^3	Sect. 5.1
17	Thermal Properties	Thermal: 1 unilluminated surface; 3 sunlit with 1 at < 30° from subsolar point	Sect. 5.3

NOTE—Abridged versions of full Level 1 Requirements are shown. Panchromatic and Spectra are abbreviated as 'Pan' and 'Spec', respectively.

2. *Geology*: *Lucy* will image the surfaces of its targets at a variety of resolutions to study their complex surface geologic processes at multiple scales. This includes investigating structures ranging from surface sub-meter blocks and regolith to large-scale cratering and mass wasting. Our goal is to measure the crater size-frequency distribution (hereafter SFD) over 2 orders of magnitude.
3. *Surface Composition and Thermal Properties*: *Lucy* will acquire resolved imagery of the asteroids' surfaces at panchromatic, color, and NIR wavelengths. The goal is to investigate possible surface variations of photometric properties. *Lucy* will also collect resolved NIR spectra from 1 to 3.6 µm to search for the presence of ices, crystalline silicates, hydrated minerals, and organics. *Lucy* will also measure the thermal inertia of it targets.
4. *Satellites and Rings*: *Lucy* will search for satellites and rings in orbit about its targets.

Each of these topics will be covered in a subsection below.

5.1 Interior Structure and Bulk Properties

The interior structure and bulk properties of *Lucy*'s targets are addressed by requirements *R.*2, 3, & 16. Together, these can be used to determine the density, one of the key diagnostics that will test hypotheses of the origin and dynamical evolution of the Trojan asteroids. *Lucy* will determine the mass of its targets from the gravitational field interaction with the spacecraft. Volume will be determined from *Lucy* imaging complemented by Earth-based measurements. Mass and volume uncertainties (±25% and ±22% respectively) are allocated to meet the science goal of determining the density to ±33% for $\rho = 1\,\text{g}/\text{cm}^3$. Estimates of the density can be used to constrain the bulk composition, porosity, and interior structure of the *Lucy* targets.

Mass: Lucy will directly measure the masses, M, of its targets using radio-Doppler tracking of the spacecraft with the X-band telecommunications system. Mass determination by radio-Doppler tracking and ranging is a powerful and precise direct measurement (Anderson 1971; Pätzold et al. 2001) and has been performed at many flybys of small bodies and satellites (Anderson et al. 1992; Andert et al. 2010; Pätzold et al. 2011, 2014). *Lucy* must target Polymele at a close-approach distance of 434 km to ensure that the mass determination meets the ±25% uncertainty allocation (*R*.16), although *Lucy* will be able to determine the orbital period of Shaun during its encounter (see discussion about the search for satellites below). The system masses for the Patroclus–Menoetius binary ($M_{sys} = 1.41 \pm 0.03 \times 10^{18}$ kg; Grundy et al. 2018) and Eurybates and its satellite Queta ($M_{sys} \approx 1.4 \times 10^{17}$ kg; Noll et al. 2020b; Brown et al. 2021) have been determined from their orbits, but *Lucy* will validate and, likely, reduce the uncertainties of these determinations.

Volume and Density: The volume of *Lucy*'s targets will be determined by obtaining panchromatic images with *Lucy*'s LOng Range Reconnaissance Imager (L'LORRI, see Weaver et al. 2023 in this collection) and its Terminal Tracking Cameras (TTCam, see Bell et al. 2023 in this collection). It is important to note, however, that volume determination from imaging is limited to the portion of the flyby when the target is resolved. As an example, this time period is about one week for Eurybates. During this time *Lucy* will obtain images of each target every 1/13 rotation and measurements of limb profiles and stereo-photogrammetric analysis will be assembled to construct an overall 3-D shape model for volume determination (*R*.2,3). A fraction of each object, determined by the spin pole orientation, may not be illuminated during the *Lucy* flybys. Leucus' slow-rotation further reduces the fraction of the object that will be sufficiently-resolved by *Lucy* to constrain the volume. As a result, *Lucy* has no requirement to measure the volume to any particular level of accuracy. Instead, we require that the observations needed to determine the volume be taken (*R*.2,3). Our comprehensive occultation program, discussed above, will be used to supplement the volume determination of the *Lucy* targets.

By measuring mass and volume of its targets, *Lucy* will determine the density of each body. We estimate that in most cases the density will be known to better than ±33% for objects with densities of $\rho = 1\,\text{g}/\text{cm}^3$ (see Mottola et al. (2024), this collection for further discussion).

5.2 Geology

Spacecraft exploration has revealed that small bodies in the Solar System have diverse and geologically complex surfaces. Resolved observations have been gathered for small bodies that span orders of magnitude in size, have variable surface compositions, and belong to distinct populations from near-Earth asteroids to trans-Neptunian objects. These data reveal complex surface geologic processes at all scales, from the generation of surface sub-meter blocks and regolith (Lauretta et al. 2019) to regional compositional variations (DeSanctis et al. 2015; Grundy et al. 2020), to large-scale cratering and mass wasting (Marchi et al. 2012; Jaumann et al. 2012; Buczkowski et al. 2016), and highly irregular shapes (Murdoch et al. 2015; Spencer et al. 2020). This multifaceted complexity hints at a range of formation and geological processes controlling landforms, shapes, and surface compositional variability. *Lucy* will probe these processes with a comprehensive set of observations.

Digital Terrain Models: 3D models representing an asteroid's terrain will provide valuable information to study the geology of the *Lucy* targets. The role of global shape models has been discussed in Sect. 5.1, also see Mottola et al. (2024), this collection. Here we focus on the more localized, higher resolution Digital Terrain Models (DTMs) that will be obtained at variable spatial and vertical resolutions and coverage depending on the specific target and flyby conditions including spacecraft velocity and close approach distance, phase angle, target pole position and spin rate. These investigations will rely on panchromatic imaging from multiple angles (*R*.4) making it possible to derive DTMs over at least 100 km^2 and at a horizontal spatial resolution less than 200 m. The anticipated vertical resolution is on the same order as the horizontal resolution. DTMs will be derived from stereo imaging obtained by L'LORRI for each primary target to study in detail local geology, including crater shapes, exposed vertical layering in crater walls, mass wasting, lineaments, and other possible landforms, such as pits resulting from the sublimation of subsurface volatiles.

Regional Mosaics: The ensemble of L'LORRI panchromatic imaging will also be used to produce regional mosaics with various degrees of resolution, and larger coverage than local DTMs (*R*.5-7). These data will enable the assessment of crater morphology and landform degradation (e.g., due to ejecta blanketing or other mass wasting processes) over an area of at least of 500 km^2 and at a resolution better than 100 m (*R*.5).

Crater morphology is diagnostic of subsurface bulk properties (e.g., material strength and stratification), and overall degradation due to impact-induced seismic shaking and regolith displacement (Marchi et al. 2015). Additional mass wasting processes may be associated with subsurface volatiles (e.g., water ice) mobilized by impact excavation/heating as observed at various locations on Ceres (Sizemore et al. 2017). *Lucy* has a requirement to investigate the variation of crater morphology as a function of latitude (*R*.5). Unlike rocky asteroids previously observed by spacecraft, the Jupiter Trojan asteroids orbit at a solar range that results in low enough near-surface temperatures to permit water ice to be stable on billion year timescales, and the latitudinal dependence of surface temperature may drive latitudinal-dependent geological crater morphology and probe existence and distribution of subsurface volatiles (Combe et al. 2019).

Regional mosaics will also be used to derive the crater size-frequency distributions both at local scale (10 km^2, craters larger than 70 m in diameter; *R*.6), and at regional scale (low resolution, 700 km^2, craters larger than 7 km in diameter; *R*.7). Impact craters are a nearly universal feature on all small bodies observed by spacecraft to date (Marchi et al. 2015). Craters enable a suite of investigations: crater population statistics constrain the surface age from the crater-retention age (e.g., Hartmann 1966), their morphologies are a function of target properties (Housen et al. 2018), and smaller craters are a proxy for the Trojan asteroid population that is too small to be observed by telescopes. Current telescopic measurements of the Trojan asteroid size-frequency distribution extend to objects with diameters $d \geq 2$ km (Yoshida and Terai 2017). Assuming a crater-to-impactor diameter ratio of 10:1, craters smaller than $d \approx 20$ km are derived from an unsampled portion of the size-frequency distribution. *Lucy*'s requirements that address the crater SFD by resolving craters across more than two orders of magnitude (from 70 m to 7 km in diameter) will provide constraints on surface crater retention ages, and past collisional evolution (Marchi et al. 2015). The shape of the SFD of the small impacting population is an important constraint for understanding the accretion process in the outer Solar System. In particular, changes in slope of the SFD is a direct result of the physical strength of these objects, and how it varies with size (Bottke et al. 2015a).

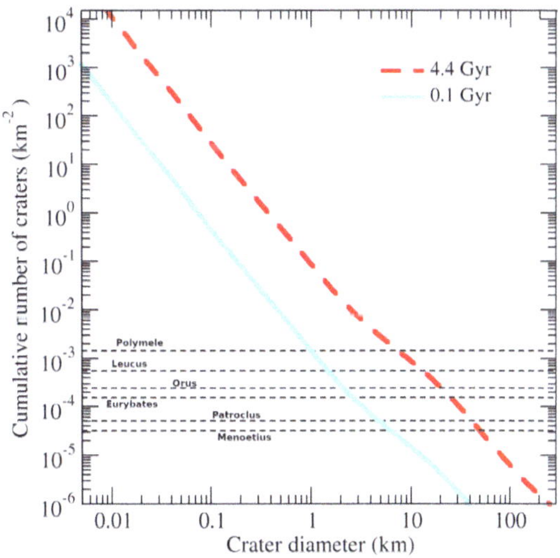

Fig. 11 The cumulative number of craters per unit area on Trojan asteroids for surfaces of 4.4 Gyr (red dashed line) and 100 Myr (blue solid line) assuming the current impact probability and velocity of the Trojans, and using the impactor size-distribution Bottke et al. (2015a) and impact rate temporal evolution O'Brien et al. (2014) from the Main Belt. We also adopt the scaling law of Holsapple and Housen (2007) for a target strength of 10^6 N/m^2. The horizontal dashed lines indicate the crater density at which there should be one crater per hemisphere for each *Lucy* target

A preliminary collisional model (Fig. 11) indicates that Trojan asteroids with a surface age as young as 100 Myr should have accumulated approximately two craters larger than 70 m per km^2 (that is, approximately 20 craters per 10 km^2). Roughly 100 $\times$ more craters are expected to have formed over the last 4.4 Gyr (not taking into account crater preservation and erasing processes). In addition, detection of 70 m craters would constrain Trojan asteroid SFD down to diameters of $\sim$ 7 m (a size scale unobservable by other means), which will enable an evaluation of the importance of non-gravitational effects on the stability of Trojans (Hellmich et al. 2019). Similar inferences of impactor populations from craters have been performed for MBAs and Cold Classical TNOs (Marchi et al. 2014; Singer et al. 2021) enabling comparisons of these distinct small body populations. These issues are further discussed by Marchi et al. (2023) in this collection.

5.3 Surface Composition and Thermal Properties

Panchromatic Imaging: During the flybys of the targets, *Lucy* will acquire resolved imagery of the asteroids' surfaces from a wide range of illumination and observation geometries. Measurements of the brightness distribution on the surface under such varying conditions will enable us to establish a photometric model for the target and to look for correlations between the model parameters and the physical properties of the surface regolith (Li et al. 2019). Spatial variations of the photometric properties will be put in the context of the morphology and geology of the targets, in order to identify possible evolutionary processes (Schröder et al. 2017). Near-closest-encounter L'LORRI images will occur at low-phase angles, which will aid in the characterization of the opposition effect (Belskaya and Shevchenko 2000), while pre- and post-encounter far-field imaging at large phase angles will help constrain the degree of unresolved roughness of the surface. Furthermore, the establishment of a photometric model will represent a prerequisite for the analysis, interpretation and comparison of heterogeneous multi-color and spectral data, described below, by allowing their representation under a reference illumination geometry. An accurate photometric model will also allow us to provide precise estimates for the albedo and phase integral of the targets. In addition to being diagnostic of the composition of the surface, those quantities are crucial for establishing the thermal balance and to assess the thermal evolution of the objects.

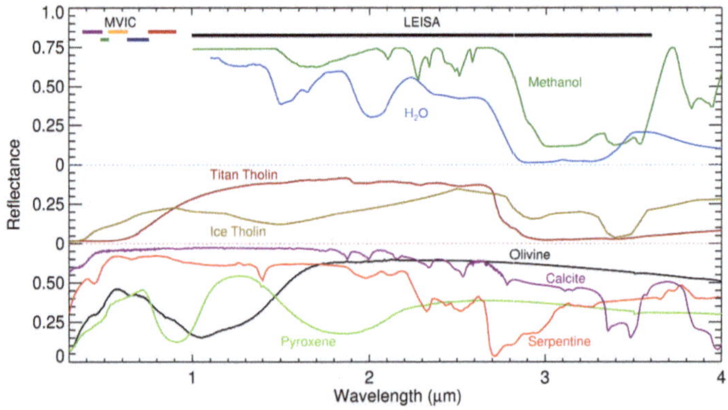

Fig. 12 The spectral range of the two components of the L'Ralph instrument (MVIC and LEISA) and spectra of species of ices, tholins and minerals which maybe visible on Trojan asteroids. Reproduced from Levison et al. (2021a)

Color Imaging: Multi-wavelength imaging in the visible and near-infrared is a proven way for investigating compositional heterogeneity and regolith properties on planetary surfaces at relatively high spatial resolution (Chapman 1996; Olkin et al. 2017; DellaGiustina et al. 2020). The Multispectral Visible Imaging Camera (MVIC, see Reuter et al. 2023 in this collection) will image the *Lucy* targets through 5 color filters (violet 377–489 nm; green 478–525 nm; orange 522–632 nm; phyllosilicate 627–758 nm; and near-IR 753–914 nm). MVIC observations will be made over a range of spatial resolutions. Low resolution observations at different rotational phases will determine the global distribution of color units. Observations at the regional ($\leq$ 1.5 km resolution) and local ($\leq$ 600 m resolution) scales will be used to search for small regions of freshly exposed materials and correlations of color with geologic features.

Some primary surface compositional units can be delineated by color imaging (Thangjam et al. 2013; Nathues et al. 2016). For example, a relevant feature for low-albedo asteroids is the absorption at 700 nm in Fe-bearing phyllosilicates (Hamilton et al. 2019; Vilas 1994). *Lucy*'s "phyllosilicate" filter is designed specifically to search for this band (Fig. 12).

The space environment can modify the surfaces of airless bodies. The spectral effects of these interactions vary due to primary surface composition and style of interaction, but generally include changes in albedo and color (e.g., Pieters and Noble 2016; Brunetto et al. 2015). MVIC color imaging can therefore be an effective means of investigating surface processing on *Lucy*'s targets. Trojan asteroids have been hypothesized to have an icy or differently processed interior overlain by an ice-free irradiation mantle (Guilbert-Lepoutre 2014; Wong and Brown 2015). If so, color maps of the surface may reveal exposures of fresh, subsurface material that can be used to constrain bulk composition (Chapman 1996). Indeed, *R*.6 and *R*.7 were specifically designed to see craters younger than 100 Myr, assuming the impact model shown in Fig. 11.

For example, as described above, Trojan asteroids fall into two distict color groupings — "red" and "less-red" groups. (Figs. 3, 4). It is notable that whereas red Trojans dominate the population at large diameters, the fraction of less-red objects increases at smaller sizes (Wong et al. 2014). Wong and Brown (2015) suggest that both color groups have similar interior colors, and that impacts on red Trojans expose their less-red interiors. If this scenario is correct, we might expect fresh surface units on the red *Lucy* targets (Orus and Leucus) to have a less-red color than surrounding terrain. If abundant subsurface ice exists (Guilbert-

Lepoutre 2014), spectral maps of fresh exposures on the *Lucy* targets will test this hypothesis (*R*.10, 11).

Near Infrared Spectroscopy: Compositions of primitive objects reflect the conditions in the solar nebula in the region in which they formed providing clues to the structure of the proto-planetary disk. Determining the compositions of Trojan asteroids is, therefore, important for understanding Solar System history. Mapping surface compositions enables investigation into potential surface processing from exposure to the space environment during an object's dynamical evolution, as well as the identification of exposed fresh material. The materials most important for understanding Trojan histories H_2O and other ices, organics, anhydrous minerals, and hydrated minerals – have diagnostic absorption features in the near infrared (NIR).

The Linear Etalon Imaging Spectral Array (LEISA) will measure NIR spectral images of *Lucy* targets from 1 to 3.6 μm with a sensitivity sufficient to detect absorptions deeper than 4% (*R*.12; see Reuter et al. 2023 in this collection). The spectral resolution of 10 nm will resolve absorption features of ices, organics, anhydrous minerals, and hydrated minerals that occur in the LEISA spectral range (*R*.13; Fig. 12). The center of the OH feature at 2.8 μm and the 3.4 μm organic absorption feature will be resolved with 10 and 20 elements respectively which are adequate for detection and characterization (De Sanctis et al. 2017). Observations when the targets do not fill the LEISA FOV will be used to search for global-scale compositional variations on the surfaces (*R*.14). Regional and local scale observations will be used to search for compositional anomalies resulting from freshly exposed materials, as a window into the interior compositions of the *Lucy* targets (*R*.15).

Thermal Infrared Observations: Surface temperatures provide important input for understanding the stability of materials detected (or not detected) on planetary surfaces (Spencer and Denk 2010; Tosi et al. 2014; Capria et al. 2014), and temperature anomalies often indicate interesting surface processes (Howett et al. 2010). Temperature maps are therefore valuable companions to compositional and geologic maps, to investigate correlations and anomalies. Observations in the thermal infrared, particularly at multiple times of day, enable determination of thermal inertia, which, in turn, gives insight in to the grain size and porosity of the regolith (Rozitis et al. 2020).

Thermal inertia estimates have been published for a few Trojan asteroids, and they are comparable to or lower than the lunar value ($\leq$100 J $m^{-2}K^{-1}s^{-1/2}$; Mueller et al. (2010)). Spectral emissivity peaks in the 10- and 20-μm spectral regions indicate fine-grained (∼1 μm) silicates, in a high porosity regolith (Emery et al. 2006; Vernazza et al. 2012), consistent with the low thermal inertias. Compacted regions of the regolith and exposures of ices could lead to regions of different thermal inertias than their surroundings. Widespread ices do not appear to be present on the surfaces of Trojan asteroids. However, Guilbert-Lepoutre (2014) finds that Trojans could support water ice ∼10 m below a dusty mantle at their equators, and as close as 10 cm to the surface in polar regions or even at the surface in isolated cold traps. If ice exists below the regolith, we may expect to see it in exposures of subsurface material on the *Lucy* targets.

The *Lucy* Thermal Emission Spectrometer (L'TES) is a point spectrometer that will measure thermal flux spectra of the *Lucy* targets over the range 7–100 μm (see Christensen et al. 2024 in this collection). From these spectra, surface temperatures will be determined with an accuracy $\leq$2 K for temperatures $\geq$75 K. L'TES is required to observe the thermal flux at $\geq$4 locations, including at least one nightside measurement and at least one observation within 30° of the sub-solar point (*R*.17).

5.4 Satellites and Rings

The discovery of (243) Ida's satellite Dactyl during the *Galileo* flyby in 1993 (Belton et al. 1996) was the first confirmed detection of a satellite for any small body other than Pluto. Since that time, there has been an explosion in the number and variety of satellites that have now been found in every small-body population (Merline et al. 2002; Margot et al. 2015; Noll et al. 2008, 2020a, see Noll et al. 2023 in this collection). Asteroids are also known to exhibit activity (Jewitt et al. 2015) and rings have been found around the TNO Haumea, the Centaur Chariklo, and possibly the Centaur Chiron (Ortiz et al. 2015; Braga-Ribas et al. 2014; Ortiz et al. 2017). Taken altogether, the possibility of satellites and other near-asteroid material must be considered during any small body flyby. Three of the *Lucy* mission targets, Patroclus, Eurybates, and Polymele, are already known to have satellites (Merline et al. 2001; Noll et al. 2020c; Buie et al. 2022), thus the need to plan for satellite observations as part of the *Lucy* mission flybys is essential.

Satellites are stable within a fraction of the Hill sphere of the system, with the stable inner third of the Hill radius sometimes referred to as the Szebehely radius (Szebehely 1967, 1978). The Hill radius is given by $R_{\rm Hill} = a(M_1/3M_\odot)^{1/3}$ where a is the heliocentric semi-major axis and M_1 and $M_\odot$ are the masses of the primary and the Sun respectively. The vast majority of known small body satellites orbit at distances from the primary of just a few percent or less of the Hill radius. At Trojan distances from the Sun and for a density of $\rho = 1\ {\rm g/cm^3}$, the Szebehely radius is equivalent to $R_{Sz} \approx 210\ r_1$ (where r_1 is the asteroid radius). The region of stable orbits extends inwards to the Roche limit at $R_{\rm Roche} \approx 2.44\ r_1$. The known rings around three small bodies all have $r_{\rm ring} \approx 3\ r_1$. *Lucy* will travel well within the Hill sphere of each Trojan target with close approach distances of $s_{\rm min} \geq 30\ r_1$.

Lucy's requirements include a search of the full Szebehely sphere for satellites with $d_{\rm sat} \geq$ 2 km. This search will be accomplished by *Lucy* using L'LORRI (*R*.17). This search will occur on either inbound or outbound from close approach depending on where the solar phase angle is lowest. The search will occur in two separate phases, a distant search and a close-in search to cover both the full Szebehely sphere and satellites or rings that are at small separations from the primary. During the encounter, there will be many images obtained for other purposes that will also be available to search for and track any satellites or other orbital material that may be present.

6 Inventory of the Instrumentation Used for Remote Sensing

The characteristics of the scientific instruments introduced above were directly derived from the *Lucy*'s Level 1 requirements. In particular, *Lucy* uses five instruments for its scientific observations:

- L'LORRI (*Lucy* LOng Range Reconnaissance Imager): This instrument provides the highest-resolution imaging for the *Lucy* mission. The instrument was built at the Johns Hopkins University Applied Physics Lab. It is a panchromatic telescope with a primary mirror with a diameter of 20.8 cm. The system has a focal length of 262 cm. Each pixel subtends 5 µrad. See Weaver et al. (2023) in this collection for a more complete description.
- L'Ralph: This instrument is a color imager and short-wavelength infrared imaging spectrometer, which was built at Goddard Spaceflight Center. The L'Ralph instrument has a 3-mirror anastigmat design f/6 with a 75 mm aperture. It has two distinct focal planes and

a wavelength dependent beamsplitter transmits light to each. As a result, it can be considered two independent instruments despite the fact that they share part of their optical train:

- MVIC (Multi-spectral Visible Imaging Camera): The MVIC focal plane consists of 6 Time Delay Integration (TDI) charge-coupled devices (CCD) on a single substrate. There are 5 color channels and 1 panchromatic channel with passbands as shown in Table 4. With a TDI CCD, signal is built up as the scene is scanned over the focal plane in coordination with the charge being transferred between rows of the detector. Each pixel subtends 29 μrad.
- LEISA (Linear Etalon Imaging Spectral Array): The LEISA's detector is covered with three linear variable filters to disperse the light from 0.95 to 3.95 μm. For each image one direction is spatial and the other is spectral. L'Ralph's scan mirror sweeps the image of the target across the LEISA detector in the spectral direction to build up a image cube which has a full spectrum of each observed surface element of the asteroid. In the spatial direction, the pixels subtend 80 μrad. The spectral resolution of LEISA is 10 nm. This resolution allows for *Lucy* to sample the 70 nm wide polycyclic aromatic hydrocarbons (PAH) absorption feature at 3.29 μm which is a class of organic molecules of interest and accessible in the LEISA spectral range.

More information about L'Ralph can be found in Reuter et al. (2023) in this collection.

- L'TES: This is a Fourier transform infrared point spectrometer built at Arizona State University. L'TES consists of a 15.2 cm diameter Cassegrain telescope, a Michelson interferometer with chemical vapor deposited diamond beamsplitter, and an uncooled, deuterated Lalanine doped triglycine sulfate (DLATGS) pyroelectric detector. It measures the thermal infrared emission from each Trojan asteroid to obtain the temperature of the asteroid's surface. The L'TES instrument collects data from 6–75 μm. The 50% encircled energy of the instrument subtends 6.5 mrad. See Christensen et al. (2024) in this collection for a more complete description.
- TTCams (Terminal Tracking Cameras): *Lucy* has two wide-field, panchromatic cameras built by Malin Space Science Systems. TTCams' main purpose is to provide on-board autonomous pointing to each Trojan near closest approach. This system is part of the spacecraft's guidance and control system and will also collect monochromatic (475–625 nm bandpass) images that address our science objective to determine the shape of the Trojan asteroids. The wide FOV of the TTCams (11° × 8.2°) allows imaging of the entire sunlit fraction of each of the Trojan asteroids during the entire flyby. The cameras have a pixel size of 2.2 μm. The telescope has a focal length of 29.7 mm and a entrance aperture of 10 mm. See Bell et al. (2023) in this collection for a more complete description.
- HGA (High-Gain Antenna): *Lucy*'s Radio Science Investigation uses the spacecraft's X-band telecommunication system to measure the Doppler shift of the spacecraft. The mass of the Trojans imparts a delta-v on the spacecraft. The concept of operations to support the Doppler shift measurements entails measuring the Doppler signature of the spacecraft before and after the flyby to determine the change in velocity.

Table 4 lists the important instrument specifications. It also provides a map linking the instruments to *Lucy*'s Level 1 science requirements listed in Table 3.

Table 4 Instrument Specifications and their Mapping to *Lucy*'s Science Requirements

	L'LORRI	L'TES	L'Ralph/MVIC	L'Ralph/LEISA	TTCam	HGA
Spectral Range (μm)	0.45 – 0.85	6 – 75	Violet: 377 – 489 Green: 478 – 525 Orange: 522 – 632 Phyllosilicate: 627 – 758 Near-IR: 753 – 914 Pan: 377 – 914	1.0 – 3.8	0.425 – 0.675	X Band
Spectral Resolution	—	10 cm^{-1}	—	10 nm	—	—
FOV (°)	0.29	0.57	8.3	1.4 × 3.4	11 × 8.2	—
Spatial Resolution (μm/px)	5.0	N/A	29	80	74	—
Links to Level 1 Requirements[a]	R.4,5,7,8 R.2,3,6	R.17	R.5,6,9–11 R.2–4,8 R.-7	R.12–15	R.2–6 R.7	R.16

[a]The requirement numbers in this column are from Table 3. Green text are for instruments that can completely accomplish the requirement. Orange text indicates the instrument partially contributes to the requirement. The purple text indicates that this instrument provides a degraded backup for the requirement.

7 Summary

The *Lucy* mission will investigate a diverse set of Trojan targets in a single 12-year mission. The mission targets span the full range of spectral types and visible and infrared albedo, diameters and rotation periods that span almost 2 orders of magnitude, and shapes that range from near-circular to highly elongated to flattened nearly-oblate spheroids. By investigating this range of properties, *Lucy* will elicit relationships that will address questions of origin and evolution of the early Solar System.

Lucy will derive the densities of its targets from radio science mass determination and volume measurements from imaging. These densities will be compared to known densities in other small body populations to identify affinities and carry out detailed compositional investigations of the surfaces of the targets. From the first resolved imaging of the Trojan asteroids, geological features will be identified on each of our Trojan asteroids. Imaging will allow an estimate of the surface age and will constrain the small impactor population from the crater SFD. The surface composition of Trojans will allow resolved mapping of color and composition units across the Trojan asteroids. Of particular interest is the potential for observations of freshly exposed materials in the crater ejecta. *Lucy*'s concept of operations and example science sequences are described by Olkin et al. (2024) in this collection. They also discuss the types of data we will obtain, including areal coverage and resolution maps. It is important to note, however, that because our goal is to directly compare diverse objects to one another, we will perform the same observations at each target as closely as possible.

To understand the impact of *Lucy* in context, it is worth noting that a total of nine Main Belt asteroids have been visited by all planetary spacecraft missions flown to date. These investigations have revolutionized our understanding of the asteroid belt and terrestrial planet formation. *Lucy*, on its own, will visit almost as many Trojan objects in a single mission. *Lucy* will do for the Trojans what has taken decades and multiple missions to accomplish for the much nearer and more accessible Main Belt asteroids.

Acknowledgement The *Lucy* mission is funded through the NASA Discovery program on contract No. NNM16AA08C.

Declarations

Competing Interests The authors declare no competing interests.

References

Anderson JD (1971) Feasibility of determining the mass of an asteroid from a spacecraft flyby. In: Gehrels T (ed) Physical studies of minor planets, NASA special publication, vol 267. NASA, p 577

Anderson JD, Armstrong JW, Campbell JK, et al (1992) Gravitation and celestial mechanics investigations with Galileo. Space Sci Rev 60(1–4):591–610. https://doi.org/10.1007/BF00216869

Andert TP, Rosenblatt P, Pätzold M, et al (2010) Precise mass determination and the nature of Phobos. Geophys Res Lett 37(9):L09202. https://doi.org/10.1029/2009GL041829

Barucci MA, Cruikshank DP, Mottola S, et al (2002) Physical properties of trojan and centaur asteroids. University of Arizona Press, Tucson, pp 273–287

Bell JF, Zhao Y, Cisneros E, et al (2023) The terminal tracking camera system on the NASA *Lucy* trojan asteroid discovery mission. Space Sci Rev 219(8):86. https://doi.org/10.1007/s11214-023-01030-5

Belskaya IN, Shevchenko VG (2000) Opposition effect of asteroids. Icarus 147(1):94–105. https://doi.org/10.1006/icar.2000.6410

Belton MJS, Mueller BEA, D'Amario LA, et al (1996) The discovery and orbit of 1993 (243)1 Dactyl. Icarus 120(1):185–199. https://doi.org/10.1006/icar.1996.0044

Bolin BT, Noll KS, Caiazzo I, et al (2023) Keck and Gemini spectral characterization of *Lucy* mission fly-by target (152830) Dinkinesh. Icarus 400:115562. https://doi.org/10.1016/j.icarus.2023.115562

Bottke WF, Brož M, O'Brien DP, et al (2015a) The collisional evolution of the main asteroid belt. In: Michel P, DeMeo FE, Bottke WF (eds) Asteroids IV. University of Arizona Press, Tucson, pp 701–724. https://doi.org/10.2458/azu_uapress_9780816532131-ch036

Bottke WF, Vokrouhlický D, Walsh KJ, et al (2015b) In search of the source of asteroid (101955) Bennu: applications of the stochastic YORP model. Icarus 247:191–217. https://doi.org/10.1016/j.icarus.2014.09.046

Bottke WF, Marschall R, Nesvorný D, et al (2023) Origin and evolution of Jupiter's trojan asteroids. Space Sci Rev 219(8):83. https://doi.org/10.1007/s11214-023-01031-4. arXiv:2312.02864 [astro-ph.EP]

Braga-Ribas F, Sicardy B, Ortiz JL, et al (2014) A ring system detected around the Centaur (10199) Chariklo. Nature 508(7494):72–75. https://doi.org/10.1038/nature13155, arXiv:1409.7259 [astro-ph.EP]

Brown ME, Levison HF, Noll KS, et al (2021) The orbit and density of the Jupiter trojan satellite system eurybates-queta. Planet Sci J 2(5):170. https://doi.org/10.3847/PSJ/ac07b0, arXiv:2106.02079 [astro-ph.EP]

Brož M, Rozehnal J (2011) Eurybates - the only asteroid family among trojans? Mon Not R Astron Soc 414(1):565–574. https://doi.org/10.1111/j.1365-2966.2011.18420.x

Brunetto R, Loeffler MJ, Nesvorný D, et al (2015) Asteroid surface alteration by space weathering processes. University of Arizona Press, Tucson, pp 597–616. https://doi.org/10.2458/azu_uapress_9780816532131-ch031

Buczkowski DL, Schmidt BE, Williams DA, et al (2016) The geomorphology of Ceres. Science 353(6303):aaf4332. https://doi.org/10.1126/science.aaf4332

Buie MW, Olkin CB, Merline WJ, et al (2015) Size and shape from stellar occultation observations of the double Jupiter trojan Patroclus and Menoetius. Astron J 149(3):113. https://doi.org/10.1088/0004-6256/149/3/113

Buie MW, Zangari AM, Marchi S, et al (2018) Light curves of *Lucy* targets: Leucus and Polymele. Astron J 155(6):245. https://doi.org/10.3847/1538-3881/aabd81

Buie M, Keeney B, Levison H, et al (2022) Shape and duplicity of *Lucy* Mission target Polymele from occultation observations. In: AAS/Division for Planetary Sciences Meeting Abstracts, p 512.03

Bus SJ, Vilas F, Barucci MA (2002) Visible-wavelength spectroscopy of asteroids. In: Asteroids III. University of Arizona Press, Tucson, pp 169–182

Capria MT, Tosi F, De Sanctis MC, et al (2014) Vesta surface thermal properties map. Geophys Res Lett 41(5):1438–1443. https://doi.org/10.1002/2013GL059026

Chapman CR (1996) S-type asteroids, ordinary chondrites, and space weathering: the evidence from Galileo's fly-bys of Gaspra and Ida. Meteorit Planet Sci 31(6):699–725. https://doi.org/10.1111/j.1945-5100.1996.tb02107.x

Christensen PR, Hamilton VE, Mehall GL, et al (2024) The *Lucy* thermal emission spectrometer (L'TES) instrument. Space Sci Rev 220(1):1. https://doi.org/10.1007/s11214-023-01029-y

Combe JP, Raponi A, Tosi F, et al (2019) Exposed H_2O-rich areas detected on Ceres with the dawn visible and infrared mapping spectrometer. Icarus 318:22–41. https://doi.org/10.1016/j.icarus.2017.12.008

de León J, Licandro J, Pinilla-Alonso N, et al (2023) Characterisation of the new target of the NASA *Lucy* mission: asteroid 152830 Dinkinesh (1999 VD57). Astron Astrophys 672:A174. https://doi.org/10.1051/0004-6361/202346278

De Sanctis MC, Ammannito E, McSween HY, et al (2017) Localized aliphatic organic material on the surface of Ceres. Science 355(6326):719–722. https://doi.org/10.1126/science.aaj2305

DellaGiustina DN, Burke KN, Walsh KJ, et al (2020) Variations in color and reflectance on the surface of asteroid (101955) Bennu. Science 370:eabc3660. https://doi.org/10.1126/science.abc3660

DeMeo FE, Carry B (2014) Solar System evolution from compositional mapping of the asteroid belt. Nature 505(7485):629–634. https://doi.org/10.1038/nature12908, arXiv:1408.2787 [astro-ph.EP]

DeMeo FE, Binzel RP, Slivan SM, et al (2009) An extension of the Bus asteroid taxonomy into the near-infrared. Icarus 202(1):160–180. https://doi.org/10.1016/j.icarus.2009.02.005

DeSanctis MC, Russell CT, Raymond CA, et al (2015) Vesta by dawn: portrait of a small world. In: IAU general assembly, p 2227768

Durda DD, Bottke WF, Enke BL, et al (2004) The formation of asteroid satellites in large impacts: results from numerical simulations. Icarus 167(2):382–396. https://doi.org/10.1016/j.icarus.2003.09.017

Emery JP, Cruikshank DP, Van Cleve J (2006) Thermal emission spectroscopy (5.2 38 μm) of three trojan asteroids with the spitzer space telescope: detection of fine-grained silicates. Icarus 182(2):496–512. https://doi.org/10.1016/j.icarus.2006.01.011

Emery JP, Burr DM, Cruikshank DP (2011) Near-infrared spectroscopy of trojan asteroids: evidence for two compositional groups. Astron J 141(1):25. https://doi.org/10.1088/0004-6256/141/1/25. arXiv:1012.1284 [astro-ph.EP]

Emery JP, Binzel RP, Britt DT, et al (2024) Surface compositions of trojan asteroids. Space Sci Rev 220(3):28. https://doi.org/10.1007/s11214-024-01060-7

Ferrais M, Jehin E, Manfroid J, et al (2021) Rotation period determination of NASA *Lucy* mission target (52246) Donaldjohanson. In: AAS/Division for Planetary Sciences Meeting Abstracts, p 306.19

Fleming HJ, Hamilton DP (2000) On the origin of the trojan asteroids: effects of Jupiter's mass accretion and radial migration. Icarus 148(2):479–493. https://doi.org/10.1006/icar.2000.6523, arXiv:astro-ph/0007296 [astro-ph]

Grav T, Mainzer AK, Bauer JM, et al (2012) WISE/NEOWISE observations of the Jovian trojan population: taxonomy. Astrophys J 759(1):49. https://doi.org/10.1088/0004-637X/759/1/49. arXiv:1209.1549 [astro-ph.EP]

Grundy WM, Noll KS, Buie MW, et al (2018) The upcoming mutual event season for the Patroclus-Menoetius trojan binary. Icarus 305:198–202. https://doi.org/10.1016/j.icarus.2018.01.009, arXiv:1903.03729 [astro-ph.EP]

Grundy WM, Bird MK, Britt DT, et al (2020) Color, composition, and thermal environment of Kuiper Belt object (486958) Arrokoth. Science 367(6481):aay3705. https://doi.org/10.1126/science.aay3705, arXiv:2002.06720 [astro-ph.EP]

Guilbert-Lepoutre A (2014) Survival of water ice in Jupiter trojans. Icarus 231:232–238. https://doi.org/10.1016/j.icarus.2013.12.014, arXiv:1401.5196 [astro-ph.EP]

Hamilton VE, Simon AA, Christensen PR, et al (2019) Evidence for widespread hydrated minerals on asteroid (101955) Bennu. Nat Astron 3:332–340. https://doi.org/10.1038/s41550-019-0722-2

Hartmann WK (1966) Martian cratering. Icarus 5(1–6):565–576. https://doi.org/10.1016/0019-1035(66)90071-6

Hellmich S, Mottola S, Hahn G, et al (2019) Influence of the Yarkovsky force on Jupiter Trojan asteroids. Astron Astrophys 630:A148. https://doi.org/10.1051/0004-6361/201834715. arXiv:1909.08584 [astro-ph.EP]

Holsapple KA, Housen KR (2007) A crater and its ejecta: an interpretation of deep impact. Icarus 187(1):345–356. https://doi.org/10.1016/j.icarus.2006.08.029

Housen KR, Sweet WJ, Holsapple KA (2018) Impacts into porous asteroids. Icarus 300:72–96. https://doi.org/10.1016/j.icarus.2017.08.019

Howett CJA, Spencer JR, Pearl J, et al (2010) Thermal inertia and bolometric Bond albedo values for Mimas, Enceladus, Tethys, Dione, Rhea and Iapetus as derived from Cassini/CIRS measurements. Icarus 206(2):573–593. https://doi.org/10.1016/j.icarus.2009.07.016

Jaumann R, Williams DA, Buczkowski DL, et al (2012) Vesta's shape and morphology. Science 336(6082):687. https://doi.org/10.1126/science.1219122

Jewitt D, Hsieh H, Agarwal J (2015) The active asteroids. University of Arizona Press, Tucson, pp 221–241. https://doi.org/10.2458/azu_uapress_9780816532131-ch012

Kary DM, Lissauer JJ (1995) Nebular gas drag and planetary accretion. II. Planet on an eccentric orbit. Icarus 117(1):1–24. https://doi.org/10.1006/icar.1995.1139

Lauretta DS, Dellagiustina DN, Bennett CA, et al (2019) The unexpected surface of asteroid (101955) Bennu. Nature 568(7750):55–60. https://doi.org/10.1038/s41586-019-1033-6

Levison HF, Marchi S, Noll K, et al (2021a) NASA's *Lucy* mission to the trojan asteroids. In: 2021 IEEE Aerospace Conference (50100), pp 1–10. https://doi.org/10.1109/AERO50100.2021.9438453

Levison HF, Olkin CB, Noll KS, et al (2021b) *Lucy* Mission to the trojan asteroids: science goals. Planet Sci J 2(5):171. https://doi.org/10.3847/PSJ/abf840

Li JY, Schröder SE, Mottola S, et al (2019) Spectrophotometric modeling and mapping of Ceres. Icarus 322:144–167. https://doi.org/10.1016/j.icarus.2018.12.038, arXiv:1812.08361 [astro-ph.EP]

Marchi S, McSween HY, O'Brien DP, et al (2012) The violent collisional history of asteroid 4 Vesta. Science 336(6082):690. https://doi.org/10.1126/science.1218757

Marchi S, Bottke WF, O'Brien DP, et al (2014) Small crater populations on Vesta. Planet Space Sci 103:96–103. https://doi.org/10.1016/j.pss.2013.05.005, arXiv:1305.6679 [astro-ph.EP]

Marchi S, Chapman CR, Barnouin OS, et al (2015) Cratering on asteroids. University of Arizona Press, Tucson, pp 725–744. https://doi.org/10.2458/azu_uapress_9780816532131-ch037

Marchi S, Bell JF, Bierhaus B, et al (2023) Surface geology of Jupiter's trojan asteroids. Space Sci Rev 219(5):44. https://doi.org/10.1007/s11214-023-00985-9

Margot JL, Pravec P, Taylor P, et al (2015) Asteroid systems: binaries, triples, and pairs. University of Arizona Press, Tucson, pp 355–374. https://doi.org/10.2458/azu_uapress_9780816532131-ch019

Marschall R, Nesvorný D, Deienno R, et al (2022) Implications for the collisional strength of Jupiter trojans from the eurybates family. Astron J 164(4):167. https://doi.org/10.3847/1538-3881/ac8d6b

Marzari F, Scholl H (1998) The growth of Jupiter and Saturn and the capture of trojans. Astron Astrophys 339:278–285

Masiero JR, Mainzer AK, Grav T, et al (2012) Preliminary analysis of WISE/NEOWISE 3-band cryogenic and post-cryogenic observations of main belt asteroids. Astrophys J Lett 759(1):L8. https://doi.org/10.1088/2041-8205/759/1/L8. arXiv:1209.5794 [astro-ph.EP]

Merline WJ, Close LM, Siegler N, et al (2001) S/2001 (617) 1. IAUC 7741: 2

Merline WJ, Weidenschilling SJ, Durda DD, et al (2002) Asteroids do have satellites. University of Arizona Press, Tucson, pp 289–312

Morbidelli A, Levison HF, Tsiganis K, et al (2005) Chaotic capture of Jupiter's trojan asteroids in the early Solar System. Nature 435(7041):462–465. https://doi.org/10.1038/nature03540

Mottola S, Marchi S, Buie MW, et al (2016) Ground-based characterization of Eurybates and Orus, two fly-by targets of the *Lucy* Discovery mission. In: AAS/Division for Planetary Sciences Meeting Abstracts #48, p 208.04

Mottola S, Hellmich S, Buie MW, et al (2020) Convex shape and rotation model of *Lucy* target (11351) Leucus from lightcurves and occultations. Planet Sci J 1(3):73. https://doi.org/10.3847/PSJ/abb942. arXiv:2009.08951 [astro-ph.EP]

Mottola S, Denk T, Marchi S, et al (2023a) Characterizing asteroid (152830) dinkinesh in preparation for the encounter with the NASA *Lucy* mission: a photometric study. Mon Not R Astron Soc 524(1):L1–L4. https://doi.org/10.1093/mnrasl/slad066

Mottola S, Hellmich S, Buie MW, et al (2023b) Shape models of *Lucy* targets (3548) Eurybates and (21900) Orus from disk-integrated photometry. Planet Sci J 4(1):18. https://doi.org/10.3847/PSJ/acaf79

Mottola S, Britt DT, Brown ME, et al (2024) Shapes, rotations, photometric and internal properties of Jupiter trojans. Space Sci Rev 220(1):17. https://doi.org/10.1007/s11214-024-01052-7

Mueller M, Marchis F, Emery JP, et al (2010) Eclipsing binary trojan asteroid Patroclus: thermal inertia from Spitzer observations. Icarus 205(2):505–515. https://doi.org/10.1016/j.icarus.2009.07.043, arXiv:0908.4198 [astro-ph.EP]

Murdoch N, Sánchez P, Schwartz SR, et al (2015) Asteroid surface geophysics. University of Arizona Press, Tucson, pp 767–792. https://doi.org/10.2458/azu_uapress_9780816532131-ch039

Nathues A, Hoffmann M, Platz T, et al (2016) FC colour images of dwarf planet Ceres reveal a complicated geological history. Planet Space Sci 134:122–127. https://doi.org/10.1016/j.pss.2016.10.017

Nesvorný D (2018) Dynamical evolution of the early Solar System. Annu Rev Astron Astrophys 56:137–174. https://doi.org/10.1146/annurev-astro-081817-052028. arXiv:1807.06647 [astro-ph.EP]

Nesvorný D, Bottke WF, Vokrouhlický D, et al (2006) Asteroid families. In: Lazzaro D, Ferraz-Mello S, Fernández JA (eds) Asteroids, comets, meteors, pp 289–299. https://doi.org/10.1017/S1743921305006800

Nesvorný D, Youdin AN, Richardson DC (2010) Formation of Kuiper Belt binaries by gravitational collapse. Astron J 140(3):785–793. https://doi.org/10.1088/0004-6256/140/3/785. arXiv:1007.1465 [astro-ph.EP]

Nesvorný D, Vokrouhlický D, Morbidelli A (2013) Capture of trojans by jumping Jupiter. Astrophys J 768(1):45. https://doi.org/10.1088/0004-637X/768/1/45. arXiv:1303.2900 [astro-ph.EP]

Nesvorný D, Vokrouhlický D, Bottke WF, et al (2018) Evidence for very early migration of the Solar System planets from the Patroclus-Menoetius binary Jupiter trojan. Nat Astron 2:878–882. https://doi.org/10.1038/s41550-018-0564-3. arXiv:1809.04007 [astro-ph.EP]

Nesvorný D, Vokrouhlický D, Alexandersen M, et al (2020a) OSSOS XX: the meaning of Kuiper Belt colors. Astron J 160(1):46. https://doi.org/10.3847/1538-3881/ab98fb. arXiv:2006.01806 [astro-ph.EP]

Nesvorný D, Vokrouhlický D, Bottke WF, et al (2020b) Very slow rotators from tidally synchronized binaries. Astrophys J Lett 893(1):L16. https://doi.org/10.3847/2041-8213/ab8311

Nesvorný D, Li R, Simon JB, et al (2021) Binary planetesimal formation from gravitationally collapsing pebble clouds. Planet Sci J 2(1):27. https://doi.org/10.3847/PSJ/abd858

Noll KS, Grundy WM, Chiang EI, et al (2008) Binaries in the Kuiper Belt. University of Arizona Press, Tucson, p 345

Noll KS, Brown ME, Levison HF, et al (2020b) (3548) Eurybates I = Queta. Minor Planet Electronic Circulars 2020-T164

Noll KS, Brown ME, Weaver HA, et al (2020c) Detection of a satellite of the trojan asteroid (3548) Eurybates—a *Lucy* Mission target. Planet Sci J 1(2):44. https://doi.org/10.3847/PSJ/abac54. arXiv:2008.01858 [astro-ph.EP]

Noll K, Grundy WM, Nesvorný D, et al (2020a) Trans-Neptunian binaries (2018). In: Prialnik D, Barucci MA, Young L (eds) The trans-Neptunian Solar System. Elsevier, Amsterdam, pp 201–224. https://doi.org/10.1016/B978-0-12-816490-7.00009-6

Noll KS, Brown ME, Buie MW, et al (2023) Trojan asteroid satellites, rings, and activity. Space Sci Rev 219(7):59. https://doi.org/10.1007/s11214-023-01001-w

O'Brien DP, Marchi S, Morbidelli A, et al (2014) Constraining the cratering chronology of Vesta. Planet Space Sci 103:131–142. https://doi.org/10.1016/j.pss.2014.05.013, arXiv:1407.3303 [astro-ph.EP]

Olkin CB, Spencer JR, Grundy WM, et al (2017) The global color of Pluto from New Horizons. Astron J 154(6):258. https://doi.org/10.3847/1538-3881/aa965b

Olkin C, Vincent M, Adam C, et al (2024) Mission design and concept of operations for the *Lucy* Mission. Space Sci Rev 220(4):47. https://doi.org/10.1007/s11214-024-01082-1

Ortiz JL, Duffard R, Pinilla-Alonso N, et al (2015) Possible ring material around centaur (2060) Chiron. Astron Astrophys 576:A18. https://doi.org/10.1051/0004-6361/201424461. arXiv:1501.05911 [astro-ph.EP]

Ortiz JL, Santos-Sanz P, Sicardy B, et al (2017) The size, shape, density and ring of the dwarf planet Haumea from a stellar occultation. Nature 550(7675):219–223. https://doi.org/10.1038/nature24051, arXiv:2006.03113 [astro-ph.EP]

Pätzold M, Häusler B, Wennmacher A, et al (2001) Gravity field determination of a comet Nucleus: Rosetta at P/Wirtanen. Astron Astrophys 375:651–660. https://doi.org/10.1051/0004-6361:20010829

Pätzold M, Andert TP, Asmar SW, et al (2011) Asteroid 21 Lutetia: low mass, high density. Science 334(6055):491. https://doi.org/10.1126/science.1209389

Pätzold M, Andert TP, Tyler GL, et al (2014) Phobos mass determination from the very close flyby of Mars Express in 2010. Icarus 229:92–98. https://doi.org/10.1016/j.icarus.2013.10.021

Pieters CM, Noble SK (2016) Space weathering on airless bodies. J Geophys Res Planets 121(10):1865–1884. https://doi.org/10.1002/2016JE005128

Reuter DC, Simon AA, Lunsford A, et al (2023) L'Ralph: a visible/infrared spectral imager for the *Lucy* Mission to the trojans. Space Sci Rev 219(8):69. https://doi.org/10.1007/s11214-023-01009-2

Roig F, Ribeiro AO, Gil-Hutton R (2008) Taxonomy of asteroid families among the Jupiter trojans: comparison between spectroscopic data and the Sloan Digital Sky Survey colors. Astron Astrophys 483(3):911–931. https://doi.org/10.1051/0004-6361:20079177. arXiv:0712.0046 [astro-ph]

Rozitis B, Emery JP, Siegler MA, et al (2020) Implications for ice stability and particle ejection from high-resolution temperature modeling of asteroid (101955) Bennu. J Geophys Res Planets 125(8):e06323. https://doi.org/10.1029/2019JE006323

Ryan EL, Sharkey BNL, Woodward CE (2017) Trojan asteroids in the Kepler campaign 6 field. Astron J 153(3):116. https://doi.org/10.3847/1538-3881/153/3/116

Schemel M, Brown ME (2021) Zwicky Transient Facility observations of trojan asteroids: a thousand colors, rotation amplitudes, and phase functions. Planet Sci J 2(1):40. https://doi.org/10.3847/PSJ/abc752. arXiv:2011.01329 [astro-ph.EP]

Schröder SE, Mottola S, Carsenty U, et al (2017) Resolved spectrophotometric properties of the Ceres surface from Dawn Framing Camera images. Icarus 288:201–225. https://doi.org/10.1016/j.icarus.2017.01.026, arXiv:1701.08550 [astro-ph.EP]

Singer KN, Greenstreet S, Schenk PM, et al (2021) Impact craters on Pluto and Charon and terrain age estimates. In: Stern SA, Moore JM, Grundy WM, et al (eds) The Pluto system after New Horizons. University of Arizona Press, Tucson, pp 121–145. https://doi.org/10.2458/azu_uapress_9780816540945-ch007, arXiv:2008.10153

Sizemore HG, Platz T, Schorghofer N, et al (2017) Pitted terrains on (1) Ceres and implications for shallow subsurface volatile distribution. Geophys Res Lett 44(13):6570–6578. https://doi.org/10.1002/2017GL073970

Spencer JR, Denk T (2010) Formation of Iapetus' extreme albedo dichotomy by exogenically triggered thermal ice migration. Science 327(5964):432. https://doi.org/10.1126/science.1177132

Spencer JR, Stern SA, Moore JM, et al (2020) The geology and geophysics of Kuiper Belt object (486958) Arrokoth. Science 367(6481):aay3999. https://doi.org/10.1126/science.aay3999, arXiv:2004.00727 [astro-ph.EP]

Stern SA, Weaver HA, Spencer JR, et al (2019) Initial results from the New Horizons exploration of 2014 MU_{69}, a small Kuiper Belt object. Science 364(6441):aaw9771. https://doi.org/10.1126/science.aaw9771, arXiv:2004.01017 [astro-ph.EP]

Szabó GM, Ivezić Ž, Jurić M, et al (2007) The properties of Jovian trojan asteroids listed in SDSS moving object catalogue 3. Mon Not R Astron Soc 377(4):1393–1406. https://doi.org/10.1111/j.1365-2966.2007.11687.x. arXiv:astro-ph/0703026 [astro-ph]

Szabó GM, Pál A, Kiss C, et al (2017) The heart of the swarm: K2 photometry and rotational characteristics of 56 Jovian trojan asteroids. Astron Astrophys 599:A44. https://doi.org/10.1051/0004-6361/201629401

Szebehely V (1967) Theory of orbits. The restricted problem of three bodies. Academic Press, New York

Szebehely V (1978) Stability of artificial and natural satellites. Celest Mech 18(4):383–389. https://doi.org/10.1007/BF01230350

Thangjam G, Reddy V, Le Corre L, et al (2013) Lithologic mapping of HED terrains on Vesta using Dawn Framing Camera color data. Meteorit Planet Sci 48(11):2199–2210. https://doi.org/10.1111/maps.12132, arXiv:1305.0953 [astro-ph.EP]

Tosi F, Capria MT, De Sanctis MC, et al (2014) Thermal measurements of dark and bright surface features on Vesta as derived from Dawn/VIR. Icarus 240:36–57. https://doi.org/10.1016/j.icarus.2014.03.017

Tsiganis K, Gomes R, Morbidelli A, et al (2005) Origin of the orbital architecture of the giant planets of the Solar System. Nature 435(7041):459–461. https://doi.org/10.1038/nature03539

Vernazza P, Delbo M, King PL, et al (2012) High surface porosity as the origin of emissivity features in asteroid spectra. Icarus 221(2):1162–1172. https://doi.org/10.1016/j.icarus.2012.04.003

Vilas F (1994) A cheaper, faster, better way to detect water of hydration on Solar System bodies. Icarus 111(2):456–467. https://doi.org/10.1006/icar.1994.1156

Vokrouhlický D, Brož M, Bottke WF, et al (2006) Yarkovsky/YORP chronology of asteroid families. Icarus 182(1):118–142. https://doi.org/10.1016/j.icarus.2005.12.010

Warner BD, Harris AW, Pravec P (2021) Asteroid Lightcurve Database (LCDB) Bundle V4.0. NASA Planetary Data System p 10. https://doi.org/10.26033/j3xc-3359

Weaver HA, Wilson JP, Conard SJ, et al (2023) The *Lucy* Long Range Reconnaissance Imager (L'LORRI). Space Sci Rev 219(8):82. https://doi.org/10.1007/s11214-023-01028-z

Wong I, Brown ME (2015) The color-magnitude distribution of small Jupiter trojans. Astron J 150(6):174. https://doi.org/10.1088/0004-6256/150/6/174. arXiv:1510.03144 [astro-ph.EP]

Wong I, Brown ME (2016) A hypothesis for the color bimodality of Jupiter trojans. Astron J 152(4):90. https://doi.org/10.3847/0004-6256/152/4/90. arXiv:1607.04133 [astro-ph.EP]

Wong I, Brown ME (2017) The bimodal color distribution of small Kuiper Belt objects. Astron J 153(4):145. https://doi.org/10.3847/1538-3881/aa60c3. arXiv:1702.02615 [astro-ph.EP]

Wong I, Brown ME, Emery JP (2014) The differing magnitude distributions of the two Jupiter trojan color populations. Astron J 148(6):112. https://doi.org/10.1088/0004-6256/148/6/112. arXiv:1408.2485 [astro-ph.EP]

Yoder CF (1979) Notes on the origin of the trojan asteroids. Icarus 40(3):341–344. https://doi.org/10.1016/0019-1035(79)90024-1

Yoshida F, Terai T (2017) Small Jupiter trojans survey with the Subaru/hyper suprime-cam. Astron J 154(2):71. https://doi.org/10.3847/1538-3881/aa7d03. arXiv:1706.10017 [astro-ph.EP]

Space Science Reviews (2023) 219:83
https://doi.org/10.1007/s11214-023-01031-4

Origin and Evolution of Jupiter's Trojan Asteroids

William F. Bottke[1] · Raphael Marschall[2] · David Nesvorný[1] · David Vokrouhlický[3]

Received: 5 July 2023 / Accepted: 20 November 2023 / Published online: 5 December 2023

Abstract
The origin of the Jupiter Trojan asteroids has long been a mystery. Dynamically, the population, which is considerably smaller than the main asteroid belt, librates around Jupiter's stable L4 and L5 Lagrange points, 60 deg ahead and behind Jupiter. It is thought that these bodies were captured into these orbits early in solar system history, but any capture mechanism must also explain why the Trojans have an excited inclination distribution, with some objects reaching inclinations of 35°. The Trojans themselves, individually and in aggregate, also have spectral and physical properties that appear consistent with many small bodies found in the outer solar system (e.g., irregular satellites, Kuiper belt objects). In this review, we assemble what is known about the Trojans and discuss various models for their origin and collisional evolution. It can be argued that the Trojans are unlikely to be captured planetesimals from the giant planet zone, but instead were once denizens of the primordial Kuiper belt, trapped by the events taking place during a giant planet instability. The Lucy mission to the Trojans is therefore well positioned to not only answer questions about these objects, but also about their place in planet formation and solar system evolution studies.

Keywords Trojans · Kuiper belt objects · Giant planet instability · Asteroid dynamics

1 Introduction

Lucy's mission to Jupiter's Trojan asteroids (hereafter Trojans) will examine one of the last unexplored small body reservoirs in the Solar System (Levison et al. 2021). By visiting at least eight Trojan bodies over a 12-year traverse, Lucy will probe the origin and evolution of these singular objects, which in turn may tell us about planetesimal and planet formation processes in the outer solar system, the effects of giant planet migration on small body populations, and how such worlds have experienced collisional evolution. The Trojans also potentially set constraints on a dramatic phase of Solar System history when the giant planets

✉ W.F. Bottke
bottke@boulder.swri.edu

1 Southwest Research Institute, Boulder, CO, USA

2 CNRS, Observatoire de la Cote d'Azur, Laboratoire J.-L. Lagrange, CS 34229, 06304 Nice Cedex 4, France

3 Institute of Astronomy, Charles University, V Holešovičkách 2, CZ-18000, Prague 8, Czech Republic

may have experienced a dynamical instability after the solar nebula had dissipated. By understanding the Trojans, we can test numerical models that suggest the giant planets formed on different orbits than they have today and that Neptune's migration through the primordial Kuiper belt injected bodies into numerous stable zones, ranging from the main asteroid belt to the irregular satellites of the giant planets to specific regions of the transneptunian populations.

At this time, researchers have limited information about the composition and physical characteristics of the Trojans and know nothing about their cratering history (see chapters by Mottola et al. and Marchi et al., this volume). Ideally, an investigation of these properties will provide vital clues concerning their provenance location (e.g., near Jupiter's orbit, or within the primordial Kuiper belt) and what the nature of the solar nebula was like at those locations when they formed. In order to interpret the data provided by Lucy's observations, we need be able to place the Trojans into a science context. Here we briefly summarize what is known of these worlds. The interested reader should also seek out two excellent comprehensive reviews of the Trojans by Marzari et al. (2002) and Emery et al. (2015) as well as the other Trojan chapters in this volume.

2 Trojan Characteristics

2.1 Orbits

Jupiter Trojans (hereafter Trojans) are populations that orbit around the L4 and L5 Lagrange points of Jupiter, which are located 60° in front of and behind Jupiter, respectively. They are contained within Jupiter's 1:1 mean motion resonance, with Jupiter having a semimajor axis of 5.2 au (Fig. 1). Details on their dynamical status can be found Marzari et al. (2002) and Emery et al. (2015). The regions themselves are relatively small, and it is unlikely that the currently observed bodies formed *in situ*. Instead, most experts argue that the Trojans were captured there during the earliest times of the Solar System, possibly when Jupiter was still forming and/or migrating. Various formation and capture scenarios will be discussed in Sect. 3.

Most Trojans reside on stable orbits that librate around L4 and L5, but a small fraction is unstable over ages shorter than the lifetime of the Solar System (Levison et al. 1997; Tsiganis et al. 2005; Di Sisto et al. 2014, 2019; Holt et al. 2020). Some unstable bodies are also sizeable; examples include (1173) Anchises, which is 100 km in diameter, and (1868) Thersites, which is 70 km in diameter (Tsiganis et al. 2000; Horner et al. 2012). These Trojans were likely captured on orbits near the stable zone boundary, and that means they will slowly diffuse out of the Trojan zone on billion-year timescales (e.g., Robutel and Gabern 2006). The remaining unstable Trojans are likely to be temporary captures from the Jupiter family comet population. The regions around L4 and L5 are dynamically "sticky", and objects that evolve to similar orbital conditions can find themselves caught for an extended time. It is also possible that some unstable Trojans reached their orbits by one of the following mechanisms: (i) the same process that captured the stable Trojans (see Sect. 3), (ii) collisions, and/or (iii) some mobility-inducing process acting over a long timescale, such as Yarkosky thermal drift or outgassing. Regardless, it can be shown that most Trojans many tens of kilometers in diameter or larger are on stable orbits, so we can presume they are survivors of the processes that created the Trojans in the first place. We can use these bodies to evaluate Trojan origin scenarios.

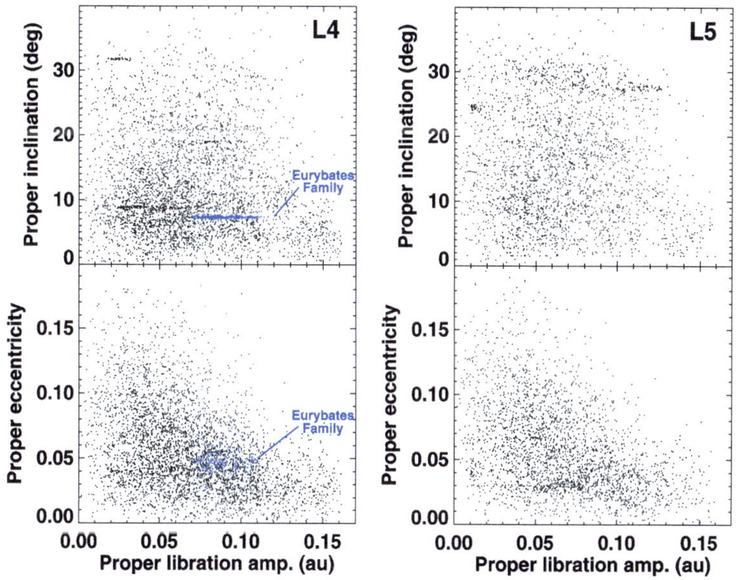

Fig. 1 Proper orbital elements of Jovian Trojans. The left two panels show the orbits of 4607 objects at the L4 libration zone, while the right two panels show the orbits of 2721 objects at the L5 libration zone (see Holt et al. 2020 for discussion of orbital elements). The Eurybates family is shown in L4 as the blue dots. The proper orbital elements, namely libration amplitude, eccentricity, inclination, of the Trojans were calculated using the methods of Milani (1993). The libration amplitude-eccentricity plots highlight the stable zone for Trojans; as objects approach the top-right corner, they reach unstable orbits and can escape

Arguably the most remarkable dynamical property of the Trojans are their inclinations, which cover a fairly uniform range from 0° to 35° in both the L4 and L5 region, with a median inclination of 10° (Fig. 1). The eccentricities of the Trojans tend to be $e < 0.15$, values required to remain within the stable L4 and L5 zones (e.g., Levison et al. 1997; Nesvorný and Dones 2002). The dynamical excitation of the Trojans presents a substantial challenge to all origin models. For example, not only must a proposed capture mechanism place comparable numbers of bodies near Jupiter's L4 and L5 zones, but it must also explain how the populations obtained high inclinations (i.e., the populations were either excited within L4/L5 in similar ways after capture or they were implanted with an excited inclination distribution). Over the last several decades, many origin scenarios have failed the test of explaining Trojan inclinations.

2.2 Population

The largest Trojan is (624) Hektor, whose diameter is $D \sim 250$ km (Marchis et al. 2014), while the smallest observed Trojans to date are likely to be a kilometer or so across (e.g., Yoshida and Terai 2017). Overall, the combined Trojan populations of L4 and L5 are considerably smaller than the population in the main asteroid belt, though how much smaller depends on how one evaluates the size frequency distributions (SFD) of each population, which do not have the same shape.

For example, both the main belt and the Trojans have a "bump" in their cumulative SFDs for $D \sim 100$ km objects (i.e., the power law slope of the SFD is substantially steeper for $D >$ 100 km bodies and is substantially shallower for $D < 100$ km objects). This bump means

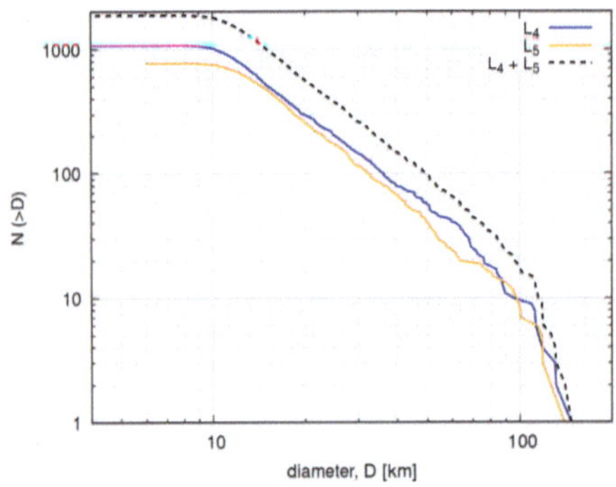

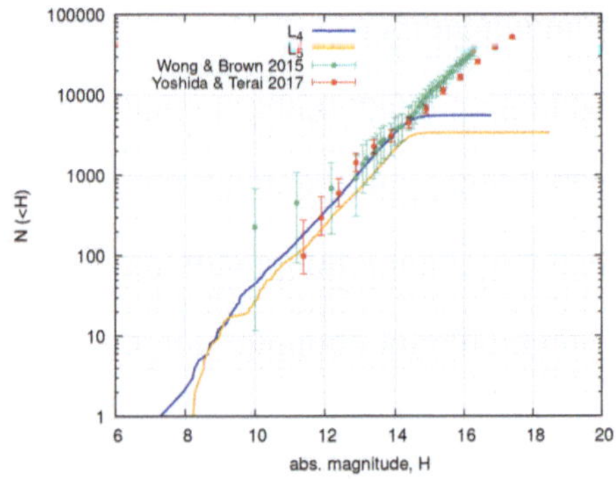

Fig. 2 The Trojan cumulative size frequency distribution (SFD) as a function of WISE diameter (left panel) and absolute magnitude (right panel) are shown for the L4 (orange) and L5 (blue) swarm (e.g., Mainzer et al. 2019; Marschall et al. 2022). The WISE SFDs are for those Trojans observed by WISE, so the curves should be interpreted in terms of shape and not absolute number. In addition, the right plot includes data from the L4 pencil beam studies of Wong and Brown (2015) and Yoshida and Terai (2017)

that most of the mass contained within the connected slopes of the SFD are in those size bodies. Indeed, it can be argued that on that basis, $D \sim 100$ km objects are a characteristic size produced by planetesimal formation processes (Bottke et al. 2005; Morbidelli et al. 2009b). Using this criterion, it can be shown that the Trojans are approximately ten times less massive than the main belt population; $\sim$20 or so bodies in the Trojans vs. $\sim$200 in the main belt (Bottke et al. 2020; Marschall et al. 2022). Other estimates suggest this ratio holds fairly well down to km-sized bodies as well; the Trojans are projected to have $\sim 10^5$ bodies with $D > 1$ km, while the asteroid belt has $\sim 10^6$ such objects (e.g., Nakamura and Yoshida 2008; Szabó et al. 2007; Bottke et al. 2020).

The relative size of the L4 and L5 Trojan populations is debated. Dynamically, there is no particular difference between the two regions that would lead to a pronounced asymmetric population by Trojan escape (e.g., Holt et al. 2020). While it is possible that a putative Trojan capture mechanism may have injected more bodies into one cloud than the other, it is not clear the population data supports this idea. The population of $D > 100$ km bodies, where most of the mass of the Trojans resides, contains 9 objects in both L4 and L5. Agreement between the populations would probably be acceptable for values of 9 ± 3. Accordingly, for the bodies least likely to be affected by collisional evolution (e.g., Bottke et al. 2005), it would appear that the two populations are comparable to one another.

If we examine $D > 10$ km bodies, the ratio of L4 to L5 objects moves to 1.4 ± 0.2 (Grav et al. 2011), values within the fractional errors given above for the $D > 100$ km bodies. These results are consistent with both populations being statistically comparable, but it is also possible that some modest asymmetry may exist (e.g., Marschall et al. 2022). This putative discrepancy will be discussed in Sect. 3 in the context of Trojan origins.

2.3 Size Distribution

The current Trojan SFD gives us crucial information about the initial Trojan population. Both the L4 and L5 swarms have similar SFDs and follow a broken power law for observed objects (Fig. 2). The largest Trojans follow cumulative power law slope $q = -5$ for $D > 100$ km bodies (e.g., Jewitt et al. 2000; Fraser et al. 2014) and a cumulative slope of approximately $q \sim -2.1$ between 10 km and 100 km (right panel of Fig. 2). The shape of the Trojan SFD for objects larger than several tens of kilometers is similar to that found among Kuiper belt objects, and it has been suggested that both have originated from the same source population (Morbidelli et al. 2009a; Fraser et al. 2014; Emery et al. 2015).

For smaller bodies, Grav et al. (2011) found a cumulative slope of $q = -2.1$ between ~10 and 100 km for both the L4 and L5 swarms, values in agreement with Jewitt et al. (2000). Pencil beam studies by several groups (Yoshida and Nakamura 2008; Wong and Brown 2016; Yoshida and Terai 2017) indicate that the same approximate slope continues to sizes smaller than 10 km. Beyond those sizes, however, the nature of the SFD becomes less clear.

Several observational studies suggest an inflection point exists in the SFD somewhere between a few kilometers and 10 kilometers, with objects smaller than the inflection point following a SFD with a shallower power law slope. For example, Yoshida and Nakamura (2008) reported a break in the SFD for absolute magnitude $H \sim 16$ in the L4 cloud. Follow up work by Wong and Brown (2016) confirmed the break but found it at $H \sim 15$. For an albedo of 0.07 this would correspond to a change in slope between 3 km and 10 km. Using HyperSuprime Cam on the Suburu telescope, Yoshida and Terai (2017) and Uehata et al. (2022) found comparable results to these studies, but also argued that a single power law of $q = -1.85 \pm 0.05$ could be fit to L4 and L5 Trojans between $13.6 < H < 17.4$, approximately 2-10 km diameter bodies, assuming an albedo of 0.07. Future observations by the Dark Energy telescope, Vera Rubin telescope, or the NEO Surveyor mission may be needed to settle the issue of what happens to the SFD of smaller Trojans.

2.4 Spectral Properties

Observations of the Trojans show that they are dominated by objects that have the broad appearance of dormant comets. Visible spectroscopy indicates Trojans have featureless spectra in visible and near-infrared wavelengths, with spectral slopes ranging from neutral to moderately red (e.g., Dotto et al. 2006; Fornasier et al. 2007; Melita et al. 2008). Using taxonomy based on main belt asteroids, Trojans mainly fall within the D- and P-type categories (e.g., Emery et al. 2015). In this nomenclature, D-types are considered "red" and the P-types are considered "less red" (Wong et al. 2014). The differences in colors may be related to where these objects formed within the solar nebula, how they evolved prior to capture within the Trojans, with some passing closer to the Sun than others, collisional evolution within the Trojan population over the last several billion years, or possibly all of these issues combined. Our discussion of Trojan colors relative to other populations (e.g., Kuiper belt) can be found in Sect. 5.3.

A small fraction of the Trojans are C-types. Most of these bodies are in the Eurybates asteroid family located within the L4 population (Fig. 1). This family will be discussed more below in Sect. 5.2. It is possible that their taxonomic type is related to their collisional disruption, with the Trojans looking different on the inside than on the outside (Wong and Brown 2016, 2017a).

Trojans also tend to have low albedos that are consistent with D- and P-type asteroids in the main asteroid belt or dormant comets (Grav et al. 2011) as well as the majority of C-complex asteroids (Masiero et al. 2013). Barucci et al. (2002) found albedos for the Trojans they observed between 0.03 and 0.07, while NEOWISE observations found albedos of 0.07 ± 0.03 across all sizes (Grav et al. 2011; see also Emery et al. 2015).

2.5 Rotation and Physical Properties

Until recently, information on the rotation rate distribution of Trojans was limited to a few tens of Trojans. The interested reader should see the review and references in Emery et al.

(2015) to find out more about the individual rotational light curve studies used to gather these data. In the section, we briefly discuss recent work determined from space-based telescopic surveys to set the context for later discussions in this chapter on Trojan formation and collisional evolution. The topics mentioned here are discussed in more detail by Mottola et al. (2024), this collection.

The spin periods for 56 Jupiter Trojans, many of them $15 < D < 50$ km, were determined using Kepler telescope photometry by Szabó et al. (2017) and Ryan et al. (2017). These data were gathered during their K2 Campaign 6 that lasted from July 14, 2015 to September 30, 2015. A crucial reason to use Kepler data is that many objects were observed for months, and as such exclude observation selection effects against slow rotators. They found that eight out of 53 Trojans, or $\sim$15%, have extremely long spin periods ($P \geq 100$ hours). A follow-up study by Kalup et al. (2021) extended their analysis of Kepler data to 101 Trojans. They showed there is a dichotomy in Trojan spin periods, with the division taking place near 100 hours; 25% had $P \geq 30$ hours, and 12% were very slow rotators with $P \geq$ 100–600 hr. Nesvorný et al. (2020a) argued that the easiest explanation for this excess was that prior to implantation within L4 or L5, these bodies were nearly equal-sized binaries that had tidally evolved toward a synchronous state. The process of capture likely led to their companions being stripped away, leaving behind a sizeable set of very slow rotators.

The Nesvorný et al. (2020a) proposal, if true, has several implications. First, equal-sized binaries may have been a common outcome of planetesimal formation wherever Trojans actually formed. Second, collisional evolution among the Trojans was not enough to substantially alter the spin state of slow rotators stripped of their companion. This scenario may explain why (11351) Leucus, a $D \sim 35$ km target of the Lucy mission (see Levison et al. 2024, this collection), has a rotation period of 445.73 hours (French et al. 2015; Buie et al. 2018; Kalup et al. 2021). Third, there may be many binaries among the Trojans today. Both Szabó et al. (2017) and Kalup et al. (2021) use the slow rotation periods and high lightcurve amplitudes of their observed Trojans to argue for a $\sim$25% rate of binaries within their samples.

As a second example, the publicly-released photometry data of the Zwicky Transient Facility was used by Schemel and Brown (2021) to obtain color, phase parameter, absolute magnitude, and amplitude of rotation measurements for 1049 Trojan asteroids. Overall, they found that the distribution of rotational light curve amplitudes is relatively constant with asteroid size, and they argued it was not distinguishable from that of main belt asteroids of similar sizes.

This result can be compared with the interpretation of observations taken from the Asteroid Terrestrial-impact Last Alert System (ATLAS). Using sparse photometric data for 863 L4 Trojans and 380 L5 Trojans, McNeill et al. (2021) argued that L4 asteroids are more elongated than the L5 asteroids. The group claimed this could be a byproduct of greater collisional evolution within the L4 swarm, which may have slightly more $D > 10$ km bodies (Grav et al. 2011). We suspect this scenario is unlikely, mainly because the collision probabilities within L4 or L5 are comparable to one another (e.g., Nesvorný et al. 2018). The discrepancy between McNeill et al. (2021) and Schemel and Brown (2021) has yet to be resolved.

3 Origin Scenarios for Trojans

3.1 Review of Trojans Origin Models

One reason that the Trojans hold our fascination is because their origin story has defied conventional explanations for decades. There is no dynamical pathway for a heliocentric body to reach a stable orbit near L4 or L5 within the context of the three-body problem (i.e., Sun, Jupiter, and an asteroid). Either the Trojans had to form *in situ*, which is unlikely, or some kind of extra perturbation was needed to capture them. This has led to a plethora of origin scenarios for the Trojans. We refer the reader to the comprehensive review of Trojan origin scenarios provided by Marzari et al. (2002) and Emery et al. (2015).

For example, a natural idea is that the Trojans are remnant planetesimals from the Jupiter zone that were captured in L4 or L5 via gas drag. As Jupiter was forming, it was proposed that gas drag from the solar nebula could have provided the kind of dissipative force needed to overcome the limitations of the three-body problem (e.g., Yoder 1979; Peale 1993). Alternatively, mutual collisions between planetesimals might have also managed to trap objects that happened to be lingering near L4 or L5 (Shoemaker et al. 1989).

Additional possibilities come from the idea that Jupiter's growth or migration would lead to a shift in the location and properties of L4 and L5, leading to capture among planetesimals that happened to be at the right place at the right time. An example of such a model would be an increase in the growth of Jupiter via accretion of planetesimals, protoplanets, and/or gas (Marzari and Scholl 1998a,b; Fleming and Hamilton 2000). Jupiter's migration in a gas disk could also lead to Trojan capture, provided the migration was smooth enough that planetesimals trapped along the way would remain trapped (e.g., Lykawka et al. 2009; Pirani et al. 2019a). It should be noted that the Pirani et al. (2019a) mechanism shows that an inward migrating Jupiter will capture more Trojans within L4 than L5. If it can be shown that the population asymmetry favoring L4 over L5 is real and statistically robust, this result would represent evidence in favor of their origin model (though the Pirani et al. (2019a) mechanism has other issues, as will discussed below).

Our expectation for these models is that the Trojans were planetesimals that were formed as a byproduct of gas and dust concentration mechanisms in the solar nebula. Given this condition, it is likely that many planetesimals would be residing near the ecliptic if they happened to be captured while the solar nebula was still in existence. This leads to an Achilles' heel for many capture models, however, in that most new Trojans would retain low inclinations once the solar nebula dissipated.

So far, no mechanism has yet been identified that can substantially increase the inclinations of captured Trojans to those matching observations. Secular resonances produced by other worlds are unable to do it, nor are protoplanets that flyby and gravitationally interact with bodies already trapped within L4/L5 (e.g., Marzari et al. 2002; Pirani et al. 2019b). It is conceivable that planetary embryos were temporarily captured within L4 and L5, thereby exciting the Trojans, but numerical simulations indicate such large bodies are unlikely to both (i) produce the appropriate inclination distributions in each cloud and (ii) escape prior to the present day (Pirani et al. 2019b). Instead, such bodies would most likely still be there and would thus be easily observable. Accordingly, the inclination distribution of the Trojans makes it difficult to accept any of the proposed capture models discussed above.

The Trojan origin models best able to explain the inclination constraint are those that assume that the planetesimals were excited prior to capture, and that the capture mechanism itself involves the migration of multiple giant planets along with Jupiter.

3.2 Review of Trojan Origin Models Involving a Giant Planet Instability

As discussed above, a good Trojan origin model requires several components:

T1) A planetesimal population that has been dynamically excited prior to capture,
T2) A mechanism that involves Jupiter's migration, since it has the ability to modify the shape of the L4 and L5 zones, and

T3) A scenario capable of placing an appropriate number of objects into L4 and L5 with the appropriate SFDs and physical/spectral properties while also satisfying other Solar System constraints.

For T1, while it is possible that planetesimals in the giant planet zone became excited during the formation of the giant planets themselves, and thus could have been captured while gas was still around, it seems likely that such a scenario would predominantly produce low inclination Trojans. This is not to say that zero Trojans were captured at this time, but something had to happen to them, otherwise this putative population would skew the current Trojans toward low inclinations. In addition, based on main belt modeling predictions and constraints, our expectation is that captured planetesimals from the giant planet zone would look much more like C-type asteroids than D- and P-type asteroids (e.g., Walsh et al. 2011).

Taken together, it seems probable that the Trojans were dynamically excited after the solar nebula had dissipated than before it was gone. This means that to satisfy T2, we need Jupiter's migration to also occur after the solar nebula had been eliminated. Finally, for T3, we need to destabilize an excited small body population that can satisfy the physical constraints for the Trojans.

3.2.1 The Giant Planet Instability

At present, the best model that can explain such behavior is characterized as the so-called "Nice model", an umbrella term for a suite of models that allow giant planet migration to occur after the elimination of the gas disk (Tsiganis et al. 2005; see Nesvorný 2018 for a review). Here we describe their basics.

After the gas disk disperses, our system of giant planets was in a different configuration than the one seen today. Gas accretion left them on nearly circular coplanar orbits between $\sim$5 and $\sim$20 au, with most/all locked in mutual mean motion resonances with one another. Beyond that zone, a massive primordial Kuiper belt existed just beyond the original orbit of Neptune, with an estimated population of $\sim$20 to 30 Earth masses (or more) that stretched well beyond 20 au. This configuration, however, eventually went unstable. The most natural explanation for the dynamical structure of the present-day Kuiper Belt is that the outermost giant planets radially migrated through the primordial Kuiper belt in a violent exchange of orbital energy and angular momentum. This led to a reorganization of the giant planets in what is now called a "giant planet instability" (e.g., Tsiganis et al. 2005; Nesvorný and Morbidelli 2012; Nesvorný 2018; Morbidelli and Nesvorný 2020).

As Neptune entered into the primordial Kuiper belt, it excited the objects residing there via resonant interactions. The subsequent migration of Neptune through this population caused most of the bodies to be flung throughout the Solar System. It led the primordial Kuiper belt to lose a factor of $\sim$1000 in population, with roughly this number of Pluto-sized bodies – and enumerable smaller bodies – scattered into the giant planet zone and/or out into a scattered disk of comets associated with Neptune. The interaction between Neptune and Pluto-sized bodies made the migration grainy rather than smooth, and this in turn affected the capture of Kuiper belt objects into Neptune resonances, explaining observations (Nesvorný and Vokrouhlický 2016; Kaib and Sheppard 2016; Lawler et al. 2019).

The number of giant planets gravitationally interacting with one another at the time of the instability is unknown. Dynamical studies suggest systems with 5 or 6 giant planets (e.g., 3 or 4 Uranus/Neptune-size bodies) have greater success reproducing dynamical constraints across the solar system than those that start with 4 giant planets (Nesvorný 2011; Nesvorný and Morbidelli 2012; Batygin et al. 2012). In the most successful cases, a Neptune-sized body interacting with Jupiter causes it to migrate slightly inward via numerous tiny jumps.

These events may help describe the dynamical structure of the asteroid belt and terrestrial planet system, and collectively are often referred to as "Jumping Jupiter" scenarios (e.g., see review in Morbidelli et al. 2015).

3.2.2 Trojan Capture During the Giant Planet Instability

From this context, we can examine several different but fairly related capture scenarios for the Trojans. For example, in the original Nice model, Morbidelli et al. (2005) proposed that gravitational interactions with refugees from the primordial Kuiper belt led Jupiter and Saturn to enter into their mutual 1:2 mean motion resonance. This resonance crossing had several effects. First, it triggered the giant planet instability and destabilized the outer solar system, with Uranus and Neptune eliminating most of the primordial Kuiper belt. This caused large numbers of objects, with a range of inclinations (see below), to be sent to the vicinity of Jupiter. Second, just after Jupiter and Saturn crossed their mutual 1:2 resonance, Jupiter's L4 and L5 regions became unstable for a short time, allowing refugee Kuiper belt objects to both enter and exit. Further giant planet migration closed the doors on the L4 and L5 regions, leaving the objects left inside those regions on stable orbits.

An issue here is that in this model, and in all models using a giant planet instability, Trojans are captured Kuiper belt objects. The model prediction is that the size distribution, colors, and physical nature of the Trojans should match what we know of the Kuiper belt and other populations that were captured at the same time by the same process, unless they are affected by subsequent evolutionary processes (e.g., collisions). This implies that Trojans are siblings of D- and P-type asteroids captured in the main asteroid belt, Hilda population, irregular satellite populations, and Neptune Trojans (Levison et al. 2009; Vokrouhlický et al. 2016; see Nesvorný 2018 for a review).

This initial version of the Nice model, and its associated Trojan capture mechanism, has increasingly become disfavored. Numerical simulations and studies of exoplanets have shown that the giant planets likely evolved into mutual mean motion resonances prior to the end of the solar nebula (Nesvorný 2018). This would suggest that the giant planet instability was not triggered when Jupiter and Saturn entered the 1:2 mean motion resonance, but rather when some of the planets broke out of resonant lock. Numerical studies show this is a common occurrence taking place shortly after the solar nebula has dissipated. In effect, the initial giant planet configuration is stable with gas, but relatively unstable without it, particularly when interacting with a large primordial Kuiper belt that may be losing bodies to the giant planet zone.

In this alternative version of the giant planet instability, giant planet encounters with one another cause them to change orbits, moving their L4 and L5 stability zones in a sudden fashion. This means that a small fraction of wandering Kuiper belt objects will have a "dynamical net" thrown over them, capturing them in stable orbits if they happen to be on the right orbits. The advantage of this model is that it better reproduces the observed distribution of Trojans, and it allows the stochastic nature of Trojan capture events to potentially produce a modest asymmetry in the populations within L4 and L5 (Emery et al. 2015). This model also favors Jupiter interacting with an extra Neptune-sized body that is eventually thrown out of the Solar System (Nesvorný 2011; Nesvorný and Morbidelli 2012; Batygin et al. 2012; Nesvorný et al. 2013; Vokrouhlický et al. 2016; Nesvorný et al. 2017a).

The probability that primordial Kuiper belt objects will be captured in the Trojan population at this time is approximately one in a million (i.e., $\sim 5 \times 10^{-7}$) (Nesvorný et al. 2013). This value is comparable to the capture probability of similar objects in the Hilda asteroid population (i.e., those in the 3:2 mean motion resonance with Jupiter) and the irregular satellites of the giant planets, but it is several times smaller than the probability of being captured in the main belt (e.g., Vokrouhlický et al. 2016). Accordingly, we would expect more D- and P-types in the main belt than the Trojans, and the largest body in the former should exceed that of the latter, both which are true (Vokrouhlický et al. 2016). Overall, these models reproduce the numbers and sizes of the largest D- and P-type asteroids in each of the aforementioned populations except the irregular satellites, which have probably experienced exceptional levels of collisional evolution (Bottke et al. 2010; Vokrouhlický et al. 2016).

This model also predicts that the Trojans, the dynamically hot Kuiper belt population, and the scattered disk are all from the same primordial Kuiper belt (Levison et al. 2008). From existing observations, it appears the SFD of the dynamically hot Kuiper belt is consistent with this idea; both show a Trojan-like like bump near 100 km and a similar shape for the observed bodies (Morbidelli et al. 2009a; Adams et al. 2014; Fraser et al. 2014). They may also be a match with the largest bodies in the scattered disk SFD, which is the source of the ecliptic comets, but observational incompleteness among scattered disk bodies and ecliptic comets make this prediction difficult to test at this time.

3.2.3 Details on the Evolution of the Primordial Kuiper Belt

The results presented above imply that the inclination distribution of the Trojans was obtained from objects that came from a dynamically excited primordial Kuiper belt that was further excited prior to capture by giant planet interactions. Accordingly, in order to interpret the nature of present-day Trojans, we need to understand the initial conditions of the primordial Kuiper belt, how and when this population became excited, and what happened to ejected objects en route to being captured in the Trojan population. On this basis, the properties of the present-day Kuiper belt, scattered disk and even ecliptic comets provide us with insights into the nature of the Trojan population, and vice versa.

Since Tsiganis et al. (2005) and Morbidelli et al. (2005), there have been many advances in our thinking about the nature and timing of the giant planet instability and the capture of the Trojans. More recent numerical simulations indicate the timing of the giant planet instability is controlled by the unknown size of the gap between Neptune's initial orbit and the primordial Kuiper belt's inner edge (Nesvorný et al. 2018). The smaller the gap, the earlier the instability. If the gap is <1.5 au, there is effectively no delay in the onset of Neptune's migration, and it would take place shortly after dissipation of the solar nebula. That might explain why the Trojans appear to have experienced limited collisional evolution, as we will discuss below.

The initial state of the primordial Kuiper belt is thought to have been dynamically cold, a necessary condition for planetesimal formation. The objects can be assumed to have had low eccentricities ($e < 0.1$) and inclinations ($i < 10^\circ$) prior to Neptune's outward migration. The disk itself stretched from near 24 au, a distance modestly outside the original semimajor axis of Neptune, to at least 30 au, the current semimajor axis of Neptune, and probably beyond. In the original Nice model, it was assumed that Neptune would migrate through the disk until it reached the outer edge and/or the spatial density of the objects dropped low enough to terminate planetesimal-driven migration (Tsiganis et al. 2005). If the former, the edge would need to be located near 30 au, but this scenario has difficulty explaining the cold classical Kuiper belt, whose outer edge is at $\sim$45-50 au.

Concentrating on the latter scenario, models can now show that the cold classicals can be explained by assuming the spatial density of objects in the primordial Kuiper belt gradually decreased away from the Sun (Nesvorný et al. 2020b). This idea is plausible, with photoevaporation potentially reducing the gas density from the outside-in and thereby limiting planetesimal formation in the cold classical region and beyond (Carrera et al. 2017). That would allow Neptune to migrate through the population until there was insufficient mass for it to move any further. From here, the only objects left outside of Neptune's orbit would be those captured in mean motion resonances with Neptune by Neptune's outward migration, destabilized objects that would eventually be ejected from the primordial Kuiper belt by Neptune perturbations (i.e., the scattered disk), destabilized objects scattered by a more eccentric Neptune that were left stranded when Neptune evolved to its current eccentricity (i.e., the detached scattered disk), and those so far away from Neptune that they were not dynamically affected by Neptune's migration.

The last population is a good fit with the properties of the cold classical Kuiper belt. It is home to objects like Arrokoth, a body that was likely formed *in situ* (Batygin et al. 2011; Parker and Kavelaars 2012; Fraser et al. 2014; McKinnon et al. 2020). Note that recent models suggest Arrokoth and other KBOs were produced when clouds of small particles experienced gravitational collapse (see Sect. 5.1). The unaffected nature of the cold classical Kuiper belt would also explain why it has a high fraction of well-separated binaries relative to the fraction in the dynamically hot Kuiper belt.

As Neptune migrated outward through the disk, it traveled slowly enough to dynamically excite the eccentricities and inclinations of objects located modestly ahead of it (Nesvorný and Vokrouhlický 2016). Most were destabilized and sent to the giant planet zone. This evolution probably explains the inclinations of the Trojan swarms. Some of the destabilized bodies from the primordial Kuiper belt reaching the giant planet zone managed to be in the right place when Jupiter was having encounters with another giant planet, and this led to their capture into many stable zones, such as the central and outer asteroid belt, the 3:2 and 1:1 mean motion resonances with Jupiter, and in distant orbits around Jupiter itself. These worlds now dominate the D- and P-type populations seen in the asteroid belt, Hildas, Trojans, and irregular satellites of Jupiter.

Finally, some objects destabilized by Neptune were sent to a scattered disk associated with Neptune's orbit; this region is now the primary source of our Centaurs and ecliptic comets (e.g., Duncan and Levison 1997; Nesvorný et al. 2017b). Other bodies were passed down the other giant planets, where they were ejected. A modest fraction of this population managed to reach Oort cloud, potentially explaining the origin of these bodies (e.g., Vokrouhlický et al. 2019), though some Oort cloud comets may also have been captured from passing stars in our primordial star cluster (Levison et al. 2010). These associations mean that the Trojans have many siblings across the solar system. Some of these bodies are still in cold storage, while others are just coming out of it as they approach the Sun. Many Trojan siblings are also collisionally evolved from their time in the excited primordial Kuiper belt.

4 Collisional and Physical Evolution of Trojans Prior to Capture

In this section, we consider the physical and collisional evolution of the Trojans. By adopting the above origin scenario, we assume that most or perhaps all of the Trojans were once members of the primordial Kuiper belt. That means many Trojans have physical histories that predate their capture in Jupiter's L4 and L5 Lagrange points. We divide this history into several stages.

4.1 Planetesimal and Binary Formation in the Primordial Kuiper Belt

In stage one, we consider the formation of Trojans as icy planetesimals in a primordial Kuiper belt that was still dynamically cold. Theoretical work and numerical simulations indicate that planetesimals were likely formed by concentrating small grains in protoplanetary disks. Specifically, the streaming instability model suggests that pebbles, defined as mm- to cm-sized particles, move toward high pressure zones in the solar nebula, and are aerodynamically collected into self-gravitating clouds (Youdin and Goodman 2005; see review by Johansen and Lambrechts 2017). When a sufficient spatial density of pebbles is reached, the particle clouds undergo gravitational collapse and form into short-lived disk structures from which planetesimals accrete. The size of the bodies formed can have a range of sizes, but the most common diameter of the largest planetesimals created is $\sim$100 km (Klahr and Schreiber 2020, 2021). They may even produce numerous equal size components that coalesce together, as has been suggested to explain the mounds on Arrokoth (Stern et al. 2023).

Numerical simulations indicate that particle/gas clouds with low angular momentum will frequently produce a single largest planetesimal with a size near $\sim$100 km, but those with higher degrees of angular momentum will often lead to the formation of equal-sized binary components orbiting one another at substantial distances (Nesvorný et al. 2019, 2021). These latter binaries are consistent with observations of many Kuiper belt binaries; the components are near 100 km in diameter, they are nearly equally sized components, they can have wide separations, and they have similar colors. These types of binaries are most common in the cold classical Kuiper belt (i.e., 40 out of 65 investigated are such binaries) (Noll et al. 2020). They are less common elsewhere, but that could be a consequence of the bodies undergoing collisional and dynamical evolution in their source populations during the giant planet instability and Neptune's migration through the primordial Kuiper belt.

These same planetesimal formation events frequently produce and eject a number of smaller bodies as well (Nesvorný et al. 2019, 2021). In some cases, the smaller objects form from flattened disks spun off from larger clusters of particles, leading to shapes that may resemble hamburgers. These kinds of origin conditions may provide a ready explanation for the flattened appearance of the two joined lobes of Arrokoth (e.g., McKinnon et al. 2020) and that of the Lucy target Polymele.

For example, in the hot classical Kuiper belt today, many binaries with a large primary also have a satellite less than half the size of the primary (though binaries with smaller equal-size components also exist) (Nesvorný and Vokrouhlický 2019). It is possible these satellites formed from impact-generated disks after their primary was struck during stage one collisions (Canup 2005; Leinhardt et al. 2010; Sekine et al. 2017). Some evidence supporting this comes from Sekine et al. (2017), who argue that high-velocity giant impacts between KBOs produce global/hemispherical darkening and reddening, thereby explaining the color variations observed among large KBOs. If true, reasons the aforementioned binaries are not found in the cold classical population may be because that population was smaller, and therefore it experienced fewer stage one collisions, or because these objects are simply harder to detect.

Our expectation is that the fraction of well-separated binaries with equal-sized components in the primordial Kuiper belt once rivaled the fraction seen in the cold classicals. Moreover, the destabilized Kuiper belt population and the hot classical Kuiper belt probably came from the same region of the primordial Kuiper belt, so we can deduce that some of these bodies were in binaries formed from impact-generated disks. The idea that many Trojans were previously in some kind of binary is consistent with observations from the Kepler spacecraft that $\sim$15% of Jupiter Trojans have very slow rotation (spin periods $>$ 100 hr;

see Sect. 2.5). Models indicate these slow rotators could have had their rotational angular momentum taken away by gravitational interactions with a satellite, with the satellite then stripped away via gravitational interactions with a giant planet prior to Jupiter capture or destroyed by collisions (Nesvorný et al. 2020a).

These ideas fit well with the diversity of Lucy's Trojan targets, of which three are known to have satellites. For example, Leucus, which is 63.8 $\times$ 36.6 $\times$ 29.6 km D-type Trojan, is a very slow rotator with a spin period of 445.68 hours (see Mottola et al. 2024, this collection). Perhaps it was one component of a stripped binary. Patroclus, which we will discuss in Sect. 4.2, and Polymele both have satellites that were probably created and/or captured during the planetesimal formation process. We consider it less likely that collisions in the primordial Kuiper belt created Queta, the small satellite of the target Eurybates (see Sect. 5.3).

These planetesimal formation mechanisms, along with accretion between planetesimals and pebble accretion, led to the formation of KBOs ranging from sub-100 km planetesimals to Pluto-sized bodies (and perhaps beyond). The net mass of the primordial Kuiper belt prior to collisional evolution was probably $\sim$20-30 Earth masses, and it contained the order of $\sim 10^3$ Pluto-sized bodies. Once created, the Pluto-sized bodies would have modestly excited the surrounding planetesimals, possibly enough to produce some degree of collisional grinding at slow impact velocities (i.e., tens to hundreds of m/s). The magnitude of stage one collisional grinding, however, is unknown, and our knowledge of how these collisions influenced the primordial Kuiper belt SFD and Kuiper belt binaries is limited.

It is also conceivable that stage one collisional evolution may be responsible for the observed cumulative power law slope of $q = -2.1$ found among observed Trojans with $D < 100$ km. The idea is that low velocity collisions in the primordial Kuiper belt may produce sufficient grinding to create a Dohnanyi-type size distribution (Dohnanyi 1969; O'Brien and Greenberg 2003; see Supplemental Fig. 4 in Nesvorný et al. 2018). The shape of this SFD would then be transferred to small body reservoirs across the solar system, including the Trojans. This scenario may be testable by examining the physical nature of Trojans (e.g., what evidence to Trojans show for very early collisional evolution history in addition to their observed craters).

4.2 Evidence for an Early Instability from the Patroclus Binary

Using the ideas presented above, we now consider an application of stage one collisional evolution. The most prominent example of an equal-sized binary in the Trojan population is (617) Patroclus and its satellite Menoetius, which we hereafter call the P-M binary. Both P-M components are larger than 100 km, and they are separated by 670 km. Located in Jupiter's L5 cloud, they will be visited by the Lucy spacecraft in 2033. As suggested above, the P-M binary likely formed in the primordial Kuiper belt as a byproduct of planetesimal formation, and its existence as an intact Trojan binary provide us with insights into the nature of the primordial Kuiper belt, a possibility recognized by Nesvorný et al. (2018). The P-M binary had to endure stage one collisional grinding in the primordial Kuiper belt between 24-30 au, collisional evolution as a destabilized binary en route to the Trojans, and disruption events in the Trojan population over the last $\sim$4.5 Gyr. Moreover, it had to avoid being stripped by encounters with a giant planet prior to capture in L5.

By putting together a collisional and dynamical model for the primordial Kuiper belt, and assuming a test population made of P-M-like binaries, Nesvorný et al. (2018) estimated the attrition rate of these bodies over time. Their results showed that if the giant planet instability occurred immediately after the dissipation of the solar nebula, and there were no stage one collisions, roughly 60% of their model P-M binaries would survive (i) stage two and stage three collisions (i.e., collisions from an excited primordial Kuiper belt and collisions from the Trojans over 4.5 Gyr, respectively) and (ii) encounters with giant planets capable of producing binary stripping. When the giant planet instability occurred at later times, allowing stage one collisions to become more important, the survival rate of the P-M binaries dropped precipitously. In other words, the earlier the giant planet instability, and the faster stage one collisions come to an end, the higher the odds that a P-M binary would survive to become a Trojan.

Ultimately, Nesvorný et al. (2018) predicted that the giant planet instability had to start within $\sim$100 Myr of the first solids, but they found the most favorable results when migration started shortly after the dissipation of the solar nebula. This argues against late migration for Neptune, which is problematic for other reasons (e.g., it is difficult to keep giant planets in a stable resonant configuration for long time periods without a gas disk, as discussed by Nesvorný (2018); late giant planet migration can produce dynamical damage to fully formed terrestrial planets; Agnor and Lin (2012); Kaib and Chambers 2016; Clement et al. 2018; Nesvorný et al. 2021). It also means there are hard limits to how much stage one collisional evolution could have occurred.

4.3 Insights into Small Trojans and KBOs from Craters on Charon and Arrokoth

At some unknown but presumably early time, Neptune entered the primordial Kuiper belt, migrated through it, and started stage two collisional evolution. Here the primordial Kuiper belt became dynamically excited, and for the first time, the objects observed by the New Horizons spacecraft, namely Pluto-Charon and Arrokoth, were capable of being hit by a large fraction of the excited primordial Kuiper belt population. Collision velocities on Pluto-Charon and Arrokoth were raised to 2.2 and 0.7 km s^{-1}, respectively (Morbidelli et al. 2021), and numerous cratering events should have taken place as the destabilized portion of the primordial Kuiper belt was scattered onto Pluto-Charon/Arrokoth-crossing orbits.

To glean insights into what the crater size distributions might be like on Trojans, it is useful to consider the smallest impactors striking Charon and Arrokoth, two worlds from the primordial Kuiper belt where crater constraints exist via New Horizons observations. Despite the potential for numerous sub-km impactors to strike either body, both Charon and Arrokoth have a curious deficiency in the number of $D < 10$ km diameter craters on their surfaces (Singer et al. 2019; Spencer et al. 2020; Robbins and Singer 2021). Charon craters follow a shallow cumulative power law slope of $q = -0.7 \pm 0.2$ for $1 < D < 10$ km craters (Singer et al. 2019; Robbins and Singer 2021), while Spencer et al. (2020) found Arrokoth craters had a slope of $q = -1.3$ for $0.2 < D < 1$ km craters. Note that a reassessment of both data sets by Morbidelli et al. (2021), which also considered the formation of a large 7 km crater on Arrokoth, argued that the power law slope of the crater size distribution was more likely to be $-1.5 < q < -1.2$, and that the projectile size distribution between 30 m and 1 km likely had $-1.2 < q < -1.0$.

Regardless, the shallow slope for a once sizeable population that is likely experienced collisional evolution is curious and intriguing, so its provenance may tell us about the physical properties of Trojans and primordial Kuiper belt objects. A similarly-shaped projectile SFD can be derived from small primary craters found on Europa, Ganymede, and Enceladus, worlds predominantly hit by comets from the scattered disk (Zahnle et al. 2003; Bierhaus et al. 2012). Morbidelli et al. (2021) argued this feature, mostly built during stage one and two collisions, was likely analogous to a similar shallow slope found in the asteroid belt. We will discuss this issue in Sect. 5, where we consider collisional evolution models for the Trojans.

5 Collisional Evolution Among the Trojans

5.1 Modeling the Collisional Evolution of the Trojans

Once the Trojans were captured in L4 and L5, they continued to undergo collisional evolution from impacting Trojans and, occasionally, from Hilda asteroids, Jupiter family comets, and nearly isotropic comets. We refer to these collisions as occurring in stage three. The nature of this evolution depends on several factors:

a. Their initial SFD at capture
b. Time of capture,
c. Collision probabilities and impact velocities of Trojans striking each other, and
d. Collisional disruption law followed by Trojans striking one another

While (a) is unknown, one can make some educated estimates as to what the SFD might have looked like and examine the amount of collisional grinding that potentially took place over the past ~4.5 Gyr. For example, the general similarity between the shapes of the Trojan and Kuiper belt SFDs for objects larger than several tens of km (see Sect. 2.3) might indicate the SFD for $D > 10$ km at the time of capture was not significantly different from the SFD today (Marschall et al. 2022). That would indicate the SFD for larger Trojans was set by some combination of planetesimal formation processes and early collisional grinding in a primordial Kuiper belt (Nesvorný et al. 2018; Nesvorný and Vokrouhlický 2019). Here we discuss how a Trojan SFD containing the combined L4 and L5 populations would evolve in the case where the observed cumulative slope of $q = -2.1$ continues to small sizes.

For (b), one can assume that the giant planet instability takes place shortly after the dissipation of the solar nebula, as discussed in Sect. 4.2. For (c), input parameters have been chosen similar to those in Nesvorný et al. (2018), namely that the intrinsic collision probability between Trojans within L4 or L5 are $P_{\rm i} = 7 \times 10^{-18}$ km^{-2} yr^{-1} and that the impact velocity $V = 4.6$ km s^{-1}. With that said, L4 objects cannot hit L5 objects, and vice versa, so the collision probability used for a single combined Trojan SFD in a collisional evolution simulation must be scaled downward by a factor of ~ 2. Note that this simulation also neglects the fact that the Hilda asteroids, Jupiter family comets, and nearly isotropic comets may also hit the Trojans, partly for simplicity but also because their influence on the overall collisional evolution of the Trojans is limited.

Concerning (d), the collisional strength of a body is commonly described by the critical impact energy Q_D^*, defined as the energy per unit target mass needed to disrupt and disperse 50% of the target (Benz and Asphaug 1999, see Davis et al. 2002). For the Trojans, the disruption law Q_D^* is unknown and therefore must be treated as a free parameter.

Using the collisional model discussed in Marschall et al. (2022), we will examine how the Trojans evolve from an initial SFD like the one discussed above when different Q_D^* laws are applied. To interpret what is happening, it is useful to first consider what we know about collisional evolution in the main asteroid belt.

Consider that in the main belt, impacts on $D > 100$ km objects create numerous fragments, which themselves disrupt as part of a collisional cascade. Collisional modeling results indicate the easiest bodies to disrupt from an energy per mass perspective are $D \sim 200$ m (e.g., Bottke et al. 2020). This means $D < 200$ m objects, which are in the strength regime, grind themselves into a Dohnanyi-shaped SFD with a cumulative slope of $q \sim -2.6$ (Dohnanyi 1969; O'Brien and Greenberg 2003). This steep population of small objects goes on to disrupt bodies between 200 m and 2 km, creating a shallow slope of $q \sim -1$ between the inflection points. In turn, the paucity of bodies between 200 m and

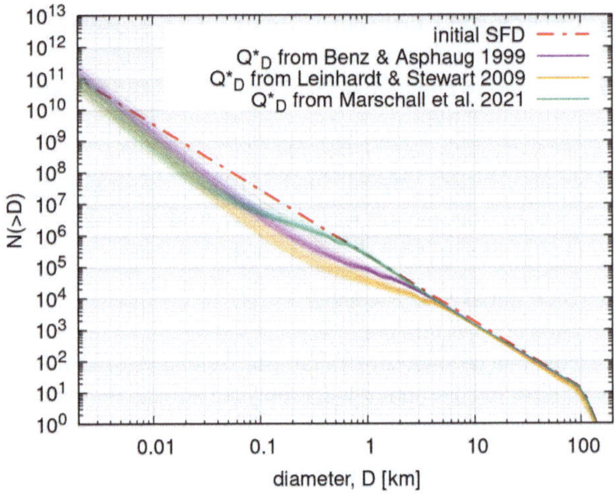

Fig. 3 Trojans size distribution after 4.5 Gy of collisional evolution for three different Q_D^* functions. The initial SFD is shown in the dashed red line. Adapted from Marschall et al. (2022)

2 km means that $D > 2$ km fragments produced by big collisions can build up over time, leading them to have a steep SFD. Overall, the models show the main belt SFD takes on a wavy shape.

Morbidelli et al. (2021) argued similar behavior would occur in the primordial Kuiper belt and its daughter populations, except there the weakest objects from an energy per mass perspective would be $D \sim 20$ m rather than 200 m. This could potentially create the shallow slope found for $D < 10$ km craters on Charon and many giant planet satellites (see last section). The issue is whether these same trends hold for Trojans, where most of their bombardment for the last 4.5 Gyr has come from their fellow captured Trojans.

In Fig. 3, the weakest objects in the Trojan disruption scaling laws were set to be between $D = 20$ to 200 m. In each case, the bodies smaller than the weakest object develop a Dohnanyi-like slope of $q = -2.6$ (Dohnanyi 1969; O'Brien and Greenberg 2003). As expected, the sizes larger than the weakest body develop a shallow slope q near -1. Overall, the behavior mimics collisional evolution in the main asteroid belt and probably the Kuiper belt/scattered disk (Bottke et al. 2005, 2015, 2020).

Though the shallow slope at intermediate sizes occurs in all cases investigated, the details of where the shallow slope begins and ends depends strongly on Q_D^*. Determining where this collisional break occurs will critically inform the nature of the Trojan disruption law. We predict the crater SFDs found on Lucy's Trojan targets may be the key data needed to understand the collisional evolution of the Trojans and other small outer solar system bodies.

5.2 The Eurybates Family in the Trojan Population

Another important feature of collisional evolution in a population is the formation of collisional families through large catastrophic impacts. Within the Trojans there is only one large collisional family: Eurybates, with (3548) Eurybates, a C-type Trojan, as its largest member family (Brož and Rozehnal 2011). The Eurybates family is well within the L4 stability region (Robutel and Gabern 2006). In inclination, the family is well defined, with most members within half a degree of 7.5° (Brož and Rozehnal 2011) (Fig. 1). From a spectral signature standpoint, the Eurybates family is also bluer than the overall L4 population. Others within the Trojans include the Hektor and Arkesilaos families (e.g. Rozehnal et al. 2016; Holt et al. 2020), but they will not be discussed further here.

The largest ten members of the Eurybates family, with diameters between 20 and 65 km, have a SFD whose power law slope is consistent with the overall slope of the Trojan population. At $D < 20$ km, the family's SFD is considerably steeper than the background population, a result indicative of a collisional family. The shape of the SFD, being diagnostic of the type of impact that formed the family, suggests that a 100 km parent body was hit by a $\sim$35 km impactor (Marschall et al. 2022). Further, the SFD indicates that the parent body had a rather weak collisional strength consistent with impact simulations of large rubble piles (Benavidez et al. 2012).

(3548) Eurybates has a km-sized satellite called Queta (Noll et al. 2020). It was likely created by disruptive collision that formed the family. The survival of Queta into the present day makes it possible to constrain the family's age, though it depends on the nature of the disruption law employed. Using the Benz and Asphaug (1999) disruption law for ice, which yields the Trojan SFD shown by the purple curves in Fig. 2, the model age for the Eurybates family is <3.7 Gy (Marschall et al. 2022). Previous family ages have resulted in a range from 1 Gyr to 4 Gyr (Brož and Rozehnal 2011; Rozehnal et al. 2016; Milani et al. 2017; Holt et al. 2020).

Though collisionally-derived satellites are a common byproduct of catastrophic disruption events, numerical models suggest they should have large eccentricities (Durda et al. 2004). Queta, with an eccentricity of only 0.125 (Brown et al. 2021), is not a typical case in that respect, though the Kozai resonance could produce eccentricity/inclination oscillations that raise the eccentricity to 0.4 (see Levison et al. 2024, this collection). With that said, its low eccentricity could also be telling us that better collisional models are needed. Regardless, Queta's formation and survival requires further explanation.

5.3 Colors of Trojans

Additional insights on Trojan evolution come from their colors and spectra. As discussed above, all Trojans tend to have low albedos. Objects with very red featureless slopes in visible and near infrared wavelengths are classified as D-type asteroids, those with modestly red slopes are P-types asteroids, and those with flat or nearly horizontal slopes are C-type asteroids (e.g., DeMeo et al. 2015). Obtaining detailed spectroscopic observations for objects more distant than Trojans can be difficult, though, so taxonomic information is limited for those worlds. For this reason, most observers have concentrated on colors, which provide a sense of the spectral slope of outer solar system worlds. Different groups have done this in different ways.

For example, one way is to use spectral slope to make color distinctions, with less red (LR) objects defined as those with spectra slopes near 5×10^{-5} Å^{-1}, red (R) objects having spectral slopes near 10×10^{-5} Å^{-1}, and very red (VR) objects having values that are considerably larger, roughly >20 to 25×10^{-5} Å^{-1} (Wong et al. 2014; Wong and Brown 2015, 2016, 2017a). Other groups use the *ugriz* magnitude system to make these color assessments (see Nesvorný et al. 2020b and references therein) or specific observation filters (Jewitt 2015; Jewitt 2018). For this discussion, we will lean on the LR, R, and VR definitions, with the boundaries between LR, R, and VR being modestly flexible. From this background, we can define what we know of various outer solar system populations.

The Trojans and Hildas have bimodal distribution of LR and R objects, and both resemble each other (Wong and Brown 2016, 2017a,b). The similarity is consistent with the idea that they have a common origin and source region. No VR objects have been identified in either population to date, an issue we will discuss below. The large asteroid families in the Trojans (i.e., Eurybates) and Hildas (i.e., Hilda, Schubert) are made up of LR objects. This suggests that collisional disruption exposes interior materials of the parent bodies which are LR.

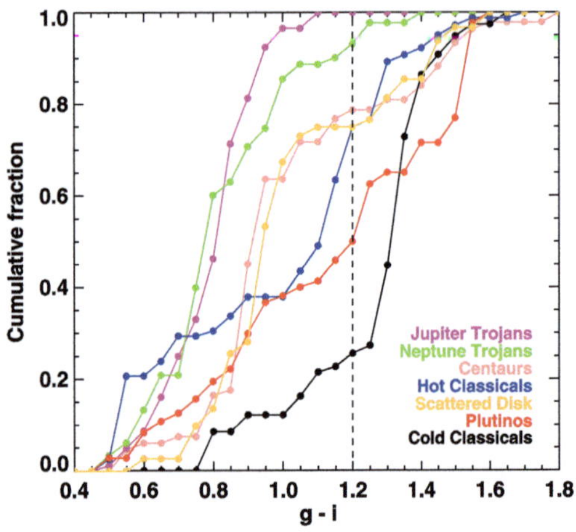

Fig. 4 Cumulative distributions of the g-i color index for many different outer solar system populations (see Sect. 5.3). Very red (VR) objects are assumed to be to the right of the dashed vertical line. Adapted from Bolin et al. (2023)

As we move further out in the solar system, the color distributions become more complex and diverse, as shown in Fig. 4, which has been adapted from Bolin et al. (2023) (see also Jewitt 2018).

- **Neptune Trojans**: These bodies have a LR/R color distribution similar to the Trojans, but include some VR objects.
- **Centaurs**: Nearly 20% of the population is comprised of VR bodies.
- **Scattered Disk**: It is the source of the Centaurs, and has color trends similar to that population.
- **Plutinos**: A 50:50 mix of R and VR objects.
- **Hot Classical**: This population has a comparable fraction of VR bodies to the Centaurs and scattered disk.
- **Cold Classicals**: They are dominated by VR objects, though a few objects are R and trend toward LR.

For this exercise, we will assume that all of the above color distributions can be explained within the context of the giant planet instability model, which provides the best available explanation of the dynamical properties of the above populations. This leads to different ways to interpret these color trends. We also must point out a caveat, namely that the majority of the Trojans and Centaurs characterized above are mainly bodies with diameters smaller than a few tens of km, while the characterized objects in more distant populations like the scattered disk, Plutinos, and various Kuiper belt populations are dominated by $D > 100$ km bodies. This size difference may mean that some comparisons below are potentially inappropriate.

Our first comment on Fig. 4 is that Jupiter and Neptune Trojans (and the Hildas, which are not plotted in Jewitt 2018 or Bolin et al. 2023) have close to the same color distribution. Dynamically, the bodies captured in the Neptune Trojans stayed far for the Sun, so that suggests that outgassing played a minimal role in producing the observed colors of either population, and that the LR objects in the Neptune Trojans were probably not produced this way.

Our second comment is that there are no VR colors in the Jupiter Trojans, but a small fraction exists in the Neptune Trojans. These data provide an update to Jewitt (2018); at that time of that publication, no VR Neptune Trojans were known. Objects with VR colors are more plentiful in the Centaurs, scattered disk, and other populations.

The dichotomy between the Jupiter and Neptune Trojans for VR objects may hold interesting information. Nesvorný et al. (2020b) showed that most of Neptune's Trojans come from 25-30 au within the primordial Kuiper belt, so placing the transition near 30 au could potentially explain why the Jupiter Trojans lack VR objects (Bolin et al. 2023). This would also imply that most VR objects are from a low mass extension of the primordial Kuiper belt between ~30-40 au, namely the zone that lies between Neptune's current orbit at 30 au and the current location of the cold classical Kuiper belt. Nesvorný et al. (2020b) also point out that having the VR objects from that zone would explain why they tend to have lower inclinations; objects destabilized by Neptune in this region do not get as excited as those between 20-30 au.

An alternative way to explain the dichotomy is to argue it is a surface veneer that goes away close to the Sun. Wong and Brown (2016, 2017a) have argued that VR colors may be produced by the sublimation of hydrogen sulfide ice at distances greater than ~15-20 au, with VR bodies then scattered to many small body reservoirs by Neptune's outward migration. The absence of VR bodies in Jupiter's Trojans and Hildas might represent the loss of surface volatiles and/or organics via solar heating close to the Sun, with the VR objects transformed into R objects. Such a transition has been observed by Jewitt (2015), with VR Centaurs gradually losing this color as they pass within 10 au. Nesvorný et al. (2020b) suggested sublimation-driven surface depletion of some organic molecules, such as NH_3, between 30-40 au might explain the R-VR transition (Brown et al. 2011).

Our third and last comment is that the role of collisional evolution in explaining these color trends has yet to be fully explored. Asteroid families in the Trojan and Hilda populations demonstrate that one way of producing LR objects is by collisional disruption (e.g., the Trojan target Eurybates). It is unclear what happens if collisions occur in other populations far from the Sun.

6 Conclusions

As this discussion shows, the origin and evolution of the Trojans are linked to the origin and evolution of planetesimals in the primordial Kuiper belt and the nature of giant planet migration. Yet, it is possible that flybys past these diverse worlds by the Lucy spacecraft will revolutionize our ideas about these bodies to such a degree that some of the suggested models discussed above will be overturned. The work done by this initial reconnaissance will lead to follow-up studies, perhaps involving landers and/or sample return missions to D- and P-types. This will potentially allow us to test the hypothesis that all such bodies derive from the primordial Kuiper belt.

Acknowledgements We thank the members of the Lucy team for their helpful comments in making this a better review article. We also thank our anonymous referee for their valuable comments that made this a better paper. The work in this paper was supported by NASA's Lucy mission through contract NNM16AA08C. The work of David Vokrouhlický was partially funded 2601 by grant 21-11058S from the Czech Science Foundation.

Declarations

Competing Interests The authors declare that they have no conflict of interest.

References

Adams ER, Gulbis AAS, Elliot JL, Benecchi SD, Buie MW, Trilling DE, Wasserman LH (2014) De-biased populations of Kuiper belt objects from the deep ecliptic survey. Astron J 148:55

Agnor CB, Lin DNC (2012) On the migration of Jupiter and Saturn: constraints from linear models of secular resonant coupling with the terrestrial planets. Astrophys J 745:143

Barucci MA, Cruikshank DP, Mottola S, Lazzarin M (2002) Physical properties of Trojan and Centaur asteroids. In: Michel P, DeMeo F, Bottke WF (eds) Asteroids IV. U. Arizona Press, pp 273–287

Batygin K, Brown ME, Fraser WC (2011) Retention of a primordial cold classical Kuiper belt in an instability-driven model of solar system formation. Astrophys J 738:13

Batygin K, Brown ME, Betts H (2012) Instability-driven dynamical evolution model of a primordially five-planet outer solar system. Astrophys J 744:L3

Benavidez PG, Durda DD, Enke BL, Bottke WF, Nesvorný D, Richardson DC, Asphaug E, Merline WJ (2012) A comparison between rubble-pile and monolithic targets in impact simulations: application to asteroid satellites and family size distributions. Icarus 219:57–76

Benz W, Asphaug E (1999) Catastrophic disruptions revisited. Icarus 142:5–20

Bierhaus EB, Dones L, Alvarellos JL, Zahnle K (2012) The role of ejecta in the small crater populations on the mid-sized Saturnian satellites. Icarus 218:602–621

Bolin BT et al (2023) Keck, Gemini, and Palomar 200-inch visible photometry of red and very-red Neptunian Trojans. Mon Not R Astron Soc 521:L29–L33

Bottke WF, Durda DD, Nesvorný D, Jedicke R, Morbidelli A, Vokrouhlický D, Levison H (2005) The fossilized size distribution of the main asteroid belt. Icarus 175:111–140

Bottke WF, Nesvorný D, Vokrouhlický D, Morbidelli A (2010) The irregular satellites. The most collisionally evolved population in the solar system. Astrophys J 139:994–1014

Bottke WF, Brož M, O'Brien DP, Campo Bagatin A, Morbidelli A, Marchi S (2015) The collisional evolution of the asteroid belt. In: Michel P, DeMeo F, Bottke WF (eds) Asteroids IV. U. Arizona Press, pp 701–724

Bottke WF, Vokrouhlický D, Ballouz R-L, Barnouin OS, Connolly HC Jr, Elder C, Marchi S, McCoy TJ, Michel P, Nolan MC, Rizk B, Scheeres DJ, Schwartz SR, Walsh KJ, Lauretta DS (2020) Interpreting the cratering histories of Bennu, Ryugu, and other spacecraft-explored asteroids. Astron J 160:14

Brown ME, Schaller EL, Fraser WC (2011) A hypothesis for the color diversity of the Kuiper belt. Astrophys J 739:L60

Brown ME, Levison HF, Noll KS, Binzel R, Buie MW, Grundy W, Marchi S, Olkin CB, Spencer J, Statler TS, Weaver H (2021) The orbit and density of the Jupiter Trojan satellite system eurybates-queta. Planet Sci J 2:170

Brož M, Rozehnal J (2011) Eurybates - the only asteroid family among Trojans? Mon Not R Astron Soc 414:565–574

Buie MW, Zangari AM, Marchi S, Levison HF, Mottola S (2018) Light curves of Lucy targets: Leucus and Polymele. Astron J 155:245

Canup RM (2005) A giant impact origin of Pluto-Charon. Science 307:546–550

Carrera D, Gorti U, Johansen A, Davies MB (2017) Planetesimal formation by the streaming instability in a photoevaporating disk. Astrophys J 839:16

Clement MS, Kaib NA, Raymond SN, Walsh KJ (2018) Mars' growth stunted by an early giant planet instability. Icarus 311:340–356

Davis DR, Durda DD, Marzari F, Campo Bagatin A, Gil-Hutton R (2002) Collisional evolution of small-body populations. In: Bottke WF, Cellino A, Paolicchi P, Binzel R (eds) Asteroids III. U. Arizona Press, pp 545–558

DeMeo FE, Alexander CMO, Walsh KJ, Chapman CR, Binzel RP (2015) The compositional structure of the asteroid belt. In: Michel P, DeMeo F, Bottke WF (eds) Asteroids IV. U. Arizona Press, pp 13–41

Di Sisto R, Ramos XS, Beaugè C (2014) Giga-year evolution of Trojans and the asymmetry problem. Icarus 243:287–295

Di Sisto RP, Ramos XS, Gallardo T (2019) The dynamical evolution of escaped Jupiter Trojan asteroids, link to other minor body populations. Icarus 319:828–839

Dohnanyi JS (1969) Collisional model of asteroids and their debris. J Geophys Res 74:2531–2554

Dotto E, Fornasier S, Barucci MA, Licandro J, Boehnhardt H, Hainaut O, Marzari F, de Bergh C, De Luise F (2006) The surface composition of Jupiter Trojans: visible and near-infrared survey of dynamical families. Icarus 183:420–434

Duncan MJ, Levison HF (1997) A scattered comet disk and the origin of Jupiter family comets. Science 276:1670–1672

Durda DD, Bottke WF, Enke BL, Asphaug E, Richardson DC, Leinhardt ZM (2004) The formation of asteroid satellites in catastrophic impacts: results from numerical simulations. Icarus 167:382–396

Emery JP, Marzari F, Morbidelli A, French LM, Grav T (2015) The complex history of Trojan asteroids. In: Michel P, DeMeo F, Bottke WF (eds) Asteroids IV. U. Arizona Press, pp 203–220

Fleming HJ, Hamilton DP (2000) On the origin of the Trojan asteroids: effects of Jupiter's mass accretion and radial migration. Icarus 148:479–493

Fornasier S, Dotto E, Hainaut O, Marzari F, Boehnhardt H, De Luise F, Barucci MA (2007) Visible spectroscopic and photometric survey of Jupiter Trojans: final results on dynamical families. Icarus 190:622–642

Fraser WC, Brown ME, Morbidelli A, Parker A, Batygin K (2014) The absolute magnitude distribution of Kuiper belt objects. Astrophys J 782:100

French LM, Stephens RD, Coley D, Wasserman LH, Sieben J (2015) Rotation lightcurves of small Jovian Trojan asteroids. Icarus 254:1–17

Grav T, Mainzer AK, Bauer J, Masiero J, Spahr T, McMillan RS, Walker R, Cutri R, Wright E, Eisenhardt PRM, Blauvelt E, DeBaun E, Elsbury D, Gautier T, Gomillion S, Hand E, Wilkins A (2011) WISE/NEOWISE observations of the Jovian Trojans: preliminary results. Astrophys J 742:40

Holt TR, Nesvorný D, Horner J, King R, Marschall R, Kamrowski M, Carter B, Brookshaw L, Tylor C (2020) Stability of Jovian Trojans and their collisional families. Mon Not R Astron Soc 495:4085–4097

Horner J, Müller TG, Lykawka PS (2012) (1173) Anchises - thermophysical and dynamical studies of a dynamically unstable Jovian Trojan. Mon Not R Astron Soc 423:2587–2596

Jewitt D (2015) Color systematics of comets and related bodies. Astron J 150:201.

Jewitt D (2018) The Trojan color conundrum. Astron J 155:56

Jewitt DC, Trujillo CA, Luu JX (2000) Population and size distribution of small Jovian Trojan asteroids. Astron J 120:1140–1147

Johansen A, Lambrechts M (2017) Forming planets via pebble accretion. Annu Rev Earth Planet Sci 45:359–387

Kaib NA, Chambers JE (2016) The fragility of the terrestrial planets during a giant-planet instability. Mon Not R Astron Soc 455:3561–3569

Kaib NA, Sheppard SS (2016) Tracking Neptune's migration history through high-perihelion resonant trans-Neptunian objects. Astron J 152:133

Kalup CE, Molnár L, Kiss C, Szabó GM, Pál A, Szakáts R, Sárneczky K, Vinkó J, Szabó R, Kecskeméthy V, Kiss LL (2021) 101 Trojans: a tale of period bimodality, binaries, and extremely slow rotators from K2 photometry. Astrophys J Suppl Ser 254:7

Klahr H, Schreiber A (2020) Turbulence sets the length scale for planetesimal formation: local 2D simulations of streaming instability and planetesimal formation. Astrophys J 901:54

Klahr H, Schreiber A (2021) Testing the jeans, toomre, and bonnor-ebert concepts for planetesimal formation: 3D streaming-instability simulations of diffusion-regulated formation of planetesimals. Astrophys J 911:9

Lawler SM, Pike RE, Kaib N, Alexandersen M, Bannister MT, Chen Y T, Gladman B, Gwyn S, Kavelaars JJ, Petit J-M, Volk K (2019) OSSOS. XIII. Fossilized resonant dropouts tentatively confirm Neptune's migration was grainy and slow. Astron J 157:253

Leinhardt ZM, Marcus RA, Stewart ST (2010) The formation of the collisional family around the dwarf planet Haumea. Astrophys J 714:1789–1799

Levison HF, Shoemaker EM, Shoemaker CS (1997) Dynamical evolution of Jupiter's Trojan asteroids. Nature 385:42–44

Levison HF, Morbidelli A, Van Laerhoven C, Gomes R, Tsiganis K (2008) Origin of the structure of the Kuiper belt during a dynamical instability in the orbits of Uranus and Neptune. Icarus 196:258–273

Levison HF, Bottke WF, Gounelle M, Morbidelli A, Nesvorný D, Tsiganis K (2009) Contamination of the asteroid belt by primordial trans-Neptunian objects. Nature 460:364–366

Levison HF, Duncan MJ, Brasser R, Kaufmann DE (2010) Capture of the Sun's Oort cloud from stars in its birth cluster. Science 329:187–190

Levison HF, Olkin CB, Noll KS, Marchi S, Bell JF, Bierhaus E, Binzel R, Bottke W, Britt D, Brown M, Buie M, Christensen P, Emery J, Grundy W, Hamilton VE, Howett C, Mottola S, Pätzold M, Reuter

D, Spencer J, Statler TS, Stern SA, Sunshine J, Weaver H, Wong I (2021) Lucy mission to the Trojan asteroids: science goals. Planet Sci J 2:171

Levison H et al (2024) Lucy science and mission overview. Space Sci Rev 220

Lykawka PS, Horner J, Jones BW, Mukai T (2009) Origin and dynamical evolution of Neptune Trojans - I. Formation and planetary migration. Mon Not R Astron Soc 398:1715–1729

Mainzer AK, Bauer JM, Cutri RM, Grav T, Kramer EA, Masiero JR, Sonnett S, Wright EL (2019) NEOWISE diameters and albedos V2.0. NASA planetary data system

Marchis F, Durech J, Castillo-Rogez J, Vachier F, Cuk M, Berthier J, Wong MH, Kalas P, Duchene G, van Dam MA, Hamanowa H, Viikinkoski M (2014) The puzzling mutual orbit of the binary Trojan asteroid (624) Hektor. Astrophys J Lett 783:L37

Marschall R, Nesvorný D, Deienno R, Wong I, Levison HF, Bottke WF (2022) Implications for the collisional strength of Jupiter Trojans from the Eurybates family. Planet Sci J 164:167

Marzari F, Scholl H (1998a) The capture of Trojans by a growing proto-Jupiter. Icarus 131:41–51

Marzari F, Scholl H (1998b) The growth of Jupiter and Saturn and the capture of Trojans. Astron Astrophys 339:278–285

Marzari F, Scholl H, Murray C, Lagerkvist C (2002) Origin and evolution of Trojan asteroids. In: Bottke WF, Cellino A, Paolicchi P, Binzel R (eds) Asteroids III. U. Arizona Press, pp 725–738

Masiero JR, Mainzer AK, Bauer JM, Grav T, Nugent CR, Stevenson R (2013) Asteroid family identification using the hierarchical clustering method and WISE/NEOWISE physical properties. Astrophys J 770:7

McKinnon WB, Richardson DC, Marohnic JC, Keane JT, Grundy WM, Hamilton DP, Nesvorný D, Umurhan OM, Lauer TR, Singer KN, Stern SA, Weaver HA, Spencer JR, Buie MW, Moore JM, Kavelaars JJ, Lisse CM, Mao X, Parker AH, Porter SB, Showalter MR, Olkin CB, Cruikshank DP, Elliott HA, Gladstone GR, Parker JW, Verbiscer AJ, Young LA, New Horizons Science Team (2020) The solar nebula origin of (486958) Arrokoth, a primordial contact binary in the Kuiper belt. Science 367:aay6620

McNeill A, Erasmus N, Trilling DE, Emery JP, Tonry JL, Denneau L, Flewelling H, Heinze A, Stalder B, Weiland HJ (2021) Comparison of the physical properties of the L4 and L5 Trojan asteroids from ATLAS data. Planet Sci J 2:6

Melita MD, Licandro J, Jones DC, Williams IP (2008) Physical properties and orbital stability of the Trojan asteroids. Icarus 195:686–697

Milani A (1993) The Trojan asteroid belt: proper elements, stability, chaos and families. Celest Mech Dyn Astron 57(1–2):59–94

Milani A, Knežević Z, Spoto F, Cellino A, Novaković B, Tsirvoulis G (2017) On the ages of resonant, eroded and fossil asteroid families. Icarus 288:240–264

Morbidelli A, Nesvorný D (2020) Kuiper belt: formation and evolution. In: Prialnik D, Barucci MA, Young L (eds) The trans-Neptunian solar system. Elsevier, Amsterdam, pp 25–59

Morbidelli A, Levison HF, Tsiganis K, Gomes R (2005) Chaotic capture of Jupiter's Trojan asteroids in the early solar system. Nature 435:462–465

Morbidelli A, Bottke WF, Nesvorný D, Levison H (2009b) Asteroids were born big. Icarus 204:558–573

Morbidelli A, Levison HF, Bottke WF, Dones L, Nesvorný D (2009a) Considerations on the magnitude distributions of the Kuiper belt and of the Jupiter Trojans. Icarus 202:310–315

Morbidelli A, Walsh KJ, O'Brien DP, Minton DA, Bottke WF (2015) The dynamical evolution of the asteroid belt. In: Michel P, DeMeo F, Bottke WF (eds) Asteroids IV. U. Arizona Press, pp 493–508

Morbidelli A, Nesvorný D, Bottke WF, Marchi S (2021) A re-assessment of the Kuiper belt size distribution for sub-kilometer objects, revealing collisional equilibrium at small sizes. Icarus 356:114256

Mottola S et al (2024) Shapes, rotations, photometric and internal properties of Jupiter Trojans. Space Sci Rev 220

Nakamura T, Yoshida F (2008) A new surface density model of Jovian Trojans around triangular libration points. Publ Astron Soc Jpn 60:293

Nesvorný D (2011) Young solar system's fifth giant planet? Astrophys J 742:L22

Nesvorný D (2018) Dynamical evolution of the early solar system. Annu Rev Astron Astrophys 56:137–174

Nesvorný D, Dones L (2002) How long-lived are the hypothetical Trojan populations of Saturn, Uranus, and Neptune? Icarus 160:271–288

Nesvorný D, Morbidelli A (2012) Statistical study of the early solar system's instability with four, five, and six giant planets. Astron J 14:117

Nesvorný D, Vokrouhlický D (2016) Neptune's orbital migration was grainy, not smooth. Astrophys J 825:94

Nesvorný D, Vokrouhlický D (2019) Binary survival in the outer solar system. Icarus 331:49–61

Nesvorný D, Vokrouhlický D, Morbidelli A (2013) Capture of Trojans by jumping Jupiter. Astrophys J 768:45, 8pp.

Nesvorný D, Roig F, Bottke WF (2017a) Modeling the historical flux of planetary impactors. Astron J 153:103

Nesvorný D, Vokrouhlický D, Dones L, Levison HF, Kaib N, Morbidelli A (2017b) Origin and evolution of short-period comets. Astrophys J 845:27

Nesvorný D, Vokrouhlický D, Bottke WF, Levison HF (2018) Evidence for very early migration of the solar system planets from the patroclus-menoetius binary Jupiter Trojan. Nat Astron 2:878–882

Nesvorný D, Li R, Youdin AN, Simon JB, Grundy WM (2019) Trans-Neptunian binaries as evidence for planetesimal formation by the streaming instability. Nat Astron 3:808–812

Nesvorný D, Vokrouhlický D, Alexandersen M, Bannister MT, Buchanan LE, Chen Y-T, Gladman BJ, Gwyn SDJ, Kavelaars JJ, Petit J-M, Schwamb ME, Volk K (2020b) OSSOS XX: the meaning of Kuiper belt colors. Astron J 160:46

Nesvorný D, Vokrouhlický D, Bottke WF, Levison HF, Grundy WM (2020a) Very slow rotators from tidally synchronized binaries. Astrophys J 893:L16

Nesvorný D, Roig FV, Deienno R (2021) The role of early giant-planet instability in terrestrial planet formation. Astron J 161:50

Noll KS, Grundy WM, Nesvorný D, Thirouin A (2020) Trans-Neptunian binaries. In: The trans-Neptunian solar system. Elsevier, Amsterdam, pp 205–224

O'Brien DP, Greenberg R (2003) Steady-state size distributions for collisional populations: analytical solution with size dependent strength. Icarus 164:334–345

Parker AH, Kavelaars JJ (2012) Collisional evolution of ultra-wide trans-Neptunian binaries. Astrophys J 744:139

Peale SJ (1993) The effect of the nebula on the Trojan precursors. Icarus 106:308–322

Pirani S, Johansen A, Bitsch B, Mustill AJ, Turrini D (2019a) Consequences of planetary migration on the minor bodies of the early solar system. Astron Astrophys 623:A169

Pirani S, Johansen A, Mustill AJ (2019b) On the inclinations of the Jupiter Trojans. Astron Astrophys 631:A89

Robbins SJ, Singer KN (2021) Pluto and Charon Impact Crater Populations: Reconciling Different Results. Planet Sci J 2:192

Robutel P, Gabern F (2006) The resonant structure of Jupiter's Trojan asteroids - I. Long-term stability and diffusion. Mon Not R Astron Soc 372:1463–1482

Rozehnal J, Brož M, Nesvorný D, Durda DD, Walsh K, Richardson DC, Asphaug E (2016) Hektor - an exceptional D-type family among Jovian Trojans. Mon Not R Astron Soc 462:2319–2332

Ryan EL, Sharkey BNL, Woodward CE (2017) Trojan asteroids in the Kepler campaign 6 field. Astron J 153:116

Schemel M, Brown ME (2021) Zwicky transient facility observations of Trojan asteroids: a thousand colors, rotation amplitudes, and phase functions. Planet Sci J 2:40

Sekine Y, Genda H, Kamata S, Funatsu T (2017) The Charon-forming giant impact as a source of Pluto's dark equatorial regions. Nat Astron 1:0031

Shoemaker EM, Shoemaker CS, Wolfe RF (1989) Trojan asteroids: populations, dynamical structure and origin of the L4 and L5 swarms. In: Binzel RP, Gehrels T, Matthews MS (eds) Asteroids II. U. Arizona Press, Tucson, pp 921–948

Singer KN, McKinnon WB, Greenstreet S, Gladman B, Bierhaus EB, Stern SA, Parker AH et al (2019) Impact craters on Pluto and Charon indicate a deficit of small Kuiper belt objects. Science 363:955–959

Spencer JR, Stern SA, Moore JM, Weaver HA, Singer KN, Olkin CB, Verbiscer AJ, McKinnon WB et al (2020) The geology and geophysics of Kuiper belt object (486958) Arrokoth. Science 367:eaay3999

Stern SA, White OL, Grundy WM, Keeney BA, Hofgartner JD, Nesvorný D, McKinnon WB, Richardson DC, Marohnic JC, Verbiscer AJ, Benecchi SD, Schenk PM, Moore JM (2023) The properties and origin of Kuiper belt object Arrokoth's large mounds. Planet Sci J 4:176

Szabó GM, Ivezić Ž, Jurić M, Lupton R (2007) The properties of Jovian Trojan asteroids listed in SDSS moving object catalogue 3. Mon Not R Astron Soc 377:1393–1406

Szabó GM, Pál A, Kiss C, Kiss LL, Molnár L, Hanyecz O, Plachy E, Sárneczky K, Szabó R (2017) The heart of the swarm: K2 photometry and rotational characteristics of 56 Jovian Trojan asteroids. Astron Astrophys 599:A44

Tsiganis K, Dvorak R, Pilat-Lohinger E (2000) Thersites: a "jumping" Trojan?. Astron Astrophys 354:1091–1100

Tsiganis K, Gomes R, Morbidelli A, Levison HF (2005) Origin of the orbital architecture of the giant planets of the solar system. Nature 435:459–461

Uehata K, Terai T, Ohtsuki K, Yoshida F (2022) Size distribution of small Jupiter Trojans in the L5 swarm. Astron J 163:213

Vokrouhlický D, Bottke WF, Nesvorný D (2016) Capture of trans-Neptunian planetesimals in the main asteroid belt. Astron J 152:39

Vokrouhlický D, Nesvorný D, Dones L (2019) Origin and evolution of long-period comets. Astron J 157:181

Walsh KJ, Morbidelli A, Raymond SN, O'Brien DP, Mandell AM (2011) A low mass for Mars from Jupiter's early gas-driven migration. Nature 475:206–209
Wong I, Brown ME (2015) The color-magnitude distribution of small Jupiter Trojans. Astron J 150:174
Wong I, Brown ME (2016) A hypothesis for the color bimodality of Jupiter Trojans. Astron J 152:90
Wong I, Brown ME (2017a) The bimodal color distribution of small Kuiper belt objects. Astron J 153:145
Wong I, Brown ME (2017b) The color-magnitude distribution of Hilda asteroids: comparison with Jupiter Trojans. Astron J 153:69
Wong I, Brown ME, Emery JP (2014) The differing magnitude distributions of the two Jupiter Trojan color populations. Astron J 148:112
Yoder CF (1979) Notes on the origin of the Trojan asteroids. Icarus 40:341–344
Yoshida F, Nakamura T (2008) A comparative study of size distributions for small L4 and L5 Jovian Trojans. Publ Astron Soc Jpn 60:297–301
Yoshida F, Terai T (2017) Small Jupiter Trojans survey with the subaru/hyper suprime-cam. Astron J 154:71
Youdin AN, Goodman J (2005) Streaming instabilities in protoplanetary disks. Astrophys J 620:459–469
Zahnle K, Schenk P, Levison H, Dones L (2003) Cratering rates in the outer solar system. Icarus 163:263–289

Space Science Reviews (2023) 219:44
https://doi.org/10.1007/s11214-023-00985-9

Surface Geology of Jupiter's Trojan Asteroids

S. Marchi[1] · J.F. Bell III[2] · B. Bierhaus[3] · J. Spencer[1]

Received: 1 June 2023 / Accepted: 5 July 2023 / Published online: 2 August 2023

Abstract
The surface geology of Jupiter's Trojan asteroids is one of the scientific investigations of the NASA *Lucy* mission. A dedicated Geology Working Group will implement these studies using primarily panchromatic and color imaging data and complement the interpretation of these data with theoretical models, such as collisional evolution models. The *Lucy* Science Team will also rely on experience and lessons learned from prior space missions, such as NASA's *NEAR*, *Dawn, OSIRIS-REx*, and *New Horizons*.

A chief goal of the Geology Working Group is to map craters and characterize their morphology across *Lucy* target's surfaces over a range of spatial resolutions. These data will be used to constrain the relative and absolute ages of terrains and their impactor size-frequency distributions. More broadly, impact-related processes such as excavation and mass wasting will inform other investigations, including geological unit mapping, stratigraphy and topography, surface composition, and internal structure. *Lucy*'s cratering data and morphology will also be used to perform comparative analyses with similar data from other small bodies across the Solar System, from Main Belt asteroids to Kuiper Belt objects. The present article provides an overview of the planned activities and methodologies of the Geology Working Group.

Keywords Trojans asteroids · Geology · Impact processes · Lucy mission

1 Introduction

The NASA *Lucy* mission will explore Jupiter's Trojan asteroids, a class of bodies co-orbital with Jupiter. The Trojan asteroids are found in two large reservoirs located at about ± 60° relative to Jupiter in its orbit around the Sun, close to the L_4 and L_5 Lagrangian Points. The L_4 and L_5 Lagrangian Points are dynamically stable locations for a restricted three-body problem (in this case, the Sun, Jupiter, asteroid), and Trojan asteroids are predicted

The Lucy Mission
Edited by Hal Levison, Simone Marchi and Keith Noll

✉ S. Marchi
marchi@boulder.swri.edu

1 Southwest Research Institute, Boulder, CO, USA

2 School of Earth and Space Exploration, Arizona State University, Tempe, AZ, USA

3 Lockeed Martin, Littleton, CO, USA

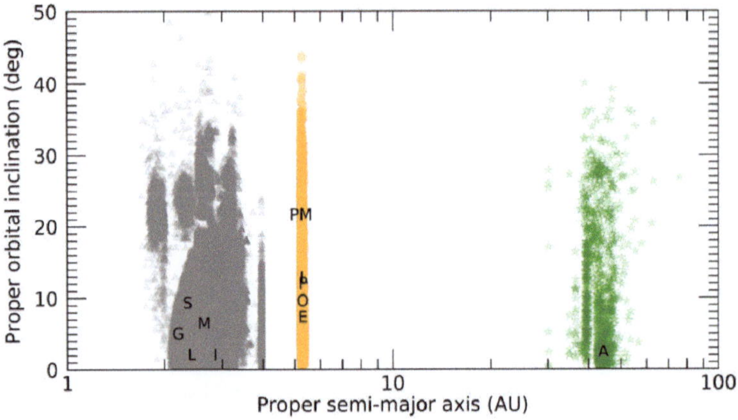

Fig. 1 Proper elements for numbered Main Belt asteroids (gray), numbered Trojan asteroids (orange), and numbered Kuiper Belt objects (green). Data from AstDyS (https://newton.spacedys.com/astdys/index.php?pc=5). Small MBAs and KBOs visited by spacecraft are indicated by their initials (Gaspra, Steins, Mathilde, Lutetia, Ida, Arrokoth). Similarly, *Lucy's* targets (Polymele, Eurybates, Orus, Leucus, Patroclus-Menoetius) are shown (note Leucus and Polymele overlap). Comets and near-Earth asteroids visited by spacecraft are not shown

to be stable over the age of the Solar System (Levison et al. 1997; Holt et al. 2021). While the origin of the Trojan asteroids is still debated (see Levison et al. 2023, this collection), there is consensus that they must have been dynamically captured early in Solar System evolution, perhaps within a few tens of Myr after formation (e.g., Nesvorny et al. 2013). Planetesimal formation models and dynamical models also indicate that Trojan asteroids probably formed within a few Myr of the origin of the Solar System and originally spanned a broad range of heliocentric distances (∼15–30 AU), but then were displaced into their current positions (∼5 AU) during the migration of the giant planets (e.g., Nesvorny et al. 2013). If such planetary migration models are correct, the Trojan asteroids could provide us with a close-up view of pristine materials that formed in the outer Solar System.

This hypothesis opens an interesting comparison with other stable classes of small bodies, such as Near-Earth objects (NEOs), Main Belt asteroids (MBAs) and Kuiper Belt objects (KBOs; Fig. 1). MBAs have heliocentric distances between the orbits of Mars and Jupiter (∼2–5 AU). Many large MBAs, especially those of the S-complex spectral type (related to ordinary chondrite meteorites), are thought to have formed at heliocentric distances within Jupiter's orbit (e.g., Bermingham and Kruijer 2022). Dynamical models also indicate that a significant fraction of MBAs, such as those with C-complex spectral types, could have formed close to the giant planets (Jupiter and Saturn) or even beyond Neptune's orbit, but were subsequently displaced into the Main Belt due to dynamical perturbations (e.g., Levison et al. 2009; Vokrouhlický et al. 2016; Raymond and Izidoro 2017). A fraction of these C-complex asteroids are the likely parent bodies of carbonaceous chondrite meteorites (e.g., Marchi et al. 2022).

KBOs are located at heliocentric distances larger than ∼30 AU. They are thought to be more volatile-rich than MBAs, including those belonging to the C-complex (e.g., Barucci et al. 2008, 2011; Brown et al. 2012), and have remained relatively unchanged since their formation. A fraction of KBOs, the so-called scattered disk and hot classical bodies, achieved their current orbits thanks to the dynamical interaction with Neptune (e.g., see review by Nesvorny 2018). The scattered disk is the source of Jupiter family comets (JFCs), objects that sporadically can reach the inner Solar System and offer opportunities for up close investigations (e.g., Nesvorny et al. 2017), such as comet 67P/Churyumov–Gerasimenko, the target of the ESA *Rosetta* mission. The cold classical disk, in contrast, is a reservoir of objects that only marginally interacted with Neptune and have remained on relatively stable orbits since their formation (e.g., Nesvorny 2018).

Space missions have explored a variety of NEOs, MBAs, and KBOs (e.g., *NEAR, Rosetta, Dawn, New Horizons, Galileo*, and others) revealing intriguing geological properties. Volatile-poor MBAs (e.g., (4) Vesta, (951) Gaspra, (243) Ida) are characterized by heavily cratered surfaces with overall morphologies resembling those of lunar terrains. The smaller sizes of MBAs compared to the Moon, however, result in distinctive features such as irregular shapes (e.g., (21) Lutetia, Ida, Gaspra), or global-scale ridges and grooves (e.g., Vesta, (433) Eros). (253) Mathilde is the only large ($\sim$50 km) MBA C-complex object with resolved images from a spacecraft (Veverka et al. 1999) and exhibits surface morphology distinct from that typically seen on S-complex bodies. Specifically, Mathilde has multiple large craters whose diameters approach the body radius. In conjunction with the low estimated bulk density ($\sim$1.3 g/cm^3, Veverka et al. 1999), Housen and Holsapple (2003) proposed that a relatively high bulk porosity explains the ability to survive these large impact events. In comparison to the NEOs and MBAs thus far explored by spacecraft, KBOs and former KBOs (i.e., JFCs) encountered so far have been seen to exhibit some surface features and compositional characteristics indicative of volatile-related processes. Examples include morphology and activity of comet 67P/Churyumov–Gerasimenko (e.g., Thomas et al. 2015) and troughs and pit chains potentially related to volatile release observed on (486958) Arrokoth (Spencer et al. 2020).

Trojan asteroids occupy intermediate heliocentric distances with respect to MBAs and KBOs. As such, they are of particular interest as they may represent a class of bodies that is not sampled in either population (e.g., distinct compositions; see Emery et al. 2023, this collection), or that have gone through radically different evolutionary processes (e.g., different collisional evolution; see Bottke et al. 2023, this collection). The comprehensive set of investigations to be performed by *Lucy*'s Geology Working Group (GWG) and broader Science Team will shed light on the origin of the Trojan asteroids, and their possible relationships with MBAs and KBOs.

In particular, the *Lucy* Level 1 science requirements have been carefully crafted to address major mission science goals, including the study of the crater populations of the targeted Trojans and their collisional evolution (Levison et al. 2021). Craters also offer an opportunity to study the near-surface stratigraphy, with implications for the composition and internal properties of Trojans (see Emery et al. 2023, this collection). For these reasons, the *Lucy* mission was designed to fly-by a set of Trojan asteroids spanning a wide range of physical properties (see Levison et al. 2023, this collection). Because of the differences in their physical properties (mass, shapes, spectral types), it is expected that their surface geology could also exhibit considerable variations. For instance, ground-based occultation data indicate that the best-fit ellipsoidal shape for *Lucy* Trojan target (15094) Polymele has major axes 27.6 km $\times$ 23.8 km $\times$ 10.0 km, with indications of a very complex, non-convex shape (Levison et al. 2023, this collection). At this time, it is not clear if this shape is due to the presence of large cavities (e.g., craters) or reflects a rubble pile nature. Notably, Polymele has a $\sim$5 km diameter satellite, whose origin is also unclear (Levison et al. 2023, this collection). *Lucy* Trojan targets (3548) Eurybates and (21900) Orus have similar and more regular shapes, with average diameters of 69.3 and 60.5 km, respectively (Mottola et al. 2023, this collection). Eurybates has a $\sim$1 km diameter satellite (Noll et al. 2020), Queta, enabling an estimate of Eurybates density to be 1.1$\pm$0.3 g/cm^3 (Brown et al. 2021). This density suggests that either Eurybates is largely made of ices, has a high porosity, or both. These properties could result in complex surface geology, affecting cratering morphology and creating features indicative of ice-loss (e.g., sublimation) processes. The *Lucy* Trojan target (11351) Leucus has a very elongated shape (major axes 60.8 km x 39.2 km x 27.8 km) and extremely long rotation period (445 hours), but we have no constraints yet on its surface geology (Mottola et al. 2020). Both elongated shape and slow rotation could be explained if Leucus was once a binary system now collapsed to a contact binary, however, several cords gathered during five occultation events did not reveal (or resolve) a contact binary shape (Mottola et al. 2020). The binary Trojan system (617) Patroclus and its companion Menoetius have a system density of $\sim$0.8 g/cm^3, thus again suggesting high ice content and/or high porosity. Patroclus and Menoetius' large diameters ($\sim$113 and $\sim$104 km, respectively; Marchis et al. 2006), and the fact that they form a roughly equal-mass binary pair similar to many found in the classical Kuiper Belt, suggest that they are primordial objects (Nesvorny et al. 2018) and thus may have accumulated significant cratering on their surfaces. In the next sections, we will review the geology of relevant objects (e.g., MBAs and KBOs), and discuss in detail *Lucy*'s planned geology-related observations.

2 The Geology of MBAs and KBOs

Spacecraft exploration of small (i.e., diameter smaller than $\sim$100-200 km) asteroids, comets, and natural satellites has revealed an impressive diversity of geologic processes on these bodies, which is perhaps surprising given their low gravity, lack of significant internal heat, and atmosphere-less surface environments. Impact cratering is generally the dominant surface modification process on such worlds, providing a way to potentially assess relative ages of surface regions (via crater areal density) as well as local relative stratigraphic relationships of geologic units. Evidence for both erosional and tectonic processes (e.g., ridges, grooves) has also been found on many of these bodies, providing additional clues about the evolution of their surfaces over time. While no specific evidence of (cryo)volcanic processes has yet been observed on the $\sim$12 small rocky asteroids or any of the small rocky planetary satellites so far encountered by spacecraft, evidence for volatile release processes has been observed on all $\sim$6 comet nuclei so far encountered. Additionally, significant geologic and compositional evidence for volcanic or cryovolcanic processes has been observed on a number of somewhat larger asteroids and planetary satellites (larger than $\sim$500 km diameter; e.g., Vesta, Ceres, Enceladus, Dione, Miranda), suggesting that the role of internal heating and potentially even differentiation begins to be much more important in the geologic evolution of bodies above that size. The relevance of these processes to smaller bodies such as the Trojan asteroids remains to be investigated.

The observed surface geology of a number of relevant small bodies can help to set expectations for the kinds of features that could be encountered during the *Lucy* mission's exploration of the Trojan asteroids. For example, spacecraft imaging of the Main Belt asteroids Ida, Mathilde, and Gaspra – all comparable in size to *Lucy's* first four Trojan asteroid targets – reveals surface morphologies dominated by the effects of impact cratering (e.g., Fig. 2a; Sullivan et al. 2002). Other small bodies such as Lutetia, Eros, Saturn's moon Phoebe, and Arrokoth – the latter two of which might have compositional similarities to Trojan asteroids – also show significant evidence for downslope movement and even "ponding" of loose, fragmental regolith materials (e.g., Fig. 2b; e.g., Robinson et al. 2002; Porco et al. 2005; Spencer et al. 2020). Evidence for throughgoing structures like grooves and ridges on Eros and the Martian moon Phobos (Fig. 2c) suggest tectonic processes and intact (but likely

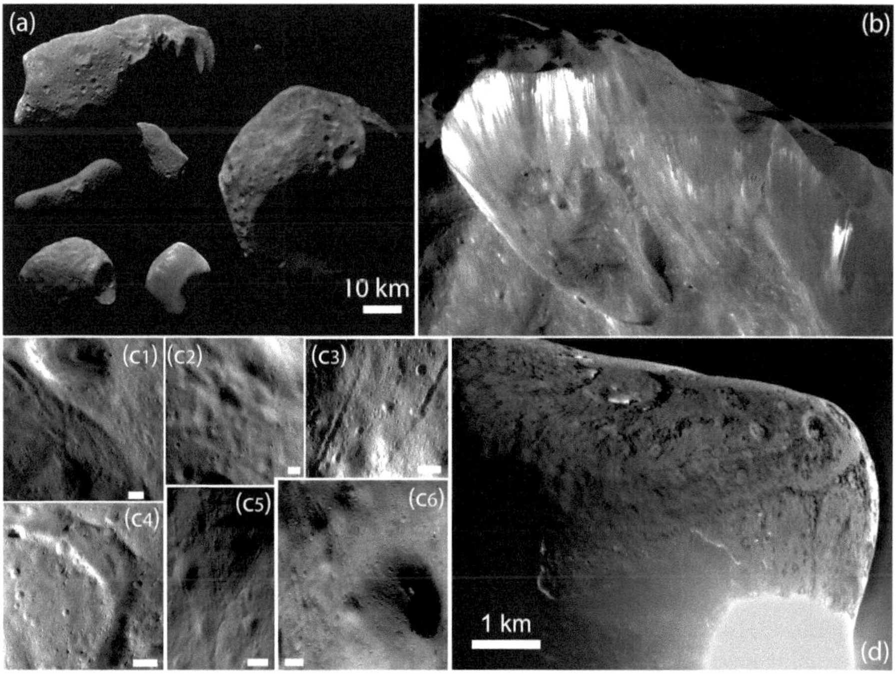

Fig. 2 A summary of surface properties of selected minor bodies. (a) Typical cratered surfaces of MBAs, NEOs, and planetary satellites generally comparable in size to many of the *Lucy* Trojan asteroid targets. Clockwise from upper left: Ida and its tiny moon Dactyl, Mathilde, Deimos, Phobos, Eros, and Gaspra (Sullivan et al. 2002), (b) Evidence for downslope regolith motion within the eastern wall of ∼45 km wide Jason crater on Saturn's ∼213 km diameter moon Phoebe. Typical streamer lengths are ∼10 km (Giesea et al. 2006); (c) Various morphological features: ridges (c1), shallow troughs (c2), pit chains (c3), flat-floored troughs (c4), fractures (c5), grooves (c6), on 34 × 11 × 11 km diameter NEO Eros (Buczkowski et al. 2008; Barnouin et al. 2013), all scale bars are 250 m; (d) Evidence for stratigraphic layering and potential flow features on the nucleus of ∼6 km diameter comet 9P/Tempel (Thomas et al. 2007). Figure 2a is from "Asteroid Geology from Galileo and NEAR Shoemaker Data" by Sullivan et al. in Asteroids III, edited by W.F. Bottke et al. © 2002 The Arizona Board of Regents. Reprinted by permission of the University of Arizona Press

fractured) rocky interiors, and also point to the potential role of tidal forces in modifying small body surfaces (e.g., Bottke et al. 1999; Robinson et al. 2002; Hurford et al. 2016). Finally, evidence for internal stratigraphic layering has been observed in spacecraft imaging of the nuclei of comets 9P/Tempel and 67P/Churyumov–Gerasimenko (Fig. 2d; e.g., Thomas et al. 2007; Giacomini et al. 2016; Birch et al. 2017). Detection of similar interior stratigraphic markers and relationships on potentially compositionally-analogous Trojan asteroids could provide unique clues about their interior structures as well as the relationship between Trojans and pristine outer Solar System cometary and KBO populations.

As noted above, cratering is one of the dominant geological processes on small bodies observed up-close so far, and we present below theoretical expectations for cratering on Trojan asteroids. The collisional environment of Trojan asteroids is similar to that of MBAs (Marschall et al. 2022). Collisional models compute an average intrinsic probability of collisions $P_i = 7\text{x}10^{-18}$ and $3\text{x}10^{-18}$ $\text{km}^{-2}\ \text{yr}^{-1}$, respectively for Trojans and MBAs. At a reference size of 1 km, the Main Belt has about 4 times more objects than the combined L_4 and L_5 Trojans. Average impact velocities are also similar, about 5 km/s for both populations (Marschall et al. 2022). Note that these parameters are significantly different from KBOs;

for instance, the impact velocities for Arrokoth are in the range 0.5–2.5 km/s, depending on the source of impactors (Morbidelli et al. 2021).

In light of these similarities between Trojan and MBA collisional properties, we derive here theoretical predictions for the Trojans' cratering history. For this purpose, we use a collisional model developed for MBAs (Marchi et al. 2009, 2012, 2016) and adapted to Trojans using the input parameters above.

The crater size is a complex function of impactor size, mass, velocity, and the mechanical properties of the target. Here we use the so-called Pi-group scaling law (e.g., Hosapple and Housen 2007) that allows computation of the transient crater diameter (D_t) as a function of impactor size (d), impact conditions and material properties:

$$D_t = kd\left[\frac{gd}{2v_\perp^2}\left(\frac{\rho}{\delta}\right)^{2\upsilon/\mu} + \left(\frac{Y}{\rho v_\perp^2}\right)^{(2+\mu)/2}\left(\frac{\rho}{\delta}\right)^{\upsilon(2+\mu)/\mu}\right]^{-\mu/(2+\mu)} \tag{1}$$

where g is the target gravitational acceleration, $v_\perp$ is the perpendicular component of the impactor velocity, δ is the projectile density, ρ and Y are the density and "cratering strength" of the target, k and μ depend on the cohesion of the target material and υ on its porosity. Therefore, the nature of the terrain affects the cratering efficiency and the functional dependence of the crater size with respect to the input parameters (e.g., impactor size and velocity).

Equation (1) models the so-called transient crater size resulting from the direct excavation and removal of target material. Large craters on planetary surfaces are expected to undergo a phase of crater modification in which highly fractured materials flow back toward the cavity. The resulting final crater (D_f) is typically from 20% to 50% larger than the transient crater in rocky targets. Here we assume a similar relationship for small bodies, and use 30% (Marchi et al. 2015). Figure 3 shows the computed crater size vs impactor size for average impact velocity for a generic 100-km Trojan asteroid. We implemented three different formulations of Eq. (1), namely a cohesive soil case for $Y = 10$, 100 kPa, and a porous case for $Y = 10$ kPa. We stress, however, that these relationships are poorly understood for highly porous materials. In addition, the strength of Trojan asteroids is not known. For reference, the strength of terrestrial basalt is of the order of 10 MPa, while dry alluvium is of the order of 50 kPa. Of note is the observation that the surface of small bodies (either comets or asteroids) are considered to be very low strength. For instance, Ballouz et al. (2020) concluded that meter-sized boulders on Bennu have a strength 0.5-1.7 MPa. These strength values are inferred from meter-scale properties, and the applicability to craters from hundreds to several km in scale is not clear.

Next, we derive a prediction for the cumulative number of craters. This requires knowledge of the evolution of the Trojan asteroid population over time, and whether other populations of objects have contributed to cratering of Trojan asteroids (for instance, interlopers from the outer Solar System). These are important details not yet fully worked out by dynamical models, and so here for simplicity we adopt a Main Belt-like impact rate over time (i.e., known as "chronology") for illustrative purposes (Fig. 4). The Main Belt chronology used here is from O'Brien et al. (2014) and has an increase on the MBA population number at 4.5 Ga of about a factor of 10 with respect to the present. We also compare our results with the observed craters on Mathilde (Chapman et al. 1999). Among the asteroids visited by spacecraft at close range, Mathilde likely has bulk properties closest to the Trojans based on its C-type spectral class, and density $\sim$1.3 g/cm^3. Mathilde's shape and geology is dominated by impact craters (Veverka et al. 1999; Thomas et al. 1999; Chapman et al. 1999). Figure 5 provides a view of Mathilde acquired by *NEAR* during approach. Large portions of

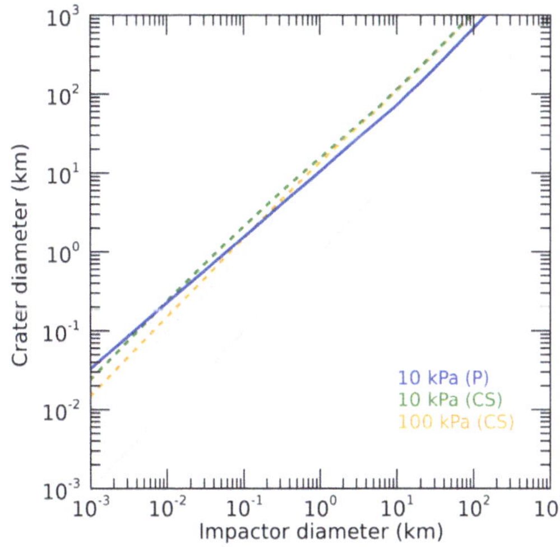

Fig. 3 Cratering scaling laws relevant for Trojan asteroids. The computed craters assume a 100-km target with a 1 g/cm^3 density (impactors have the same density). Scaling laws are from Eq. (1) for cohesive soils (CS) and porous materials (P). The material parameters are: $\nu = 0.4$, $\mu = 0.41$, $k = 1.03$ (CS); and $\nu = 0.4$, $\mu = 0.4$, $k = 0.725$ (P) are from Holsapple and Housen (2007). Gray lines indicate craters with diameters equal to 1x and 10x the impactor size

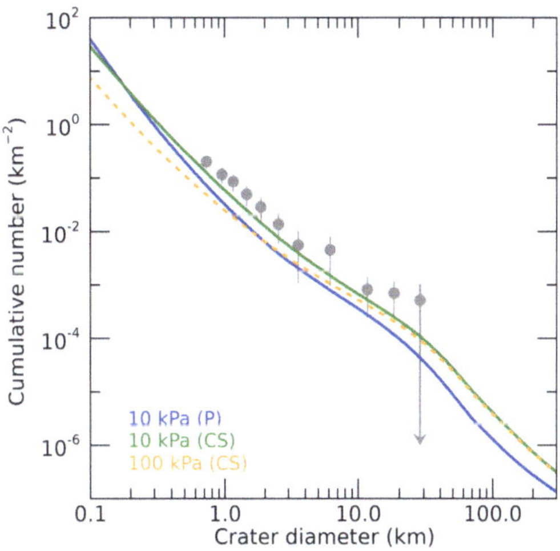

Fig. 4 Model crater production functions for a generic 100-km Trojan asteroid. All curves are for 4 Ga. We assumed a MBA-like decay in the impact rate over time, intrinsic probability and impact velocities as described in the text. The crater size-frequency distribution of asteroid Mathilde (gray points) is also reported for comparison (see text)

the object from both views are hidden in the depressions of large craters. The crater population seen in the illuminated surfaces is close to saturation (Chapman et al. 1999; Marchi et al. 2015), except for the walls of the giant craters. The relative paucity on the walls suggests downslope movement erases craters that form in those regions. Given the very slow rotation rate of Mathilde (17.4 days, Mottola et al. 1995), the $\sim$1 cm/s^2 surface acceleration for Mathilde (Thomas et al. 1999) is not modulated by rotation, and slope failure of material under such low acceleration implies weak cohesive strength. The low strength of the surface material is countered by high bulk porosity (Veverka et al. 1999; Thomas et al. 1999), which is expected to give Mathilde its ability to survive multiple large impacts (Housen and Holsapple 2003) whose resulting craters have diameters comparable to Mathilde's mean radius.

From this comparison, we find that if the Trojans have cratering similar to Mathilde, then either the Trojan population was much more numerous back in time as we assumed for the Main Belt, or other impactor populations contributed to cratering. Alternatively, if the primordial Trojan population was less numerous than assumed, then Trojans likely have

Fig. 5 Main belt asteroid Mathilde, as seen by *NEAR*. The large depressions are interpreted to be impact craters. Credit: https://photojournal.jpl.nasa.gov/catalog/PIA02477

significantly fewer craters compared to Mathilde. We stress that this analysis is speculative and based on a Main Belt-like chronology, and it will be updated in preparation for *Lucy*'s flybys. For further details on this topic, we refer to Marchi et al. (2023).

Another interesting comparison is with KBO Arrokoth, the target of the *New Horizons* mission (Spencer et al. 2020). As discussed by Morbidelli et al. (2021), cratering on Arrokoth indicates that the impactor SFD has a cumulative slope $q \sim -1.1$ in the size range ∼0.2–2 km, similar to MBAs. Arrokoth cratering shows that a similar slope is likely to hold down to impactor sizes ∼0.03 km, while MBAs are much steeper ($q \sim -2.7$) in the size range ∼0.03–0.2 km. Thus, a prediction of Morbidelli et al. (2021) is that cratering on Trojans may be significantly shallower than Mathilde for craters diameter in the size range ∼0.6–4 km (based on the scaling cratering scaling laws discussed above; see Fig. 3). As we shall discuss next, this is well within the observable resolution of the *Lucy* mission.

3 Lucy's Geology Investigation

The bulk of *Lucy*'s science investigation is carried out by five Working Groups (Geology, Surface Composition, Interior and Bulk Properties, Satellites and Rings, Activity), while two additional Working Groups coordinate *Lucy*'s cruise science and Earth-based observations. Here we report the Level 1 mission science requirements that are relevant to the Geology Working Group:

Requirement 4.1.2, Global Imaging for Shape Models and Geological Units:

Lucy shall obtain panchromatic images, spaced by 1/25 to 1/13 of a rotation, over a full target rotation. The target shall not overfill the field of view of the imagers during these observations. The resulting images shall be used to determine the global distribution of geological units and create shape models.

Requirement 4.1.3, Global Imaging for Shape Models and Geological Units:

Lucy shall obtain, during encounter, panchromatic images at a series of phase angles covering the sunlit surface in total, and separated by between 15° and 25°. The target shall not overfill the field of view of the imagers during these observations. The resulting images shall be used to determine the global distribution of geological units and create shape models.

Requirement 4.1.4, Stereoscopic Imaging for Elevation Models:

Lucy shall obtain panchromatic images of an area > 100 km^2 with a sub-spacecraft characteristic resolution $\leq$ 200 m at two emission angles (angle at the surface between the surface normal and the spacecraft) that are appropriate for the construction of stereoscopic imagery. The resulting images shall be used to produce stereoscopic images and elevation models.

Requirement 4.1.5, Landform Degradation at Different Latitudes:

Lucy shall obtain panchromatic images that cover a total area $\geq$ 500 km^2 or 50% of the available sunlit surface (whichever is less) and sample regions from the equator to latitudes of at least 60° from the equator, at a characteristic resolution $\leq$ 100 m. The resulting images shall be used to study how the structure of craters varies with latitude.

Requirement 4.1.6, Impact Crater Size Distribution:

Lucy shall obtain panchromatic images of an area $\geq$ 700 km^2 or 80% of the available sunlit surface (whichever is less) that are capable of resolving craters larger than 7 km in diameter. The resulting images shall be used to determine the surface number density of the observed craters.

Requirement 4.1.7, Impact Crater Size Distribution:

Lucy shall obtain panchromatic images of an area $\geq$ 10 km^2 that are capable of resolving craters larger than 70 m in diameter. The resulting images shall be used to determine the surface number density of the observed craters.

The surface geology Level 1 mission requirements will be accomplished by a combination of imaging data taken by different instruments, and in particular panchromatic imaging from L'LORRI, MVIC panchromatic and color imaging, and panchromatic imaging from the TTCam tracking camera (Olkin et al. 2023; Bell et al. 2023; Emery et al. 2023, this collection). Note that all the spatial requirements are meant to be based on resolution elements, that is, 3 native pixels for L'LORRI, 2 pixels for MVIC, 2 pixels for TTCam.

For each Lucy target, we have developed a robust and comprehensive timeline for data acquisition. As an example, we report here the simulated panchromatic imaging data to be acquired at Polymele, the smallest of *Lucy*'s primary targets (Fig. 6). We stress that this is based on current flyby trajectory planning, and the final imaging sequence is subject to change. For this simulation, we assumed a close approach distance of 434 km, a flyby speed of 6 km/s, and an approach phase angle of ~80°. We approximated Polymele's shape using an ellipsoid with major axes 28 km × 28 km × 11 km (about 784 km^2 hemispheric area), a rotation period of 11.5 hours (Mottola 2023, this collection), and we assumed a pole orientation perpendicular to the plane defined by the Sun and spacecraft velocity vector.

In this specific example, Requirement 4.1.2 is achieved by L'LORRI observations during the inbound trajectory from a range from 160,000 to 33,000 km, or from 7 to 1.5 hours prior to close approach. These data, in addition to selected TTCam images near close approach, will be used to achieve Requirement 4.1.3. TTCam images near close approach (range −13,000 to +13,000 km) will be used to achieve Requirement 4.1.4, while MVIC panchromatic images with a similar range will be used to achieve Requirement 4.1.5. TTCam images at a range of about +10,000 km will achieve Requirement 4.1.6, while LORRI imaging at a range of about −1200 km will achieve the resolution and coverage requirement for Requirement 4.1.7.

Additional data will be used to support and expand *Lucy*'s geological investigations, in particular MVIC color imaging. Planning the acquisition of these data is the responsibility

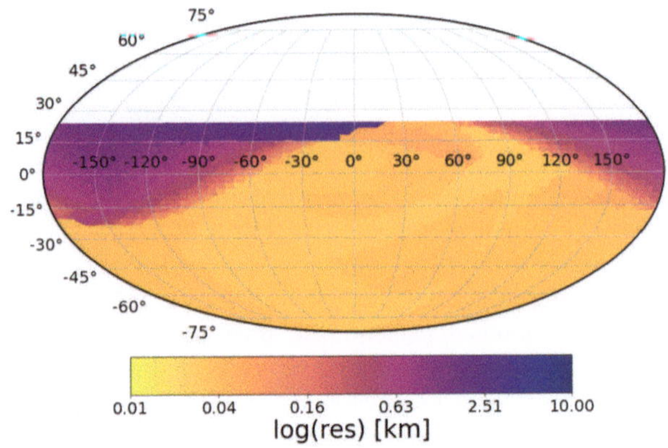

Fig. 6 *Lucy*'s panchromatic imaging of Polymele. The resolution coverage map shows the higher resolution hemisphere imaged near close approach, while the rest of the imaged surface is from a larger range. Note that the northern pole is not visible for the assumed pole orientation

of the Surface Composition Working Group, and includes the goal of obtaining a global color map of each target's surface (Emery et al. 2023, this collection).

The general structure of the data acquisition plan presented here for Polymele will be repeated for all of *Lucy*'s targets, with modifications as needed to accommodate specific flyby conditions. For instance, the planned close approach distance of about 434 km for Polymele is needed to measure the mass of the object with the required accuracy (Mottola et al. 2023, this collection). Because *Lucy's* other targets are significantly more massive than Polymele, the close approach distance can be larger. A larger close approach distance (e.g., 1000 km) allows for a more relaxed timeline for additional observations to be performed, while still achieving all the Level 1 mission requirements.

An alternative example is provided by the largest of *Lucy*'s targets, the Patroclus-Menoetius binary system (Fig. 7). The nominal close approach distance is 1075 km (with respect to Menoetius), a flyby speed of 8.8 km/s, and an approach phase angle of about 56°. The binary orbital period is 103 hours (Marchis et al. 2006), assumed to be the same as each object's rotation period, and we adopted a pole orientation perpendicular to the mutual orbital plane, with a declination of $-23°$ with respect to the plane defined by the Sun and spacecraft velocity vector. Similar to the Polymele encounter, the Level 1 requirements are met with a combination of L'LORRI, MVIC panchromatic and TTCam imaging. A peculiarity of the binary system is that the highest resolution observations (Requirement 4.1.7) taken near close approach require Lucy to slew its instrument platform a few times back and forth between Patroclus and Menoetius, adding significant complexities to the observation sequence.

4 Conclusions

Lucy's observations are carefully planned for maximum scientific return. Not only the nominal observational sequences guarantee fulfillment of the Level 1 science requirements, but the science return is also significantly expanded by adding more observations when possible.

The investigations pertinent to the *Lucy* mission's Geology Working Group cover a broad range of topics that include terrain morphologies, stratigraphic relationships, cratering, and overall asteroid shapes. *Lucy's* diversity of targets provides unique opportunities for comparative studies among Trojan asteroids, and with other populations of small bodies visited at close range, including near-Earth asteroids, Main Belt asteroids, and Kuiper Belt objects. Of particular interest is the range of physical properties of *Lucy's* targets, across different

Fig. 7 *Lucy*'s panchromatic imaging coverage of Patroclus and Menoetius (bottom and top panel, respectively)

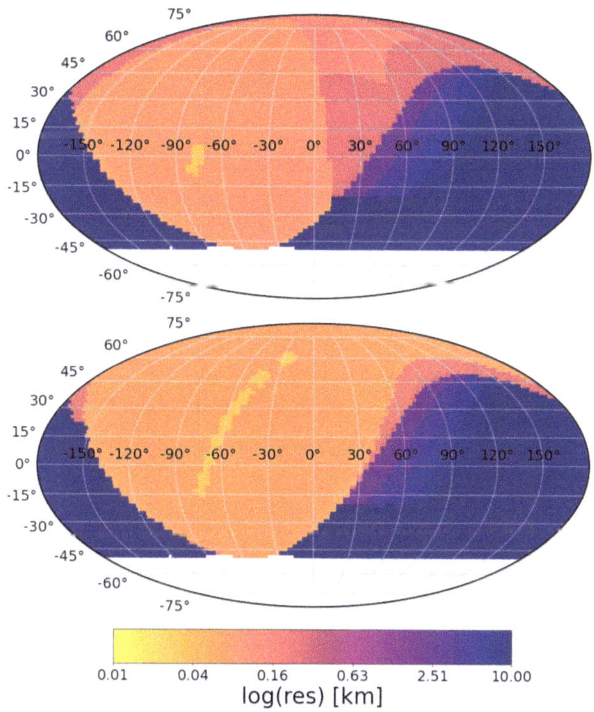

spectral types (C-, D- P-types), masses (a factor of 100), and rotational periods (a factor of 50).

While the physical properties of Trojan asteroids are poorly understood, *Lucy* observations offer a unique opportunity to study at close range objects that may have formed in the outer Solar System and that could be volatile-rich. Because collisional rates within Trojans are similar to MBAs (Marshall et al. 2022) the *Lucy* encounters will provide a multi-body data set to compare with MBAs and to assess the similarities and differences of the two populations. A fundamental aspect of the *Lucy* mission is to address whether its targets are primordial or collisional fragments, thus helping to significantly constrain models of the origin of the Jupiter Trojans.

An interesting application of the methods described in this article is offered by Eurybates, the largest remnant of a collisional family. While the age of the family is not well constrained (Marshall et al. 2022), Eurybates' surface could be significantly less cratered than other similarly sized Trojans, including Orus. For this reason, the comparison of crater SFDs between Eurybates and Orus may reveal whether the shape of the impactor population SFD has evolved over time. In addition, Patroclus and Menoetius could inform us about the timing of their formation. Our expectation is that this binary system is primordial (e.g., Nesvorny et al. 2018), and their cratering ages could confirm or reject this expectation.

The investigations described here by the *Lucy* mission's Geology Working Group will directly address some of these fundamental open questions in planetary science, and synergies with other working groups will provide an unprecedented understanding of Jupiter's Trojan asteroids.

Declarations

Competing Interests The authors declare no competing interests.

Publisher's Note Springer Nature remains neutral with regard to jurisdictional claims in published maps and institutional affiliations.

References

Ballouz R-L, Walsh KJ, Barnouin OS, DellaGiustina DN, Asad MA, Jawin ER, Daly MG, Bottke WF, Michel P, Avdellidou C, Delbo M, Daly RT, Asphaug E, Bennett CA, Bierhaus EB, Connolly HC, Golish DR, Molaro JL, Nolan MC, Pajola M, Rizk B, Schwartz SR, Trang D, Wolner CWV, Lauretta DS (2020) Bennu's near-Earth lifetime of 1.75 million years inferred from craters on its boulders. Nature 587:205–209. https://doi.org/10.1038/s41586-020-2846-z. 2020

Barnouin OS, Daly M, Mazarouei S, Ernst CM, Buczkowski DL, Michel P (2013) The geology of 433 Eros and 25143 Itokawa: observational constraints on asteroid disruption and reaccretion. In: 8th workshop of catastrophic disruption in the Solar System, Hawaii, June 24-27, 2013

Barucci MA, Brown ME, Emery JP, Merlin F, Boehnhardt H, Cruikshank DP, Morbidelli A, Dotson R (2008) Composition and surface properties of transneptunian objects and centaurs. In: Barucci MA et al (eds) The Solar System beyond Neptune, University of Arizona Press, Tucson, p 143–160

Barucci MA, Alvarez-Candal A, Merlin F, Belskaya IN, de Bergh C, Perna D, DeMeo F, Fornasier S (2011) New insights on ices in Centaur and Transneptunian populations. Icarus 214:297–307. https://doi.org/10.1016/j.icarus.2011.04.019

Bell et al. (2023) Space Sci Rev 219

Bermingham K, Kruijer T (2022) Isotopic constraints on the formation of the Main Belt. In: Marchi S, Raymond C, Russell C (eds) Vesta and Ceres: insights from the Dawn Mission for the origin of the Solar System (Cambridge Planetary Science). Cambridge University Press, Cambridge, pp 212–226. https://doi.org/10.1017/9781108856324.018

Birch SPD et al (2017) Geomorphology of comet 67P/Churyumov–Gerasimenko. Mon Not R Astron Soc 469:S50–S67. https://doi.org/10.1093/mnras/stx1096

Bottke WF Jr, Richardson DC, Michel P, Love SG (1999) 1620 Geographos and 433 Eros: shaped by planetary tides? Astron J 117:1921–1928

Bottke et al. (2023) Space Sci Rev 219

Brown ME, Schaller EL, Fraser WC (2012) Water ice in the Kuiper Belt. Astron J 143:146. https://doi.org/10.1088/0004-6256/143/6/146

Brown ME, Levison HF, Noll KS, Binzel R, Buie MW, Grundy W, Marchi S, Olkin CB, Spencer J, Statler TS, Weaver H (2021) The orbit and density of the Jupiter Trojan satellite system Eurybates–Queta. Planet Sci J 2:170. https://doi.org/10.3847/PSJ/ac07b0

Buczkowski DL, Barnouin-Jha OS, Prockter LM (2008) 433 Eros lineaments: global mapping and analysis. Icarus 193:39–52. https://doi.org/10.1016/j.icarus.2007.06.028

Chapman CR, Merline WJ, Thomas P (1999) Cratering on Mathilde. Icarus 140:28–33. https://doi.org/10.1006/icar.1999.6119

Emery et al. (2023) Space Sci Rev 219

Giacomini L et al (2016) Geologic mapping of the Comet 67P/Churyumov–Gerasimenko's northern hemisphere. Mon Not R Astron Soc 462:S352–S367. https://doi.org/10.1093/mnras/stw2848

Giesea B, Neukum G, Roatsch T, Denk T, Porco CC (2006) Topographic modeling of Phoebe using Cassini images. Planet Space Sci 54:1156–1166. https://doi.org/10.1016/j.pss.2006.05.027

Holsapple KA, Housen KR (2007) A crater and its ejecta: an interpretation of Deep Impact. Icarus 187:345–356. https://doi.org/10.1016/j.icarus.2006.08.029

Holt TR, Horner J, Nesvorný D, King R, Popescu M, Carter BD, Tylor CCE (2021) Astrocladistics of the Jovian Trojan swarms. Mon Not R Astron Soc 504:1571–1608. https://doi.org/10.1093/mnras/stab894

Housen KR, Holsapple KA (2003) Impact cratering on porous asteroids. Icarus 163:102–119. https://doi.org/10.1016/S0019-1035(03)00024-1

Hurford TA, Asphaug E, Spitale JN, Hemingway D, Rhoden AR, Henning WG, Bills BG, Kattenhorn SA, Walker M (2016) Tidal disruption of Phobos as the cause of surface fractures. J Geophys Res 121:1045–1065. https://doi.org/10.1002/2015JE004943

Levison HF, Shoemaker EM, Shoemaker CS (1997) Dynamical evolution of Jupiter's Trojan asteroids. Nature 385:42–44. https://doi.org/10.1038/385042a0

Levison HF, Bottke WF, Gounelle M, Morbidelli A, Nesvorný D, Tsiganis K (2009) Contamination of the asteroid belt by primordial trans-Neptunian objects. Nature 460:364–366. https://doi.org/10.1038/nature08094

Levison HF, Olkin CB, Noll KS, Marchi S, Bell JF III, Bierhaus E, Binzel R, Bottke W, Britt D, Brown M, Buie M, Christensen P, Emery J, Grundy W, Hamilton VE, Howett C, Mottola S, Pätzold M, Reuter D, Spencer J, Statler TS, Stern SA, Sunshine J, Weaver H, Wong I (2021) Lucy mission to the Trojan asteroids: science goals. Planet Sci J 2:171. https://doi.org/10.3847/PSJ/abf840

Levison et al. (2023) Space Sci Rev 219

Marchi S, Mottola S, Cremonese G, Massironi M, Martellato E (2009) A new chronology for the Moon and Mercury. Astron J 137:4936–4948. https://doi.org/10.1088/0004-6256/137/6/4936

Marchi S, McSween HY, O'Brien DP, Schenk P, De Sanctis MC, Gaskell R, Jaumann R, Mottola S, Preusker F, Raymond CA, Roatsch T, Russell CT (2012) The violent collisional history of asteroid 4 Vesta. Science 336:690–694. https://doi.org/10.1126/science.1218757

Marchi S, Chapman CR, Barnouin OS, Richardson JE, Vincent J-B (2015) Cratering on asteroids. In: Michel P, DeMeo FE, Bottke WF (eds) Asteroids IV. University of Arizona Press, Tucson. https://doi.org/10.2458/azu_uapress_9780816532131-ch037

Marchi S, Ermakov AI, Raymond CA, Fu RR, O'Brien DP, Bland MT, Ammannito E, De Sanctis MC, Bowling T, Schenk P, Scully JEC, Buczkowski DL, Williams DA, Hiesinger H, Russell CT (2016) The missing large impact craters on Ceres. Nat Commun. https://doi.org/10.1038/ncomms12257

Marchi S, Raymond C, Russell C (2022) Epilogue: the renaissance of Main Belt asteroid science. In: Marchi S, Raymond C, Russell C (eds) Vesta and Ceres: insights from the Dawn Mission for the origin of the Solar System (Cambridge Planetary Science). Cambridge University Press, Cambridge, pp 262–264. https://doi.org/10.1017/9781108856324.021

Marchi S, Nesvorný D, Vokrouhlický D, Bottke WF, Levison H (2023) A crater chronology for the Jupiter's Trojan asteroids. Astron J (in press)

Marchis F, Hestroffer D, Descamps P, Berthier J, Bouchez AH, Campbell RD, Chin JCY, van Dam MA, Hartman SK, Johansson EM, Lafon RE, Mignant DL, de Pater I, Stomski PJ, Summers DM, Vachier F, Wizinovich PL, Wong MH (2006) A low density of 0.8 g cm-3 for the Trojan binary asteroid 617 Patroclus. Nature 439:565–567. https://doi.org/10.1038/nature04350

Marschall R, Nesvorný D, Deienno R, Wong I, Levison HF, Bottke WF (2022) Implications for the collisional strength of Jupiter trojans from the Eurybates family. Astron J 164:167. https://doi.org/10.3847/1538-3881/ac8d6b

Morbidelli A, Nesvorny D, Bottke WF, Marchi S (2021) A re-assessment of the Kuiper Belt size distribution for sub-kilometer objects, revealing collisional equilibrium at small sizes. Icarus 356:114256. https://doi.org/10.1016/j.icarus.2020.114256

Mottola S et al (1995) The slow rotation of 253 Mathilde. Planet Space Sci 43:1609–1613

Mottola S, Hellmich S, Buie MW, Zangari AM, Marchi S, Brown ME, Levison HF (2020) Convex shape and rotation model of Lucy target (11351) Leucus from lightcurves and occultations. Planet Sci J 1:73. https://doi.org/10.3847/PSJ/abb942

Mottola S, Hellmich S, Buie MW, Zangari AM, Stephens RD, Di Martino M, Proffe G, Marchi S, Olkin CB, Levison HF (2023) Shape models of Lucy targets (3548) Eurybates and (21900) Orus from disk-integrated photometry. Planet Sci J 4:18. https://doi.org/10.3847/PSJ/acaf79

Mottola et al. (2023) Space Sci Rev 219

Nesvorný D (2018) Dynamical evolution of the early Solar System. Annu Rev Astron Astrophys 56:137–174

Nesvorný D, Vokrouhlický D, Morbidelli A (2013) Capture of trojans by jumping Jupiter. Astrophys J 768:45. https://doi.org/10.1088/0004-637X/768/1/45

Nesvorný D, Vokrouhlický D, Dones L, Levison HF, Kaib N, Morbidelli A (2017) Origin and evolution of short-period comets. Astrophys J 845:27. https://doi.org/10.3847/1538-4357/aa7cf6

Nesvorný D, Vokrouhlický D, Bottke WF, Levison HF (2018) Evidence for very early migration of the Solar System planets from the Patroclus–Menoetius binary Jupiter Trojan. Nat Astron 2:878–882. https://doi.org/10.1038/s41550-018-0564-3

Noll KS, Brown ME, Weaver HA, Grundy WM, Porter SB, Buie MW, Levison HF, Olkin C, Spencer JR, Marchi S, Statler TS (2020) Detection of a satellite of the Trojan asteroid (3548) Eurybates—a Lucy mission target. Planet Sci J 1:44. https://doi.org/10.3847/PSJ/abac54

O'Brien DP, Marchi S, Morbidelli A, Bottke WF, Schenk PM, Russell CT, Raymond CA (2014) Constraining the cratering chronology of Vesta. Planetary and Space Science 103:131–142. https://doi.org/10.1016/j.pss.2014.05.013

Olkin et al. (2023) Space Sci Rev 219

Porco CC et al (2005) Cassini imaging science: initial results on Phoebe and Iapetus. Science 307:1237–1242. https://doi.org/10.1126/science.1107981

Raymond SN, Izidoro A (2017) The empty primordial asteroid belt. Sci Adv 3:e1701138. https://doi.org/10.1126/sciadv.1701138

Robinson MS, Thomas PC, Veverka J, Murchie SL, Wilcox BB (2002) The geology of 433 Eros. Meteorit Planet Sci 37:1651–1684

Spencer JR et al (2020) The geology and geophysics of Kuiper Belt object (486958) Arrokoth. Science. https://doi.org/10.1126/science.aay3999

Sullivan RJ, Thomas PC, Murchie SL, Robinson MS (2002) Asteroid geology from Galileo and NEAR Shoemaker data. In: Bottke WF Jr, Cellino A, Paolicchi P, Binzel RP (eds) Asteroids III. University of Arizona Press, Tucson, pp 331–350

Thomas PC, Veverka J, Bell JF, Clark BE, Carcich B, Joseph J, Robinson M, McFadden LA, Malin MC, Chapman CR, Merline W, Murchie S (1999) Mathile: size, shape, and geology. Icarus 140:17–27. https://doi.org/10.1006/icar.1999.6121

Thomas PC, Veverka J, Belton MJS, Hidy A, A'Hearn MF, Farnham TL, Groussin O, Li J-Y, McFadden LA, Sunshine J, Wellnitz D, Lisse C, Schultz P, Meech KJ, Delamere WA (2007) The shape, topography, and geology of tempel 1 from Deep Impact observations. Icarus 187:4–15. https://doi.org/10.1016/j.icarus.2006.12.013

Thomas N, Sierks H, Barbieri C, Lamy PL, Rodrigo R, Rickman H, Koschny D, Keller HU, Agarwal J, A'Hearn MF, Angrilli F, Auger A-T, Barucci MA, Bertaux J-L, Bertini I, Besse S, Bodewits D, Cremonese G, Da Deppo V, Davidsson B, De Cecco M, Debei S, El-Maarry MR, Ferri F, Fornasier S, Fulle M, Giacomini L, Groussin O, Gutierrez PJ, Güttler C, Hviid SF, Ip W-H, Jorda L, Knollenberg J, Kramm J-R, Kührt E, Küppers M, La Forgia F, Lara LM, Lazzarin M, Moreno JJL, Magrin S, Marchi S, Marzari F, Massironi M, Michalik H, Moissl R, Mottola S, Naletto G, Oklay N, Pajola M, Pommerol A, Preusker F, Sabau L, Scholten F, Snodgrass C, Tubiana C, Vincent J-B, Wenzel K-P (2015) The morphological diversity of comet 67P/Churyumov-Gerasimenko. Science 347:aaa0440. https://doi.org/10.1126/science.aaa0440

Veverka J, Thomas P, Harch A, Clark B, Bell JF, Carcich B, Joseph J, Murchie S, Izenberg N, Chapman C, Merline W, Malin M, McFadden L, Robinson M (1999) NEAR encounter with asteroid 253 Mathilde: overview. Icarus 140:3–16. https://doi.org/10.1006/icar.1999.6120

Vokrouhlický D, Bottke WF, Nesvorný D (2016) Capture of trans-neptunian planetesimals in the main asteroid belt. Astron J 152:39. https://doi.org/10.3847/0004-6256/152/2/39

Space Science Reviews (2024) 220:17
https://doi.org/10.1007/s11214-024-01052-7

Shapes, Rotations, Photometric and Internal Properties of Jupiter Trojans

Stefano Mottola[1] · Daniel T. Britt[2] · Michael E. Brown[3] · Marc W. Buie[4] · Keith S. Noll[5] · Martin Pätzold[6]

Received: 8 August 2023 / Accepted: 29 January 2024 / Published online: 12 February 2024

Abstract
The Jupiter Trojans, being trapped around the stable L4 and L5 Jupiter Lagrangian points, are thought to be more primitive than the Main Belt asteroids. They are believed to have originated from a range of heliocentric distances in the trans-Neptunian region, to have subsequently been scattered inwards, and finally captured in their current location during the phase of Giant Planet migration. As a consequence, their bulk composition is expected to reflect that of the protoplanetary disk at the time and location of their formation. The photometric properties of Trojans appear to have a bi-modal distribution. A few Trojans have been discovered to be binary systems, suspected contact binaries, or to possess moonlets, which has revealed consistently low bulk densities (around 1×10^3 kg m^{-3}) for those systems. Those estimates, together with the presence of a spin barrier between 4 and 4.8 h rotation period, suggest that low densities are a general property of the population, similar to that of cometary nuclei.

Current Trojan physical properties provide clues that relate to their formation that can, in turn, be traced back to the origin of the solar system. We review here our current knowledge on the physical properties of Trojans and the methods used for their determinations. Most of these methods are based on Earth-bound observations, and are limited by the large distance to these objects. The next breakthrough will be made possible by the Lucy mission, which, by visiting several Trojans during a tour through both clouds, will address many open questions and probably raise new ones. The combination of the ground truth for select objects provided by Lucy with the context view given by the Earth-bound observations will result in powerful synergy.

✉ S. Mottola
stefano.mottola@dlr.de

[1] Institute of Planetary Research, DLR, Rutherfordstr. 2, 12489, Berlin, Germany

[2] University of Central Florida, 12354 Research Parkway, Orlando, FL, 32826, USA

[3] Division of Geological and Planetary Sciences California Institute of Technology, Pasadena, CA, 91125, USA

[4] Southwest Research Institute, 1050 Walnut St., Boulder, CO, 80302, USA

[5] NASA Goddard Space Flight Center, 8800 Greenbelt Rd., Greenbelt, MD, 20771, USA

[6] Abteilung Planetenforschung, Rheinisches Institut für Umweltforschung an der Universität zu Köln, Aachenerstrasse 209, 50931 Cologne, Germany

Keywords Lucy Mission · Jupiter Trojans · shape inversion · photometry · internal properties

1 Introduction

Jupiter Trojans are a population of bodies trapped at the L4 and L5 Jovian Lagrangian points. The 1:1 mean-motion resonance with Jupiter makes their current orbits stable over the lifetime of the Solar System (Emery et al. 2015). The most likely formation scenario of the Trojan clouds involves inward scattering of planetesimals from different zones of the trans-Neptunian region and subsequent capture in the 1:1 resonance with Jupiter. This capture has probably taken place during the migration phase of the Giant Planets during which Jupiter has abruptly changed its semimajor axis due to one or multiple encounters with an ice giant (Morbidelli et al. 2009; Nesvorný et al. 2013; Bottke et al. 2023). This "jumping Jupiter" scenario can potentially explain the current eccentricities of Jupiter and Saturn, and account for the observed number asymmetries that characterize the L4 and L5 Trojan clouds. Due to their dynamical isolation, Jupiter Trojans represent relatively unprocessed samples from the respective formation regions, and their study can help constrain current models about the origin and evolution of the Solar System. The NASA Lucy mission, by visiting several targets in both Trojan clouds, will explore their diversity in terms of composition and physical properties (Levison et al. 2021, 2024).

Due to their distance from Earth, current knowledge about the physical properties of Trojans is still sparse, but has been rapidly progressing in the past decade, mainly due to advances in observational techniques and equipment. In this chapter we review the state of the current knowledge about Jupiter Trojans in terms of their rotations, shapes, masses and densities.

2 Rotation Properties

As of today, reliable spin rates are known for about 240 Trojans with a level of completeness of about 98% down to a diameter of 50 km (Asteroid Lightcurve Database (LCDB), Warner et al. 2009).

Recently, rotation properties of Trojans have been extracted from data from the K2, TESS and Gaia missions (Szabó et al. 2017; Ryan et al. 2017; Pál et al. 2020; Kalup et al. 2021; Ďurech and Hanuš 2023). These spacecraft data, which benefit from uninterrupted observation opportunities covering several weeks – thereby helping reduce the bias that affects ground-based observations and favors short rotation periods (Marciniak et al. 2018) – have considerably increased the size of the sample and confirmed and extended the earlier results.

Previous works (Mottola et al. 2014; French et al. 2015) have noted, based on data available at that time, that the rotation properties of Trojans are different from those of the Main Belt. It was also realized that the distribution of the rotation frequency of Trojans deviates in a statistically significant way from that of a 3-D Maxwellian distribution – the expected end state for a perfectly relaxed population of interacting bodies with random orientation of their spin axes (Harris and Burns 1979). In particular, Trojans show a substantial excess of slow rotators (see Fig. 1) and appear to lack fast-spinning objects, compared to Main Belt Asteroids (MBAs), Mars Crossers (MCs) and Near Earth Objects (NEOs).

Several mechanisms have been proposed for explaining the excess of slow rotators among Trojans. One of those invokes the YORP effect (Rubincam 2000), which is known to effectively act on Main Belt and Near-Earth asteroids. However, Kalup et al. (2021) estimated

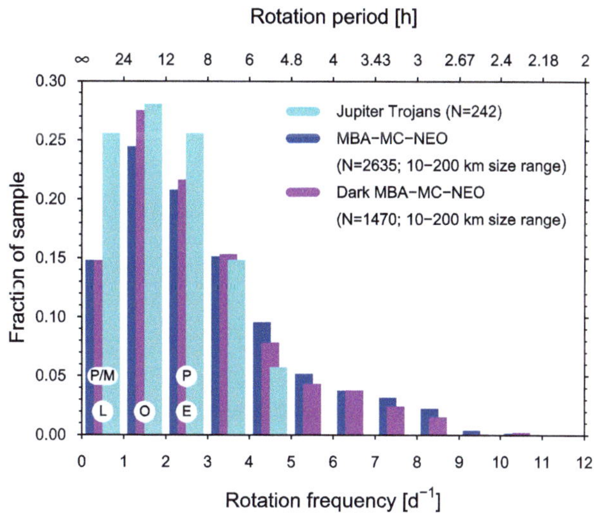

Fig. 1 Distribution of the rotation frequency of Jupiter Trojans and of MBAs, MCs and NEOs in the size range 10 - 200 km. 'Dark' refers to objects belonging to the C, P and D taxonomic groups. The letters mark the location in the histogram of the Lucy Trojan targets (P/M= Patroclus-Menoetius; L=Leucus; O=Orus; P=Polymele; E=Eurybates). Data from LCDB (Warner et al. 2009), updated 2023 April

that the typical YORP spin-down time scale for Trojans with a diameter larger than 10 km is $\geq$ 1 Gyr, which, being much longer than the typical times for spin randomization due to mutual collisions, makes it an unviable mechanism, at least for all but the smallest objects. Another possible explanation is that localized pockets of volatiles in subsurface reservoirs might be sporadically activated, either by minor impacts, or by periodic variations in the heliocentric range, thereby producing a net torque that could spin up or spin down the body (Mottola et al. 2014). A third explanation is that the slow-rotating Trojans were born as similar-size tight binaries – possibly in the transneptunian formation region – and were eventually slowed down due to tidal evolution until they locked in an orbit-synchronous rotation, analogously to the Patroclus-Menoetius system (Nesvorný et al. 2020). Such a system could eventually become either separated, due to impacts or planetary perturbations, or remain intact. In any case, the excess of slow rotators is an interesting feature of the Trojan population that is deserving of additional study. Lucy will contribute to understanding slow rotators by visiting three: the Patroclus-Menoetius synchronous binary system and (11351) Leucus (Mottola et al. 2020), for which no companion has yet been discovered. Also, Lucy will visit the MBA (52246) Donaldjohanson, which will enable a comparison between slow rotators in different environments.

The fastest spin for Trojans for which reliable determinations exist on the LCDB corresponds to a rotation period of about 4.8 h for objects larger than about 10 km (see Fig. 2). Recently, (Chang et al. 2021) conducted a dedicated survey with the Subaru telescope and determined the spin rates of 53 small (2 km $\leq D \leq$ 40 km) Trojans. Three of these objects (with a size between 3.5 and 5 km) appear to spin faster than 4.8 h, with periods ranging from 4.0 h to 4.3 h. These objects, however, were previously unknown and their classification as Trojans is based only on a short-arc determination. It would be therefore desirable to recover those objects and confirm their orbital classification and their rotation periods. Determining the location of the rotational breakup barrier – as in the case of the $\sim$ 2.2 h limit for MBAs, MCs and NEOs – is important because it places a strong constraint on the bulk density of the population, under the assumption that the bodies are cohesionless rubble piles held together by gravity (see Sect. 6). Trojans larger than about 2 km likely belong to this category since they are above the transition zone from "strength-dominated" to "gravity-dominated" bodies, which is thought to occur around a diameter of about 200-400 m (Benz

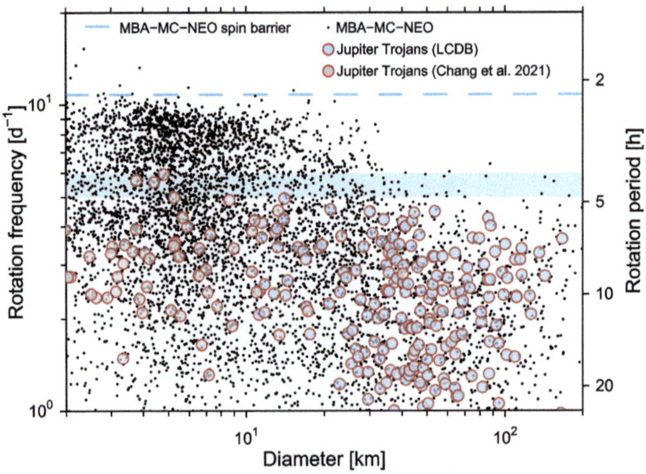

Fig. 2 Scatter plot of the rotation frequency of asteroids vs. their diameter. The horizontal dashed line corresponds to the spin barrier for MBAs, MCs and NEOs at a rotation period of about 2.2 h. Trojan asteroids appear to have a spin barrier between 4 and 4.8 h (shaded area). Data from Chang et al. (2021) and LCDB (Warner et al. 2009), updated 2023 April

and Asphaug 1999). It is to be expected, however, that as soon as the rotations for Trojans in the sub-100-m size range will become measurable, some of those objects, which are likely to be monoliths, will display very short rotation periods.

The distribution of the spin axes of asteroids can provide clues on their formation mechanisms and post-accretion evolution. For example, the tendency for a polarization of the spin axis directions towards the Ecliptic poles has been established for MBAs smaller than 30 km (Hanuš et al. 2016), which has been interpreted as the result of the YORP effect (Rubincam 2000). Also, Cibulková et al. (2016), by determining the spin axis directions of a few thousands of MBAs, have found an asymmetry in the distribution of the Ecliptic longitude of their poles, for which no satisfactory explanation exists to-date.

Until very recently, spin axis orientations for Trojans were only known for about a dozen of objects. As of writing, however, new results based on dense and sparse photometric surveys have been reported in Hanuš et al. (2023). From a sample of 82 objects, the authors find that the distribution of the spin vectors of Trojans in the sky is roughly isotropic. However, the authors also find that the distribution of the orbital obliquity – the angle between the spin vector of an object and its orbital plane – shows an excess of prograde rotators, with a ratio of prograde-to-retrograde of about 1.5. Although there is a 10% probability that the observed asymmetry be due to chance – due to the relatively small size of the sample – their result seems to be compatible with the current understanding of a Trojan formation mechanism driven by streaming instability in a cold trans-Neptunian protoplanetary disk (Youdin and Goodman 2005), which also predicts an excess of primordial prograde rotators. The asymmetry of spins observed for Trojans is smaller than that predicted by numerical simulations of the streaming instability scenario, which is interpreted by Hanuš et al. (2023) as being due to spin randomization during the post-accretion collisional evolution phase.

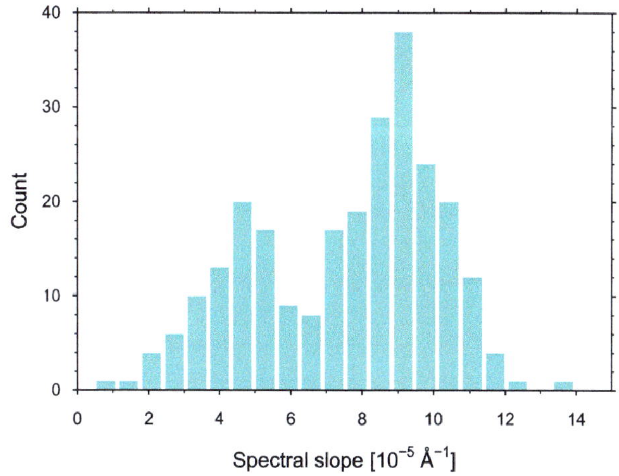

Fig. 3 The distribution of spectral slopes for 254 Trojan asteroids shows a bi-modal distribution. Modified after Wong et al. (2014)

3 Photometric Properties

Studying the photometric properties of planetary surfaces holds the potential of understanding physical properties of their regolith, such as surface porosity, roughness, particle albedo and size distribution. Our current understanding of the photometric properties of Trojans, however, is still quite limited. This is due to the low apparent brightness of most of the objects and to the restricted range of illumination geometries from which they can be observed Earth-bound. As a result, we have accurate and well sampled measurements for a few Trojans, and a large number of less accurate, sparse and serendipitous measurements from systematic observation surveys. The techniques used are disk-integrated photometry, mostly in the form of lightcurves and phase curves, spectrophotometry and spectroscopy.

In addition to providing clues about the location and conditions of formation regions, and of the surface evolutionary history, knowledge of the photometric properties enables the computation of quantities as the albedo and phase integral, which are fundamental for, e.g., computing the thermal balance of the bodies. Compared to MBAs, Trojans have a narrow variety of taxonomic types, being mostly limited to the dark P and D types, and some rare C types (Grav et al. 2012). In turn, the P and D types are very rare in the inner Main Belt. The colors of Trojans mostly cluster in two major groups (Wong et al. 2014; Emery et al. 2024): red and less red (see Fig. 3).

A growing body of observations (Shevchenko et al. 2012; Mottola et al. 2020, 2023) shows that their phase curves are characterized by a very small or absent opposition effect - the non-linear brightness surge usually present on atmosphereless bodies at small phase angles. As a consequence, many Trojans have phase curves with a linear trend down to sub-degree phase angles (Slyusarev and Belskaya 2014). The interpretation of this phenomenon is very likely linked to the relative contributions of the two effects causing the opposition effect: shadow hiding and coherent backscatter. The limited data set, however, does not allow firm answers, with different scholars reaching opposite conclusions (Schaefer et al. 2010; Shevchenko et al. 2012). It is also apparent that the HG photometric system (Bowell et al. 1989), formerly adopted by the IAU and now deprecated, fails to describe the shape of the phase curves of Trojans. On the other hand, the more recently IAU-adopted H, G_1, G_2 system (Muinonen et al. 2010) provides a very good match to Trojans, at least in the limited phase angle range spanned by ground-based observations. Studying the photometric properties of the Trojans, and putting them in the context of other Solar System objects such as transneptunian objects, comets and giant planets irregular satellites, can help resolve the question of their formation region.

Although less accurate for individual objects, studies of a large number of Trojans through the analysis of serendipitous observations by large-scale surveys provide statistically solid results about the ensemble properties of the population. Schemel and Brown (2021) have analyzed data from the Zwicky Transient Facility (ZTF) survey and established a correlation between colors and phase curves. Working in the HG-System, they found that for phase angles smaller than 12°, redder objects have systematically higher values of G, while the less-red objects have values closer to the canonical $G = 0.15$ value. Schemel and Brown (2021) suggest that the correlation between color and phase parameter is less likely to be linear, but rather a function of two separate average phase parameters for the two separate color classes of Trojans. They find an average phase parameter values of $G = 0.14 \pm 0.05$ for the less-red cluster of Trojans, which generally corresponds to objects classified as P-type, and $G = 0.22 \pm 0.07$ for the red population, which generally corresponds to objects classified as D-type. Objects classified as C-type have colors only slightly bluer than the less-red population and do not appear to be a clearly distinguishable sub-group, but based on the color correlation they would have phase parameters slightly smaller than those of the less-red population.

4 Shapes

The shapes of asteroids can provide information about the formation mechanisms of these bodies, about the processes that controlled their evolution, and about their physical characteristics. Due to their large geocentric distance, however, even the largest Trojans are beyond, or at the limit of the resolution of current ground-based and Earth co-orbital direct-imaging instruments. A telltale case is (624) Hektor – the largest Trojan – for which, already in the late seventies, Hartmann and Cruikshank (1978) predicted, based on its exceptional maximum lightcurve amplitude of 1.23 mag, a dumbbell shape. Further hypotheses included a close or contact binary. To date, however, a solid description of Hektor's shape is still elusive. Attempts to detect the bilobed nature of Hektor were made by Storrs et al. (1999) with the Hubble Space Telescope (HST) in direct imaging, and by Tanga et al. (2003) with the HST Fine Guidance Sensors in interferometry mode. Neither study, however, could distinguish between the different hypotheses for the shape. Adaptive Optics (AO) observations of Hektor were performed in the near IR by Marchis et al. (2014) between 2006 and 2011 with the 10 m Keck telescope. Those observations seem to show a bilobed shape in some frames, but do not allow distinguishing between a contact or a close binary.

For essentially all Trojans, the most readily available source of information about their shape is given today by photometric lightcurves. The maximum observed amplitude is a first indication about the elongation of the body. Early studies of a limited sample of the largest Trojans (Hartmann et al. 1988) seemed to suggest that Trojans show in average larger lightcurve amplitudes than a sample of similarly sized MBAs. Further data and a bias analysis (Binzel and Sauter 1992) seemed to confirm significantly larger amplitudes only for Trojans larger than 90 km. Later studies, however, based on a much larger sample of Trojans (Mottola et al. 2011), showed no anomalously high amplitudes, suggesting no major differences in the elongation of the MBAs and in the Trojans in the size range 50 - 150 km. More recently, (Schemel and Brown 2021) have compared the amplitudes of Trojans and MBAs in the size range 20 km to 100 km and also found no statistically significant differences, though the large uncertainties could hide important distinctions between these populations which have yet to be observed.

Close and contact binaries, discussed in more detail by Noll et al. (2023) in this collection, are likely to display above-average amplitudes. For this reason, several studies have searched for binary candidates by identifying objects with large amplitudes in lightcurve surveys (Mann et al. 2007; Sonnett et al. 2015; Ryan et al. 2017). Two large-amplitude Trojans have been identified as likely close or contact binaries based on the incompatibility of their slow-rotation with a dynamically-controlled shape for a fluid-like or rubble-pile object (Mann et al. 2007). Of the survey objects with more sparsely sampled lightcurves, some tens of percent also display high amplitudes and slow rotations, thereby constituting similar contact-binary candidate systems (Sonnett et al. 2015; Ryan et al. 2017). The question remains, however, as to how many of these candidates are really contact binaries, since a large amplitude is not a sufficient condition for binarity (see e.g. Harris and Warner 2020).

A more certain proof for the presence of a binary system and a means for determining its geometry is the detection in the lightcurves of mutual events between well-separated eclipsing companions. The Patroclus-Menoetius system is such a configuration where the two components are orbit-synchronous and undergo periodic mutual event seasons (Buie et al. 2015; Grundy et al. 2018; Wong and Brown 2019; Berthier et al. 2020). Pinilla-Alonso et al. (2022) used observations of seven multiple mutual events occurred in 2017-2018 to identify evidence for a large deviation from a spheroidal shape. Combining these results with those by Buie et al. (2015), suggests that the feature, possibly a large impact basin, must be near the south pole of Menoetius.

Whenever photometric observations are available over a wide range of viewing and illumination geometries, it is possible to reconstruct a 3D convex approximation of the body (Kaasalainen et al. 2001, 2002). This is a robust and established technique that enables the simultaneous determination of the orientation of the body and its "convex hull" shape. Although solutions allowing nonconvexities are also possible, those are in general non-unique if based on lightcurve data alone (Ďurech and Kaasalainen 2003; Viikinkoski et al. 2017). Data fusion with other observational techniques as radar imaging, adaptive optics observations and stellar occultations, however, can help remove the ambiguity in the determination of non-convexities (Viikinkoski et al. 2015). Of those techniques, however, only stellar occultations can be currently successfully applied to Trojans, due to their large geocentric distance.

One of the advantages of the convex inversion technique is that in addition to conventional, densely sampled lightcurve data, it can use also sparse photometric data, as those serendipitously acquired by recent ground- and space-based photometric surveys (Kaasalainen 2004). Although shape and rotation models solely based on sparse data can produce spurious results, the large number of objects for which sparse data are available enables statistical studies to be made and to compare the global properties of different populations. As an example, recent studies have exploited the serendipitous photometric data acquired by the *Gaia* mission (Gaia Collaboration et al. 2018) and the ATLAS photometric survey (Tonry et al. 2018) to derive ellipsoidal or coarse convex shape models of asteroids, among which are also about a dozen of Trojans (Ďurech et al. 2019, 2020). Further results, consisting of convex shape models for additional 71 Trojans have been reported recently (Hanuš et al. 2023; Ďurech and Hanuš 2023).

As mentioned earlier, one complementary technique for the determination of the shape of asteroids that is applicable to the Lucy targets and to Jupiter Trojans in general, is the observation of stellar occultations, which, in recent years, has greatly advanced thanks to the introduction of affordable and high-performance focal plane detectors, the advent of precise

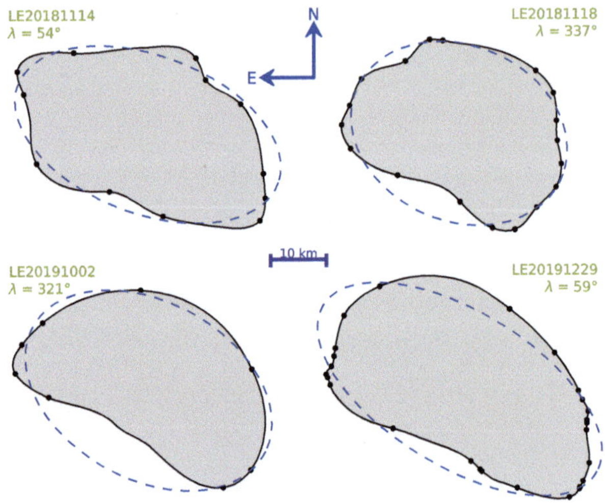

Fig. 4 Shape constraints for Leucus from occultations. The date and subearth longitude is shown for each dataset. All are projected on the plane of the sky at the same scale as observed with no extra geometric corrections. For each event, the formal best-fit ellipse is shown with dotted blue lines. The solid curve that is filled in is a spline curve drawn through the points. The agreement between the ellipse and actual profile is very poor. Occultation data are from Buie et al. (2021)

GPS time recording and the availability of accurate all-sky, dense astrometric catalogs. This technique produces a snapshot of the sky projection of the object at the time of the occultation, by sampling its silhouette at discrete positions defined by the chords subtended by the observing stations onto the body. The number of stations on the occultation path defines the density of the silhouette sampling. For this reason, a well sampled object profile calls for the coordinated deployment of a large number of mobile observing stations and the computation of accurate shadow-path predictions (Buie et al. 2015, 2021). This is a field that greatly benefits from the amateur-professional connection, as demonstrated by recent large-scale deployment efforts (Keeney et al. 2022), which, among other things, have resulted in the discovery of a satellite of the Lucy target Polymele.

A common assumption when assessing the shape of a small body from occultation data is to start with the convenience of a tri-axial ellipsoid. The projection of such an ellipsoid is an ellipse and the best-fit ellipse shape is often reported. This ellipse is commonly used both for astrometry derived from the occultation and for a first determination of the geometric albedo. A further approach is to approximate the profile with a continuous function, which can provide a better description of the limb shape, compared to an ellipse. As an example, Fig. 4 shows a collection of profiles for the Lucy target, (11351) Leucus. The data presented in this figure are exactly as reported by Buie et al. (2021). In addition to the fitted ellipses from that work, a spline curve is drawn that satisfies the constraints provided by the occultation data. Although not producing unique profiles, this representation can help identify major features in the body's shape.

Similarly to lightcurve inversion, stellar occultations cannot uniquely resolve concavities on the target, in the sense defined by Viikinkoski et al. (2017). However, if a sufficient number of occultation cross-sections are available, they can describe the shape of those non-convex objects that go under the class of *tangent-covered bodies* or TCBs (Kaasalainen 2011). As an example, these shapes include individually-convex binary systems, dumbbell shapes, banana-shaped objects as (433) Eros or (25143) Itokawa, but not concavities as the large, deep craters on (253) Mathilde.

By obtaining a dense and regular sampling of the limb of a particular target from well-observed occultations at different aspect geometries it is in principle possible to reconstruct the 3D object shape and its spin orientation from occultation data alone. However, it is in general more effective to combine occultation and lightcurve data into one global solution providing pole direction, shape, absolute size and albedo. Depending on the accuracy and sampling of the data, the shape can be retrieved in its convex (e.g. Ďurech et al. 2011, Mottola et al. 2020, 2023) or nonconvex approximation (Carry et al. 2010; Viikinkoski et al. 2017). Extensive occultation and photometric lightcurve campaigns enabling the application of these methods have been performed and are currently planned for all of the Lucy targets. Lucy, in turn, will return data that will provide an essential set of ground-truth observations from which to further guide our ability to interpret future occultation and photometric data from the rest of the Trojan population.

5 Mass Determination

There are three main methods to determine the mass of a small body in the solar system: 1) by the observation of spacecraft velocity perturbations during a flyby, 2) by observing the gravitational effect on a perturbed asteroid by a planet or a large asteroid, 3) by the astrometric observation of the orbit of a satellite about an asteroid. One fourth possibility is given by the detection of the change of the orbital semi-major axis of an object due to the Yarkovsky nongravitational acceleration (Vokrouhlický et al. 2015) by measuring its astrometric position over many decades. This method has been successfully applied to several NEOs to determine their physical properties, among which their mass. However, since the magnitude of the Yarkovsky drift scales as the Solar irradiation flux and as the inverse of the object's size, its application to Trojan asteroids is currently impractical.

5.1 Spacecraft Flyby

The most precise way to determine masses and, if possible, gravity fields of planetary bodies is the radio tracking of an interplanetary spacecraft flying-by or orbiting the body of interest. The motion of the spacecraft is perturbed by the attracting force of the body acting on the spacecraft, which causes an extra Doppler shift contribution proportional to the projected perturbed velocity vector to be added to the transmitted radio carrier frequency. Essentially, all fundamental information on planetary masses, gravity fields, and internal structure and composition available today was gained by radio tracking experiments.

The very first mass determination of a small body by radio tracking was successfully performed by the NEAR spacecraft for asteroid (253) Mathilde in 1997 (Yeomans et al. 1997) followed by a first determination for NEO (433) Eros during a flyby in 1998 before NEAR eventually entered Eros' orbit in 2000 (Yeomans et al. 1999). Further successful mass determinations of small bodies were conducted by the Japanese orbiter Hayabusa for (25143) Itokawa; by the European comet chaser Rosetta for (21) Lutetia in 2010 (Pätzold et al. 2011); and during the orbital phase around comet 69P/Churyumov-Gerasimenko by Rosetta from 2014 to 2016 (Pätzold et al. 2016, 2019). Those determinations were followed by the gravity investigations by Dawn at Vesta and Ceres (Konopliv et al. 2018) and by OSIRIS-REx at (101955) Bennu – a potentially hazardous asteroid for the Earth (Milani et al. 2009).

The attracting force of a planetary body perturbs the trajectory and velocity of a spacecraft flying-by if the body has a sufficiently large mass and the spacecraft is sufficiently close to that body. The radio signal transmitted by the spacecraft will show a total Doppler shift Δf of the radio carrier frequency f_0 proportional to the radial components of the relative spacecraft velocity $v_{s/c,r}$ and the relative perturbed spacecraft velocity $\Delta v_{flyby,r}$ during the flyby

$$\Delta f = -f_0 \cdot \frac{v_{s/c,r}}{c} - f_0 \cdot \frac{\Delta v_{flyby,r}}{c} \tag{1}$$

$$= -f_0 \cdot \frac{(v_{s/c} + \Delta v_{flyby}) \cdot \cos(\alpha)}{c} \tag{2}$$

where α is the angle between the spacecraft velocity vector $\vec{v_{s/c}}$ and the line-of-sight to Earth, and c is the speed of light. The unperturbed spacecraft velocity $v_{s/c}$ and its Doppler shift can be predicted with high precision for the time of the flyby by the integration of the equation of motion starting with an unperturbed spacecraft state vector many hours before closest approach and considering other non-gravitational forces acting on the spacecraft. The total perturbed spacecraft velocity and Doppler shift Δv_{flyby} and Δf_{flyby}, which are the differences between the pre- and post-encounter values of the respective quantities (measured many hours before and after closest approach), are

$$\Delta v_{flyby} = -2 \cdot \frac{G \cdot M}{v_{s/c} \cdot d} \tag{3}$$

$$\Delta f_{flyby} = -2 \cdot \frac{f_0}{c} \cdot \frac{G \cdot M}{v_{s/c} \cdot d} \cdot \sin(\alpha) \tag{4}$$

where d is the flyby distance. For a two-way radio link, where the uplink signal is transmitted by the ground station, received by the spacecraft and transponded back coherently to the ground station, the two equations above have to be multiplied by a factor of two. The coherent two-way radio link is commonly used in order to take advantage of the superior noise performance of the ground station equipment. The product $G \cdot M$ is derived from the perturbed Doppler shift Δf_{flyby}, which is directly observed, and the known flyby geometry and spacecraft velocity (Pätzold et al. 2001, 2010, 2011, 2014; Andert et al. 2010; Tyler et al. 2008). In the case that the flyby is performed with a binary asteroid or an asteroid with a moon, as Lucy will do for the Trojans Eurybates and its moonlet Queta (Noll et al. 2020) and Polymele and its satellite, provisionally nicknamed Shaun (Buie et al. 2021), both in 2027, and the binary system Patroclus-Menoetius in 2032, the product $G \cdot M$ will represent the total system mass, should it not be possible to track the spacecraft at the time of closest approach.

In order to be able to resolve a sufficiently strong Doppler shift Δf_{flyby} from the background frequency noise, Eq. (4) suggests that the relative flyby velocity $v_{s/c}$ of the spacecraft shall be slow, and that the spacecraft's flyby distance to the target shall be small. The background frequency noise is dominated by the contribution from the propagation of the radio signal though the turbulent solar wind plasma, which induces random phase changes on the radio carrier ("solar plasma noise"). The geometry of the flyby is another important factor for the accuracy of the mass determination: even if $G \cdot M$ is large, the flyby velocity slow and the distance to the object is close, the perturbed velocity vector is still projected into the line-of-sight (LOS) to Earth. If the angle α between the velocity vector of the spacecraft and the LOS is close to the radial direction to Earth, the sine term in Eq. (4) decreases a strong velocity perturbation to a small Doppler shift Δf_{flyby}. The selection of an optimal flyby geometry within the mission constraints makes flyby planning a challenge.

Assuming that the Doppler shift as observed by the ground station could be corrected for other gravitational and non-gravitational forces acting on the spacecraft, Eq. (4) represents

Table 1 Lucy Trojan flyby parameters

Target	r (au)	Δ (au)	Effective diameter (km)	Solar elongation (°)	Fly-by velocity (km s^{-1})	α (°)	Fly-by distance (km)	Doppler shift (MHz)	Nominal GM (10^6 m^3 s^{-2})
Eurybates	5.671	5.623	69	87.5	5.8	108	750 ± 3.0	168 ± 2.2	$9.2 \pm 1.5\%$
Polymele	5.688	6.148	21	58.6	6.0	109	434 ± 3.0	13 ± 2.3	$0.32 \pm 19\%$
Leucus	5.649	4.891	41	135.2	5.9	69	1000 ± 3.0	21 ± 2.0	$2.4 \pm 5.2\%$
Orus	5.342	6.245	61	22.3	7.1	57	1000 ± 3.0	65 ± 2.3	$4.9 \pm 3.9\%$
Patroclus/ Menoetius	5.384	5.275	137	90.9	8.1	63	1223 ± 5.0	835 ± 2.1	$90 \pm 0.5\%$

r and Δ represent the heliocentric and geocentric distance, respectively. α is the angle between the spacecraft trajectory and the direction to Earth. The nominal GM value is computed for an assumed bulk density of 10^3 kg m^{-3}. The effective diameter is defined as the spherical volume-equivalent diameter. The effective diameter of the Patroclus-Menoetius system corresponds to the sum of the volumes of the two bodies.

the Doppler shift as a response from the attracting force of the asteroid. The following is an estimate of the mass determination accuracy based on Eq. (4)

$$\left(\frac{\sigma_{GM}}{G\cdot M}\right)^2 = \left(\frac{\sigma_{\Delta f_{flyby}}}{\Delta f_{flyby}}\right)^2 + \left(\frac{\sigma_d}{d}\right)^2 + \left(\frac{\sigma_{v_{s/c}}}{v_{s/c}}\right)^2 + (\sigma_\alpha \cdot \tan\alpha)^2 \tag{5}$$

where $\left(\frac{\sigma_{GM}}{G\cdot M}\right)$ is the fractional accuracy of $G \cdot M$, σ_d is the uncertainty in the flyby distance d, $\sigma_{v_{s/c}}$ is the error in the spacecraft velocity and σ_α is the uncertainty in the geometry angle α. In the case of the Lucy mission, σ_d is estimated by the navigation team as about ±3 km, $\sigma_{v_{s/c}}$ is in the order of ± 0.1 mm s^{-1} and σ_α is typically $< 0.1°$.

Values for Δf, d, $v_{s/c}$ and α for the five Lucy Trojan flybys are listed in Table 1. $\sigma_{\Delta f}$ is the frequency noise as measured in the ground station receiver. It is a combination of the thermal noise of the equipment, the phase noise of the hydrogen maser in the DSN ground station and the plasma noise from the propagation of the radio signal through the turbulent solar wind. The latter dominates the total measured noise $\sigma_{\Delta f}$. The potential solar wind plasma noise was estimated from an analysis of X-band tracking of the Rosetta spacecraft on its way to rendezvous comet 69P/Churyumov-Gerasimenko at 5 au and from the New Horizons spacecraft on its way to Jupiter. A noise range was extrapolated as a function of geocentric distance out to 6 au and as a function of the Sun-Earth-Probe (SEP) angle to the solar disk (Table 1) for each flyby. It is evident that the fractional uncertainty of $G \cdot M$ is dominated by the first two terms in Eq. (5) and amounts to the values in the last column of Table 1. The Lucy project has put a requirement on the Trojan mass determination to be better than 25% (Levison et al. 2021), based on the assumed effective diameters listed in Table 1. Deviations from these assumptions will affect the actual mass accuracy.

5.2 Astrometric Mass Determinations

A further method for estimating the mass of an asteroid is the determination of its gravitational influence on another asteroid or even a planet. As Standish (2000) emphasized, precise mass estimates of the largest asteroids are of fundamental importance, as they currently represent the limiting factor for the accuracy of planetary ephemerides. Baer et al. (2011) presented a list of 60 asteroids with masses estimated from asteroid-asteroid perturbations, from asteroids with moons or binaries and from spacecraft radio tracking during flybys. A direct comparison of the mass of (21) Lutetia from asteroid-asteroid perturbations (Baer and Chesley 2008; Baer et al. 2011) and the Rosetta flyby in 2010 (Pätzold et al. 2011), however, showed that Lutetia's astrometric mass was overestimated by 50% (against a quoted error of 30%) compared to the highly precise Rosetta flyby mass (with a quoted error of 1%). This inconsistency shows that the method was subjected to a considerable bias. A more recent study by Fienga et al. (2020) attempts to improve the accuracy of perturbation-based mass determinations of asteroids by using prior information about their physical properties and a random Monte Carlo sampling. The comparison of their results with spacecraft-based mass determinations appears promising, with a higher accuracy and smaller systematics than previous methods. A further promising development is anticipated from the Gaia mission, which is expected to improve the orbital accuracy of a large number of asteroids by several orders of magnitude. This improvement, in addition to a corresponding improvement in planetary ephemeris, is expected to result in accurate mass determinations for a few hundreds of asteroids (Mouret et al. 2008).

5.3 Binary Asteroids and Asteroids with a Moon

A direct method for determining the mass of an asteroid is applicable when a satellite or binary companion is identified (Noll et al. 2023). By determining the period T, and semimajor axis a of the satellite, the system mass can be computed using Kepler's third law:

$$M_{sys} = 4\pi^2 a^3 / GT^2$$

Among the Trojans, three systems have had their mass determined in this way. Marchis et al. (2014) determined the mass of (624) Hektor to be $7.9 \pm 1.4 \times 10^{18}$ kg from analysis of the orbit of the small satellite Skamandrios using Keck AO imaging. The mass of the (617) Patroclus-Menoetius binary is well-determined at $M_{sys} = 1.41 \pm 0.03 \times 10^{18}$ kg with the mutual orbit measured with a combination of HST imaging, Keck AO, and astrometry from a stellar occultation (Grundy et al. 2018). Lastly, the mass of (3548) Eurybates and its small satellite, Queta, is now known to be $M_{sys} = 0.151 \pm 0.003 \times 10^{18}$ kg from the satellite's orbit, measured with HST (Brown et al. 2021). The recent discovery by the Lucy Team of a satellite orbiting Polymele (Buie et al. 2022) will probably also result in an accurate mass determination for this system.

High resolution imaging is the most important factor in deriving precise relative astrometry of the primary and satellite needed to determine the orbit. For very small satellites, multiple exposure depths may be required to determine the position of both the primary and secondary. The orbital period may be improved by making repeated observations over longer timescales. Also, by observing the satellite orbit at multiple epochs, the changing geometry of the orbit as viewed from the Earth allows for better constraints of orbital elements other than the period, including the semimajor axis. Over long time scales, non-Keplerian effects may become evident and may be used to obtain the masses of individual objects in a multi-component system (Ragozzine and Brown 2009).

Mass determinations, whether by spacecraft flyby or satellite orbit, inform the determination of density, as discussed in the next section.

6 Densities and Internal Properties

6.1 Current Knowledge About the Density of Trojans from Remote Sensing

Density is a fundamental property of materials that is a function both of composition and structure. For asteroids, bulk density determinations or estimates range from lows of 0.5×10^3 kg m^{-3} and less, about half that of water ice, to as much as 5×10^3 kg m^{-3} or more (Britt et al. 2002; Carry 2012; Scheeres et al. 2015). Among small body populations more similar to Trojans are low-albedo asteroids with satellites in the Outer Main Belt, mostly C-types; these systems have densities of roughly 1.3×10^3 kg m^{-3} (Margot et al. 2015). In the Kuiper Belt, densities skew lower than 1×10^3 kg m^{-3} for binary systems with components smaller than $\sim$500 km in diameter (Grundy et al. 2019). In both of these cases, pore spaces, either macro- or microscopic, must account for a significant fraction of the total volume for the bulk density to be consistent with the density of the compositionally most-likely solids.

There are three Trojan systems for which the mass has been determined, as noted above. The mass uncertainties are $\pm 18\%$ for Hektor and only about $\pm 2\%$ for both Patroclus and Eurybates. The other necessary component for the calculation of the bulk density, the volume, is less well-determined and is usually the limiting factor in density determination (e.g. Carry 2012). The bilobed shape of the primary component in the Hektor system presents a major challenge in determining its volume. Uncertainties in the shape lead to nearly a factor of two difference in density estimates from 1320$\pm$ 530 kg m^{-3} (from Marchis et al. (2014) as corrected by Descamps 2015) to 2430 $\pm$ 350 kg m^{-3} (Descamps 2015). For Eurybates, the effective diameter was initially estimated from radiometric observations, and more recently with results from stellar occultations (Keeney et al. 2021). Combined with lightcurve data, a shape model can be used to predict the volume (Mottola et al. 2023). Brown et al. (2021) used a volume of $V = 0.14 \pm 0.04 \times 10^{15}$ m^3 to arrive at a Eurybates bulk density of $\rho = 1100 \pm 300$ kg m^{-3}. For the Patroclus-Menoetius binary Buie et al. (2015) determined the shape of both components from a stellar occultation in 2013 as well as a more recent series of events in 2022 (Keeney et al. 2022) combined with lightcurve data. From the 2013 occultation data and multi-epoch lightcurves, an ellipsoidal shape model with a total volume for both components of $V = 1.36 \times 10^{15}$ m^3 was determined. This volume, combined with the more accurate system mass from Grundy et al. (2018), yields a density of ρ = 1040 kg m^{-3}. Unfortunately, Buie et al. (2015) do not cite uncertainties for their shape model, but given the evidence for possible large cratering on Menoetius presented in their work and by Pinilla-Alonso et al. (2022), it is almost certain that the density uncertainty for this system is dominated by the volume uncertainty.

The spin barrier for a population, as shown in Fig. 2, represents the critical rotation rate at which a particle on the equator of the spinning body experiences an apparent centrifugal acceleration that exactly balances the acceleration of gravity (Pravec and Harris 2000). The spin barrier is only meaningful in a statistical sense – if a population that can be assumed to consist of similar objects exhibits a minimum period, it implies that there is a maximum density. In the Main Belt, there is a limit near 2.2 hours which implies a maximum density of 3×10^3 kg m^{-3}. For the Trojans, that limit appears to fall between 4 - 4.8 hours, leading to a maximum density of $\rho_{max} \approx 0.9 \times 10^3$ kg m^{-3} for strengthless rubble-pile Trojans (Ryan et al. 2017; Chang et al. 2021). This limit is marginally consistent with the densities of Patroclus and Eurybates and suggests that the higher density range for Hektor would be not typical for the Trojans as a family.

Overall, low densities would appear to be the norm for Trojans, a result that implies high porosity and also favors ice-rich bulk composition as discussed next. What the Lucy mission will contribute to the determination of density is the possibility of achieving a more accurate, if not first, system mass and, crucially, a much better volume determination for the Trojans it encounters. If densities can be determined for even a subset of the encountered targets with uncertainties of a few percent, compared to the current tens of percent, it will allow many currently unanswerable questions about composition and structure to be addressed.

6.2 Mineralogies of Trojans and Implications for Their Density and Internal Structure

The mineralogy of the Trojan asteroids is something of an enigma. It is dynamically unlikely that we have any samples of Trojans in the meteorite collections (Bottke et al. 2012; Morbidelli et al. 2010). What clues we do have are based on the typical Trojan low albedo, flat or red spectral slope their low bulk density, and their location outside the frost line in the solar system. The timing of accretion from dust into planets was largely driven by the density of available material.

In the inner regions of the solar system where nebular material densities and temperatures were highest, accretion occurred early and was dominated by high temperature minerals such as silicates and metal, while being depleted in volatiles. The heat from the decay of the short-lived radioactive elements turned the early accreting asteroids into melted balls of liquid rock and metal, or somewhat later accreting asteroids into thermally metamorphosed rocks that were also depleted in volatiles (Hutchison 2004).

Further out in the solar system past the "frost line", lower material densities meant later accretion and produced bodies with lower amounts of short-lived radioactive elements and larger amounts of water ice and organics – the carbonaceous chondrites. Relatively volatile-depleted carbonaceous chondrites like the CVs and the COs have bulk mineralogies similar to ordinary chondrites, but also accreted some volatiles and carbonaceous organics along with enough radioactive elements to produce modest thermal metamorphism.

Accreting later and probably farther out in the solar system are the volatile-rich carbonaceous chondrites, the CM, CR and CI carbonaceous chondrites (Papike 2018). The original mineralogy of these asteroids included silicates such as abundant olivine and pyroxene, along with carbonaceous organics and water ice. They also incorporated modest amounts of radioactive elements, enough to melt accreted ice and drive aqueous metamorphism. The melted ice interacted with the abundant olivine, converting the olivine into the hydrated phyllosilicate serpentine along with a number of other hydrated clay minerals. This had the effect of locking some of the original water into the serpentine crystal structure in the form of hydroxyls. The amount of aqueous alteration, the degree of hydration, the degree of metamorphism, and the local chemical conditions varied hugely in these meteorites and thus the parent bodies (Hutchison 2004; Desch et al. 2018; Hamilton et al. 2019).

Comets probably had a similar accretion history as volatile-rich carbonaceous chondrites in that they accreted a similar mix of high-temperature silicates (olivine and pyroxene), ices, and carbonaceous organics, but with very low amounts of short-lived radioactive isotopes. As a result, they never underwent significant heating or metamorphism and retained their accretional ices including some proportion of supervolatiles, ices such as CO_2, CO, and CH_4 and low-temperature organics (Sitko et al. 2011; Wooden et al. 2017).

Where did the Trojans fit into this accretionary continuum between fully differentiated bodies and almost unprocessed objects like comets? Based on the spectra and physical properties of Trojans there are two possibilities. First, and most likely, the Trojans may be very analogous to comets. They would essentially represent unprocessed outer nebular material and would be under-dense, volatile-rich, and organic rich. The dusty portions of Trojans would be composed of mixtures of crystalline and amorphous silicates, primarily olivine and pyroxene, along with metal sulfides, carbonates, and perhaps a few clays. The actual proportions of volatiles and dusty components will likely be highly variable since comets can vary in this fraction by a factor of five.

The other alternative is that Trojans accreted with some non-negligible portion of radioactive elements and were altered like CI and CM carbonaceous chondrites. In this scenario the dusty portion to be dominated by the aqueous alteration product serpentine and other hydrated clays. This seems less likely because the process of aqueous alteration would likely compress and consolidate the altered bodies and Trojans exhibit very low-density characteristic of high volatile contents and very little consolidation.

7 Outstanding Questions and Lucy Contributions

With its tour through both Trojan clouds the Lucy mission will help address a number of outstanding questions about the origin and evolution of the Trojans and inform the current Solar System formation models. Through the direct measurement of the volume –via imaging– and of the mass –via the radio science experiment– Lucy will determine the bulk density for all of the Trojan targets, a quantity that bears information about the mineralogy and the structural properties of the bodies. By measuring this quantity for objects of different sizes and belonging to all different taxonomic types present in the Trojan regions, Lucy will look for possible correlations between density, composition and structure. Direct measurements of shape and topography will enable the search for expressions of shape evolution processes as erosion by collision or by sublimation and will enable the application of comparative methods with other bodies previously visited by spacecraft. Accurate measurement of disk-integrated and disk-resolved photometric properties will be possible for observing geometries that are diagnostic and not accessible from Earth. These allow analyses that will complement the geomorphology studies and support the classification of the properties of the surface regolith. Furthermore, they will support the search for regions of possible cometary activity.

Acknowledgements This work has made use of data from the Asteroid Lightcurve Database (Warner et al. 2009).

Funding Open Access funding enabled and organized by Projekt DEAL. Research at the DLR was funded by the DLR Programmatik Raumfahrtforschung und -technologie through the grant Q5 2474029 Lucy. MP is funded by the Bundesministerium für Wirtschaft und Energie BMWi, Berlin, via the Deutsche Raumfahrtagentur, Deutsches Zentrum für Luft- und Raumfahrt, Bonn-Oberkassel, under grant 50 OW 2102.

Declarations

Competing Interests The authors declare that they have no competing interests.

References

Andert TP, Rosenblatt P, Pätzold M et al (2010) Geophys Res Lett 37:L09202. https://doi.org/10.1029/2009GL041829

Baer J, Chesley SR (2008) Celest Mech Dyn Astron 100:27. https://doi.org/10.1007/s10569-007-9103-8

Baer J, Chesley SR, Matson RD (2011) Astron J 141:143. https://doi.org/10.1088/0004-6256/141/5/143

Benz W, Asphaug E (1999) Icarus 142:5. https://doi.org/10.1006/icar.1999.6204

Berthier J, Descamps P, Vachier F et al (2020) Icarus 352:113990. https://doi.org/10.1016/j.icarus.2020.113990

Binzel RP, Sauter LM (1992) Icarus 95:222. https://doi.org/10.1016/0019-1035(92)90039-A

Bottke WF, Vokrouhlický D, Minton D et al (2012) Nature 485:78. https://doi.org/10.1038/nature10967

Bottke WF, Marschall R, Nesvorný D, Vokrouhlický D (2023). Space Sci Rev 219:83. https://doi.org/10.1007/s11214-023-01031-4

Bowell E, Hapke B, Domingue D et al (1989) In: Binzel RP, Gehrels T, Matthews MS (eds) Asteroids II. University of Arizona Press, Tucson, pp 524–556

Britt DT, Yeomans D, Housen K, Consolmagno G (2002) In: Bottke WF, Cellino A, Paolicchi P, Binzel RP (eds) Asteroids III. University of Arizona Press, Tucson, pp 485–500

Brown ME, Levison HF, Noll KS et al (2021) Planet Sci J 2:170. https://doi.org/10.3847/PSJ/ac07b0

Buie MW, Olkin CB, Merline WJ et al (2015) Astron J 149:113. https://doi.org/10.1088/0004-6256/149/3/113

Buie MW, Keeney BA, Strauss RH et al (2021) Planet Sci J 2:202. https://doi.org/10.3847/PSJ/ac1f9b

Buie MW et al (2022) AAS Division of Planetary Science meeting. Bull Am Astron Soc 54(8):2022n8i512p03

Carry B (2012) Planet Space Sci 73:98. https://doi.org/10.1016/j.pss.2012.03.009

Carry B, Dumas C, Kaasalainen M et al (2010) Icarus 205:460. https://doi.org/10.1016/j.icarus.2009.08.007

Chang C-K, Chen Y-T, Fraser WC et al (2021) Planet Sci J 2:191. https://doi.org/10.3847/psj/ac13a4

Cibulková H, Ďurech J, Vokrouhlický D, Kaasalainen M, Oszkiewicz DA (2016) Astron Astrophys 596:A57. https://doi.org/10.1051/0004-6361/201629192

Descamps P (2015) Icarus 245:64. https://doi.org/10.1016/j.icarus.2014.08.002

Desch SJ, Kalyaan A, Alexander CMO (2018) Astrophys J Suppl Ser 238:11. https://doi.org/10.3847/1538-4365/aad95f

Ďurech J, Hanuš J (2023) Astron Astrophys 675:A24. https://doi.org/10.1051/0004-6361/202345889

Ďurech J, Kaasalainen M (2003) Astron Astrophys 404:709. https://doi.org/10.1051/0004-6361:20030505

Ďurech J, Kaasalainen M, Herald D et al (2011) Icarus 214:652. https://doi.org/10.1016/j.icarus.2011.03.016

Ďurech J, Hanuš J, Vančo R (2019) Astron Astrophys 631:A2. https://doi.org/10.1051/0004-6361/201936341

Ďurech J, Tonry J, Erasmus N et al (2020) Astron Astrophys 643:A59. https://doi.org/10.1051/0004-6361/202037729

Emery JP, Marzari F, Morbidelli A, French LM, Grav T (2015) In: Bottke WF, DeMeo FE, Michel P (eds) Asteroids IV. University of Arizona Press, Tucson, pp 203–220. https://doi.org/10.2458/azu_uapress_9780816532131-ch011

Emery JP, Binzel RP, Britt DT et al (2024). Space Sci Rev 220

Fienga A, Avdellidou C, Hanuš J (2020) Mon Not R Astron Soc 492:589. https://doi.org/10.1093/mnras/stz3407

French LM, Stephens RD, Coley D, Wasserman LH, Sieben J (2015) Icarus 254:1. https://doi.org/10.1016/j.icarus.2015.03.026

Gaia Collaboration, Spoto F, Tanga P et al (2018) Astron Astrophys 616:A13. https://doi.org/10.1051/0004-6361/201832900

Grav T, Mainzer AK, Bauer JM, Masiero JR, Nugent CR (2012) Astrophys J 759:49. https://doi.org/10.1088/0004-637X/759/1/49

Grundy W, Noll K, Buie M, Levison H (2018) Icarus 305:198. https://doi.org/10.1016/j.icarus.2018.01.009

Grundy WM, Noll KS, Buie MW et al (2019) Icarus 334:30. https://doi.org/10.1016/j.icarus.2018.12.037

Hamilton VE, Simon AA, Christensen PR et al (2019) Nat Astron 3:332. https://doi.org/10.1038/s41550-019-0722-2

Hanuš J, Ďurech J, Oszkiewicz DA et al (2016) Astron Astrophys 586:A108. https://doi.org/10.1051/0004-6361/201527441

Hanuš J, Vokrouhlický D, Nesvorný D et al (2023) Astron Astrophys 679:A56. https://doi.org/10.1051/0004-6361/202346022

Harris AW, Burns JA (1979) Icarus 40:115. https://doi.org/10.1016/0019-1035(79)90058-7

Harris A, Warner BD (2020) Icarus 339:113602. https://doi.org/10.1016/j.icarus.2019.113602

Hartmann WK, Cruikshank DP (1978) Icarus 36:353. https://doi.org/10.1016/0019-1035(78)90114-8

Hartmann WK, Binzel RP, Tholen DJ, Cruikshank DP, Goguen J (1988) Icarus 73:487. https://doi.org/10.1016/0019-1035(88)90058-9
Hutchison R (2004) Meteorites. Cambridge University Press, Cambridge
Kaasalainen M (2004) Astron Astrophys 422:L39. https://doi.org/10.1051/0004-6361:20048003
Kaasalainen M (2011) Inverse Probl Imaging 5:37. https://doi.org/10.3934/ipi.2011.5.37
Kaasalainen M, Torppa J, Muinonen K (2001) Icarus 153:37. https://doi.org/10.1006/icar.2001.6674
Kaasalainen M, Mottola S, Fulchignoni M (2002) In: Bottke WF, Cellino A, Paolicchi P, Binzel RP (eds) Asteroids III. University of Arizona Press, Tucson, pp 139–150
Kalup CE, Molnár L, Kiss C et al (2021) Astrophys J Suppl Ser 254:7. https://doi.org/10.3847/1538-4365/abe76a
Keeney B, Buie M, Kaire M et al (2021) In: AGU fall meeting abstracts, vol 2021, pp P32B–03
Keeney B, Buie M, Levison H (Lucy Occultations Team) (2022) In: AAS/Division for planetary sciences meeting abstracts, Vol. 54, 512.04
Konopliv AS, Park RS, Vaughan AT et al (2018) Icarus 299:411. https://doi.org/10.1016/j.icarus.2017.08.005
Levison HF, Olkin CB, Noll KS et al (2021) Planet Sci J 2:171. https://doi.org/10.3847/PSJ/abf840
Levison HF, Marchi S, Noll KS, Olkin CB, Statler TS (2024). Space Sci Rev 220
Mann RK, Jewitt D, Lacerda P (2007) Astron J 134:1133. https://doi.org/10.1086/520328
Marchis F, Ďurech J, Castillo-Rogez J et al (2014) Astrophys J 783:L37. https://doi.org/10.1088/2041-8205/783/2/l37
Marciniak A, Bartczak P, Müller T et al (2018) Astron Astrophys 610:A7. https://doi.org/10.1051/0004-6361/201731479
Margot JL, Pravec P, Taylor P, Carry B, Jacobson S (2015) In: Bottke WF, DeMeo FE, Michel P (eds) Asteroids IV. University of Arizona Press, Tucson, pp 355–374. https://doi.org/10.2458/azu_uapress_9780816532131-ch019
Milani A, Chesley SR, Sansaturio ME et al (2009) Icarus 203:460. https://doi.org/10.1016/j.icarus.2009.05.029
Morbidelli A, Brasser R, Tsiganis K, Gomes R, Levison HF (2009) Astron Astrophys 507:1041. https://doi.org/10.1051/0004-6361/200912876
Morbidelli A, Brasser R, Gomes R, Levison HF, Tsiganis K (2010) Astron J 140:1391. https://doi.org/10.1088/0004-6256/140/5/1391
Mottola S, Di Martino M, Erikson A et al (2011) Astron J 141:170. https://doi.org/10.1088/0004-6256/141/5/170
Mottola S, Di Martino M, Carbognani A (2014) Mem Soc Astron Ital Suppl 26:47
Mottola S, Hellmich S, Buie MW et al (2020) Planet Sci J 1:73. https://doi.org/10.3847/PSJ/abb942
Mottola S, Hellmich S, Buie MW et al (2023) Planet Sci J 4:18. https://doi.org/10.3847/PSJ/acaf79
Mouret S, Hestroffer D, Mignard F (2008) Planet Space Sci 56:1819. https://doi.org/10.1016/j.pss.2008.02.026
Muinonen K, Belskaya IN, Cellino A et al (2010) Icarus 209:542. https://doi.org/10.1016/j.icarus.2010.04.003
Nesvorný D, Vokrouhlický D, Morbidelli A (2013) Astrophys J 768:45. https://doi.org/10.1088/0004-637X/768/1/45
Nesvorný D, Vokrouhlický D, Bottke WF, Levison HF, Grundy WM (2020) Astrophys J 893:L16. https://doi.org/10.3847/2041-8213/ab8311
Noll KS, Brown ME, Weaver HA et al (2020) Planet Sci J 1:44. https://doi.org/10.3847/PSJ/abac54
Noll KS, Brown ME, Grundy WH (2023). Space Sci Rev 219:59. https://doi.org/10.1007/s11214-023-01001-w
Pál A, Szakáts R, Kiss C et al (2020) Astrophys J Suppl Ser 247:26. https://doi.org/10.3847/1538-4365/ab64f0
Papike JJ (ed) (2018) Planetary Materials. De Gruyter, Berlin, Boston. https://doi.org/10.1515/9781501508806
Pätzold M, Wennmacher A, Häusler B et al (2001) Astron Astrophys 370:1122. https://doi.org/10.1051/0004-6361:20010244
Pätzold M, Andert TP, Häusler B et al (2010) Astron Astrophys 518:L156. https://doi.org/10.1051/0004-6361/201014325
Pätzold M, Andert TP, Asmar SW et al (2011) Science 334:491. https://doi.org/10.1126/science.1209389
Pätzold M, Andert TP, Tyler GL et al (2014) Icarus 229:92. https://doi.org/10.1016/j.icarus.2013.10.021
Pätzold M, Andert T, Hahn M et al (2016) Nature 530:63. https://doi.org/10.1038/nature16535
Pätzold M, Andert TP, Hahn M et al (2019) Mon Not R Astron Soc 483:2337. https://doi.org/10.1093/mnras/sty3171
Pinilla-Alonso N, Popescu M, Licandro J et al (2022) Planet Sci J 3:267. https://doi.org/10.3847/PSJ/ac9f11

Pravec P, Harris AW (2000) Icarus 148:12. https://doi.org/10.1006/icar.2000.6482
Ragozzine D, Brown ME (2009) Astron J 137:4766. https://doi.org/10.1088/0004-6256/137/6/4766
Rubincam DP (2000) Icarus 148:2. https://doi.org/10.1006/icar.2000.6485
Ryan EL, Sharkey BNL, Woodward CE (2017) Astron J 153:116. https://doi.org/10.3847/1538-3881/153/3/116
Schaefer MW, Schaefer BE, Rabinowitz DL, Tourtellotte SW (2010) Icarus 207:699. https://doi.org/10.1016/j.icarus.2009.11.031
Scheeres DJ, Britt D, Carry B, Holsapple KA (2015) In: Bottke WF, DeMeo FE, Michel P (eds) Asteroids IV. University of Arizona Press, Tucson, pp 745–766. https://doi.org/10.2458/azu_uapress_9780816532131-ch038
Schemel M, Brown ME (2021) Planet Sci J 2:40. https://doi.org/10.3847/psj/abc752
Shevchenko VG, Belskaya IN, Slyusarev IG et al (2012) Icarus 217:202. https://doi.org/10.1016/j.icarus.2011.11.001
Sitko ML, Lisse CM, Kelley MS et al (2011) Astron J 142:80. https://doi.org/10.1088/0004-6256/142/3/80
Slyusarev IG, Belskaya IN (2014) Sol Syst Res 48:139. https://doi.org/10.1134/S0038094614020063
Sonnett S, Mainzer A, Grav T, Masiero J, Bauer J (2015) Astrophys J 799:191. https://doi.org/10.1088/0004-637x/799/2/191
Standish M (2000) In: Johnston KJ, McCarthy DD, Luzum BJ, Kaplan GH (eds) IAU colloq. 180: Towards models and constants for sub-microarcsecond astrometry, p 120
Storrs A, Weiss B, Zellner B et al (1999) Icarus 137:260. https://doi.org/10.1006/icar.1999.6047
Szabó GM, Pál A, Kiss C et al (2017) Astron Astrophys 599:A44. https://doi.org/10.1051/0004-6361/201629401
Tanga P, Hestroffer D, Cellino A et al (2003) Astron Astrophys 401:733. https://doi.org/10.1051/0004-6361:20030032
Tonry JL, Denneau L, Heinze AN et al (2018) Publ Astron Soc Pac 130:064505. https://doi.org/10.1088/1538-3873/aabadf
Tyler GL, Linscott IR, Bird MK et al (2008) Space Sci Rev 140:217. https://doi.org/10.1007/s11214-007-9302-3
Viikinkoski M, Kaasalainen M, Ďurech J (2015) Astron Astrophys 576:A8. https://doi.org/10.1051/0004-6361/201425259
Viikinkoski M, Hanuš J, Kaasalainen M, Marchis F, Ďurech J (2017) Astron Astrophys 607:A117. https://doi.org/10.1051/0004-6361/201731456
Vokrouhlický D, Bottke WF, Chesley SR, Scheeres DJ, Statler TS (2015) In: Bottke WF, DeMeo FE, Michel P (eds) Asteroids IV. University of Arizona Press, Tucson, pp 509–531. https://doi.org/10.2458/azu_uapress_9780816532131-ch027
Warner BD, Harris AW, Pravec P (2009) Icarus 202:134. https://doi.org/10.1016/j.icarus.2009.02.003
Wong I, Brown ME (2019) Astron J 157:203. https://doi.org/10.3847/1538-3881/ab18f4
Wong I, Brown ME, Emery JP (2014) Astron J 148:112. https://doi.org/10.1088/0004-6256/148/6/112
Wooden DH, Ishii HA, Zolensky ME (2017) Philos Trans R Soc Lond Ser A 375:20160260. https://doi.org/10.1098/rsta.2016.0260
Yeomans DK, Barriot JP, Dunham DW et al (1997) Science 278:2106. https://doi.org/10.1126/science.278.5346.2106
Yeomans DK, Antreasian PG, Cheng A et al (1999) Science 285:560. https://doi.org/10.1126/science.285.5427.560
Youdin AN, Goodman J (2005) Astrophys J 620:459. https://doi.org/10.1086/426895

Publisher's Note Springer Nature remains neutral with regard to jurisdictional claims in published maps and institutional affiliations.

Space Science Reviews (2024) 220:28
https://doi.org/10.1007/s11214-024-01060-7

Surface Compositions of Trojan Asteroids

Joshua P. Emery[1] · Richard P. Binzel[2] · Daniel T. Britt[3] · Michael E. Brown[4] · Carly J.A. Howett[5,6] · Audrey C. Martin[3] · Mario D. Melita[7,8] · Ana Carolina Souza-Feliciano[3,9] · Ian Wong[10]

Received: 16 June 2023 / Accepted: 4 March 2024 / Published online: 25 March 2024

Abstract
The Jupiter Trojan asteroids are a key population for understanding the chemical and dynamical evolution of the Solar System. Surface compositions of Trojans, in turn, provide crucial information for reconstructing their histories. NASA's *Lucy* mission will soon complete the first spacecraft reconnaissance of this population. This review summarizes the current state of knowledge of Trojan surface compositions and looks ahead to expected advances in that knowledge from *Lucy*. Surface compositions of Trojans remain uncertain due to a relative lack of diagnostic absorption features, though dedicated observations have begun to provide some clues to compositions. Trojans have uniformly low albedos, with a population average of $\sim$5.3%, and red spectral slopes at ultraviolet, visible, and near-infrared wavelengths. A bimodality of spectral slopes has been detected and confirmed across all these wavelengths, and the ratio of "less-red" to "red" Trojans increases with decreasing size. A broad absorption at $\sim$3.1 μm in some less-red Trojans may indicate the presence of N-H bearing material. Mid-infrared emissivity spectra reveal the presence of fine-grained anhydrous silicates on the surfaces. The meteorite collection contains no identifiable analogs to Trojan asteroids. Among small body populations, some Main Belt asteroids, comets, irregular satellites, and Centaurs provide reasonable spectral matches, supporting some genetic relationships among some members of these groups. The cause of the observed spectral properties remains uncertain, but recent suggestions include a combination of volatile ice sublimation and space weathering or a combination of impact gardening and space weathering. The *Lucy* mission will provide detailed compositional analysis of (3548) Eurybates, (15094) Polymele, (11351) Leucus, (21900) Orus, and (617) Patroclus-Menoetius, a suite of targets that sample the diversity among the Trojan population along several dimensions. With these flybys, the *Lucy* mission is poised to resolve many of the outstanding questions regarding Trojan surface compositions, thereby revealing how the Trojans formed and evolved and providing a clearer view of Solar System history.

Keywords Trojan asteroids · Asteroids · Lucy · Composition · Spectroscopy

1 Introduction

The Solar System is ever-changing, and changes were particularly rapid and dramatic in the early stages of its evolution. Spurred by observations of diverse planetary systems around

Extended author information available on the last page of the article other stars as well as open questions about the Solar System, various styles and episodes of planet migration have been modeled for the Solar System. Small bodies (asteroids and comets) and their distribution across the Solar System are tracers of the original composition and thermochemical conditions in their formation regions, dynamical events (such as planet migration), and physical effects from interactions with the space environment. What we learn about the Solar System, the only planetary system we can study *in situ*, illuminates how other planetary systems evolve and how many of those, and which ones, may foster life.

The Trojan asteroids share Jupiter's orbit around the Sun, trapped in the L4 (leading) and L5 (trailing) stable Lagrange regions at $\sim$5.28 AU. These $1:1$ mean motion resonant orbits are stable over the lifetime of the Solar System (Levison et al. 1997). Marzari and Scholl (1998) showed that objects would be captured from the local population into the Lagrange regions as Jupiter grew. However, this capture mechanism does not explain the observed inclination distribution of Trojan asteroids. More recent works have alternatively proposed capture later in Solar System history, as part of the orbital migration of the giant planets in the Nice model (Tsiganis et al. 2005). Morbidelli et al. (2005) originally suggested that as Jupiter and Saturn smoothly migrate into a mean motion resonance with each other, any existing bodies are removed from the Lagrange regions, and as they migrate out of the resonance, a new population is captured (so-called chaotic capture). Nesvorný et al. (2013) demonstrated that bodies can also be trapped in Lagrange regions when Jupiter "jumps" (i.e., changes semi-major axis quickly) during close scattering of ice giant planets (so-called jump capture, as part of the jumping-Jupiter version of the Nice model) (Morbidelli et al. 2009). In any version of the Nice model, the captured Trojan population is dominated by bodies that were scattered from the Kuiper Belt, predicting that present-day Trojan asteroids originated in the Kuiper Belt.

Surface compositions of Trojan asteroids provide crucial information for reconstructing their histories. Nebular conditions in their formation region(s) set original compositions (e.g., silicate mineralogy, ices, and organics). Trojan asteroids are thought to be primitive, meaning their compositions have not been significantly altered since formation. Nevertheless, any internal heating that may have modified bulk compositions would be controlled by the original composition. Any potential surface modification by space weathering would also depend on the original surface composition as well as Trojan dynamical histories. Detailed compositional analysis of the suite of minerals, ices, and organics on the surface and in the near subsurface enables the unraveling of the complex influences to help establish evolutionary timelines. Comparisons with analog populations also provide key insights into present-day compositions and evolutionary histories of Trojan asteroids.

Several reviews of the surface compositions of Trojan asteroids have been published previously (Dotto et al. 2008; Slyusarev and Belskaya 2014; Emery et al. 2015). Here, we build on those reviews with a focus on advances in knowledge since their publication. We begin with an overview of the current state of knowledge of Trojan surface compositions, including insights gained from spacecraft and telescopic observations of other small body populations, and a discussion of potential mechanisms that can alter their surface compositions over time. We then describe outstanding questions regarding the surface compositions of Trojans and discuss how NASA's *Lucy* mission to provide the first up-close investigation of Trojans will address those questions.

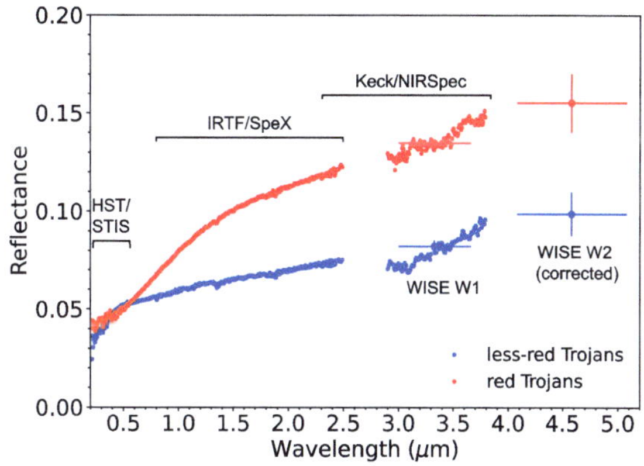

Fig. 1 Average reflectance spectra of less-red and red Trojans from 0.2 to 5.0 μm, compiled from Emery et al. (2011), Grav et al. (2012), Brown (2016), and Wong et al. (2019a). The two subpopulations show distinct near-UV, optical, and near-infrared continuum slopes. Note the weak 3.1 μm absorption that is primarily observed on less-red objects

2 Observational Constraints on Surface Composition

2.1 Albedos

Compared to other populations of small bodies in the solar system, the albedos of the Trojan asteroids appear remarkably uniform and low. Analysis of thermal emission from the Wide-field Infrared Space Explorer (WISE) mission of 476 Trojan asteroids showed a mean V-band albedo of 6.9% — with only a small spread — for the largest, well-measured objects (Grav et al. 2012). Correcting these measurements using updated absolute magnitudes (Romanishin and Tegler 2018) drops the true albedos to a mean of only 5.3%. These low values are similar to those seen in dark cometary nuclei (Lamy et al. 2004).

While the largest Trojan asteroids appear uniformly low albedo, there is some evidence that smaller Trojans can have higher albedos. Fernández et al. (2009) measured thermal flux of 44 small (5 to 24 km) Trojans using the *Spitzer Space Telescope* and found that these small bodies had statistically higher albedos than larger Trojans. The WISE data show an upward trend of albedo for smaller diameters, but this trend can plausibly be attributed solely to the larger uncertainties for the smaller objects, known bias in absolute magnitudes for small objects, and the positive bias of albedo. Nonetheless, a small number of small well-measured WISE Trojans appear to have higher-than-typical albedos. Simpson et al. (2022) observed a subset of the WISE high-albedo objects with ALMA and also found that some ~20-30 km Trojans have albedos as high at 13%. They hypothesized that these smaller objects contain evidence of recent catastrophic impacts which can — perhaps only briefly — give access to fresh interior material before it has been extensively space weathered.

2.2 Reflectance Spectrophotometry in the Ultraviolet, Visible, and Near-Infrared

Average reflectance spectra of Trojans from 0.2 to 5.0 μm are shown in Fig. 1. The earliest efforts to characterize the surface composition of Trojans through visible reflectance spectroscopy revealed featureless spectra with positive spectral slopes comparable to D-type asteroids in the outer Main Belt (Gradie and Veverka 1980). Subsequent visible spectroscopy throughout the following decades targeted a significant fraction of the large Trojans ($D > 50$ km), uncovering a wide range of spectral slopes — from neutral to moderately red — while continuing to yield no incontrovertible molecular absorptions (Jewitt and Luu 1990; Vilas et al. 1993; Bendjoya et al. 2004; Lazzaro et al. 2004; Fornasier et al. 2004; Dotto et al. 2006; Fornasier et al. 2007; Melita et al. 2008). Similarly, near-infrared

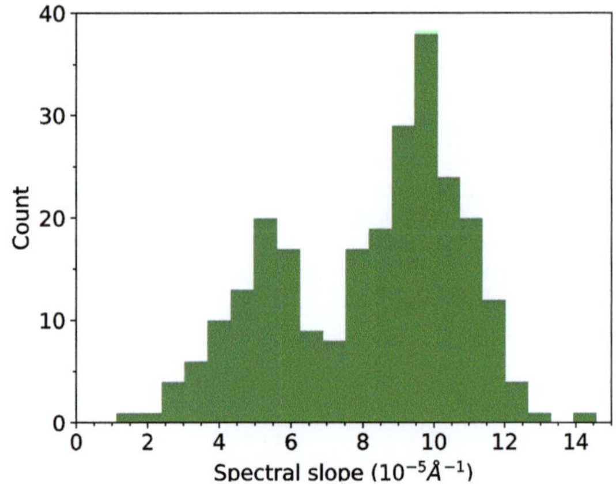

Fig. 2 Histogram of visible spectral slope for 293 Trojan asteroids observed by SDSS, showing the observed bimodality in color

spectroscopy through 2.5 μm has hitherto failed to detect any spectral features (Luu et al. 1994; Dumas et al. 1998; Emery and Brown 2003; Dotto et al. 2006; Yang and Jewitt 2007; De Luise et al. 2010; Emery et al. 2011; Marsset et al. 2014). Particularly notable is the lack of silicate and water ice features in the near-infrared spectra, which are otherwise quite common throughout the middle and outer Solar System. The absence of these prominent absorptions in spectra of Trojans has been a persistent mystery, given that water ice and silicates are expected to comprise much of the bulk composition of Trojans.

The advent of all-sky photometric surveys has greatly enriched our knowledge of Trojan reflectance spectra by enabling the study of population-level trends. The Sloan Digital Sky Survey (SDSS) imaged hundreds of Trojans in five photometric bands: u' (360 nm), g' (470 nm), r' (620 nm), i' (750 nm), and z' (890 nm). Analyses of the visible spectral slope distribution derived from SDSS photometry demonstrated that the Trojans have a bimodal color distribution (Fig. 2) (Szabó et al. 2007; Roig et al. 2008; Wong et al. 2014). The bimodality is attested in both the L4 and L5 swarms and does not appear to be correlated with any orbital parameter (Fornasier et al. 2007; Melita et al. 2008; Emery et al. 2011). Likewise, broadband near-infrared reflectance measurements obtained by the WISE mission in the W1 band (3.4 μm) showed a robust bifurcation among the largest Trojans in the sample (Grav et al. 2012). Further evidence of bimodal spectral characteristics comes from 0.7–2.5 μm spectroscopy of 68 Trojans observed using the Infrared Telescope Facility (IRTF; Fig. 1) (Emery and Brown 2003; Emery et al. 2011): the derived broadband near-infrared colors are clustered into two groups in $J - K$ vs. $0.85 - J$ space.

Crucially, when examining Trojans that are shared among the SDSS, WISE, and IRTF datasets, the respective bimodal classifications are consistent; i.e., objects belonging to the redder of the two visible color groups also have redder near-infrared colors and higher W1 reflectance, and vice versa (Wong et al. 2014). This observation indicates that the Trojan population as a whole contains two subpopulations, whose members differ systematically with respect to several spectral properties across the visible and near-infrared wavelength range and therefore must possess distinct surface compositions and/or regolith properties. These subpopulations are conventionally referred to as less-red and red. The *Lucy* mission targets are representative of the two subpopulations. (617) Patroclus-Menoetius and (15094) Polymele are in the less-red group, whereas (11351) Leucus and (21900) Orus are in the red group. (3548) Eurybates is an exception, with published spectral slopes from neutral (compatible with C-type asteroids) (Fornasier et al. 2007) to less-red (Souza-Feliciano et al. 2020; Emery et al. 2011).

The large body of visible colors provided by SDSS has allowed for the luminosity/size distributions of the less-red and red subpopulations to be characterized independently. The less-red and red luminosity distributions were shown to differ substantially. Specifically, the slope of the distributions are strongly divergent at intermediate sizes ($H \gtrsim 10$ or $D \lesssim 50$ km), with the fraction of less-red objects monotonically increasing with increasing H (decreasing size) (Wong et al. 2014). A follow-up wide-field photometric survey of small Trojans corroborated this finding by extending the attested trend to objects with diameters in the 1–10 km range and demonstrating that less-red objects dominate the Trojan population at the smallest sizes (Wong and Brown 2015). If Trojan albedos turn out to be systematically higher at smaller sizes (see Sect. 2.1), the observation of an increasing less-red fraction at small sizes would not change, though the size at which the transition from red-dominated to less-red dominated would be at a somewhat smaller size than noted in Wong and Brown (2015). Previously, Fornasier et al. (2007) found no trend in spectral slope from visible spectra of 142 Trojans in collisional families with $D > 16$ km. It is likely that the smaller sample size was not sufficient to resolve the trend in this size range observed in Wong and Brown (2015). Another notable finding concerns the Eurybates family — the only robustly attested collisional family in the Trojans (Brož and Rozehnal 2011): the collisional fragments do not display the color bimodality seen in the background population and are exclusively less-red (Fornasier et al. 2007). These two peculiar observations have broad implications for our understanding of Trojan surface composition (see Sect. 4).

Recently, spectroscopic characterization of Trojans has been extended into the near-ultraviolet (NUV). Reflectance spectra spanning 220–550 nm were obtained using the Hubble Space Telescope (HST) for six large Trojans, three from each color subpopulation (Wong et al. 2019a). The subpopulation-averaged ultraviolet spectra show distinct spectral shapes: namely, less-red objects (as characterized by visible color) have steeper NUV spectral slopes than red objects. Figure 1 summarizes the current state of our understanding of Trojan reflectance spectra from 0.2 to 5.0 μm. The subpopulation-averaged HST and IRTF spectra are combined with the averaged visible spectra from available literature sources, as well as the WISE broadband reflectances at W1 (3.4 μm) and W2 (4.6 μm). The differences in spectral shape between the less-red and red subpopulations are most evident at wavelengths less than ~2 μm. Humes et al. (2022) observed four of the *Lucy* targets (Patroclus, Eurybates, Leucus, and Orus) in the UV with HST. They did not find the same slope-trend reversal as Wong et al. (2019a), but they report a UV reflectance "bump" near 0.35 μm not detected in the Wong et al. (2019a) data. They interpret the UV bump as Rayleigh scattering by sub-micron-sized opaque grains.

2.3 3–5 μm Spectra

Ground-based reflectance spectroscopy of these faint bodies is difficult beyond 2.5 μm and has only been obtained for a handful of the brightest objects (Cruikshank et al. 2001; Emery and Brown 2003; Brown 2016), though WISE observations obtained photometry in the W1 (3-4 μm) and W2 (4-5 μm) bands of many objects (Grav et al. 2012). The spectra and WISE photometry show that the reflectance of both the less-red and red Trojans continues to rise approximately linearly through at least 5 μm (Fig. 1).

Cruikshank et al. (2001) and Emery and Brown (2003) report no absorptions in this wavelength range outside of the noise in the spectra. Emery and Brown (2004) use those non-detections to place limits of a few percent on abundances of water ice and hydrated silicates on Trojan surfaces and slightly higher limits on complex organic molecules. They also suggest that the lack of strong organic features indicates that the red VNIR slopes cannot be due to organics, as previously suggested.

Brown (2016) report an absorption at 3.1 μm in several large Trojans, with a strong anti-correlation with color. Absorptions in these regions are often attributed to fine-grained water ice, but such grains would quickly be depleted on the surfaces of these objects, so unusual circumstances such as recent collisions would be required for water ice to be present. The appearance of this feature on all of the less-red objects observed makes this explanation unlikely. The absorption could perhaps be explained as part of an N-H stretch feature at 3.13 μm. Beyond the 3.1-μm absorption, additional absorptions appear, consistent with organic materials that might be expected on objects such as these. In the ground-based data, these appear strongest on the least-red of these objects. Unfortunately, with the low signal-to-noise of these data, little definitive can be said about the species responsible for these absorptions beyond 3 μm.

2.4 Mid-Infrared Emission Spectroscopy

The first Mid-Infrared (MIR; 5-30 μm) spectra of Trojan asteroids looked at (624) Hektor, (1172) Aneas, and (911) Agamemnon taken with the Spitzer Space Telescope (SST) (Emery et al. 2006). The spectra reveal a broad 10-μm emissivity feature with a spectral contrast of roughly 10-15% along with two smaller and broader peaks near 19 and 24 μm (Fig. 3). The spectra of these three Trojans are indicative of fine-grained crystalline silicates, such as olivine and pyroxene. The shape and spectral contrast bear a resemblance to that of comets, for which optically thin comae result in strong 10-μm emission features when comprised of fine-grained ($\lesssim$ 10 μm) disperse silicates. Trojans do not have comae (e.g. Perna et al. 2018b), so researchers hypothesize that the emissivity peaks arise from a fluffy (i.e, highly porous) regolith of fine grain silicates, or from silicates suspended in a transparent matrix (Emery et al. 2006; Vernazza et al. 2012; Yang et al. 2013). Through comparisons of the SST spectra to laboratory spectra of olivine and pyroxene mixed with potassium bromide salt to simulate regolith porosity, Martin et al. (2022, 2023a) conclude that the regolith porosity on Hektor is in excess of 80%. Using T-Matrix and Hapke spectral modeling, Lowry et al. (2022) obtain similar results, including a likely olivine-rich silicate component. Martin and Emery (2023) present MIR spectra of 11 additional Trojans measured with SST, and all show similar emissivity peaks indicative of fine-grained silicates. The red spectral group have larger spectral contrasts in the 10-μm pleateau than the less-red group.

Subsequent MIR observations of the Patroclus-Menoetius system with the SST revealed spectra that look similar to that of Hektor (Mueller et al. 2010). The Patroclus spectrum has a well-defined 10-μm feature, though the spectral contrast is about half that of Hektor's 10-μm feature, and a perceptible feature near 20 μm. As with the other three Trojans, Patroclus likely contains fine-grain crystalline silicates on its surface. Though Hektor belongs to the spectrally red group, and Patroclus belongs to the spectrally less-red group, their similar MIR spectra suggest a compositional similarity, at least for the silicate component.

In addition to the crystalline silicates on Trojan surfaces observed in the MIR, there are likely contributions from amorphous and opaque materials. Patroclus and Aneas have rounded 10-μm emissivity features, as opposed to a trapezoidal 10-μm feature seen in Hektor's spectrum (Kelley et al. 2017; Martin 2022; Martin and Emery 2023). Rounded emission features can be explained by amorphous silicates or hydrated silicate spectra (Lisse et al. 2006; Harker et al. 2007; Kelley and Wooden 2009; Izawa et al. 2021). No indications of hydration have been detected at either 0.7 μm or ∼3 μm on Trojans, but amorphous silicates would be expected on bodies that formed in the outer Solar System (e.g. Henning 2010). The MIR spectral contrasts of all four Trojans are much less than that of comets or laboratory studies of analog material (Izawa et al. 2021; Martin et al. 2022, 2023a). Additional

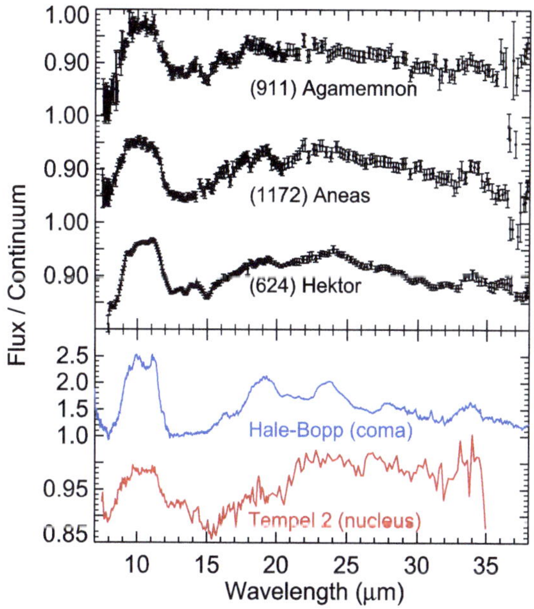

Fig. 3 MIR emissivity spectra of (1172) Aneas, (911) Agamemnon, and (624) Hektor, from Emery et al. (2006). Spectra of the coma of comet Hale-Bopp (blue; Crovisier et al. 1997) and the nucleus of comet Tempel 2 (red; Kelley et al. 2017) are shown for comparison (see Sect. 3.3). The Trojan and Tempel 2 nucleus spectra are shown at the same scale, but note the significant difference in scale factor for Hale-Bopp

material, such as amorphous carbon or opaques (e.g., FeS), reduce the spectral contrasts of emission features in the MIR and are featureless in the VNIR (e.g. Hanner et al. 1994; Wooden et al. 2004; Wooden 2008; Woodward et al. 2015). Non-compositional factors may also contribute to the decreased spectral contrast, such as larger grain sizes or lower surface porosity compared to comets.

2.5 Thermophysical Properties

Thermal inertias based on thermophysical modeling of thermal infrared observations have been reported for only a few Trojan asteroids. Fernández et al. (2003) measured simultaneous visible and multi-band infrared (12.5 and 20.8 μm) fluxes of 9 Trojan asteroids, from which they derived an average beaming parameter of 0.94. Arguing that such a low beaming parameter indicates a "slow rotator" thermal model ($\Theta < 1$, where $\Theta = \Gamma\sqrt{\omega}/\epsilon\sigma T_{ss}^3$, Γ is thermal inertia, ω is rotation rate, ϵ is bolometric emissivity, σ is the Stefan-Boltzmann constant, and T_{ss} is sub-solar temperature) applies to the Trojans, they computed an upper limit to the thermal inertia of $\sim 50\ \mathrm{J\ m^{-2}\ K^{-1}\ s^{-1/2}}$, similar to the Moon and large Main Belt asteroids. Applying similar arguments to the results for WISE observations of Trojans reported by Grav et al. (2012), in which beaming parameters for nearly all of the 478 Trojans reported are <1, would suggest that low thermal inertias are common among Trojan asteroids.

Emery et al. (2006) mention that thermophysical models of infrared spectra (5.2 to 38 μm) of the three Trojans in Fig. 3 provide upper limits of $\sim 50\ \mathrm{J\ m^{-2}\ K^{-1}\ s^{-1/2}}$ on the thermal inertias of those three large Trojans. Mueller et al. (2010) report thermal inertias of 21 ± 14 and $6.4 \pm 1.6\ \mathrm{J\ m^{-2}\ K^{-1}\ s^{-1/2}}$ for the two components of the 617 Patroclus binary from heating and cooling curves measured with SST infrared spectral observations of shadowing during mutual events. Horner et al. (2012) computed a thermal inertia of 25 to $100\ \mathrm{J\ m^{-2}\ K^{-1}\ s^{-1/2}}$ (3σ confidence range) for the dynamically unstable Trojan 1173 Anchises from thermophysical modeling of IRAS, AKARI, and WISE thermal photometry.

These low thermal inertia values reported for Trojan asteroids suggest dusty, porous surfaces, consistent with the thermal infrared emissivity spectra discussed in Sect. 2.4. Such low thermal inertias appear common on moons of the giant planets, Centaurs, and Kuiper Belt Objects (e.g. Müller et al. 2020, and refs therein) and indicate that fine-grained regoligths are common in the outer Solar System.

3 Analogs

3.1 Meteorites

Meteorite analogs provide useful insight into the surface composition of asteroids. Interestingly, there are no meteorite analogs that have been confidently identified for the enigmatic Trojans. The Tagish Lake Meteorite was initially touted as a spectral analog for D-type asteroids based on the relatively steep VNIR spectral slope, although it displays a 3-μm absorption feature (e.g. Hiroi et al. 2001) not observed in spectra of Trojans. Emery et al. (2006) and Vernazza et al. (2013) found that the MIR spectrum of Tagish Lake meteorite powder (Hiroi et al. 2001; Roush 2003) is a poor match to the emission spectra of the large Trojans Hektor, Aneas, and Agamemnon, and Vernazza et al. (2013) also demonstrated that the VNIR reflectance spectrum of Tagish Lake is also a poor match to VNIR Trojan spectra. Despite the incompatibility between Tagish Lake and Trojans, Vernazza et al. (2013) found a good spectral match at all wavelengths between Tagish Lake and the Main Belt D-type asteroid 368 Haidea, indicating compositional variability among D-type asteroids. Searches within meteorite spectral databases such as ASTER,[1] which includes carbonaceous chondrites, ordinary chondrites, and enstatite chondrites, among others, return no good matches to the observed Trojan emissivity spectra (Emery et al. 2006).

To broaden the possibilities in identifying meteorite analogs for Trojan asteroids, Vernazza et al. (2012) measured chondrite powder diluted with KBr in the MIR and found similarities to the Hektor emission spectra. KBr salt is transparent in the MIR, which results in the diluted meteorite grains appearing well separated in this wavelength regime. Similar experiments were later performed with additional meteorites such as Tagish Lake and ordinary chondrites (Izawa et al. 2021), as well as terrestrial olivine and pyroxene minerals (Martin et al. 2022, 2023a). Regardless of Tagish Lake dilution, its spectra deliver a poor match to any Trojan emission spectra. However powdered ordinary chondrites and powdered olivine spectra produce similar spectra to the Trojans, particularly Hektor, when diluted with KBr.

3.2 Main Belt and Near-Earth Asteroids

In plotting the heliocentric distribution of asteroid taxonomic classes Gradie and Tedesco (1982) showed the predominance of low albedo and red-sloped P- and D-type asteroids in the outer belt and Trojan populations. Ongoing spectral surveys have solidified these findings, showing that strong representations of these classes are also included in the Cybeles (7:4 resonance at 3.5 AU) and the Hildas (3:2 resonance at 4 AU) (Grav et al. 2012; DeMeo and Carry 2013; Wong and Brown 2017b). Nevertheless, D-types have been detected among small members of the inner and middle Main Belt as well (e.g. Lazzaro et al. 2004; DeMeo et al. 2014). Bourdelle de Micas et al. (2022) also found 3 D/T types among relatively large (D∼30 to 50 km), likely primordial planetesimals in the inner Main Belt (∼4% of their

[1] http://speclib.jpl.nasa.gov/

measured sample). The identification of D-types across the Main Belt suggests scattering of D-types into the innermost parts of the Solar System at some point during its evolution. In their spectral modeling of a large data set of D-type spectra, Gartrelle et al. (2021) found that Main Belt D-types appear to contain a higher abundance of opaque minerals than Trojans.

Low albedo and D-type members are also well known in the near-Earth population (Fernández et al. 2001; Binzel et al. 2004; Perna et al. 2018a). The majority of these objects are very highly correlated with Jovian Tisserand parameter values less than 3, indicating a dynamical link and origin associated with Jupiter Family Comets (see Sect. 3.3). However, not all D-types found in the near-Earth and Mars-crossing populations have a viable dynamical link to the outer belt or Jupiter Family comet source regions (Carry et al. 2016; Binzel et al. 2019). The existence of NEA D-types sourced from the inner and middle Main Belt emphasizes that although D-types are predominantly located in the outer belt and Trojan swarms, some D-type material has been scattered more broadly.

The only low-albedo Main Belt object that has been visited by a spacecraft is the dwarf planet Ceres. Although Ceres is a differentiated C-complex body with a disk-averaged VNIR spectrum that is very different from Trojan spectra and absorption features in the 3–4-μm region that are also distinct from Trojans (e.g. Rivkin et al. 2022), insights from its evolution have some relevance to understanding Trojans. NASA's *Dawn* mission found an average surface composition on Ceres rich in hydrated and ammoniated clays and carbonates (De Sanctis et al. 2015; Ammannito et al. 2016), indicating widespread heating and alteration of the original materials. The presence relatively large amounts of nitrogen, as expressed in the widespread occurrence of ammoniated clays, has been attributed to an outer Solar System origin for Ceres (e.g. Ribeiro de Sousa et al. 2022), though the nitrogen may be consistent with typical CM/CI chondrite bulk composition (McSween et al. 2018). In any case, the *Dawn* results highlight that searches for nitrogen-bearing materials on Trojans are important for comparisons with other bodies and for assessing formation region. Spectral mapping by the *Dawn* mission also revealed significant surface heterogeneity not indicated by disk-integrated spectra, including substantial exposures of H_2O ice, complex organic material, and salts, providing important clues to and constraints on interior composition and passed geologic activity (e.g. Combe et al. 2016; De Sanctis et al. 2017; de Sanctis et al. 2018; De Sanctis et al. 2020). By analogy, spectral mapping of Trojans is likely to provide a clearer view of interior compositions and any past activity than disk-integrated spectra alone.

3.3 Comets

The Nice Model predicts that Trojans are dynamically linked to outer Solar System Kuiper belt objects (KBOs) and are therefore related to Jupiter Family Comets (JFCs; Morbidelli et al. 2005). JFCs, such as 9P/Tempel, have comae made primarily of H_2O, CO_2, and sub-micron sized silicate grains (Farnham et al. 2007; Feaga et al. 2007; Harker et al. 2007). Ices have not been observed on Trojan surfaces (Emery and Brown 2003; Yang and Jewitt 2007; Brown 2016), but the observed silicates can be useful for comparisons to comets.

Typically, silicate-rich comae are dominated by amorphous phases (Wooden 2002; Kelley and Wooden 2009), although crystalline silicates have been detected (e.g. Harker et al. 2005; Sugita et al. 2005; Watanabe 2004; Lisse et al. 2006; Harker et al. 2007). The silicate phase (i.e., amorphous vs. crystalline) is also instructive for understanding the condition and region of formation. The crystalline fraction of silicates (f_{cryst}) is typically used in comet research to investigate high-temperature processing that formed dust in the inner disk of the solar nebula (Bockelée-Morvan et al. 2002; Ciesla 2007; Hughes and Armitage 2010). In general, minerals that formed in the inner solar nebula were transported outward and mixed with amorphous dust grains prior to forming a comet (Charnoz and Morbidelli 2007).

Another parameter of silicate composition that can be used to compare Trojans and comets is the Mg/(Mg+Fe) value of olivine and pyroxene, which has been used to constrain the structure and evolution of circumstellar dust (Hanner et al. 1994; Tielens et al. 1998; Gail and Sedlmayr 1999). Silicates in the solar nebula trended from more Mg-rich crystalline silicates in the inner Solar System to Fe-rich silicates in the outer Solar System. In the MIR, as Mg/(Mg+Fe) decreases, emissivity peaks associated with olivine and pyroxene shift from shorter to longer wavelengths (Koike et al. 1993; Jäger et al. 1998; Chihara et al. 2002; Koike et al. 2003).

Trojans' highly porous surfaces, evidenced by strong emissivity peaks (see Sect. 2.4), coupled with red-sloped reflectance spectra (see Sect. 2.2) and low thermal inertia (see Sect. 2.5), makes comets a useful analog, despite Trojans' lack of comae (Perna et al. 2018b). Generally, comet nuclei fall under D-type classification in the VNIR (see Sect. 2.2), and, in the MIR, comet nuclei emission spectra bear similarities to Trojans (Kelley et al. 2017) (Fig. 3). The fine-grained material on the surface of comet nuclei is primarily made of Mg-rich silicates, intermediate Mg/Fe silicates, and carbonaceous and metal sulfide material (e.g. Frank et al. 2014; Wooden 2002). Currently, the JFC 10P/Tempel 2 is the prime cometary spectral analog to Hektor (Kelley et al. 2017), suggesting that other red (D-type) Trojans should have a similar silicate makeup of fine-grained Mg-rich and Mg/Fe mixed silicates.

The JFC 67P/Churyumov-Gerasimenko was observed in exquisite detail by the European Space Agency (ESA) *Rosetta* spacecraft. Despite being an active comet, disk-averaged spectra of the nucleus do not show H_2O, though local exposures of H_2O are observed in color and spectral mapping data (e.g. Pommerol et al. 2015; De Sanctis et al. 2015; Barucci et al. 2016; Fornasier et al. 2016; Deshapriya et al. 2018; Fornasier et al. 2023), and H_2O was exposed in boulders contacted by the Philae lander (O'Rourke et al. 2020). Observed exposures range in size from about 1 to over 1500 m^2 and have lifetimes from a few hours (for frosts) to more than two years. After sublimation, these regions became indistinguishable from the surrounding dark and red terrain. The many observations of H_2O exposures and their distribution indicate that H_2O is abundant below the surface (e.g. Fornasier et al. 2016, 2023; Oklay et al. 2017). These observations are consistent with *Deep Impact* spacecraft observations of the nucleus of the JFC 9P/Tempel 1 (Sunshine et al. 2006) and suggest that Trojans could have H_2O beneath their surfaces, even though no spectral signatures have been detected in disk-integrated telescopic data. Localized exposures of CO_2 ice were also detected in resolved spectra of 67P (Filacchione et al. 2016), highlighting the possibility of more volatile ices also being masked by dusty regoliths.

Spectra of the low-albedo material covering the nucleus of 67P show a broad ~3.1-μm absorption that has been interpreted as ammonium salts (Poch et al. 2020). This feature is distinct from the ammoniated clay feature detected on Ceres, but still indicates significant nitrogen, highlighting again the importance of searches for nitrogen-bearing materials on Trojans. Complex organic molecules also likely help explain the broad ~2.9- to 3.6-μm absorption on 67P (e.g. Capaccioni et al. 2015; Quirico et al. 2016), and aliphatic organics have been detected in the dark material (Raponi et al. 2020).

3.4 Irregular Satellites

The Nice Model predicts that the irregular satellites of the giant planets were captured in the same scattering event and from the same Kuiper Belt source population as the Trojan asteroids (Morbidelli et al. 2005). Graykowski and Jewitt (2018) find that the average visible colors of the irregular satellites of all four giant planets are statistically consistent with each other and that they lack the ultra-red colors seen among the Kuiper Belt, though the small number of Uranian and Neptunian irregulars weaken the statistics for those groups. The average colors of Jupiter and Saturnian irregulars are consistent with the average colors of Trojan asteroids, though the Jupiter irregulars contain more neutral-sloped (C-type) surfaces than the Trojans. Bottke et al. (2010) argue that the irregular satellite populations were once much larger and have been extensively ground down from collisions. If irregular satellites and Trojans have less-red or neutral interiors (e.g. Wong and Brown 2016) (see Sect. 4.1), this larger abundance of neutral-sloped surfaces may be a result of the enhanced collisional environment. Interestingly, Graykowski and Jewitt (2018) point out that the similarity of average colors of the irregulars at different distance from the Sun (and therefore different surface temperatures) and distinct color distributions from KBOs argue against a thermal process at 5 AU causing the difference in colors between Trojans and their presumed KBO source population.

Our closest view of what a Jupiter Trojan asteroid would look like *before* being heated to the temperatures of the Jupiter region comes from Saturn's irregular satellite Phoebe. Phoebe was presumably captured from solar orbit early in the history of the Solar System and now resides on a distant inclined orbit about Saturn. Phoebe is only a little smaller than Hektor, the largest Jupiter Trojan, so could be a model for a less-heated version of a large Trojan.

Phoebe's albedo of about 8% (Buratti et al. 2008) is relatively low, but a bit higher than the average for Trojans. Phoebe's average VNIR spectrum is spectrally neutral (Owen et al. 1999), distinct from those of Trojans, though some regions have slightly red spectral slopes (Buratti et al. 2008). Near infrared spectroscopy reveals, distinct absorption features due to water ice on the surface of Phoebe (Owen et al. 1999; Clark et al. 2005). Images and spectra from Cassini reveal that Phoebe's surface is dominated by large impact basins (Porco et al. 2005) and that the water ice spectral features, though present everywhere, are strongest in these regions of impact (Clark et al. 2005; Buratti et al. 2008; Fraser and Brown 2018). Overall, the surface appears to be water ice poor but with impacts revealing an icier interior. While water ice is not stable for long periods of time at the current temperatures of the Trojans, the overall low albedo with high albedo patches showing on Phoebe in regions of impacts is similar to what is inferred from the presence of higher albedo small Trojans that might have had recent impacts (see Sect. 2.1).

3.5 Centaurs

Centaurs are an enigmatic population of minor bodies with orbits that traverse the giant planet region. Dynamical studies have shown that these objects are inward-scattered KBOs and are characterized by orbital stability lifetimes on the order of 1–10 Myr (Tiscareno and Malhotra 2003; Horner et al. 2004). Given their formation location in the outer Solar System and relatively close present-day heliocentric distances, Centaurs have attracted a great deal of interest as proxies for studying the more distant KBOs and may represent bodies in a transition from KBO-like to Trojan-like surfaces.

Over the past few decades, observational studies have uncovered several similarities between Centaurs and Jupiter Trojans. The most salient common trait is color bimodality. Centaurs have been shown to display bifurcations in both visible (Peixinho et al. 2003; Tegler et al. 2003, 2008, 2016) and near-infrared (Delsanti et al. 2006) colors. In the following, we refer to the two color classes in the Centaurs and KBOs as red and very-red. Notably, the color centers of the two Centaur sub-populations lie at redder spectral slope values than the two color groups in the Trojans. Red Trojans have similar slopes as red Centaurs, and very red Centaurs have no analog in the Trojan population. Given the dynamical history of Centaurs, it had long been suggested that this color bimodality is indicative of two broad color classes in the wider KBO population (Fraser and Brown 2012; Peixinho et al. 2012, 2015; Tegler et al. 2016), a scenario that has been confirmed by recent photometric survey observations of similarly-sized KBOs (Wong and Brown 2017a; Schwamb et al. 2019).

Another important similarity between Centaurs and Trojans is very low geometric albedos. However, whereas the less-red and red Trojans have similar albedo distributions, the Centaur sub-populations differ significantly in their reflectivity: the median albedos of the red and very-red groups are 5% (similar to the average for Trojans) and 11%, respectively (Duffard et al. 2014; Romanishin and Tegler 2018; Müller et al. 2020). Median albedos of most dynamical groups of KBOs are higher than those of Trojans (Müller et al. 2020). One exception is the Scattered Disk Objects (SDOs), which have a median albedo of $\sim$6%, though the spread of albedos is much higher among the SDOs. If Trojans are scattered KBOs, it would not be surprising that their albedos are systematically lower from loss of ices that are common on KBOs. Likewise, the two Centaur sub-populations are characterized by discrepant inclination distributions, with the very-red objects having lower inclinations on average (Tegler et al. 2008; Romanishin and Tegler 2018). The emissivity spectrum of the red Centaur Asbolus is very similar to those of red Trojans (Barucci et al. 2008), suggesting similar surface texture and silicate mineralogy (see Sect. 2.4).

With their origins in the outer Solar System, Centaurs are expected to be rich in volatile ices. Near-infrared spectroscopy of several Centaurs, including Chiron, Pholus, Thereus, and Chariklo, has revealed absorption signatures of water ice (Brown and Koresko 1998; Cruikshank et al. 1998; Romon-Martin et al. 2003; Licandro and Pinilla-Alonso 2005; Barkume et al. 2008; Stansberry et al. 2008; Duffard et al. 2014). In addition, methanol ice has been detected on four objects, including Pholus and Crantor (Cruikshank et al. 1998; Stansberry et al. 2008; Duffard et al. 2014). The volatile-rich nature of Centaurs is also manifested by the peculiar subset of objects that show cometary activity (Jewitt 2009). Comae have been observed on Centaurs at heliocentric distances greater than 12 AU (Tegler et al. 2006; Bauer et al. 2008; Jewitt 2009, 2015) — far beyond the region where water ice sublimates — which indicates that the activity is primarily driven by a different physical process, perhaps the crystallization of amorphous ice or sublimation of more volatile ices (Prialnik et al. 1995; Jewitt 2009; Guilbert-Lepoutre 2012). Notably, the color distribution of active Centaurs is skewed toward bluer colors when compared to the inactive population (Jewitt 2009, 2015; Wong et al. 2019b), suggesting that the activity may be altering the primordial colors of Centaurs. Analogous bluing of colors due to activity was observed on the nucleus of comet 67P by the *Rosetta* mission (e.g. Fornasier et al. 2016, 2023; Ciarniello et al. 2022).

The advent of dynamical instability models of solar system evolution and the proposed outer Solar System origin of Trojans have made comparisons between Centaurs and Trojans particularly apropos. Numerical models simulating the dynamical instability show that some planetesimals from the primordial trans-Neptunian protoplanetary disk were initially scattered onto highly-eccentric Centaur-like orbits before being captured into resonance by Jupiter to become Trojans (Morbidelli et al. 2005; Nesvorný et al. 2013). Therefore, detailed study of Centaurs may further our understanding of how the combined effects of increased solar irradiation and cometary activity could have affected the surface properties of Trojans in the early Solar System.

4 Compositional Interpretation and Surface Evolution

4.1 Theoretical and Observational Considerations

Of the observational constraints discussed in Sect. 2, the color bimodality is perhaps the most consequential observational constraint on compositional models of Trojans. Broadly speaking, the disparate spectral properties of less-red and red Trojans suggest one of two possible scenarios: (1) less-red and red Trojans were formed in different locations and have distinct bulk compositions, or (2) all Trojans were initially formed in the same environment and accreted an identical bulk composition, with the bifurcation in spectral properties arising from different sublimation and evolutionary histories that affect only the ices retained in the near surface and the surface regolith. Current models of Solar System evolution largely align with the second scenario, with the Trojans forming in the outer Solar System alongside the progenitors of the present-day Kuiper belt and Centaurs, before being scattered inward and captured into resonance during an episode of dynamical instability sometime after the dispersal of the protoplanetary disk. It follows that in order to reconcile the observed surface properties of Trojans with these dynamical models, a mechanism must be invoked to create two distinct surface chemistries from a common bulk composition.

One hypothesis posits that volatile ice sublimation and space weathering may be the primary drivers behind the observed Trojan color bimodality (Wong and Brown 2016) and also behind the color and spectral diversity in the Kuiper Belt (e.g. Schaller and Brown 2007). Within the framework of dynamical instability models of Solar System evolution, all bodies that formed in the planetesimal disk beyond the primordial orbits of the ice giants would have accreted a roughly cometary composition, i.e., rocky material and water ice, with a significant volume of other volatile ices such as carbon dioxide, methanol, and ammonia. When the protoplanetary disk dissipated, these objects were exposed to direct insolation, resulting in sublimation of volatile ices in the outermost layers. After millions of years of exposure, the strongly temperature-dependent, and by extension heliocentric-distance-dependent, sublimation rates produced distinct surface ice compositions across the primordial icy planetesimal disk, while keeping the interior bulk composition unchanged (e.g. Schaller and Brown 2007, Steckloff et al. 2021).

Simple sublimation modeling has demonstrated that hydrogen sulfide ice (H_2S) is conditionally unstable to sublimation loss at these heliocentric distances and could therefore be the critical volatile ice species responsible for the development of the color bimodality. Objects in the planetesimal disk would have been divided into two groups — the objects situated closer to the Sun experienced higher surface temperatures and became depleted of H_2S on their surfaces, while objects farther out retained H_2S. The retention of H_2S, which is known to induce a strong reddening upon irradiation (Carlson et al. 2007), would contribute additional reddening relative to the case where H_2S is absent. The result of space weathering on these two distinct surfaces is the development of a color bimodality among objects in the primordial planetesimal disk: less-red objects without surface H_2S, and red objects with surface H_2S. The subsequent dynamical instability spread these bodies throughout the Solar System to become the present-day Trojans, KBOs, and Centaurs, which thereby inherited the color bimodality. Other simple ices also redden upon irradiation (e.g., CH_3OH and CH_4), but their volatilities are such that their sublimation lines would not have bifurcated the original Kuiper Belt like H_2S would.

This hypothesis explains the color bimodality attested across all middle and outer solar system minor body populations. In addition, it naturally accounts for the less-red Eurybates collisional family, as well as the broader trend of increasing less-red fraction with decreasing size. Since the physical and chemical processes responsible for the bifurcation in color

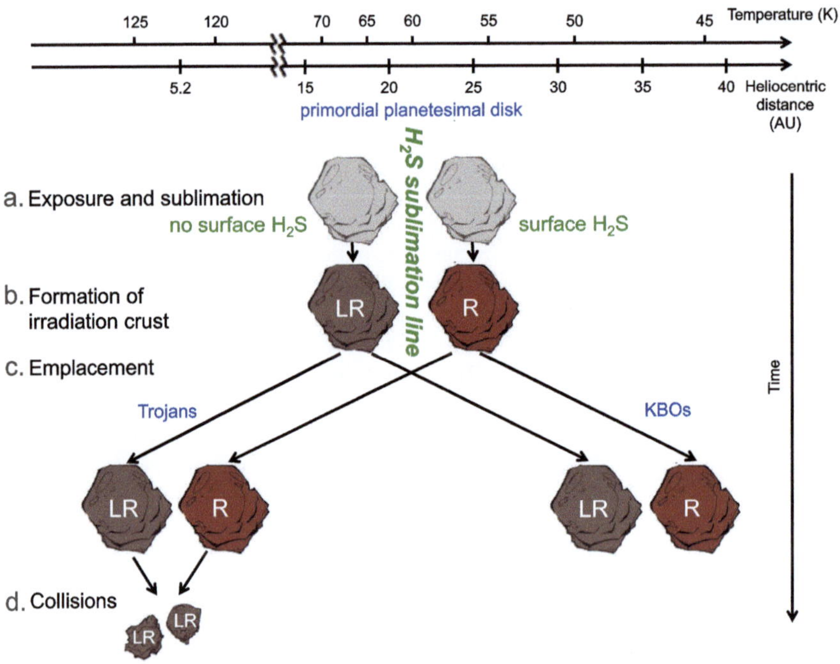

Fig. 4 Schematic diagram illustrating the H_2S ice sublimation hypothesis proposed by Wong and Brown (2016) to explain the color bimodality of Jupiter Trojans and KBOs. Primordial planetesimals that were situated on two sides of the H_2S sublimation line attained different levels of reddening depending on the loss or retention of H_2S ice on their surfaces

are confined to a thin outer irradiated shell, collisions readily destroy this crust, exposing the pristine interior material. At the much higher surface temperatures of the present-day Trojans, any remaining volatile ices would instantaneously sublimate away. In particular, H_2S, the reddening species responsible for the development of red objects, would be depleted in all cases, and subsequent irradiation of the fragments would only produce less-red colors. The various features of the H_2S ice sublimation hypothesis are illustrated in Fig. 4.

4.2 Laboratory Studies

Motivated by the aforementioned H_2S sublimation hypothesis (Wong and Brown 2016), a laboratory study was carried out with a focus on the irradiation of volatile ices of relevance to the Jupiter Trojans (Mahjoub et al. 2016, 2017; Poston et al. 2018). Samples consisting of water, ammonia, and methanol ices, with or without the inclusion of H_2S ice were subjected to high-energy electron bombardment at temperatures representative of the primordial planetesimal disk and then heated to Trojan temperatures. The experimental methodology was designed to simulate the initial space weathering of Trojans in the outer Solar System and the subsequent thermophysical evolution that transpired during the transition to their current environment.

Notably, the results of the experiments reproduce many of the general observed features of Trojan asteroids. Both samples reddened and darkened upon electron bombardment, accompanied by the formation of an irradiation mantle on the surface. The sample with H_2S attained a steeper spectral slope throughout the visible and near-infrared, consistent with the hypothesized role of sulfur irradiation chemistry in creating the redder colors of red Trojans. In addition, spectral slope was reduced in both samples upon subsequent heating to Trojan-like temperatures, suggesting that secondary thermal processing of the surface regolith upon emplacement into their present-day orbits may have led to the more neutral colors of Trojans when compared to Centaurs and KBOs.

Detailed compositional analysis of the irradiation residues revealed the production of short-chain sulfur allotropes S_2, S_3, and S_4, which have been previously shown to induce strong reddening in laboratory samples. These allotropes are unstable to heating and can further polymerize to form longer sulfur chains such as S_8 that are characterized by more neutral visible spectra. High-resolution reflectance spectroscopy of the samples also uncovered a series of narrow absorptions at wavelengths longer than 3 μm, including several sulfur-bearing molecules (e.g., OCS, CS_2, and SO_2) only detected in the sample with H_2S. Future high-precision spectroscopic observations with facilities like the *James Webb Space Telescope* (JWST) may be able to probe for some these diagnostic absorption features, which will definitively test the sulfur hypothesis for the Trojan color bimodality.

4.3 Space Weathering and Collisional Processes

Even if a genetic relationship with the hot classical KBOs is established, differences between their surface properties are to be expected due to the different evolution of the two populations in their different environments. The major processes responsible for modifying the surfaces are space weathering (e.g., irradiation by solar ions, galactic cosmic rays, electrons, UV, and X-rays) and/or physical collisions with other minor bodies or micrometeorites. How a surface is altered by these processes depends on chemical composition, the rate of the process, and on the location in the Solar System.

The most studied space-weathered airless body is the Moon, where the most prominent effect is the formation of nano-phase iron (npFe). Inclusion of npFe in glassy rims coating surface grains causes the reflectance spectra to become redder (i.e., increasing the spectral slope) and lowers the albedo (see for example Hapke 2001). In the case of S-type asteroids, as parent bodies of ordinary chondrite meteorites, a reddening trend has also been established (Binzel et al. 2004). These trends have been validated in the laboratory with similar results for different weathering agents, since both ion irradiation and laser irradiation/ablation experiments produce on the silicates reddening and darkening spectral trends (Hapke 1965; Moroz et al. 1996; Sasaki et al. 2001; Strazzulla et al. 2005; Marchi et al. 2005; Brunetto et al. 2006; Loeffler et al. 2008, 2009; Fu et al. 2012). On darker objects, such as C-type asteroids, the evidence is more controversial – weak reddening trends have been observed by some investigators (Lazzarin et al. 2006), while other studies have found a blueing effect (Nesvorný et al. 2005; Lantz et al. 2013; Vernazza et al. 2013).

A review of space weathering experiments performed on terrestrial and extraterrestrial materials can be found in Brunetto et al. (2015), which can be briefly summarized as follows. On high albedo materials, such as olivine, both ion and laser irradiation produce strong spectral reddening and darkening. Similar, but weaker, effects occur for pyroxenes. On low albedo terrestrial bitumens, a very strong spectral flattening and a weak brightening trend is produced. On organic residues of C-rich ices, a spectral reddening and darkening is produced due to the formation of refractory molecular solid materials, and a similar trend is observed for organics covering silicates. Irradiation experiments of carbonaceous chondrite meteorites show that the CV and CO groups darken and redden, like ordinary chondrites, olivine, and pyroxene, whereas CI and CM groups brighten and become bluer. It is unclear whether the different behaviors are caused by organic content (low vs high), silicate composition

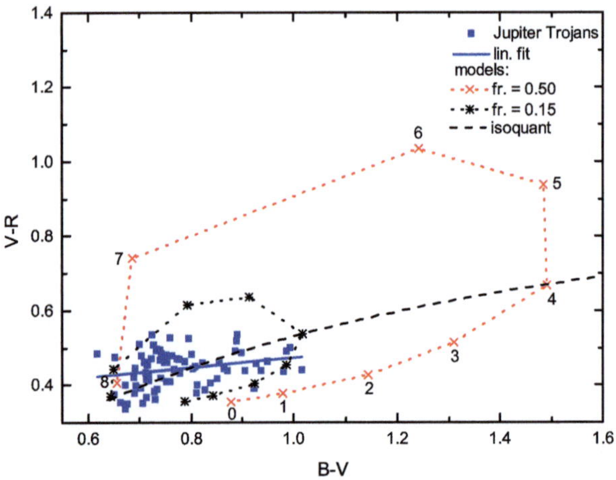

Fig. 5 V-R vs. B-V color indexes of the Jupiter Trojans from the MBOSS database. Color change of the model Trojan for different surface fraction of the "polystyrene phase". The isoquant that better reassembles the linear fit of the data is plotted

(anhydrous vs hydrated), or chondrule abundance (relatively higher vs lower) (Lantz et al. 2017).

Several attempts have been made to assess the combined effect of space weathering and the rejuvenation of the surface by impact gardening. It has been noted that water ice is very short lived on the surface of a Jovian-Trojan due to rapid sublimation (Guilbert-Lepoutre 2014). Melita et al. (2009) postulated a bulk composition in which fresh sub-surface material is spectrally red dust. Under very specific circumstances, they found that the timescale for resurfacing the entire surface of the Trojan asteroids is similar to that of the flattening of the red spectrum of the new dust by solar-proton irradiation. Under such circumstances, the surfaces of most Trojan asteroids might be composed mainly by lowly irradiated (red) dust. This model would predict that smaller bodies, which are more likely to have suffered recent impacts, would be predominantly red-sloped, opposite to the trend described in Sect. 2.2.

As an analog to silicate grains coated in organic material, Kaňuchová et al. (2010) investigated the effect of irradiation of polystyrene-coated olivine. Polystyrene exhibits a progressive spectral reddening and then darkening during irradiation. At high irradiation doses, the reddening reaches a maximum, while the sample continues to darken and starts to flatten. Interestingly, the spectral shape of the irradiated material at different doses can reproduce some asteroid spectra. On the basis of these experimental results, a reddening-blueing loop has been proposed by Kaňuchová et al. (2012): Irradiated organics, initially exhibiting a flat spectrum and a high albedo, are progressively reddened while the albedo decreases. At very high doses, the spectra are flattened (blueing) while the albedo continues to decrease. A fully weathered organic will then have a flat and dark spectrum. In Melita et al. (2015), a model is constructed in which the surface is composed of two phases: one with constant albedo and spectral slope and one that evolves in the color-color diagram due to irradiation as polystyrene deposited on olivine, described in Kaňuchová et al. (2010). The color evolves due to irradiation along the reddening-blueing loop described by Kaňuchová et al. (2010), with a magnitude that is controlled by the relative initial amounts of the two components (see Fig. 5). The Trojan color distribution is best represented by a fraction of the organic-rich component of $\sim$15% and an albedo of 0.04 for the other component. However, the timescale of surface rejuvenation implied by this model is unreallistically short, of only 100 yr.

The added contribution of cratering leads to irradiated less-red objects becoming red objects, once again through resurfacing, thereby preventing all the red Trojans from turning into less-red objects of lower albedo. However, the characteristic collisional timescale and, correspondingly, the timescale of resurfacing increase with decreasing asteroid size because of the decreasing number of impactors that can reset the surface (Melita et al. 2015), while the rate of irradiation is the same for all objects, as mentioned earlier. Therefore, the resurfacing of Trojans through cratering becomes less effective at returning irradiated less-red objects to red objects when one goes to smaller sizes. This would lead to a relative excess of less-red objects at faint magnitudes, but it would imply a lower albedo for highly processed small objects, which does not coincides with the observations. In any case, these models based on the evolution of polystyrene deposited on olivine (Kaňuchová et al. 2010) under cosmic irradiation would not predict a bimodal distribution of colors but a smooth gradient.

5 Synthesis and Outstanding Questions

Dedicated telescopic observations, as discussed in Sect. 2, have begun to reveal the nature of this pivotal population. Generally, Trojans are dark, reddish objects with featureless spectra from the near-ultraviolet through 2.5 μm. Although albedos are uniformly clustered around 5%, VNIR spectral slopes are bimodal, with the ratio of less-red to red slopes increasing with decreasing size. The lack of any discernible water ice absorption between 1 and 4 μm places strong upper limits on the water ice abundance in the uppermost layers of the regolith. The spectral feature at ~3.1 μm observed in spectra of some less-red Trojans has an ambiguous interpretation, as it can be attributed to either fine-grained water frost or a N–H vibrational absorption. MIR thermal emission spectra of Trojans show spectral features indicative of fine-grained silicates in a highly porous (i.e., "fluffy") regolith.

Despite these important discoveries, several fundamental questions remain regarding the surface compositions of Trojan asteroids that confound a full understanding of their origin and evolution. Whereas fine-grained silicates have been detected, the mineralogy, chemistry, and crystalline fraction of these silicates, and whether these properties vary significantly among the population, remain uncertain. On primitive bodies, these silicate properties are indicative of nebular formation region. It is also important to establish the hydration state of silicates on Trojan asteroids in order to assess whether these bodies have undergone past heating events, either geothermal or impact-related.

The abundance and composition of organic materials is another significant unknown. The red spectral slopes of Trojans were initially interpreted to be due to abundant organic materials, but no clear organic absorptions have been detected. Characterization of organic molecules and comparisons to those on surfaces of KBOs will inform the origin and surface processing undergone by Trojans. The detection of N-H-bearing material is intriguing, and it will be illuminating to determine if that material is similar to the materials detected on Ceres or on comet 67P Churyumov-Gerasimenko. Of particular interest will be assessing whether that material is indicative of an outer Solar System origin, as suggested for Ceres and 67P.

The absence of H_2O absorptions from all Trojan spectra is something of a mystery. Bulk densities derived from binary orbits among Trojans are close to 1 g cm^{-3}, suggesting a significant fraction of interior ice. Thermal stability studies show that H_2O ice can be stable within 1 to 10 m of the surface of Trojans (Guilbert-Lepoutre 2014). Spacecraft spectra of the surfaces of active comets have shown that interior ice is effectively masked, even while sublimation is occurring, by a dusty surface layer (see Sect. 3.3). Nevertheless, with high quality published NIR spectra of dozens of Trojans, one might expect to have seen some recent exposure of interior ice from recent large impacts, even considering that Earth-based telescopic measurements yield spatial resolutions that are just hemispherical averages. It remains to be demonstrated whether interior ice is present and was lost from the surface by sublimation.

One of the biggest current open questions about Trojan asteroids is the cause of the observed color bimodality. The two scenarios in Sect. 4 discuss different possible paths, one due to different formation regions and the other due to surface processing, for a fundamental chemical difference between the red and less-red material. Determining the compositions of the two spectral groups is critical for distinguishing these scenarios. A specific question to assess the Wong and Brown (2015) scenario is whether the red material shows a sulfur-rich composition, and a question related to the Melita et al. (2015) model is whether either spectral class is consistent with the organic material suggested in that model. On the other hand, it is not even clear whether the bimodality is due to a chemical difference between the red and less-red material, or if a physical difference in the regoliths (e.g., grain size, regolith porosity) could cause the bimodality. A key to understanding the bimodality is determining whether fresh material is red or less-red, or if the fresh material itself is bimodal in composition.

It is also critical to determine how Trojan surfaces evolve due to space weathering. Many analyses invoke different styles, pathways, and timescales for surface alteration due to exposure to the space environment, but there are few constraints. It is not known whether one spectral group evolves from the other, or if the two surface types evolve differently. Along with determining surface composition of fresh and evolved materials, understanding space weathering requires determining the rate of alteration by irradiation compared to the rate of collisional resurfacing for Trojans of different size. Along with assessing the current ongoing space weathering evolution, it is necessary to unravel how Trojan surfaces were altered as they migrated from their presumed Kuiper Belt origin region to their present location. Trojans exhibit several observational differences from KBOs, Centaurs, and irregular satellites, and it is necessary to determine whether these differences can be explained by the changing thermal and irradiation environment that they experienced due to their orbital migration to Jupiter's Lagrange regions.

A theme that runs through all these open questions relates to potential regional scale variation in surface compositions on individual Trojans. The Trojan population as a whole exhibits less heterogeneity than almost any other Solar System small body population, yet Trojans are almost always inferred to have surfaces that are in some ways distinct from their interiors, due to a combination of sublimation, irradiation, and impact gardening. If this distinction between interior and surface is correct, individual Trojans should show regional and local scale variations in composition from exposures of fresh material and perhaps from differing irradiation doses of surfaces exposed for different amounts of time. If geothermal and/or impact-related hydration has occurred, one might also expect those processes to lead to regional scale variations in composition. Since telescopic observations observe entire hemispheres, it is difficult to assess surface exposures, unless a hemispheric-scale event has occurred. All told, deciphering surface compositions of Trojan asteroids is clearly dependent on spacecraft observations providing high spatial resolution of surface heterogeneities. With spatial resolution providing geologic context, we can hope to unravel which (if any) of the currently envisioned processes are active in the outer solar system.

6 How the Lucy Mission Will Address These Outstanding Questions

The *Lucy* mission is designed to address these outstanding questions related to surface composition and thereby reveal not only the histories of Trojan asteroids, but key aspects of the evolution of the Solar System (Levison et al. 2021; Olkin et al. 2021). Levison et al. (2021) presents the Level 1 requirements of the *Lucy* mission and describes the broad science motivation for these requirements. The flyby targets sample the different Trojan spectral classes, sizes, binarity, rotation rate, shape, etc, enabling investigation of the compositional diversity of Trojans along multiple dimensions (see Table 3 and Fig. 7 of Levison et al. 2021), satisfying requirement 1 (R1) of the mission. As mentioned in Sect. 2.2, Orus and Leucus are in the red spectral group, Patroclus-Menoetius and Polymele are in the less-red group, and Eurybates is the largest member of a collisional family with neutral to less-red spectral slopes. Patroclus-Menoetius is a nearly equal-sized binary, and Eurybates and Polymele both have small moons. Rotation periods range from 8 hrs (Eurybates) to 445 hrs (Leucus), and shapes derived from light curve and occulation analyses (e.g. Mottola et al. 2023; Buie et al. 2021; Pinilla-Alonso et al. 2022) range from spheroidal (Patroclus-Menoetius) to elongate with significant topography (Leucus).

Instrumentation on the *Lucy* spacecraft is optimized for detecting materials than are critical for understanding the origin and evolution of Trojans (R12 and R13 of the mission). The Linear Etalon Imaging Spectral Array (LEISA) mapping spectrometer spectral range (1.0–3.6 μm), spectral resolution ($\Delta\lambda \sim 0.01$ μm), and sensitivity will enable detection of ices, organics, salts, and crystalline silicates of differing compositions (Olkin et al. 2021; Reuter et al. 2023). On approach and departure, low spatial resolution spectra will be obtained of each target over a full rotation period, spaced no more coarsely than 1/6 of a rotation period (R14) in order to map global-scale variations on the targets. Nearer the close approach, LEISA's spatial resolution of $\sim$40 m at a distance of 500 km will map compositional heterogeneity, including fresh exposures of sub-surface materials (mission requirement R14).

The Multi-spectral Visible Imaging Camera (MVIC) will image the targets through 5 color filters (violet 377–489 nm; green 478–525 nm; orange 522–632 nm; phyllosilicate 627–758 nm; and NIR 753–914 nm) (Olkin et al. 2021; Reuter et al. 2023). The phyllosilicate filter is designed to search for the 0.7 μm absorption feature in Fe-bearing phyllosilicates. Approach and departure low spatial resolution images covering a full rotation spaced no more coarsely than 1/6 of a rotation period will map global-scale color variations (mission requirement R9). During the flybys, MVIC's higher spatial resolution ($\sim$15 m at a distance of 500 km) will enable color imaging that covers 80% or $\geq$700 km^2 of each target capable of resolving features $\leq$1.5 km in size, to characterize regional-scale compositional variations (mission requirement R10). Nearer to close approach, MVIC will obtain color images covering $\geq$150 km^2 capable of resolving features $\leq$600 m in size, to characterize regional-scale compositional variations (mission requirement R11), including freshly exposed materials (Olkin et al. 2021; Reuter et al. 2023).

The highest spatial resolution images from the Lucy LOng Range Reconnaissance Imager (L'LORRI) panchromatic visible camera ($\sim$3 m at a distance of 500 km) will provide necessary geologic context for interpreting compositional heterogeneity and an understanding of the crater and resurfacing histories of Trojans (Olkin et al. 2021; Weaver et al. 2023). Measurements of thermal emission by the Lucy Thermal Emission Spectrometer (L'TES) point spectrometer (6–50 μm; spatial resolution[2] $\sim$3.6 km at a distance of 500 km) at multiple times of day, including at least one night time spot, will provide insights into surface structure and volatile stability (mission requirement R17) (Olkin et al. 2021; Christensen et al. 2024). For the warmest parts of the surfaces of each target, the SNR of the L'TES

[2] 85% encircled energy.

images may be high enough to search for and interpret spectral features (e.g., of surface silicates), but thermal emission spectroscopy is not a requirement of the *Lucy* mission.

The majority of questions regarding the current surface composition of Trojans will be directly answered by spectra measured by LEISA. The abundance and composition of organics will be assessed primarily the 3–3.6 μm segment of measured spectra. Aliphatic organics have diagnostic features at ~3.4 and 3.5 μm, aromatics have features near 3.3 μm, and specific composition and structural details affect the shapes and positions of these features. N-H-bearing materials detected other Solar System objects (e.g., ammoniated clays on Ceres and ammonium salts on comet 67P) are identified by relatively broad absorptions near 3.1 μm. Potential N-H features have been detected in ground-based spectra of some less-red Trojans, albeit with fairly low S/N, and LEISA spectra will enable more detailed compositional analyses of these features to determine whether the materials are similar to those on Ceres or 67P or distinct from either. Hydrated minerals, diagnostic of past geothermal heating and aqueous alteration, have strong features near 2.7–2.9 μm, the exact position of which is indicative of composition and therefore degree of alteration. The MVIC filter centered near 0.7 μm is designed to search for Fe-bearing hydrated silicates. Crystalline anhydrous silicates, such as olivine and pyroxene, exhibit broad absorptions near 1 and 2 μm. H_2O ice has a strong, broad fundamental absorption near 3 μm as well as weaker overtone and combination absorptions at 1.5 and 2.0 μm that, together, will enable very sensitive searches for H_2O as well as characterization of the phase of any detected.

The overriding benefit of sending spacecraft to explore the Trojan asteroids is getting spatially resolved data. The *Lucy* mission will obtain rotationally-resolved global color and spectral data of the Trojan targets on approach and/or departure in order to assess regional-scale color and compositional variations. Color and compositional variations at regional scales will provide important constraints on large-scale geologic and internal thermal processes. Closer in, higher resolution color and spectral data will be obtained of smaller regions to investigate local variations, particularly those that might be due to exposures of fresh subsurface material. Analysis of compositional variations at different spatial scales is critical for addressing several of the most important outstanding questions. A search for H_2O ice in fresh exposures is important for answering whether H_2O was originally present in the bulk composition and was lost from the surface layer due to sublimation. Though the spatial resolutions expected for LEISA and MVIC during the Trojan flybys would not resolve the smallest H_2O exposures seen by *Rosetta* on the nucleus of comet 67P, it is sufficient to identify the larger exposures. Comparing the colors and compositions of fresh, subsurface material to surface layers exposed to the space environment will directly address how the surfaces of these bodies are space weathered. It is critical to understand whether the fresh material corresponds to the average red Trojans, the average less-red Trojans, or something else. Compositional signatures of weathered versus unweathered material will provide an understanding of the chemical processes involved in space weathering. Searching for sulfur-bearing materials in the fresh exposures will provide a key test of the Wong and Brown (2016) hypothesis for the color evolution of Trojans. Compositional and color analyses of surface and subsurface materials are also key to understanding the processing that may have occurred as Trojans migrated from the Kuiper Belt, answering why they currently look so different from KBOs.

Another key aspect of the design of the *Lucy* mission for addressing the outstanding compositional questions is that the spacecraft will flyby targets of different spectral classes and sizes. Comparison of the detailed color and compositional analyses at different spatial scales for the red (Orus and Leucus), less-red (Patroclus-Menoetius and Polymele), and neutral-sloped (Eurybates) Trojans is crucial for understanding whether the different classes have different original compositions or are just weathered differently. An important focus will be determining how similar (or different) freshly exposed material on the different spectral classes is, and what the specific compositions of those materials are. The compositions of fresh material on the different bodies should trace back to formation region(s).

Information gained about surface geology (e.g. Marchi et al. 2023), internal properties (e.g. Mottola et al. 2024), activity (e.g. Noll et al. 2023), and collisional history (e.g. Bottke et al. 2023) will also provide important context for interpreting surface compositions. For example, collisional resurfacing rates on Trojans of different size control, compared to space weathering rates, control whether disk-integrated colors of large Trojans or small Trojans should appear fresher. This information, in turn, informs how the population-level trends of surface colors (see Sect. 4.1) should be interpreted. Mass and density determinations enabled by the close *Lucy* flybys, in particular, will provide important constraints on bulk composition to be compared with the surface composition. The suite of observations to be obtained by the *Lucy* mission analyzed together will revolutionize our understanding of the Trojan asteroids and the Solar System.

A JWST cycle 1 GO program (2574) has measured NIR and MIR spectra of all *Lucy* Trojan targets. The Near-InfraRed Spectrograph (NIRSpec) obtained 1.7 to 5.1 μm spectra of all 5 targets with the medium resolution grism. The Mid-InfraRed Instrument (MIRI) has obtained 5 to 28 μm spectra with the medium resolution grism of all 5 targets. These *JWST* data are much higher quality than possible from ground-based or even previous space-based facilities. The NIRSpec data show a ∼3.1-μm feature in the less-red targets, similar to that reported by Brown (2016), as well as some possible organic absorptions near 3.4 μm, which are deepest in the red targets (Wong et al. 2023). Eurybates also shows an absorption near 4.2 μm similar to features seen in the dark material of the large, regular moons of Jupiter and Saturn and attributed to complexed CO_2. The MIRI data show 10-μm emissivity plateaus in all objects, consistent with previous MIR observations of Trojans (Martin et al. 2023b). Detailed analysis of these high quality spectra will provide the standard by which *Lucy* spectral data will be analyzed.

7 Conclusions

The Trojan asteroids remain poorly understood in terms of composition, yet those compositions would provide key constraints on Solar System evolution. Trojan's featureless VNIR spectra have long confounded attempts to investigate compositions. Broad absorptions near 3.1 μm in spectra of several less-red Trojans may be due to N-H-bearing materials, which could suggest an outer Solar System origin. MIR emissivity peaks near 10 and 20 μm indicate the presence of fine-grained silicates that appear to be similar in composition to cometary silicates. There are no good Trojan analogs among the meteorite collection, prohibiting similar insights that meteorites provide for many Main Belt asteroids. Placing Trojans in context with Main Belt asteroids, comets, irregular satellites of the giant planets, and Centaurs indicates some potential genetic relationships among at least some members of these disparate groups.

Bimodal spectral groups among the Trojans and the changing fractions of these groups with object size provide important clues to the evolution of Trojans. Hypotheses that rely solely on space weathering to explain colors fail to predict the spectral bimodality. One model that fits current data, put forth in Wong and Brown (2016), posits that the difference between the two groups is not original composition, but the stability of sulfur-bearing ices at different heliocentric distances in the early Kuiper Belt. NASA's *Lucy* mission is poised to reveal the first views of the surfaces of Trojan asteroids. The detailed measurements of ices, organics, and silicates at different spatial scales on the diverse set of *Lucy* targets will answer the question of what the surface and bulk compositions of Trojans are, how the Trojans formed and evolved, and what their evolution tells us about Solar System history.

Acknowledgements We are grateful to the reviewer for comments and suggestions that improved the manuscript.

Declarations

Competing Interests The authors declare no competing interests.

References

Ammannito E, DeSanctis MC, Ciarniello M et al (2016) Distribution of phyllosilicates on the surface of Ceres. Science 353(6303):aaf4279. https://doi.org/10.1126/science.aaf4279

Barkume KM, Brown ME, Schaller EL (2008) Near-infrared spectra of centaurs and Kuiper belt objects. Astron J 135(1):55–67. https://doi.org/10.1088/0004-6256/135/1/55

Barucci MA, Brown ME, Emery JP et al (2008) Composition and surface properties of transneptunian objects and centaurs. In: Barucci MA, Boehnhardt H, Cruikshank DP et al (eds) The solar system beyond Neptune. University of Arizona Press, Tucson, pp 143–160

Barucci MA, Filacchione G, Fornasier S et al (2016) Detection of exposed H_2O ice on the nucleus of comet 67P/Churyumov-Gerasimenko. As observed by Rosetta OSIRIS and VIRTIS instruments. Astron Astrophys 595:A102. https://doi.org/10.1051/0004-6361/201628764. arXiv:1609.00551 [astro-ph.EP]

Bauer JM, Choi YJ, Weissman PR et al (2008) The large-grained dust coma of 174P/Echeclus. Planet Space Sci 120(866):393. https://doi.org/10.1086/587552

Bendjoya P, Cellino A, Di Martino M et al (2004) Spectroscopic observations of Jupiter Trojans. Icarus 168(2):374–384. https://doi.org/10.1016/j.icarus.2003.12.004

Binzel RP, Rivkin AS, Stuart JS et al (2004) Observed spectral properties of near-Earth objects: results for population distribution, source regions, and space weathering processes. Icarus 170(2):259–294. https://doi.org/10.1016/j.icarus.2004.04.004

Binzel RP, DeMeo FE, Turtelboom EV et al (2019) Compositional distributions and evolutionary processes for the near-Earth object population: results from the MIT-Hawaii near-Earth object spectroscopic survey (MITHNEOS). Icarus 324:41–76. https://doi.org/10.1016/j.icarus.2018.12.035. arXiv:2004.05090 [astro-ph.EP]

Bockelée-Morvan D, Gautier D, Hersant F et al (2002) Turbulent radial mixing in the solar nebula as the source of crystalline silicates in comets. Astron Astrophys 384(3):1107–1118. https://doi.org/10.1051/0004-6361:20020086

Bottke WF, Nesvorný D, Vokrouhlický D et al (2010) The irregular satellites: the most collisionally evolved populations in the solar system. Astron J 139(3):994–1014. https://doi.org/10.1088/0004-6256/139/3/994

Bottke WF, Marschall R, Nesvorný D et al (2023) Origin and evolution of Jupiter's Trojan asteroids. Space Sci Rev 219(8):83. https://doi.org/10.1007/s11214-023-01031-4. arXiv:2312.02864 [astro-ph.EP]

Bourdelle de Micas J, Fornasier S, Avdellidou C et al (2022) Composition of inner main-belt planetesimals. Astron Astrophys 665:A83. https://doi.org/10.1051/0004-6361/202244099

Brož M, Rozehnal J (2011) Eurybates - the only asteroid family among Trojans? Mon Not R Astron Soc 414(1):565–574. https://doi.org/10.1111/j.1365-2966.2011.18420.x. arXiv:1109.1109 [astro-ph.EP]

Brown ME (2016) The 3-4 μm spectra of Jupiter Trojan asteroids. Astron J 152(6):159. https://doi.org/10.3847/0004-6256/152/6/159. arXiv:1606.03013 [astro-ph.EP]
Brown ME, Koresko CD (1998) Detection of water ice on the centaur 1997 CU_{26}. Astrophys J Lett 505(1):L65–L67. https://doi.org/10.1086/311593
Brunetto R, Romano F, Blanco A et al (2006) Space weathering of silicates simulated by nanosecond pulse UV excimer laser. Icarus 180(2):546–554. https://doi.org/10.1016/j.icarus.2005.10.016
Brunetto R, Loeffler MJ, Nesvorný D et al (2015) Asteroid surface alteration by space weathering processes. Univ Ariz Press, pp 597–616. https://doi.org/10.2458/azu_uapress_9780816532131-ch031
Buie MW, Keeney BA, Strauss RH et al (2021) Size and shape of (11351) Leucus from five occultations. Planet Sci J 2(5):202. https://doi.org/10.3847/PSJ/ac1f9b
Buratti BJ, Soderlund K, Bauer J et al (2008) Infrared (0.83 5.1 μm) photometry of Phoebe from the Cassini visual infrared mapping spectrometer. Icarus 193(2):309–322. https://doi.org/10.1016/j.icarus.2007.09.014
Capaccioni F, Coradini A, Filacchione G et al (2015) The organic-rich surface of comet 67P/Churyumov-Gerasimenko as seen by VIRTIS/Rosetta. Science 347(6220):aaa0628. https://doi.org/10.1126/science.aaa0628
Carlson RW, Kargel JS, Douté S et al (2007) Io's surface composition. In: Lopes RMC, Spencer JR (eds) Io after Galileo: a new view of Jupiter's volcanic moon, Springer, Berlin, p 193. https://doi.org/10.1007/978-3-540-48841-5_9
Carry B, Solano E, Eggl S et al (2016) Spectral properties of near-Earth and Mars-crossing asteroids using Sloan photometry. Icarus 268:340–354. https://doi.org/10.1016/j.icarus.2015.12.047. arXiv:1601.02087 [astro-ph.EP]
Charnoz S, Morbidelli A (2007) Coupling dynamical and collisional evolution of small bodies. II. Forming the Kuiper belt, the Scattered Disk and the Oort Cloud. Icarus 188(2):468–480. https://doi.org/10.1016/j.icarus.2006.11.018
Chihara H, Koike C, Tsuchiyama A et al (2002) Compositional dependence of infrared absorption spectra of crystalline silicates I. Mg-Fe pyroxenes. Astron Astrophys 391(1):267–273. https://doi.org/10.1051/0004-6361:20020791
Christensen PR, Hamilton VE, Mehall GL et al (2024) The Lucy Thermal Emission Spectrometer (L'TES) Instrument. Space Sci Rev 220(1):1. https://doi.org/10.1007/s11214-023-01029-y
Ciarniello M, Fulle M, Raponi A et al (2022) Macro and micro structures of pebble made cometary nuclei reconciled by seasonal evolution. Nat Astron 6:546–553. https://doi.org/10.1038/s41550-022-01625-y
Ciesla FJ (2007) Dust coagulation and settling in layered protoplanetary disks. Astrophys J 654:159–162
Clark RN, Brown RH, Jaumann R et al (2005) Compositional maps of Saturn's moon Phoebe from imaging spectroscopy. Nature 435(7038):66–69. https://doi.org/10.1038/nature03558
Combe JP, McCord TB, Tosi F et al (2016) Detection of local H_2O exposed at the surface of Ceres. Science 353(6303):aaf3010. https://doi.org/10.1126/science.aaf3010
Crovisier J, Leech K, Bockelee-Morvan D et al (1997) The spectrum of comet Hale-Bopp (C/1995 O1) observed with the infrared space observatory at 2.9 astronomical units from the sun. Science 275:1904–1907. https://doi.org/10.1126/science.275.5308.1904
Cruikshank DP, Roush TL, Bartholomew MJ et al (1998) The composition of centaur 5145 Pholus. Icarus 135(2):389–407. https://doi.org/10.1006/icar.1998.5997
Cruikshank DP, Dalle Ore CM, Roush TL et al (2001) Constraints on the composition of Trojan asteroid 624 Hektor. Icarus 153(2):348–360. https://doi.org/10.1006/icar.2001.6703
De Luise F, Dotto E, Fornasier S et al (2010) A peculiar family of Jupiter Trojans: the eurybates. Icarus 209(2):586 590. https://doi.org/10.1016/j.icarus.2010.04.024, arXiv.1004.4180 [astro-ph.EP]
De Sanctis MC, Ammannito E, Raponi A et al (2015) Ammoniated phyllosilicates with a likely outer solar system origin on (1) Ceres. Nature 528(7581):241–244. https://doi.org/10.1038/nature16172
De Sanctis MC, Capaccioni F, Ciarniello M et al (2015) The diurnal cycle of water ice on comet 67P/Churyumov-Gerasimenko. Nature 525(7570):500–503. https://doi.org/10.1038/nature14869
De Sanctis MC, Ammannito E, McSween HY et al (2017) Localized aliphatic organic material on the surface of Ceres. Science 355(6326):719–722. https://doi.org/10.1126/science.aaj2305
de Sanctis MC, Ammannito E, Carrozzo FG et al (2018) Ceres's global and localized mineralogical composition determined by dawn's visible and infrared spectrometer (VIR). Meteorit Planet Sci 53(9):1844–1865. https://doi.org/10.1111/maps.13104
De Sanctis MC, Ammannito E, Raponi A et al (2020) Fresh emplacement of hydrated sodium chloride on Ceres from ascending salty fluids. Nat Astron 4:786–793. https://doi.org/10.1038/s41550-020-1138-8
Delsanti A, Peixinho N, Boehnhardt H et al (2006) Near-infrared color properties of Kuiper belt objects and centaurs: final results from the ESO large program. Astron J 131(3):1851–1863. https://doi.org/10.1086/499402

DeMeo FE, Carry B (2013) The taxonomic distribution of asteroids from multi-filter all-sky photometric surveys. Icarus 226(1):723–741. https://doi.org/10.1016/j.icarus.2013.06.027. arXiv:1307.2424 [astro-ph.EP]

DeMeo FE, Binzel RP, Carry B et al (2014) Unexpected D-type interlopers in the inner main belt. Icarus 229:392–399. https://doi.org/10.1016/j.icarus.2013.11.026. arXiv:1312.2962 [astro-ph.EP]

Deshapriya JDP, Barucci MA, Fornasier S et al (2018) Exposed bright features on the comet 67P/Churyumov-Gerasimenko: distribution and evolution. Astron Astrophys 613:A36. https://doi.org/10.1051/0004-6361/201732112

Dotto E, Fornasier S, Barucci MA et al (2006) The surface composition of Jupiter Trojans: visible and near-infrared survey of dynamical families. Icarus 183(2):420–434. https://doi.org/10.1016/j.icarus.2006.02.012

Dotto E, Emery JP, Barucci MA et al (2008) De troianis: the Trojans in the planetary system. In: Barucci MA, Boehnhardt H, Cruikshank DP et al (eds) The solar system beyond Neptune. Univ Ariz Press, p 383

Duffard R, Pinilla-Alonso N, Santos-Sanz P et al (2014) "TNOs are cool": a survey of the trans-Neptunian region. XI. A Herschel-PACS view of 16 centaurs. Astron Astrophys 564:A92. https://doi.org/10.1051/0004-6361/201322377. arXiv:1309.0946 [astro-ph.EP]

Dumas C, Owen T, Barucci MA (1998) Near-infrared spectroscopy of low-albedo surfaces of the solar system: search for the spectral signature of dark material. Icarus 133(2):221–232. https://doi.org/10.1006/icar.1998.5927

Emery JP, Brown RH (2003) Constraints on the surface composition of Trojan asteroids from near-infrared (0.8-4.0 μm) spectroscopy. Icarus 164(1):104–121. https://doi.org/10.1016/S0019-1035(03)00143-X

Emery JP, Brown RH (2004) The surface composition of Trojan asteroids: constraints set by scattering theory. Icarus 170(1):131–152. https://doi.org/10.1016/j.icarus.2004.02.004

Emery JP, Cruikshank DP, Van Cleve J (2006) Thermal emission spectroscopy (5.2-38 μm) of three Trojan asteroids with the spitzer space telescope: detection of fine-grained silicates. Icarus 182(2):496–512. https://doi.org/10.1016/j.icarus.2006.01.011

Emery JP, Burr DM, Cruikshank DP (2011) Near-infrared spectroscopy of Trojan asteroids: evidence for two compositional groups. Astron J 141(1):25. https://doi.org/10.1088/0004-6256/141/1/25. arXiv:1012.1284 [astro-ph.EP]

Emery JP, Marzari F, Morbidelli A et al (2015) The Complex History of Trojan Asteroids. In: Michel P, DeMeo FE, Bottke WF (eds) Asteroids IV. University of Arizona Press, Tucson, pp 203–220. https://doi.org/10.2458/azu_uapress_9780816532131-ch011

Farnham TL, Wellnitz DD, Hampton DL et al (2007) Dust coma morphology in the deep impact images of comet 9P/Tempel 1. Icarus 191(2 Suppl.):146–160. https://doi.org/10.1016/j.icarus.2006.10.038

Feaga LM, A'Hearn MF, Sunshine JM et al (2007) Asymmetries in the distribution of H_2O and CO_2 in the inner coma of comet 9P/Tempel 1 as observed by Deep Impact. Icarus 190(2):345–356. https://doi.org/10.1016/j.icarus.2007.04.009

Fernández YR, Jewitt DC, Sheppard SS (2001) Low albedos among extinct comet candidates. Astrophys J Lett 553(2):L197–L200. https://doi.org/10.1086/320689. arXiv:astro-ph/0104478 [astro-ph]

Fernández YR, Sheppard SS, Jewitt DC (2003) The albedo distribution of Jovian Trojan asteroids. Astron J 126(3):1563

Fernández YR, Jewitt D, Ziffer JE (2009) Albedos of small Jovian Trojans. Astron J 138(1):240–250. https://doi.org/10.1088/0004-6256/138/1/240. arXiv:0906.1786 [astro-ph.EP]

Filacchione G, Raponi A, Capaccioni F et al (2016) Seasonal exposure of carbon dioxide ice on the nucleus of comet 67P/Churyumov-Gerasimenko. Science 354(6319):1563–1566. https://doi.org/10.1126/science.aag3161

Fornasier S, Dotto E, Marzari F et al (2004) Visible spectroscopic and photometric survey of L5 Trojans: investigation of dynamical families. Icarus 172(1):221–232. https://doi.org/10.1016/j.icarus.2004.06.015

Fornasier S, Dotto E, Hainaut O et al (2007) Visible spectroscopic and photometric survey of Jupiter Trojans: final results on dynamical families. Icarus 190(2):622–642. https://doi.org/10.1016/j.icarus.2007.03.033. arXiv:0704.0350 [astro-ph]

Fornasier S, Mottola S, Keller HU et al (2016) Rosetta's comet 67P/Churyumov-Gerasimenko sheds its dusty mantle to reveal its icy nature. Science 354(6319):1566–1570. https://doi.org/10.1126/science.aag2671

Fornasier S, Hoang HV, Fulle M et al (2023) Volatile exposures on the 67P/Churyumov-Gerasimenko nucleus. Astron Astrophys 672:A136. https://doi.org/10.1051/0004-6361/202245614. arXiv:2302.11424 [astro-ph.EP]

Frank D, Zolensky M, Le L (2014) Olivine in terminal particles of stardust aerogel tracks and analogous grains in chondrite matrix. Geochim Cosmochim Acta 142:240–259. https://doi.org/10.1016/j.gca.2014.05.037

Fraser WC, Brown ME (2012) The Hubble wide field camera 3 test of surfaces in the outer solar system: the compositional classes of the Kuiper belt. Astrophys J 749(1):33. https://doi.org/10.1088/0004-637X/749/1/33. arXiv:1202.0827 [astro-ph.EP]

Fraser WC, Brown ME (2018) Phoebe: a surface dominated by water. Astron J 156(1):23. https://doi.org/10.3847/1538-3881/aac213. arXiv:1803.04979 [astro-ph.EP]

Fu X, Zou Y, Zheng Y et al (2012) Effects of space weathering on diagnostic spectral features: results from He^+ irradiation experiments. Icarus 219(2):630–640. https://doi.org/10.1016/j.icarus.2012.03.009

Gail HP, Sedlmayr E (1999) Mineral formation in stellar winds I. Condensation sequence of silicate and iron grains in stationary oxygen rich outflows. Astron Astrophys 347:594–616

Gartrelle GM, Hardersen PS, Izawa MRM et al (2021) Illuminating the dark side of the asteroid population: visible near-infrared (0.7-2.45 μm) surface mineralogy modeling of D-type asteroids using Shkuratov theory. Icarus 354:114043. https://doi.org/10.1016/j.icarus.2020.114043

Gradie J, Tedesco E (1982) Compositional structure of the asteroid belt. Science 216(4553):1405–1407. https://doi.org/10.1126/science.216.4553.1405

Gradie J, Veverka J (1980) The composition of the Trojan asteroids. Nature 283(5750):840–842. https://doi.org/10.1038/283840a0

Grav T, Mainzer AK, Bauer JM et al (2012) WISE/NEOWISE observations of the Jovian Trojan population: taxonomy. Astrophys J 759(1):49. https://doi.org/10.1088/0004-637X/759/1/49. arXiv:1209.1549 [astro-ph.EP]

Graykowski A, Jewitt D (2018) Colors and shapes of the irregular planetary satellites. Astron J 155(4):184. https://doi.org/10.3847/1538-3881/aab49b. arXiv:1803.01907 [astro-ph.EP]

Guilbert-Lepoutre A (2012) Survival of amorphous water ice on centaurs. Astron J 144(4):97. https://doi.org/10.1088/0004-6256/144/4/97

Guilbert-Lepoutre A (2014) Survival of water ice in Jupiter Trojans. Icarus 231:232–238. https://doi.org/10.1016/j.icarus.2013.12.014. arXiv:1401.5196 [astro-ph.EP]

Hanner MS, Lynch DK, Russell RW (1994) The 8-13 micron spectra of comets and the composition of silicate grains. Astrophys J 425:274–285

Hapke B (1965) Effects of a simulated solar wind on the photometric properties of rocks and powders. Ann NY Acad Sci 123:711–721. https://doi.org/10.1111/j.1749-6632.1965.tb20395.x

Hapke B (2001) Space weathering from Mercury to the asteroid belt. J Geophys Res 106(E5):10039–10074. https://doi.org/10.1029/2000JE001338

Harker DE, Woodward CE, Wooden DH (2005) The dust grains from 9P/Tempel 1 before and after the encounter with Deep Impact. Science 310(5746):278–280. https://doi.org/10.1126/science.1119143

Harker DE, Woodward CE, Wooden DH et al (2007) Gemini-N mid-IR observations of the dust properties of the ejecta excavated from comet 9P/Tempel 1 during Deep Impact. Icarus 190(2):432–453. https://doi.org/10.1016/j.icarus.2007.03.008

Henning T (2010) Cosmic silicates. Annu Rev Astron Astrophys 48:21–46. https://doi.org/10.1146/annurev-astro-081309-130815

Hiroi T, Zolensky ME, Pieters CM (2001) The Tagish Lake meteorite: a possible sample from a D-type asteroid. Science 293(5538):2234–2236. https://doi.org/10.1126/science.1063734

Horner J, Evans NW, Bailey ME (2004) Simulations of the population of centaurs - I. The bulk statistics. Mon Not R Astron Soc 354(3):798–810. https://doi.org/10.1111/j.1365-2966.2004.08240.x. arXiv:astro-ph/0407400 [astro-ph]

Horner J, Müller TG, Lykawka PS (2012) (1173) Anchises - thermophysical and dynamical studies of a dynamically unstable Jovian Trojan. Mon Not R Astron Soc 423(3):2587–2596. https://doi.org/10.1111/j.1365-2966.2012.21067.x. arXiv:1204.1388 [astro-ph.EP]

Hughes AL, Armitage PJ (2010) Particle transport in evolving protoplanetary disks: implications for results from stardust. Astrophys J 719(2):1633–1653. https://doi.org/10.1088/0004-637X/719/2/1633. arXiv:1007.1989

Humes OA, Thomas CA, Emery JP et al (2022) Ultraviolet spectroscopy of Lucy mission targets with the Hubble space telescope. Planet Sci J 3(8):190. https://doi.org/10.3847/PSJ/ac8059. arXiv:2207.11390 [astro-ph.EP]

Izawa MR, King PL, Vernazza P et al (2021) Salt – a critical material to consider when exploring the solar system. Icarus 359(October 2020):114328. https://doi.org/10.1016/j.icarus.2021.114328

Jäger C, Molster FJ, Dorschner J et al (1998) Steps toward interstellar silicate mineralogy IV. The crystalline revolution. Astron Astrophys 339:904–916

Jewitt D (2009) The active centaurs. Astron J 137(5):4296–4312. https://doi.org/10.1088/0004-6256/137/5/4296. arXiv:0902.4687 [astro-ph.EP]

Jewitt D (2015) Color systematics of comets and related bodies. Astron J 150(6):201. https://doi.org/10.1088/0004-6256/150/6/201. arXiv:1510.07069 [astro-ph.EP]

Jewitt DC, Luu JX (1990) CCD spectra of asteroids. II - The Trojans as spectral analogs of cometary nuclei. Astron J 100:933–944. https://doi.org/10.1086/115572

Kaňuchová Z, Baratta GA, Garozzo M et al (2010) Space weathering of asteroidal surfaces. Influence on the UV-Vis spectra. Astron Astrophys 517:A60. https://doi.org/10.1051/0004-6361/201014061

Kaňuchová Z, Brunetto R, Melita M et al (2012) Space weathering and the color indexes of minor bodies in the outer solar system. Icarus 221(1):12–19. https://doi.org/10.1016/j.icarus.2012.06.043

Kelley MS, Wooden DH (2009) The composition of dust in Jupiter-family comets inferred from infrared spectroscopy. Planet Space Sci 57(10):1133–1145. https://doi.org/10.1016/j.pss.2008.11.017

Kelley MS, Woodward CE, Gehrz RD et al (2017) Mid-infrared spectra of comet nuclei. Icarus 284:344–358. https://doi.org/10.1016/j.icarus.2016.11.029

Koike C, Shibai H, Tuchiyama A (1993) Extinction of olivine and pyroxene in the mind- and far-infrared. Mon Not R Astron Soc 264(3):654–658

Koike C, Chihara H, Tsuchiyama A et al (2003) Compositional dependence of infrared absorption spectra of crystalline silicate II. Natural and synthetic olivines. Astron Astrophys 399(3):1101–1107. https://doi.org/10.1051/0004-6361:20021831

Lamy PL, Toth I, Fernández YR et al (2004) The sizes, shapes, albedos, and colors of cometary nuclei. In: Festou MC, Keller HU, Weaver HA (eds) Comets II. University of Arizona Press, Tucson, pp 223–264

Lantz C, Clark BE, Barucci MA et al (2013) Evidence for the effects of space weathering spectral signatures on low albedo asteroids. Astron Astrophys 554:A138. https://doi.org/10.1051/0004-6361/201321593

Lantz C, Brunetto R, Barucci MA et al (2017) Ion irradiation of carbonaceous chondrites: a new view of space weathering on primitive asteroids. Icarus 285:43–57. https://doi.org/10.1016/j.icarus.2016.12.019

Lazzarin M, Marchi S, Moroz LV et al (2006) Space weathering in the main asteroid belt: the big picture. Astrophys J Lett 647(2):L179–L182. https://doi.org/10.1086/507448

Lazzaro D, Angeli CA, Carvano JM et al (2004) S^3OS^2: the visible spectroscopic survey of 820 asteroids. Icarus 172(1):179–220. https://doi.org/10.1016/j.icarus.2004.06.006

Levison HF, Shoemaker EM, Shoemaker CS (1997) Dynamical evolution of Jupiter's Trojan asteroids. Nature 385:42–44

Levison HF, Olkin CB, Noll KS et al (2021) Lucy mission to the Trojan asteroids: science goals. Planet Sci J 2(5):171. https://doi.org/10.3847/PSJ/abf840

Licandro J, Pinilla-Alonso N (2005) The inhomogeneous surface of centaur 32522 Thereus (2001 PT_{13}). Astrophys J Lett 630(1):L93–L96. https://doi.org/10.1086/491469

Lisse CM, VanCleve J, Adams AC et al (2006) Spitzer spectral observations of the deep impact ejecta. Science 313(5787):635–640. https://doi.org/10.1126/science.1124694

Loeffler MJ, Baragiola RA, Murayama M (2008) Laboratory simulations of redeposition of impact ejecta on mineral surfaces. Icarus 196(1):285–292. https://doi.org/10.1016/j.icarus.2008.02.021

Loeffler MJ, Dukes CA, Baragiola RA (2009) Irradiation of olivine by 4 keV He^+: simulation of space weathering by the solar wind. J Geophys Res, Planets 114(E3):E03003. https://doi.org/10.1029/2008JE003249

Lowry VC, Donaldson Hanna KL, Ito G et al (2022) T-matrix and Hapke modeling of the thermal infrared spectra of Trojan asteroids and (944) Hidalgo: implications for their regolith particle size and porosity. Planet Sci J 3(7):181. https://doi.org/10.3847/PSJ/ac7a30

Luu J, Jewitt D, Cloutis E (1994) Near-infrared spectroscopy of primitive solar system objects. Icarus 109(1):133–144. https://doi.org/10.1006/icar.1994.1081

Mahjoub A, Poston MJ, Hand KP et al (2016) Electron irradiation and thermal processing of mixed-ices of potential relevance to Jupiter Trojan asteroids. Astrophys J 820(2):141. https://doi.org/10.3847/0004-637X/820/2/141

Mahjoub A, Poston MJ, Blacksberg J et al (2017) Production of sulfur allotropes in electron irradiated Jupiter Trojans ice analogs. Astrophys J 846(2):148. https://doi.org/10.3847/1538-4357/aa85e0

Marchi S, Brunetto R, Magrin S et al (2005) Space weathering of near-Earth and main belt silicate-rich asteroids: observations and ion irradiation experiments. Astron Astrophys 443(3):769–775. https://doi.org/10.1051/0004-6361:20053525

Marchi S, Bell JF, Bierhaus B et al (2023) Surface geology of Jupiter's Trojan asteroids. Space Sci Rev 219(5):44. https://doi.org/10.1007/s11214-023-00985-9

Marsset M, Vernazza P, Gourgeot F et al (2014) Similar origin for low- and high-albedo Jovian Trojans and Hilda asteroids? Astron Astrophys 568:L7. https://doi.org/10.1051/0004-6361/201424105. arXiv:1407.7016 [astro-ph.EP]

Martin AC (2022) Mid-infrared spectral studies of Jovian Trojan asteroids and the effects of regolith porosity. PhD thesis, Northern Arizona University. https://www.proquest.com/dissertations-theses/mid-infrared-spectral-studies-Jovian-trojan/docview/2708280124/se-2

Martin AC, Emery JP (2023) MIR spectra and analysis of Jovian Trojan asteroids. Planet Sci J 4(8):153. https://doi.org/10.3847/PSJ/aced0c

Martin AC, Emery JP, Loeffler MJ (2022) Spectral effects of regolith porosity in the mid-IR - forsteritic olivine. Icarus 378:114921. https://doi.org/10.1016/j.icarus.2022.114921

Martin AC, Emery JP, Loeffler MJ (2023a) Spectral effects of regolith porosity in the mid-IR - pyroxene. Icarus 397:115507. https://doi.org/10.1016/j.icarus.2023.115507

Martin AC, Emery JP, Wong I et al (2023b) JWST MIRI-MRS observations of Lucy's Trojan targets. In: Asteroids, comets, meteors 2023, p 2465

Marzari F, Scholl H (1998) Capture of Trojans by a growing proto-Jupiter. Icarus 131:41–51

McSween HY, Emery JP, Rivkin AS et al (2018) Carbonaceous chondrites as analogs for the composition and alteration of Ceres. Meteorit Planet Sci 53(9):1793–1804. https://doi.org/10.1111/maps.12947

Melita MD, Licandro J, Jones DC et al (2008) Physical properties and orbital stability of the Trojan asteroids. Icarus 195(2):686–697. https://doi.org/10.1016/j.icarus.2008.01.004

Melita MD, Strazzulla G, Bar-Nun A (2009) Collisions, cosmic radiation and the colors of the Trojan asteroids. Icarus 203(1):134–139. https://doi.org/10.1016/j.icarus.2009.04.024. arXiv:0906.4130 [astro-ph.EP]

Melita MD, Kaňuchová Z, Brunetto R et al (2015) Space weathering and the color-color diagram of Plutinos and Jupiter Trojans. Icarus 248:222–229. https://doi.org/10.1016/j.icarus.2014.09.050

Morbidelli A, Levison HF, Tsiganis K et al (2005) Chaotic capture of Jupiter's Trojan asteroids in the early solar system. Nature 435(7041):462–465. https://doi.org/10.1038/nature03540

Morbidelli A, Brasser R, Tsiganis K et al (2009) Constructing the secular architecture of the solar system. I. The giant planets. Astron Astrophys 507(2):1041–1052. https://doi.org/10.1051/0004-6361/200912876

Moroz LV, Fisenko AV, Semjonova LF et al (1996) Optical effects of regolith processes on S-asteroids as simulated by laser shots on ordinary chondrite and other mafic materials. Icarus 122(2):366–382. https://doi.org/10.1006/icar.1996.0130

Mottola S, Hellmich S, Buie MW et al (2023) Shape models of Lucy targets (3548) Eurybates and (21900) Orus from disk-integrated photometry. Planet Sci J 4(1):18. https://doi.org/10.3847/PSJ/acaf79

Mottola S, Britt D, Brown ME et al (2024) Shapes, rotations, photometric and internal properties of Jupiter Trojans. Space Sci Rev 220:17. https://doi.org/10.1007/s11214-024-01052-7

Mueller M, Marchis F, Emery JP et al (2010) Eclipsing binary Trojan asteroid patroclus: thermal inertia from spitzer observations. Icarus 205(2):505–515. https://doi.org/10.1016/j.icarus.2009.07.043. arXiv:0908.4198

Müller T, Lellouch E, Fornasier S (2020) Trans-Neptunian objects and centaurs at thermal wavelengths. In: Prialnik D, Barucci MA, Young L (eds) The trans-Neptunian solar system. Elsevier, Amsterdam, pp 153–181. https://doi.org/10.1016/B978-0-12-816490-7.00007-2

Nesvorný D, Jedicke R, Whiteley RJ et al (2005) Evidence for asteroid space weathering from the sloan digital sky survey. Icarus 173(1):132–152. https://doi.org/10.1016/j.icarus.2004.07.026

Nesvorný D, Vokrouhlický D, Morbidelli A (2013) Capture of Trojans by jumping Jupiter. Astrophys J 768(1):45. https://doi.org/10.1088/0004-637X/768/1/45. arXiv:1303.2900 [astro-ph.EP]

Noll KS, Brown ME, Buie MW et al (2023) Trojan asteroid satellites, rings, and activity. Space Sci Rev 219(7):59. https://doi.org/10.1007/s11214-023-01001-w

Oklay N, Mottola S, Vincent JB et al (2017) Long-term survival of surface water ice on comet 67P. Mon Not R Astron Soc 469:S582–S597. https://doi.org/10.1093/mnras/stx2298

Olkin CB, Levison HF, Vincent M et al (2021) Lucy mission to the Trojan asteroids: instrumentation and encounter concept of operations. Planet Sci J 2(5):172. https://doi.org/10.3847/PSJ/abf83f. arXiv:2104.04575 [astro-ph.IM]

O'Rourke L, Heinisch P, Blum J et al (2020) The Philae lander reveals low-strength primitive ice inside cometary boulders. Nature 586(7831):697–701. https://doi.org/10.1038/s41586-020-2834-3

Owen TC, Cruikshank DP, Dalle Ore CM et al (1999) NOTE: detection of water ice on Saturn's satellite Phoebe. Icarus 139(2):379–382. https://doi.org/10.1006/icar.1999.6116

Peixinho N, Doressoundiram A, Delsanti A et al (2003) Reopening the TNOs color controversy: centaurs bimodality and TNOs unimodality. Astron Astrophys 410:L29–L32. https://doi.org/10.1051/0004-6361:20031420. arXiv:astro-ph/0309428 [astro-ph]

Peixinho N, Delsanti A, Guilbert-Lepoutre A et al (2012) The bimodal colors of centaurs and small Kuiper belt objects. Astron Astrophys 546:A86. https://doi.org/10.1051/0004-6361/201219057. arXiv:1206.3153 [astro-ph.EP]

Peixinho N, Delsanti A, Doressoundiram A (2015) Reanalyzing the visible colors of centaurs and KBOs: what is there and what we might be missing. Astron Astrophys 577:A35. https://doi.org/10.1051/0004-6361/201425436. arXiv:1502.04145 [astro-ph.EP]

Perna D, Barucci MA, Fulchignoni M et al (2018a) A spectroscopic survey of the small near-Earth asteroid population: peculiar taxonomic distribution and phase reddening. Planet Space Sci 157:82–95. https://doi.org/10.1016/j.pss.2018.03.008, arXiv:1803.08953

Perna D, Bott N, Hromakina T et al (2018b) Rotationally-resolved spectroscopy of Jupiter Trojans (624) Hektor and (911) Agamemnon. Mon Not R Astron Soc 475(1):974–980. https://doi.org/10.1093/mnras/stx3341

Pinilla-Alonso N, Popescu M, Licandro J et al (2022) Detection of the irregular shape of the southern limb of Menoetius from observations of the 2017-2018 Patroclus–Menoetius mutual events. Planet Sci J 3(12):267. https://doi.org/10.3847/PSJ/ac9f11

Poch O, Istiqomah I, Quirico E et al (2020) Ammonium salts are a reservoir of nitrogen on a cometary nucleus and possibly on some asteroids. Science 367(6483):aaw7462. https://doi.org/10.1126/science.aaw7462. arXiv:2003.06034 [astro-ph.EP]

Pommerol A, Thomas N, El-Maarry MR et al (2015) OSIRIS observations of meter-sized exposures of H_2O ice at the surface of 67P/Churyumov-Gerasimenko and interpretation using laboratory experiments. Astron Astrophys 583:A25. https://doi.org/10.1051/0004-6361/201525977

Porco CC, Baker E, Barbara J et al (2005) Cassini imaging science: initial results on Phoebe and Iapetus. Science 307(5713):1237–1242. https://doi.org/10.1126/science.1107981

Poston MJ, Mahjoub A, Ehlmann BL et al (2018) Visible near-infrared spectral evolution of irradiated mixed ices and application to Kuiper belt objects and Jupiter Trojans. Astrophys J 856(2):124. https://doi.org/10.3847/1538-4357/aab1f1

Prialnik D, Brosch N, Ianovici D (1995) Modelling the activity of 2060 Chiron. Mon Not R Astron Soc 276(4):1148–1154. https://doi.org/10.1093/mnras/276.4.1148

Quirico E, Moroz LV, Schmitt B et al (2016) Refractory and semi-volatile organics at the surface of comet 67P/Churyumov-Gerasimenko: insights from the VIRTIS/Rosetta imaging spectrometer. Icarus 272:32–47. https://doi.org/10.1016/j.icarus.2016.02.028

Raponi A, Ciarniello M, Capaccioni F et al (2020) Infrared detection of aliphatic organics on a cometary nucleus. Nat Astron 4:500–505. https://doi.org/10.1038/s41550-019-0992-8. arXiv:2009.14476 [astro-ph.EP]

Reuter DC, Simon AA, Lunsford A et al (2023) L'Ralph: a visible/infrared spectral imager for the Lucy mission to the Trojans. Space Sci Rev 219(8):69. https://doi.org/10.1007/s11214-023-01009-2

Ribeiro de Sousa R, Morbidelli A, Gomes R et al (2022) Dynamical origin of the dwarf planet Ceres. Icarus 379:114933. https://doi.org/10.1016/j.icarus.2022.114933, arXiv:2202.09238 [astro-ph.EP]

Rivkin AS, Emery JP, Howell ES et al (2022) The nature of low-albedo small bodies from 3 μm spectroscopy: one group that formed within the ammonia snow line and one that formed beyond it. Planet Sci J 3(7):153. https://doi.org/10.3847/PSJ/ac7217. arXiv:2205.09166 [astro-ph.EP]

Roig F, Ribeiro AO, Gil-Hutton R (2008) Taxonomy of asteroid families among the Jupiter Trojans: comparison between spectroscopic data and the sloan digital sky survey colors. Astron Astrophys 483(3):911–931. https://doi.org/10.1051/0004-6361:20079177. arXiv:0712.0046 [astro-ph]

Romanishin W, Tegler SC (2018) Albedos of centaurs, Jovian Trojans, and Hildas. Astron J 156(1):19. https://doi.org/10.3847/1538-3881/aac210

Romon-Martin J, Delahodde C, Barucci MA et al (2003) Photometric and spectroscopic observations of (2060) Chiron at the ESO Very Large Telescope. Astron Astrophys 400:369–373. https://doi.org/10.1051/0004-6361:20021890

Roush T (2003) Estimated optical constants of the Tagish Lake meteorite. Meteorit Planet Sci 38(3):419–426. https://doi.org/10.1111/j.1945-5100.2003.tb00277.x

Sasaki S, Nakamura K, Hamabe Y et al (2001) Production of iron nanoparticles by laser irradiation in a simulation of lunar-like space weathering. Nature 410(6828):555–557

Schaller EL, Brown ME (2007) Volatile loss and retention on Kuiper belt objects. Astrophys J Lett 659(1):L61–L64. https://doi.org/10.1086/516709

Schwamb ME, Fraser WC, Bannister MT et al (2019) Col-OSSOS: the colors of the outer solar system origins survey. Astrophys J Suppl Ser 243(1):12. https://doi.org/10.3847/1538-4365/ab2194. arXiv:1809.08501 [astro-ph.EP]

Simpson AM, Brown ME, Schemel MJ et al (2022) An ALMA search for high-albedo objects among the midsized Jupiter Trojan population. Astron J 164(1):23

Slyusarev IG, Belskaya IN (2014) Jupiter's Trojans: physical properties and origin. Sol Syst Res 48(2):139–157. https://doi.org/10.1134/S0038094614020063

Souza-Feliciano AC, De Prá M, Pinilla-Alonso N et al (2020) Analysis in the visible range of NASA Lucy mission targets: Eurybates, Polymele, Orus and Donaldjohanson. Icarus 338:113463. https://doi.org/10.1016/j.icarus.2019.113463, arXiv:1907.11451 [astro-ph.EP]

Stansberry J, Grundy W, Brown M et al (2008) Physical properties of Kuiper belt and centaur objects: constraints from the spitzer space telescope. In: Barucci MA, Boehnhardt H, Cruikshank DP et al (eds) The solar system beyond Neptune. University of Arizona Press, Tucson, p 161

Steckloff JK, Lisse CM, Safrit TK et al (2021) The sublimative evolution of (486958) Arrokoth. Icarus 356:113998. https://doi.org/10.1016/j.icarus.2020.113998, arXiv:2007.12657 [astro-ph.EP]

Strazzulla G, Dotto E, Binzel R et al (2005) Spectral alteration of the Meteorite Epinal (H5) induced by heavy ion irradiation: a simulation of space weathering effects on near-Earth asteroids. Icarus 174(1):31–35. https://doi.org/10.1016/j.icarus.2004.09.013

Sugita S, Ootsubo T, Kadono T et al (2005) Subaru telescope observations of deep impact. Science 310(5746):274–278. https://doi.org/10.1126/science.1119091

Sunshine JM, A'Hearn MF, Groussin O et al (2006) Exposed water ice deposits on the surface of comet 9P/Tempel 1. Science 311(5766):1453–1455. https://doi.org/10.1126/science.1123632

Szabó GM, Ivezić Ž, Jurić M et al (2007) The properties of Jovian Trojan asteroids listed in SDSS moving object catalogue 3. Mon Not R Astron Soc 377(4):1393–1406. https://doi.org/10.1111/j.1365-2966.2007.11687.x. arXiv:astro-ph/0703026 [astro-ph]

Tegler S, Consolmagno G, Romanishin W (2006) Comet 174P/Echeclus. IAU Circ 8701:1

Tegler SC, Romanishin W, Consolmagno GJ (2003) Color patterns in the Kuiper belt: a possible primordial origin. Astrophys J Lett 599(1):L49–L52. https://doi.org/10.1086/381076

Tegler SC, Bauer JM, Romanishin W et al (2008) Colors of centaurs. In: Barucci MA, Boehnhardt H, Cruikshank DP et al (eds) The solar system beyond Neptune. University of Arizona Press, Tucson, p 105

Tegler SC, Romanishin W, Consolmagno GJ et al (2016) Two color populations of Kuiper belt and centaur objects and the smaller orbital inclinations of red centaur objects. Astron J 152(6):210. https://doi.org/10.3847/0004-6256/152/6/210

Tielens AGGM, Waters LBFM, Molster FJ et al (1998) Circumstellar silicate mineralogy. Astrophys Space Sci 255:415–426. https://doi.org/10.1023/A:1001585120472

Tiscareno MS, Malhotra R (2003) The dynamics of known Centaurs. Astron J 126(6):3122–3131. https://doi.org/10.1086/379554, arXiv:arXiv:astro-ph/0211076 [astro-ph]

Tsiganis K, Gomes R, Morbidelli A et al (2005) Origin of the orbital architecture of the giant planets of the solar system. Nature 435(7041):459–461. https://doi.org/10.1038/nature03539

Vernazza P, Delbo M, King PL et al (2012) High surface porosity as the origin of emissivity features in asteroid spectra. Icarus 221(2):1162–1172. https://doi.org/10.1016/j.icarus.2012.04.003

Vernazza P, Fulvio D, Brunetto R et al (2013) Paucity of Tagish Lake-like parent bodies in the Asteroid Belt and among Jupiter Trojans. Icarus 225(1):517–525. https://doi.org/10.1016/j.icarus.2013.04.019

Vilas F, Larson SM, Hatch EC et al (1993) CCD reflectance spectra of selected asteroids. II. Low-albedo asteroid spectra and data extraction techniques. Icarus 105(1):67–78. https://doi.org/10.1006/icar.1993.1111

Watanabe JI (2004) Meteor streams and comets. Earth Moon Planets 95(1-4):49–61. https://doi.org/10.1007/s11038-005-9045-7

Weaver HA, Wilson JP, Conard SJ et al (2023) The Lucy Long Range Reconnaissance Imager (L'LORRI). Space Sci Rev 219(8):82. https://doi.org/10.1007/s11214-023-01028-z

Wong I, Brown ME (2015) The color-magnitude distribution of small Jupiter Trojans. Astron J 150(6):174. https://doi.org/10.1088/0004-6256/150/6/174. arXiv:1510.03144 [astro-ph.EP]

Wong I, Brown ME (2016) A hypothesis for the color bimodality of Jupiter Trojans. Astron J 152(4):90. https://doi.org/10.3847/0004-6256/152/4/90. arXiv:1607.04133 [astro-ph.EP]

Wong I, Brown ME (2017a) The bimodal color distribution of small Kuiper belt objects. Astron J 153(4):145. https://doi.org/10.3847/1538-3881/aa60c3. arXiv:1702.02615 [astro-ph.EP]

Wong I, Brown ME (2017b) The color-magnitude distribution of Hilda asteroids: comparison with Jupiter Trojans. Astron J 153(2):69. https://doi.org/10.3847/1538-3881/153/2/69. arXiv:1701.00367 [astro-ph.EP]

Wong I, Brown ME, Emery JP (2014) The differing magnitude distributions of the two Jupiter Trojan color populations. Astron J 148(6):112. https://doi.org/10.1088/0004-6256/148/6/112. arXiv:1408.2485 [astro-ph.EP]

Wong I, Brown ME, Blacksberg J et al (2019a) Hubble ultraviolet spectroscopy of Jupiter Trojans. Astron J 157(4):161. https://doi.org/10.3847/1538-3881/ab0e00. arXiv:1903.03576 [astro-ph.EP]

Wong I, Mishra A, Brown ME (2019b) Photometry of active centaurs: colors of dormant active centaur nuclei. Astron J 157(6):225. https://doi.org/10.3847/1538-3881/ab1b22. arXiv:1904.09255 [astro-ph.EP]

Wong I, Brown M, Emery J et al (2023) JWST observations of the Lucy flyby targets: new spectroscopic constraints on Jupiter Trojan composition. In: Asteroids, Comets, Meteors Conference 2023, p 2556

Wooden DH (2002) Comet grains: their IR emission and their relation to ISM grains. Earth Moon Planets 89(1):247–287. https://doi.org/10.1023/A:1021515023679

Wooden DH (2008) Cometary refractory grains: interstellar and nebular sources. Space Sci Rev 138:75–108. https://doi.org/10.1007/s11214-008-9424-2

Wooden DH, Woodward CE, Harker DE (2004) Discovery of crystalline silicates in comet C/2001 Q4 (NEAT). Astrophys J 612(1):L77–L80. https://doi.org/10.1086/424593

Woodward CE, Kelley MSP, Harker DE et al (2015) SOFIA infrared spectrophotometry of comet C/2012 K1 (Pan-STARRS). Astrophys J 809(2):181. https://doi.org/10.1088/0004-637X/809/2/181. arXiv:1508.00288 [astro-ph.EP]

Yang B, Jewitt D (2007) Spectroscopic search for water ice on Jovian Trojan asteroids. Astron J 134(1):223–228. https://doi.org/10.1086/518368

Yang B, Lucey P, Glotch T (2013) Are large Trojan asteroids salty? An observational, theoretical, and experimental study. Icarus 223(1):359–366. https://doi.org/10.1016/j.icarus.2012.11.025. arXiv:1211.3099

Publisher's Note Springer Nature remains neutral with regard to jurisdictional claims in published maps and institutional affiliations.

Authors and Affiliations

Joshua P. Emery[1] · Richard P. Binzel[2] · Daniel T. Britt[3] · Michael E. Brown[4] · Carly J.A. Howett[5,6] · Audrey C. Martin[3] · Mario D. Melita[7,8] · Ana Carolina Souza-Feliciano[3,9] · Ian Wong[10]

✉ J.P. Emery
joshua.emery@nau.edu

1 Northern Arizona University, Flagstaff, AZ, USA

2 Massachusetts Institute of Technology, Cambridge, MA, USA

3 University of Central Florida, Orlando, FL, USA

4 California Institute of Technology, Pasadena, CA, USA

5 University of Oxford, Oxford, UK

6 Planetary Science Institute, Tucson, AZ, USA

7 CONICET-Universidad de Buenos Aires, Buenos Aires, Argentina

8 Universidad Nacional de La Plata, La Plata, Argentina

9 Florida Space Institute, Orlando, FL, USA

10 NASA Goddard Space Flight Center, Greenbelt, MD, USA

Space Science Reviews (2023) 219:59
https://doi.org/10.1007/s11214-023-01001-w

Trojan Asteroid Satellites, Rings, and Activity

Keith S. Noll[1] · Michael E. Brown[2] · Marc W. Buie[3] · William M. Grundy[4] · Harold F. Levison[3] · Simone Marchi[3] · Catherine B. Olkin[3,5] · S. Alan Stern[3] · Harold A. Weaver[6]

Received: 3 July 2023 / Accepted: 9 September 2023 / Published online: 12 October 2023

Abstract

The *Lucy* mission will encounter five Jupiter Trojans during its mission with three of the five already known to be multiple systems. These include a near-equal-mass binary, a small and widely separated satellite, and one intermediate-size satellite system. This chapter reviews the current state of knowledge of Trojan asteroid satellites in the context of similar satellite systems in other small body populations. The prospects for the detection of additional satellites as well as other near-body phenomena are considered. The scientific utility of satellites makes their observation with *Lucy* an important scientific priority for the mission.

Keywords Asteroids · Trojan asteroids · Satellites · Rings

1 Introduction

The existence of asteroid satellites was, for many years, a purely speculative topic that had yielded hints, but no incontrovertible evidence, despite numerous searches (Weidenschilling et al. 1989). But, starting with the serendipitous discovery of S/(243) I Ida by the Galileo spacecraft (Belton et al. 1996), there was a rapid convergence of multiple observational and theoretical developments that led to the paradigm-shifting realization that satellites exist in essentially all small body populations (Merline et al. 2002; Richardson and Walsh 2006; Noll et al. 2008), including the Jupiter Trojans (Table 1).

Satellites and binary systems exist over a wide range of relative sizes and separations, but, when scaled to the region of gravitational influence given by the Hill radius, regularities

✉ K.S. Noll
keith.s.noll@nasa.gov

1 NASA Goddard Spaceflight Center, 8800 Greenbelt Rd., Code 693.0, Greenbelt, MD 20771, USA

2 California Institute of Technology, Pasadena, CA 91125, USA

3 Southwest Research Institute, Boulder, CO 80302, USA

4 Lowell Observatory, Flagstaff, AZ 86001, USA

5 Muon Space, Mountain View, CA 94043, USA

6 Johns Hopkins Applied Physics Laboratory, Laurel, MD 20723, USA

Table 1 Known or Suspected Trojan Satellites and Binaries

Object	d_2/d_1 (%)	a/r_{Hill} (%)	method	discovery
(617) Patroclus	92	1.7	imaging	Merline et al. (2001)
(640) Hektor	4.8	1.1	imaging	Marchis et al. (2006b)
(911) Agamemnon	*≥3*	*≥3*	occultation	Timerson et al. (2013)
(2207) Antenor	-	-	lightcurve	Stephens et al. (2018)
(3548) Eurybates	1.9	11.3	imaging	Noll et al. (2020b)
(15094) Polymele	*≥20*	*≥3*	occultation	Buie et al. (2022)
(16974) Iphthime	76	1	imaging	Noll et al. (2016)
(17365) Thymbraeus	*∼100*	*contact*	lightcurve	Mann et al. (2007)
(29314) Eurydamas	*∼100*	*contact*	lightcurve	Mann et al. (2007)

Estimated relative sizes and separations of components. Uncertain values are shown in italics

are evident (see Fig. 1). Widely separated systems with satellites at more than ten percent of the Hill radius are known, but the majority of satellites cluster closer to their primaries, at distances of a few percent of the Hill radius or less. Relative sizes also span the range from near-equal size binaries down to small satellites at observational limits. During the flybys, *Lucy* will be able to identify $d \geq 2$ km objects over the full Szebehely sphere, *i.e.* the inner third of the Hill sphere where satellites can have stable orbits.

Satellites are produced in several ways - systems may be primordial, the product of collisions and subsequent capture, or can arise from rotational fission. The three different formation processes are partially correlated with size and can be grouped by the total angular momentum of the system (Pravec and Harris 2007). Primordial binaries are a natural and expected outcome of the streaming instability which results in a gravitationally-bound collapse (Nesvorný et al. 2010, 2019). Kuiper Belt objects, especially the Cold Classicals, are the best examples of likely primordial objects that formed by co-accretion in the nebula and were not subsequently disrupted. Large Trojan and Outer Main Belt asteroids with similar-size binary components, including contact binaries, may also have formed in this way.

A second ubiquitous process that can result in the formation of satellites is collisional evolution. Collisions can result in one or more satellites over a wide range of primary size and relative component size (Durda et al. 2004). Collisions also play an important role in driving intermittent activity (Jewitt et al. 2015) and could be a source mechanism for transient rings or other orbital debris.

Finally, small asteroids can split by fission after being spun up by solar radiation pressure via YORP (see Vokrouhlický et al. 2015). This is the likely origin for most Near-Earth Asteroids (NEA) with satellites as well as the majority of small, rapidly rotating Main Belt asteroid systems (Margot et al. 2015). It remains to be seen whether this mechanism might be important for Trojans, both because few small Trojans are known and because YORP is much less effective at 5 AU (Kalup et al. 2021).

Satellite discoveries provide great scientific opportunities. The relative size, composition, and orbit properties of a satellite are all important constraints on its origin and the dynamical and collisional history of the system. Satellite orbits can be used to determine system mass which can then be combined with volume determinations to yield densities. The *Lucy* encounters provide a unique opportunity to search for satellites in orbits and size scales inaccessible to current Earth-based telescopes. Therefore, the search for satellites during Trojan encounters is a scientific priority for the *Lucy* mission.

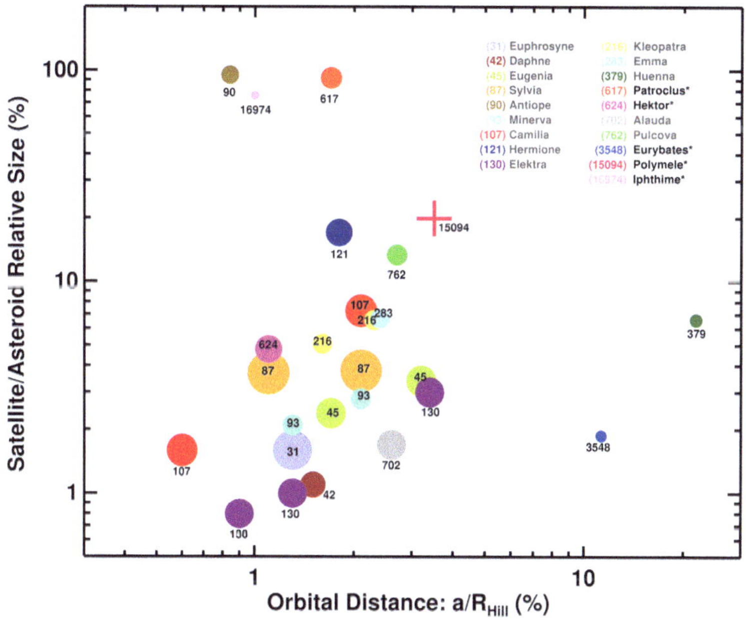

Fig. 1 The relative size and orbital distance of satellites of large, primitive, low albedo asteroids (C, P, and D spectral types) in the Main Belt (MB) and Trojans having primaries with volume-equivalent diameters $d_{veq} \geq 50$ km are compared (plus Polymele with $d_{veq} = 19$ km). Five Trojan (denoted by bold with * in the legend) and fourteen Main Belt systems are shown. The symbol size is proportional to the effective diameter of the system except for Polymele which is shown as a red cross because of the uncertainty in both the satellite size and separation (see text). Five of the nineteen systems shown consist of two satellites and one system has three; the orbital properties of each satellite relative to its primary are shown for these systems

Rings and other circum-body material may be considered as special cases of the more general topic of satellites. Rings are likely collisional in origin, but might also arise from other forms of mass loss. Of the four known small bodies with rings, none are Trojans. Asteroids are also known to exhibit activity driven by a range of processes including impact ejection and disruption, rotational instabilities, thermal fracture, and sublimation of volatile ices (Jewitt et al. 2015). To date, no such activity has been observed among the Trojans, but, observational biases may account for the lack of detection. Taken altogether, the possibility of rings and near-asteroid phenomena must be considered during any small body flyby.

In this chapter we focus primarily on the Jupiter Trojans, and specifically the targets of the *Lucy* mission. However, to understand the Trojans in context it will be important to look at other populations as well, especially carbonaceous (low-albedo, C-, X-, and D-type) asteroids in the Outer Main Belt that potentially share similar origin and evolutionary tracks with the Trojans.

2 Earth-Based Observations

The vast majority of known satellites of small bodies (see Johnston 2019 for compilation) have been discovered from ground-based and Earth-orbiting observatories using one or more distinct observational techniques. Search methodologies that have yielded satellite detections include direct imaging, stellar occultations, and lightcurve analysis. All of these observing techniques have been used in searches of Trojans, but only a small number of satellites or binary systems have been identified. It remains to be seen whether this represents a fundamental difference in this population compared to other small body populations, or, more likely, is an artifact of observational biases, as described below.

2.1 Direct Imaging

The most straightforward method by which to identify satellites is with direct imaging. The relevant parameters governing detectability are the angular separation of the bodies, their relative brightness, and the mutual orbit period and orientation. The sub-arcsecond angular scale of most known small-body satellite systems and the occurrence of large brightness differences between primary and secondary means that observations from the Hubble Space Telescope (HST) or from large ground-based telescopes with adaptive optics are the most productive search techniques. Such searches in the Main Belt and the Kuiper Belt have yielded the detection of many satellite and binary systems (Noll et al. 2008; Margot et al. 2015; Noll et al. 2020a).

Imaging searches of Jupiter Trojans have led to the discovery of four satellites/binaries (Table 1). Two systems were detected from the ground with adaptive optics (AO) systems - (617) Patroclus has a near-equal mass binary companion, Menoetius (Merline et al. 2001; Buie et al. 2015) and (624) Hektor is a probable contact binary with a small satellite, Skamandrios (Marchis et al. 2006b, 2014). Two additional systems have been found with HST - (16974) Iphthime was found to be a partially-resolved, near-equal mass binary in a search for possible synchronous binaries among Trojans with long rotation periods (Noll et al. 2016), and (3548) Eurybates was found to have a small, widely-separated satellite, Queta, in a deep search of the *Lucy* mission targets (Noll et al. 2020b; Brown et al. 2021). Other searches have had mixed results: a search of 35 moderate-size Trojans with HST's Advanced Camera for Surveys High Resolution Camera did not find any companions, while ground-based observations of a "few dozen" larger Trojans discovered only one, Patroclus (Merline et al. 2001, 2007). Marchis et al. (2006a) reported observations of 20 Trojans brighter than 18th mag in R-band (approximately $d > 30$ km) with only the detection of the satellite of Hektor (Marchis et al. 2006b). Unfortunately, the negative searches of Trojans from ground-based AO systems are not well-documented - neither target lists nor observing details are available - so the significance of the non-detections is difficult to evaluate. In hindsight, however, the non-detections are not especially constraining because, even with Hubble or ground-based AO observations, the expected angular separations of satellites falls below detection limits for all but the largest objects and/or unusually wide satellite separations, even in very deep images. As shown in Fig. 1, the satellite of Eurybates, Queta, is relatively wide compared to the orbits of comparable Main Belt objects, with the exception of the very wide satellite of (379) Huenna. At equivalent diameters of 69.8 km and 57.3 km respectively, the Trojans Eurybates and Iphthime are smaller than all of the Main Belt asteroids where satellites have been directly imaged from Earth-based observatories. As the Hill radius shrinks proportionally with less massive systems, closer and/or smaller satellites are rendered undetectable, even with the superior resolution and contrast of HST.

Main Belt (MB) asteroids with satellites are useful for bounding the kinds of systems the *Lucy* spacecraft might encounter. Among MB asteroids of similar spectral types (Tholen types C-, P-, and D-) approximately 5% of objects with diameters of $d \geq 85$ km have been found to have satellites (Fig. 1). It is worth noting that, if these MB systems were moved to the same distance as the Trojans, many would be undetectable from Earth-based AO or HST. As this figure shows, for this subset of asteroids, satellites are mostly found at a few percent or less of the Hill radius with more weakly-bound, distant satellites being rare. We also note the high prevalence of multiple satellite systems - five of the MB asteroids host two satellites each and a sixth, (130) Elektra, previously known to have two satellites, has recently been found to have a third (Berdeu et al. 2022). The MB asteroids with multiple satellites are also the largest bodies in collisional families, a correlation that is highly suggestive of a collisional origin for these satellite systems. Likewise, among the Trojans, we note that Eurybates is the largest collisional family known and similarly conclude that its satellite Queta likely originated from the family-forming event. Orbital analysis and photometric observations point to another potential catastrophic family in the Trojans, that of (4709) Ennomos (Nesvorný et al. 2015; Wong and Brown 2023), which, however, has yet to be searched for potential satellites. Thus, based on the prevalence and architecture of known satellite systems, the detection of close-in satellites and multiple satellite systems must be considered in planning the *Lucy* flybys.

2.2 Stellar Occultations

Stellar occultations also have the potential to identify satellites. Timerson et al. (2013) summarize previous reports of possible visual detections and present evidence of a secondary event recorded in an occultation of the Trojan (911) Agamemnon. The projected sky-plane separation was 278 km and the diameter of the satellite was estimated to be 5 km. A satellite this small and close to its primary is not detectable by direct imaging with AO systems or by using HST, leaving stellar occultation as the only means to detect it. With sparse chord sampling, the probability of serendipitous detection is low and suggests that close-in satellites may not be rare among the Trojans.

In preparation for the *Lucy* flybys, there has been an intensive campaign to observe stellar occultations of the *Lucy* mission targets (Buie et al. 2021; Keeney et al. 2021, 2022) with the primary goal of refining size and shape estimates needed to plan the spacecraft encounters. Unexpectedly, during an occultation by Polymele on 27 March, 2022, observers on two adjacent chords detected a secondary event corresponding to a 5-6 km object at a projected separation of 204 km (Buie et al. 2022). The satellite was observed again on 03 February, 2023 in another stellar occultation, this time with a massive deployment of over 90 telescopes tightly spaced across the track, intentionally arranged to redetect the satellite (Buie et al. 2023; Levison et al. 2023). As the discovery of Polymele's satellite demonstrates, intensive observations with close spacing of telescopes across the shadow path can be a successful technique for the discovery of satellites. However, the logistical challenges of this method have, so far, limited its widespread application.

2.3 Lightcurves

Satellites can also be found by detecting the occurrence of mutual events or from the analysis of the lightcurve shape, amplitude, and period (Tedesco 1979; Pravec et al. 2002). The appearance of multiple, asynchronous periods in a lightcurve is one indicator that can be interpreted as evidence of a binary system (Mottola and Lahulla 2000; Pravec et al. 2006). The analysis of lightcurve periods for populations can also reveal patterns that may be interpreted as evidence of binaries. In particular, Trojans show a marked excess of slow rotators (French et al. 2015; Ryan et al. 2017; Kalup et al. 2021) which may be attributable to tidally synchronous binaries (Nesvorný et al. 2020). At the other extreme, there is an apparent spin barrier at a $P \sim$ 4-5 hr (Pravec and Harris 2000; Melita et al. 2010; Ryan et al. 2017; Chang et al. 2021) signaling a maximum rotation rate that is consistent with the observed low bulk densities for the Patroclus and Eurybates (Buie et al. 2015; Noll et al. 2020b). Further acceleration of bodies rotating near the spin barrier leads to the formation of binaries by fission.

The mechanism for acceleration of rotation rates at Trojan distances is not clear as solar radiation that powers the YORP effect in the Main Asteroid Belt is significantly reduced; instead, collisions may play a dominant role.

Large amplitude, slow rotators have been interpreted as being consistent with close- or contact-binary configurations (Leone et al. 1984; Sheppard and Jewitt 2004). Mann et al. (2007) identified two Trojans, (17365) Thymbraeus and (29314) Eurydamas, as likely contact binaries where they also note that the primary component of Hektor meets this criterion as well. Sonnett et al. (2015) used sparse photometric data from WISE to identify 34 Trojans with large amplitude lightcurves out of a sample of 953 Trojans as well as 48 of 554 Hildas. The high fraction of candidates is consistent with other lines of evidence that suggest close binaries may be a relatively common feature of the Trojans, particularly at smaller diameters.

When asteroid satellite systems are suitably aligned, mutual events are observable and can yield the relative sizes of the components in addition to the orbital period. The observation of mutual events accounts for a significant fraction of systems known in the Main Belt and NEA populations because it can be carried out with telescopes with modest apertures. Stephens et al. (2018), Stephens and Warner (2019) report the observation of five or more possible mutual events in the lightcurve of (2207) Antenor observed from Jan. 23-May 22, 2018 and another in March 2019. They were, however, unable to extract a consistent orbital period from the observed events and so are unable to estimate the size of the secondary. The utility of this technique remains mostly unrealized for Trojans, primarily because they are significantly fainter than their Main Belt counterparts owing to their larger heliocentric distances and because the smaller-diameter asteroids that account for the majority of Main Belt and NEA systems remain largely unidentified among the Trojans.

2.4 Orbit Determination and Tidal Evolution

Once satellites are detected by any of the methods described above, the next observational goal is to establish the mutual orbit. This can be accomplished with repeated detections of the satellite with sufficiently accurate relative astrometry to allow for orbit solutions. In practical terms this usually requires at least four additional detections but can require many more (Grundy 2012). In cases where the number of detections is insufficient to uniquely determine an orbit, other orbital constraints can be applied. One of the most useful is the state of tidal evolution of the system which can potentially constrain the orbital eccentricity. In cases where the shape of the primary deviates significantly from spherical, the mutual orbit plane can also sometimes be inferred.

Mutual tides in binary systems lead naturally to a synchronous, circular-orbit end state. The time scale for eccentricity evolution is very strongly dependent on the semimajor axis of the system, a, scaled by the radius of the satellite, R_2, with $\tau_e \propto (a/R_2)^5$. The timescale is also directly proportional to Q/k, the tidal quality factor scaled by the tidal Love number (Goldreich and Sari 2009). However, because tidal dissipation for porous, rubble-pile-like structures is complex and poorly constrained, the tidal synchronization formulae are most useful in predicting the functional sensitivity, but less so in determining absolute timescales. Fortunately, empirical data from known binary systems can be used to constrain tidal evolution timescales. Kuiper Belt binaries span the transition from eccentric to circularized (Noll et al. 2020a) and show a very regular behavior, transitioning from circularized orbits to random eccentricities at $a/R_{avg} \approx 55$, where R_{avg} is the average of the binary component radii (Nesvorný et al. 2020). The trend is similar, but less clearcut, among the Main Belt systems shown in Fig. 1. This may be due to the prevalence of multi-satellite systems where additional orbital resonances come into play.

Table 2 Orbits of Trojan Satellite Systems

Secondary	a (km)	e	T (days)	a/R_{avg}	reference
(617) I Menoetius	688.4 ±4.7	0.0043 ± 0.0049	4.28270±0.00007	12.6	Grundy et al. (2018)*
(640) I Skamandrios	623.5 ±10	0.31 ± 0.03	2.9651 ± 0.0003	9.5	Marchis et al. (2014)
(3548) I Queta	2321.2 ± 35	0.098 ± 0.009	82.47± 0.06	131	Brown et al. (2021)**
S/2022 (15094) 1	*≥204*	*0*	*14-17*	*34*	Buie et al. (2022)
S/2016 (16974) 1	*≥205*	*0*	3.288 ± 0.016	*10.1*	Noll et al. (2016)

Estimated or assumed values shown in italics. The satellites of (15094) Polymele and (16974) Iphthime do not have established orbits. The eccentricity of both, however, is assumed to be near zero due to the expected short tidal evolution timescales in both cases. The orbital period of (16974) Ipthime is assumed to be the same as the observed lightcurve period Mottola et al. (2011). The range of orbital periods for Polymele's satellite is for density from 700-1000 kg/m^3 (Levison et al. 2023) *Unconstrained orbit fit **Updated with additional observations obtained in December 2022

Among the Trojans (see Table 2), the Patroclus-Menoetius binary, with $a/R_{avg} \approx 6$, is fully circularized and synchronous (Grundy et al. 2018). Queta orbits Eurybates at $a/R_{avg} \approx 131$, a distance where it would not be expected to be circularized, consistent with its observed eccentricity of $e = 0.125 \pm 0.009$ (Brown et al. 2021). If Polymele's recently discovered satellite's observed separation of 204 km represents its semimajor axis, it would have $a/R_{avg} \approx 34$, in the range where it would also be expected to be tidally evolved on timescales of order 200 Myr or less. Absent a recent formation or other factors, its yet-to-be-determined orbit is likely to be nearly circular. The close binary Iphthime should be tidally evolved, and multi-epoch HST observations are consistent with this (Noll & Grundy, in preparation). The orbit of Hektor's satellite stands out as an exception. Despite having $a/R_{avg} \approx 9.5$, its orbit is substantially eccentric at $e \sim 0.3$. The satellite orbit is also inclined relative to the primary's equator with $i \sim 50$ (Marchis et al. 2014). It is possible, however, that both unusual features may result from forcing by the elongated, bilobed Hektor primary (Jiang et al. 2018).

Besides mutual tides of the primary/satellite system, it is also important to consider orbital evolution from orbital perturbations caused by the Sun. For example, for the Eurybates-Queta system, a Kozai resonance could force coupled oscillations of the eccentricity and inclination over surprisingly short timescales of ≈ 500 years (Brown et al. 2021). The precise mutual orbit that will be established by the *Lucy* flyby will set the stage for testing whether or not this resonance comes into play in this system.

2.5 Rings and Activity

Rings have been found or have been suggested to be present around four small bodies, the Kuiper Belt objects Haumea and Quaoar, and the Centaurs Chariklo and, possibly, Chiron (Ortiz et al. 2015; Morgado et al. 2023; Braga-Ribas et al. 2014; Ortiz et al. 2017). The evidence for rings comes from stellar occultations that are sensitive to the largest and most optically thick examples of rings. Whether similar structures exist around smaller and less active bodies such as the Trojans is currently a matter for speculation that can be directly addressed by the *Lucy* flybys.

Activity is an umbrella term that covers a wide range of potentially time-variable sources of material found in proximity of any given small body. In addition to satellites and rings, asteroids may intermittently release material from the surface as the result of collisions or the activation of subsurface volatiles. So-called Main Belt comets (MBC) are bodies with asteroid-like orbits that show comet-like characteristics activated by a collision (Jewitt et al. 2015). No examples of this phenomenon have been detected among the Trojans despite the fact that the Trojans are likely to be compositionally similar to MBCs. The lack of detected activity may be the result of their colder surface temperatures and correspondingly lower sublimation rates, or due to the lower collision rates in the Trojans, or simply due to observational biases. It is worth noting that Chiron and a handful of other Centaurs are known to exhibit episodic activity (e.g., Jewitt 2009) despite the fact that their surfaces are generally even colder than Trojan surfaces and collision rates are also likely lower. Centaur activity might suggest that Trojans are more depleted in near-surface volatiles than either MBCs or Centaurs.

3 Observations of Lucy Targets

Given the widespread evidence of satellites, rings, and activity in asteroid populations, along with the diagnostic scientific value of satellite dynamical and physical properties, the search for and study of these phenomena was identified as one of the top scientific priorities of the *Lucy* mission (Levison et al. 2021). *Lucy*'s close flybys of five Trojan systems puts it in a unique position to make observations that are difficult or impossible to make in any other way. Beforehand, there are ground- and space-based observations that can help optimize *Lucy*'s flyby sequences. At all steps of the process leading up to the flybys, careful planning is of paramount importance.

Satellite-focused observations of the *Lucy* targets can be divided into two main categories - searches for currently unknown satellites, rings, or other near-asteroid phenomena - and observations of currently known satellites. We discuss each below.

3.1 Searches for Unknown Satellites

Lucy will search for currently unknown satellites and rings or other orbital debris with direct imaging. How best to conduct an imaging search involves tradeoffs between sensitivity, coverage, angular resolution and timing. *Lucy* has a requirement to be capable of conducting a search for a 2 km diameter or larger object with an albedo of $p = 0.04$ or greater anywhere in the stable portion of the Hill sphere (Levison et al. 2021). Because such a search covers such a large scale in distance and, at the same time, requires fine resolution and high contrast close to the primary, the search is broken up into two phases as described below.

The volume of space that must be searched is defined by the region around each target where satellites and rings can have stable orbits. This is usually parameterized by the Hill radius where $R_{Hill} = a\ (m_1/m_{Sun})^{1/3}$ where a is the semimajor axis of the binary and m_1 is the mass of the primary; the Hill radius thus scales with the effective radius of the primary. Hamilton and Burns (1991) showed that some orbits, particularly retrograde orbits, can be exist at separations up to $\sim$ 1/2 R_{Hill}, however, there are few known examples of such weakly bound satellites, possibly because the lifetime of any such systems is short. Most orbit at smaller fractions of the Hill radius. On the other hand, the existence of asteroid pairs (Pravec et al. 2019) - formerly bound small asteroids that now orbit the Sun independently - attests to the fact that some very wide systems must exist at times, if only briefly. Fortunately, satellites at such wide separations from the primary can be found with Earth-based observatories such as HST or Keck at the required sensitivity or better. Thus, the Szebehely radius, 1/3 R_{Hill}, where satellites are most stable, is the region of interest for satellite searches by *Lucy* (Table 3, Fig. 2).

Table 3 Szebehely Radius and Angular Size for *Lucy* Targets

Object	Mass[a] (10^{15} kg)	r_{sz} (10^3 km)	range[b] (10^6 km)	t±CA[c] days
(3548) Eurybates	151(3)	6.93	2.52	5.12
(15094) Polymele	21.1(1)	2.18	0.79	1.47
(11351) Leucus	40.4(4)	3.73	1.36	3.21
(21900) Orus	50.8(8)	6.91	2.51	4.68
(617) Patroclus	1410(29)	13.9	5.01	11.2

Uncertainty in final digit(s) shown in parentheses [a]Eurybates and Patroclus use system masses derived from satellite orbits (Noll et al. 2020b; Grundy et al. 2018), other objects use assumed density of 1000 kg^{-3} and effective diameters of 21, 34, and 63 km for Polymele, Leucus, and Orus respectively [b]The range at which the angular size of the Szebehely sphere equals 5.5 mrad, the size of the L'LORRI field of view [c]Time from closest approach when Szebehely sphere exceeds the L'LORRI field of view

As detailed in Table 4, deep searches around *Lucy* mission targets have been carried out with HST at two or more epochs and already constrain the presence of distant satellites at separations encompassing the full Szebehely sphere (Noll et al. 2018). The sensitivity of these HST observations exceed that needed to reach the $d = 2$ km requirement. However, at separations of $s < 0.5$ arcsec, where the background is determined by the PSF of the primary, the sensitivity to km-scale satellites rapidly declines. Equal-diameter binaries are detectable at significantly smaller separations of $s > 0.07$ arcsec. Intermediate cases fall between these extremes, although it is worth noting that the satellite of Polymele (plotted in Fig. 1) would not be detectable with HST due to the small diameter of the primary and consequent small angular scale of the Hill sphere. Thus, a complete search satisfying the mission requirement will rely on observations to be made by the *Lucy* spacecraft during each flyby.

Satellite searches with *Lucy* can be attempted once a potential satellite would be bright enough to be detectable with one of the *Lucy* instruments. L'LORRI is the most sensitive instrument for this measurement and will be the primary instrument used for searches. As shown in Fig. 3, L'LORRI has sufficient sensitivity to meet this requirement for approximately one month before or after close approach (CA). This leaves a large window when observations can be scheduled while optimizing coverage and resolution. L'LORRI has two observing modes – in 1 × 1 mode each pixel is read out for the highest resolution. In 4 × 4 mode, individual pixels are binned to larger superpixels resulting in higher sensitivity at the expense of resolution. Detection limits for specified exposures times in the two modes are shown in Fig. 3. Observations from *Lucy* will switch from deep exposures in the 4 × 4 mode to shorter exposures in 1 × 1 mode with when the expected brightness of a satellite crosses the detection threshold for 1 × 1 mode. The timing of this switchover varies depending on the size of the target and details of the encounter geometry (see Fig. 3).

In order to differentiate a satellite from a background source, there must be detectable motion of the *Lucy* target and any potential bound satellites relative to distant stars. The apparent rate of Trojan targets as seen from the *Lucy* spacecraft is a function of both the satellite's orbital motion and the spacecraft trajectory. Thus, a successful search strategy requires repeated observations over multiple days while at large distances with increasingly smaller time steps between observations nearer to close approach as the viewing geometry changes more rapidly. *Lucy* will carry out its search using just such a cadence.

Lucy encounters its Trojan targets at phase angles ranging from 63° to 126°, with the phase angle of egress complementing that of the ingress. The lower phase angle leg is the most favorable for satellite searches because the greater fraction of illuminated surface as

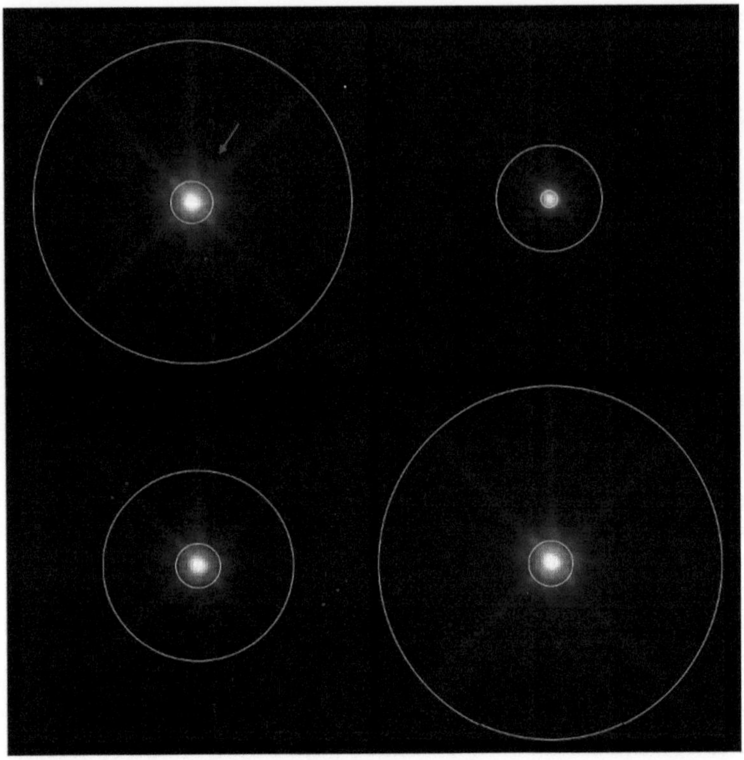

Fig. 2 Four ~5 arcsec square regions are shown - clockwise from the upper left, Eurybates, Polymele, Orus, and Leucus. Images are median combined WFC3 images shown in a logarithmic stretch at a common scale with the central region saturated to bring out faint detail in the wings of the PSF. Pixels are ~40 mas on a side. The Szebehely radius for a density of $\rho = 1500$ kg/m^3 and the geocentric distance at the time of observation is shown in green. The smaller blue circle centered on the target has a radius of 1000 km for Eurybates, Orus, and Leucus, 400 km for Polymele. The red arrow points at Eurybates' 1.2 ± 0.4 km satellite Queta, barely visible at this stretch (Noll et al. 2020a). Other small image artifacts are not repeatable in individual frames

Table 4 HST Satellite Search of *Lucy* Targets

Object	dates		Object	dates	
Donaldjohanson	09 Jan	2021	Orus	07 Aug	2018
Eurybates[a]	12 Sep	2018		08 Aug	"
	14 Sep	"		11 Sep	"
Polymele	13 Sep	2018		12 Sep	"
	14 Sep	"		13 Sep	"
Leucus	17 Jun	2017		20 Sep	"
	30 Aug	2018	Patroclus	13 Feb	2018
	01 Sep	"		15 Feb	"

[a] Additional observations of Eurybates were obtained following the discovery of Queta as described in Noll et al. (2020b) and Brown et al. (2021)

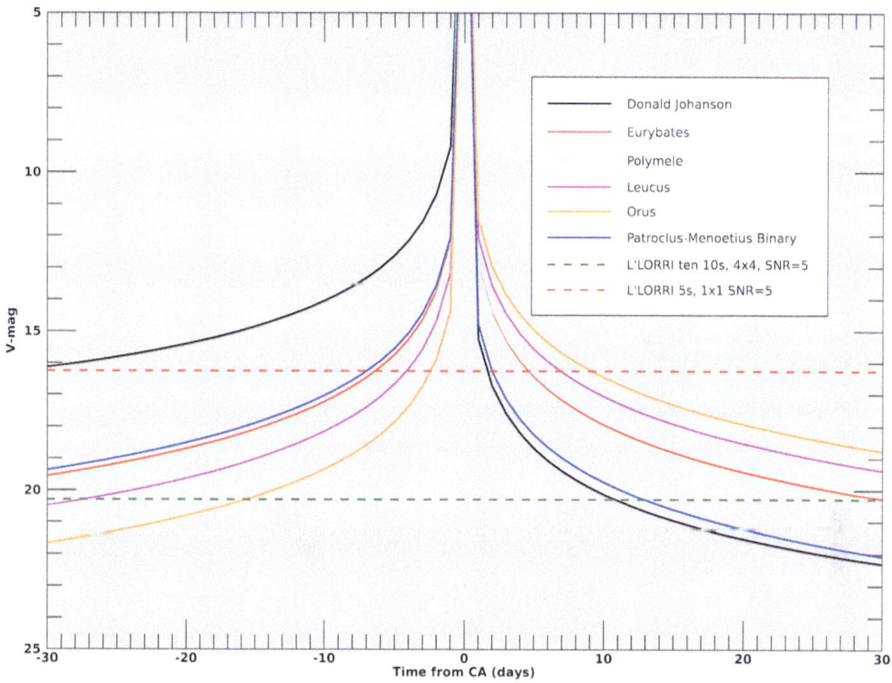

Fig. 3 The brightness of a 2 km diameter satellite as seen by *Lucy* as a function of time from close approach is shown. The satellites are assumed to have an albedo of p = 0.04 and a phase function constant of 0.04 mag/deg. Dashed curves show detection limits for ten 10 s exposures using a 4 × 4 L'LORRI superpixel (green dashed) and a single 5 s exposure in 1 × 1 pixel mode (red dashed)

seen from the spacecraft results in a brighter target. The low phase angle leg occurs on ingress for Eurybates, Polymele, and Patroclus. At Orus and Leucus, the phase angle is lowest on egress. The prime satellite search will be scheduled for the optimal leg of the encounter. However, in order to provide backup redundancy, a smaller set of satellite search observations will also be scheduled for the less-favorable encounter leg. An added benefit of this built-in redundancy will be improved orbit determination for any existing or newly discovered satellites.

3.2 Observations of Known Satellites

Three of the *Lucy* targets are known to have a companion. One of these, the near-equal-mass Patroclus-Menoetius binary, was chosen as a mission target, in part, because it was known to be a binary that shares characteristics with many primordial binary systems in the Kuiper Belt (Buie et al. 2015; Levison et al. 2021). The satellites of Eurybates and Polymele were discovered after the selection of the *Lucy* mission as a result of the observations undertaken in preparation for the mission (Noll et al. 2018, 2020b; Brown et al. 2021; Buie et al. 2022). During the flybys, observations of each of these companions will be planned as part of the encounter observing sequence.

The state of knowledge of binary or satellite orbit is the primary driver of what observations can be carried out during encounters. For the Patroclus-Menoetius binary, which *Lucy* will fly by in March 2033, the orbital phase of the binary pair must be precisely known in order for both binary components to be observable during the close approach. The last planned

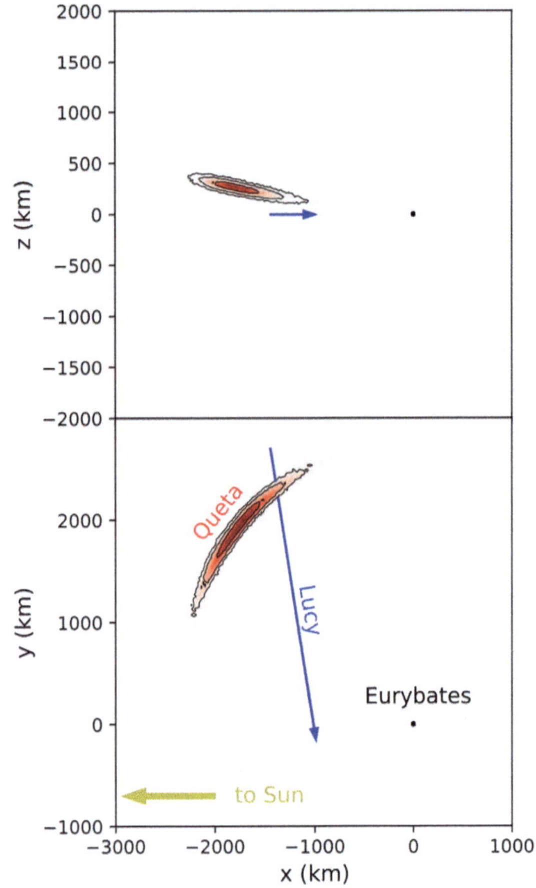

Fig. 4 Fig. 3 from Brown et al. (2021) is reproduced here showing the Eurybates flyby trajectory of the *Lucy* spacecraft and the predicted location of the satellite Queta. The top panel show the view in the flyby plane (near-ecliptic) looking away from the Sun. The lower panel shows the view looking down from the plus-z axis (near the north ecliptic pole) with the Sun to the left. In both, the motion of the *Lucy* spacecraft from t-500 s to close approach is shown by the blue arrow. The 1-, 2-, and 3-σ error ellipses for Queta's position are shown

deep space maneuver in mid-2029 will set *Lucy* on its trajectory to the binary and lock in the timing of the encounter, thus, the state of knowledge of the binary orbit at that time is critical. In order to recover the lost phasing of the binary Grundy et al. (2018) completed a series of observations using HST that successfully refined the orbit to sufficient precision to allow predictions for mutual events that occurred from 2017 to 2019 (see chapter by Mottola et al. in this volume for discussion of the use of mutual events as a probe of object shapes). The timing of mutual events, in principle, offer a significant advantage in constraining the mutual orbit compared to imaging observations. The next mutual event season for the Patroclus-Menoetius binary will take place in 2024. Stellar occultations that detect both binary components are another technique that can address this problem (Keeney et al. 2022). Event prediction and planning that enable additional observations will be needed up to the 2033 encounter to ensure the complex *Lucy* encounter with this binary object can be successfully completed.

During the flyby of the Eurybates system, *Lucy* will have a very close encounter with the satellite Queta as shown in Fig. 4. However, because *Lucy* is passing Queta on the anti-sunward side, it is unlikely that *Lucy*'s instrument platform range of motion will allow for an observation near the time of closest approach. Observations will be planned when *Lucy* is at a larger separation where the geometry is more favorable and where Queta's positional uncertainty can be covered with the L'LORRI field of view or by a small mosaic. Additional pre-encounter astrometric observations of Queta can further reduce its positional uncertainty enabling observations at closer range and higher spatial resolution.

The orbit of Polymele's satellite is the least well-known of the three, having been detected only twice. However, because of the inherent precision of the two stellar occultation detections, one or two additional detections could be enough to yield an orbit, especially if the orbit has near-zero eccentricity. Either another stellar occultation detection or early detections from *Lucy* during approach may be enough to allow for a late target update that will enable resolved observations of the satellite during the flyby. An important added benefit of tracking the orbit of Polymele's satellite will be to obtain an independent measure of the system mass. Mass determination from the radio science spacecraft system will be carried out for all of the *Lucy* targets. But because Polymele is the smallest Trojan primary that *Lucy* will encounter, the mass determination by this method will likely be less precise than can be obtained from the satellite orbit.

3.3 Searches for Rings and Activity

Lucy will search for evidence of Trojan rings, other dust assemblages, and for Trojan activity on each of its flybys, primarily using high-phase-angle imaging to search for coma particulates. *Lucy* will also use its L'Ralph MVIC instrument Violet filter channel to search for coma gas due to activity; MVIC's Violet filter bandpass includes the 388.3 nm CN resonance line that is often used to trace gas activity in comets. The unusually high fluorescence efficiency (g-factor) of CN makes this species detectable in even minute quantities of gas. Apart from observations *Lucy* will make on its specific flyby targets, it would also be useful to conduct groundbased search for activity using CN filters and for both activity and for rings using stellar occultations on a wide variety of Trojans.

Lucy will have a unique opportunity to search close-in to the bodies it will encounter. The region of stable satellite orbits extends inwards to the Roche limit at $R_{Roche} \approx 2.44\ r_1$ where r_1 is the radius of the parent body, much closer than can be accessed by any ground- or space-based observatory (with the exception of stellar occultation observations). Thus, there exists a significant region where satellites could exist that is largely unexplored. Rings are also situated at small separations from their parent body. The known rings around three of the four asteroids have $r_{ring} \approx 3\ r_1$ (Ortiz et al. 2015; Braga-Ribas et al. 2014; Ortiz et al. 2017) while the recently identified ring around Quaoar is at $r_{ring} \approx 7.4\ r_1$ (Morgado et al. 2023). Satellites are typically found at $a \approx 5 - 15\ r_1$ with a few found closer or farther. *Lucy* will fly by well within the Hill sphere of each Trojan target with close approach distances of $s_{min} \geq 20\ r_1$ giving it the opportunity to probe portions of the Hill sphere that are otherwise difficult or impossible to access. We note that *Lucy*'s flybys at distances of $\sim$ 450-1100 km are likely too distant to be able to observe phenomena such as the flurries of cm-scale particles that were seen episodically by OSIRIS-REx at Bennu (Lauretta et al. 2019). However, the appearance of unexpected activity is a possibility that nevertheless must be considered when planning *Lucy* observations.

3.4 Main Belt Asteroid Encounters

The *Lucy* mission will also fly by two Main Belt asteroids on the way to the Trojans (152830) Dinkinesh and (52246) Donaldjohanson. While these objects are not part of the formal science requirements and are primarily intended to test critical spacecraft systems, observations will nonetheless allow for a number of scientific investigations, including a search for satellites. Interestingly, Donaldjohanson has an extremely high amplitude lightcurve of 1.7 mag and a long period of 251 hr (Levison et al. 2021), consistent with a possible synchronous binary or contact binary configuration. A single epoch of deep HST imaging was obtained near lightcurve maximum on 09 January 2021, but did not resolve a detached binary. This is, however, only a weak constraint because the most likely angular separation for a synchronous secondary would be below HST's resolution limit. The HST observations were unfortunately obtained away from opposition due to a guide star failure during the initial attempt resulting in a reduced spatial resolution. *Lucy* imaging during a successful encounter with Donaldjohanson will directly constrain the configuration of this object. Dinkinesh is a much smaller target with a likely diameter near 1 km and a 0.39 ± 0.02 mag lightcurve amplitude with a period of 52.67 ± 0.04 hours (Mottola et al. 2023). Imaging observations taken as part of optical navigation will also be useful for identifying any possible satellites.

4 Summary

As the number of known satellites of asteroids has grown, their importance and utility in understanding the formation and evolution of small-body populations, and the solar system more generally, has become apparent. Satellites are known in almost every small body population despite observational limitations that leave significant portions of the Hill sphere out of reach of Earth-based searches. Because of their distance and relatively small sizes compared to Main Belt and Kuiper Belt objects, the Jupiter Trojans are one of the least-well explored of the major small body populations. Despite this, satellites have been detected directly and there are indications of many more in the population as a whole. The *Lucy* mission will encounter five Trojans, with at least three of them having a binary companion or a satellite. Two Main Belt asteroids will also be encountered on the way to the Trojans and one of these is a candidate contact binary. The flybys will provide opportunities for close-up studies of these objects, and may reveal the presence of more satellites that have remained undetected. *Lucy* will also be able to search for rings and activity - two less-common, but possible phenomena. Up to now, the only small bodies with satellites that have been studied *in situ* by spacecraft are Ida and its satellite Dactyl in the Main Belt, the Pluto-Charon binary with its retinue of smaller satellites, and Arrokoth, a possible primordial contact binary. *Lucy* will study objects that may include both primordial binaries and collisional satellite systems, thereby significantly expanding our knowledge of this aspect of the planetary system.

Funding The *Lucy* mission is funded through the NASA Discovery program on contract No. NNM16AA08C.

Declarations

Competing Interests The authors declare that they have no competing interests.

References

Belton MJS, Mueller BEA, D'Amario LA et al (1996) Icarus 120:185. https://doi.org/10.1006/icar.1996.0044

Berdeu A, Langlois M, Vachier F (2022) Astron Astrophys 658:L4. https://doi.org/10.1051/0004-6361/202142623

Braga-Ribas F, Sicardy B, Ortiz JL et al (2014) Nature 508:72. https://doi.org/10.1038/nature13155

Brown ME, Levison HF, Noll KS et al (2021) Planet Sci J 2:170. https://doi.org/10.3847/PSJ/ac07b0

Buie MW, Olkin CB, Merline WJ et al (2015) Astron J 149:113. https://doi.org/10.1088/0004-6256/149/3/113

Buie MW, Keeney BA, Strauss RH et al (2021) Planet Sci J 2:202. https://doi.org/10.3847/PSJ/ac1f9b

Buie M, Keeney B, Levison H, Olkin C, Lucy Occultations Team (2022) In: AAS/Division for planetary sciences meeting abstracts, vol 54, 512.03

Buie M, Keeney B, Levison H (Lucy Occultations Team) (2023) In: Asteroids, comets, meteors conference 2023, p 2442

Chang C-K, Chen Y-T, Fraser WC et al (2021) Planet Sci J 2:191. https://doi.org/10.3847/PSJ/ac13a4 arXiv: 2107.06685

Durda DD, Bottke WF, Enke BL et al (2004) Icarus 167:382. https://doi.org/10.1016/j.icarus.2003.09.017

French LM, Stephens RD, Coley D, Wasserman LH, Sieben J (2015) Icarus 254:1. https://doi.org/10.1016/j.icarus.2015.03.026

Goldreich P, Sari R (2009) Astrophys J 691:54. https://doi.org/10.1088/0004-637X/691/1/54

Grundy WM (2012) In: Arenou F, Hestroffer D (eds) Orbital couples: pas de deux in the solar system and the Milky Way, pp 13–17

Grundy WM, Noll KS, Buie MW, Levison HF (2018) Icarus 305:198. https://doi.org/10.1016/j.icarus.2018.01.009

Hamilton DP, Burns JA (1991) Icarus 92:118. https://doi.org/10.1016/0019-1035(91)90039-V

Jewitt D (2009) Astron J 137:4296. https://doi.org/10.1088/0004-6256/137/5/4296

Jewitt D, Hsieh H, Agarwal J (2015) The active asteroids. In: Bottke WF, DeMeo FE, Michel P (eds) Asteroids IV. University of Arizona Press, Tucson, pp 221–241. https://doi.org/10.2458/azu_uapress_9780816532131-ch012

Jiang Y, Baoyin H, Li H (2018) Adv Space Res 61:1371. https://doi.org/10.1016/j.asr.2017.12.011

Johnston WR (2019) Binary minor planets compilation v3.0. NASA Planetary Data System. https://doi.org/10.26033/bb68-pw96

Kalup CE, Molnár L, Kiss C et al (2021) Astrophys J Suppl Ser 254:7. https://doi.org/10.3847/1538-4365/abe76a

Keeney B, Buie M, Kaire M et al (2021) In: AGU fall meeting abstracts, vol 2021, P32B–03

Keeney B, Buie M, Levison H (Lucy Occultations Team) (2022) In: AAS/Division for planetary sciences meeting abstracts, vol 54, 512.04

Lauretta DS, Dellagiustina DN, Bennett CA et al (2019) Nature 568:55. https://doi.org/10.1038/s41586-019-1033-6

Leone G, Paolicchi P, Farinella P, Zappala V (1984) Astron Astrophys 140:265

Levison H, Buie M, Keeney B, Mottola S (Lucy Occultations Team) (2023) In: Asteroids, comets, meteors conference 2023, p 2184

Levison HF, Olkin C, Noll K, Marchi S, Bell JF, Bierhaus E et al (2021) Planet Sci J 2:171. https://doi.org/10.3847/PSJ/abf840

Mann RK, Jewitt D, Lacerda P (2007) Astron J 134:1133. https://doi.org/10.1086/520328

Marchis F, Berthier J, Wong MH et al (2006a) In: AAS/Division for planetary sciences meeting abstracts, vol 38, 65.07

Marchis F, Wong MH, Berthier J et al (2006b) IAU Circ 8732:1

Marchis F, Durech J, Castillo-Rogez J et al (2014) Astrophys J Lett 783:L37. https://doi.org/10.1088/2041-8205/783/2/L37

Margot JL, Pravec P, Taylor P, Carry B, Jacobson S (2015) Asteroid systems: binaries, triples, and pairs. In: Bottke WF, DeMeo FE, Michel P (eds) Asteroids IV. University of Arizona Press, Tucson, pp 355–374. https://doi.org/10.2458/azu_uapress_9780816532131-ch019

Melita MD, Duffard R, Williams IP et al (2010) Planet Space Sci 58:1035. https://doi.org/10.1016/j.pss.2010.03.009

Merline WJ, Close LM, Siegler N et al (2001) IAU Circ 7741:2

Merline WJ, Weidenschilling SJ, Durda DD et al (2002) Asteroids do have satellites. Bottke Jr WF, Cellino A, Paolicchi P, Binzel RP (eds) Asteroids III. University of Arizona Press, Tucson, pp 289–312

Merline WJ, Tamblyn PM, Dumas C et al (2007) In: AAS/Division for planetary sciences meeting abstracts #39, 60.09

Morgado BE, Sicardy B, Braga-Ribas F et al (2023) Nature 614:239. https://doi.org/10.1038/s41586-022-05629-6
Mottola S, Lahulla F (2000) Icarus 146:556. https://doi.org/10.1006/icar.2000.6421
Mottola S, Di Martino M, Erikson A et al (2011) Astron J 141:170. https://doi.org/10.1088/0004-6256/141/5/170
Mottola S, Denk T, Marchi S et al (2023) Mon Not R Astron Soc. https://doi.org/10.1093/mnrasl/slad066
Nesvorný D, Youdin AN, Richardson DC (2010) Astron J 140:785. https://doi.org/10.1088/0004-6256/140/3/785
Nesvorný D, Brož M, Carruba V (2015) Identification and dynamical properties of asteroid families. In: Bottke WF, DeMeo FE, Michel P (eds) Asteroids IV. University of Arizona Press, Tucson, pp 297–321. https://doi.org/10.2458/azu_uapress_9780816532131-ch016
Nesvorný D, Li R, Youdin AN, Simon JB, Grundy WM (2019) Nat Astron 3:808. https://doi.org/10.1038/s41550-019-0806-z
Nesvorný D, Vokrouhlický D, Bottke WF, Levison HF, Grundy WM (2020) Astrophys J Lett 893:L16. https://doi.org/10.3847/2041-8213/ab8311
Noll KS, Grundy WM, Chiang EI, Margot JL, Kern SD (2008) Binaries in the Kuiper Belt. In: Barucci MA, Boehnhardt H, Cruikshank DP, Morbidelli A (eds) The solar system beyond Neptune. University of Arizona Press, Tucson, p 345
Noll KS, Grundy WM, Ryan EL, Benecchi SD (2016) In: Lunar and planetary science conference, p 2632
Noll K, Grundy W, Buie M, Levison HF, Marchi S (2018) In: AAS/Division for planetary sciences meeting abstracts #50, 217.04
Noll K, Grundy WM, Nesvorný D, Thirouin A (2020a) Trans-neptunian binaries (2018) Elsevier, Amsterdam, pp 201–224. https://doi.org/10.1016/B978-0-12-816490-7.00009-6
Noll KS, Brown ME, Weaver HA et al (2020b) Planet Sci J 1:44. https://doi.org/10.3847/PSJ/abac54
Ortiz JL, Duffard R, Pinilla-Alonso N et al (2015) Astron Astrophys 576:A18. https://doi.org/10.1051/0004-6361/201424461
Ortiz JL, Santos-Sanz P, Sicardy B et al (2017) Nature 550:219. https://doi.org/10.1038/nature24051
Pravec P, Harris AW (2000) Icarus 148:12. https://doi.org/10.1006/icar.2000.6482
Pravec P, Harris AW (2007) Icarus 190:250. https://doi.org/10.1016/j.icarus.2007.02.023
Pravec P, Harris AW, Michalowski T (2002) Asteroid rotations. In: Bottke Jr WF, Cellino A, Paolicchi P, Binzel RP (eds) Asteroids III. University of Arizona Press, Tucson, pp 113–122
Pravec P, Scheirich P, Kušnirák P et al (2006) Icarus 181:63. https://doi.org/10.1016/j.icarus.2005.10.014
Pravec P, Fatka P, Vokrouhlický D et al (2019) Icarus 333:429. https://doi.org/10.1016/j.icarus.2019.05.014
Richardson DC, Walsh KJ (2006) Annu Rev Earth Planet Sci 34:47. https://doi.org/10.1146/annurev.earth.32.101802.120208
Ryan EL, Sharkey BNL, Woodward CE (2017) Astron J 153:116. https://doi.org/10.3847/1538-3881/153/3/116
Sheppard SS, Jewitt D (2004) Astron J 127:3023. https://doi.org/10.1086/383558
Sonnett S, Mainzer A, Grav T, Masiero J, Bauer J (2015) Astrophys J 799:191. https://doi.org/10.1088/0004-637X/799/2/191
Stephens RD, Warner BD (2019) Minor Planet Bull 46:315
Stephens RD, Pravec P, Kuèáková H et al (2018) Minor Planet Bull 45:341
Tedesco EF (1979) Science 203:905. https://doi.org/10.1126/science.203.4383.905
Timerson B, Brooks J, Conard S et al (2013) Planet Space Sci 87:78. https://doi.org/10.1016/j.pss.2013.08.015
Vokrouhlický D, Bottke WF, Chesley SR, Scheeres DJ, Statler TS (2015) The Yarkovsky and YORP effects. In: Bottke WF, DeMeo FE, Michel P (eds) Asteroids IV. University of Arizona Press, Tucson, pp 509–531. https://doi.org/10.2458/azu_uapress_9780816532131-ch027
Weidenschilling SJ, Paolicchi P, Zappala V (1989) Do asteroids have satellites? In: Binzel RP, Gehrels T, Matthews MS (eds) Asteroids II. University of Arizona Press, Tucson, pp 643–658
Wong I, Brown ME (2023) Astron J 165:15. https://doi.org/10.3847/1538-3881/ac9eb3

Space Science Reviews (2025) 221:94
https://doi.org/10.1007/s11214-025-01190-6

The Lucy Spacecraft

Matthew Cox[1] · **Cavan Cuddy**[1] · **Jeremiah Schneider**[1] · **Tom Kennedy**[1] · **Jonathan Taylor**[1] · **Greg Bollendonk**[1]

Received: 2 April 2025 / Accepted: 24 June 2025 / Published online: 14 October 2025

Abstract
The Lucy spacecraft is developed from a combination of heritage components used on other NASA deep space missions, combined with a set of newly developed hardware specific to Lucy's mission, most critically the solar arrays. These components are configured into a spacecraft capable of launching on the least expensive Atlas launch vehicle, deploying into a power-safe configuration, executing the high-precision Trojan asteroid encounters, and surviving the 12-year mission timeline.

1 Introduction

Prior to the Lucy Mission, no spacecraft in NASA's Discovery Program had travelled more than 2.72 AU from the Sun (Blume 2005) and no solar powered spacecraft had ever travelled beyond the orbit of Jupiter (Bolton et al. 2017). In order to survey the Jupiter Trojan asteroids (Levison et al. 2025), Lucy required a spacecraft with the necessary power and propulsive capability to support operation in the outer solar system, while minimizing launch mass. Lockheed Martin, having a long history of planetary spacecraft configurations and hardware to choose from, drew on its expansive inventory of qualified deep space hardware and software elements to assemble the Lucy spacecraft system. Table 1 shows the most critical components and their development origin.

At launch, the spacecraft mass was 1500 kg, of which 712 kg is spacecraft dry mass, 50 kg is instrument dry mass, 438 kg is fuel, 297 kg is oxidizer, and 3 kg is helium pressurant.

2 Spacecraft Configuration

The Lucy spacecraft is comprised of five major elements, each accommodating a specific operational purpose. These five elements are assembled independently, and then combined and tested during the Assembly and Test phase, lasting the final 17 months of development prior to launch (Table 2).

Figure 1 thru 3, below, show the primary elements and overall size of the Lucy spacecraft.

Technical data about the Lucy Spacecraft is extracted from the Lucy Critical Design Review, conducted in October 2019.

✉ G. Bollendonk

[1] Lockheed Martin Space, Littleton, CO, USA

Table 1 Sources of Lucy spacecraft heritage components

Component/Subsystem	Heritage
Structure	2001 Mars Odyssey Spacecraft (Saunders et al. 2004)
Bi-Propellant Propulsion System	2001 Mars Odyssey Spacecraft (Saunders et al. 2004)
Electrical Solar Array Design	2011 Juno Mission to Jupiter (Bolton et al. 2017)
Mechanical Solar Array Design	2007 Mars Phoenix Lander Mission (Smith et al. 2008)
High Gain Antenna	2013 Mars MAVEN Mission (Jakosky et al. 2015)
Core Avionics	2011 Juno Mission to Jupiter (Bolton et al. 2017)
Instrument Pointing Platform	2013 Mars MAVEN Mission (Jakosky et al. 2015)
Instrument Pointing Platform Gimbal	2016 GOES-R Weather Spacecraft
Guidance, Navigation, Control Components	Commercial Products
Propulsion Components	Commercial Products
Flight Software	Lockheed Martin Incremental Development over 20+ years of deep space missions

Table 2 Lucy spacecraft subassemblies

Element	Location	Purpose
Propulsion Module	Lower Half of Main Structure	Contain Propulsion Subsystem, Distribute Launch Loads
Equipment Module	Upper Half of Main Structure	Contain avionics and harnessing
Instrument Pointing Platform	Platform on top of Main Structure	Contain science instruments and navigation cameras, track Trojan asteroids during encounters
7.3 m Deployable Solar Arrays	Appendages on either side of main structure	Provide solar power throughout mission
2 m High Gain Antenna	Fixed dish antenna on one side of main structure	Provide high-rate communication to earth

3 Solar Arrays

Lucy utilizes "Ultraflex" solar arrays developed and manufactured by Northrop Grumman. The solar arrays deliver about 500 watts of power at the far sun range of the Trojan asteroid encounters (5.7 earth distances from the sun) and weigh 176 kg.

The solar cells are mounted on a mesh material formed by a group of composite spars radially connected to a rotating hub. After launch, the mechanisms rotate the assembly away from the spacecraft body, deploy the spars around the hub, and tension and latch the mesh elements into stiff solar panels.

Figure 4, below, illustrates the solar array deployment sequence.

3.1 Power Subsystem

The Lucy power subsystem consists of three main elements: Power Generation, Power Storage, and Power Distribution.

Power generation is provided by the most prominent feature of the Lucy spacecraft, two large 7.3 meter diameter solar arrays attached to the spacecraft's main body. Lucy will venture farther from the sun than any other solar-powered vehicle in history. Although the solar

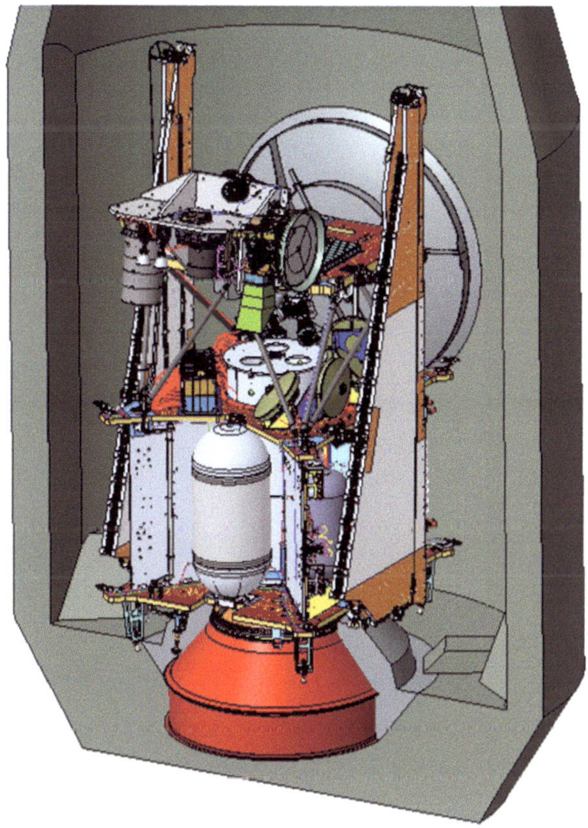

Fig. 1 Lucy Launch Configuration (inside Atlas-V Fairing)

arrays are capable of producing up to 18,000 Watts in Earth orbit (enough to power several homes), at Jupiter's orbit, Lucy will only generate 500 Watts (enough to almost power a hair dryer) due to the reduced solar irradiance at that distance. Solar irradiance, defined as power per unit area received from the Sun, reduces by the inverse square of the distance ($1/D^2$) from the Sun, so power diminishes rapidly as we travel into deep space at nearly 6 times the distance of Earth from the Sun. Lucy instruments, electronics, and thermal system have been carefully designed to consume as little power as possible in order to meet mission needs without breaking the 500 Watt power budget.

During certain mission activities, including launch and the Trojan asteroid encounters, the solar arrays cannot be continuously pointed at the sun, requiring stored energy to be supplied. When this occurs the battery comes into action. The single Lucy battery is a light-weight design which uses an internally redundant small cell format (slightly larger than AA sized cells) strung up in an 8-series and 16-parallel string configuration to provide a nominal bus voltage of 32 V and an energy storage capacity of 44 amp-hours. With this much energy the battery can power the spacecraft on its own for about 3 hours.

In order to get power from the solar arrays to the batteries and electronic loads, two electronics boxes are utilized, the Power Distribution and Power Unit (PDDU) and the Pyro and Propulsion Unit (PAPU).

The PDDU consists of circuit card assemblies to manage solar array and battery power, switch on/off power to instruments, heater and other various loads, collect telemetry from

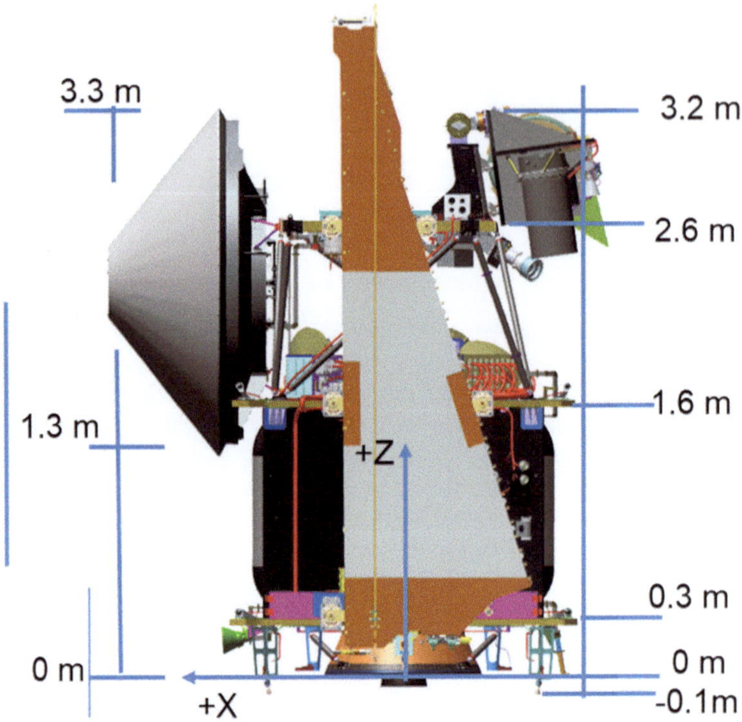

Fig. 2 Lucy Launch Configuration Dimensions

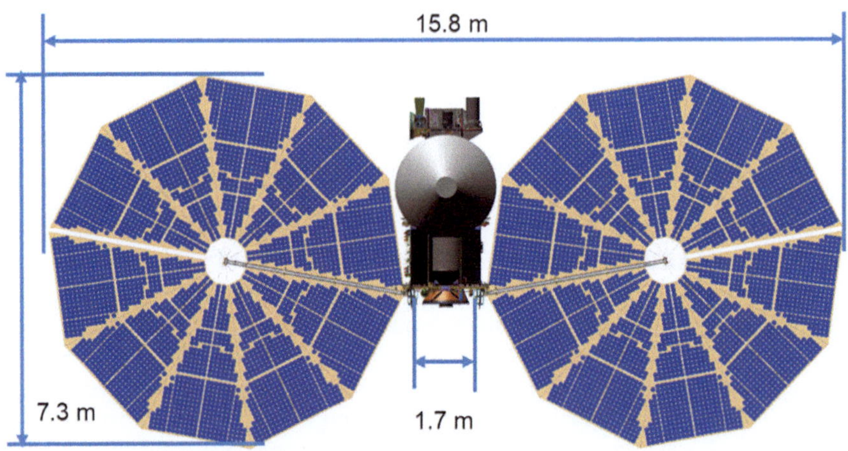

Fig. 3 Lucy In-Flight Configuration and Dimensions

the spacecraft, maintain housekeeping power for the PDDU and PAPU, and to provide a main power and grounding bus for the entire spacecraft.

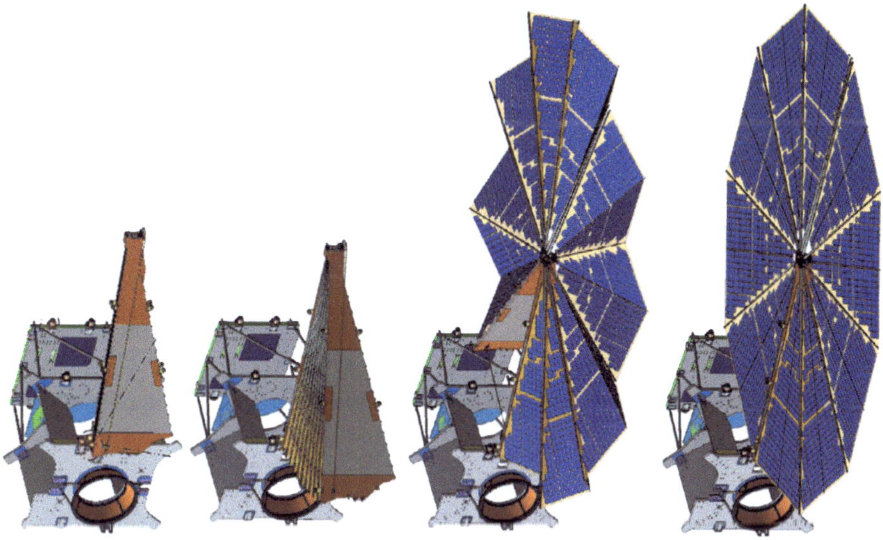

Fig. 4 Lucy Solar Array Deployment Sequence from stowed to fully deployed

The PAPU consists of four circuit card assemblies to provide energy to the pyrotechnic devices and valve control for the propulsion subsystem when commanded by the flight software.

Both the PDDU and PAPU have a high amount of heritage as versions of the units were used on other NASA missions including Juno, MAVEN, and OSIRIS-REx.

4 Instrument Pointing Platform

The Instrument Pointing Platform (IPP) provides volume accommodation and pointing of the three science instruments, L'Lorri, L'Ralph and L'TES, as well as the Terminal Tracking Camera (TTCam) assemblies that GN&C use to track the asteroid. The IPP is constructed as a carbon fiber sandwich panel, which is mounted to a two axis gimbal to steer the platform's pointing.

The primary design consideration for the platform structure was co-alignment of each instrument. Instrument and IPP interface design had to account for thermal distortion, mounting pattern clearance and overall machining tolerances. The instruments are co-aligned to within about 0.5 milliradian. This angle is equivalent to the angle between the top and bottom of a silver dollar coin, when viewed from 100 yards away.

The science instrument boresights must remain steadily pointed at the asteroid targets while Lucy flies by. To maximize observational flexibility, the gimbals are used to point the instruments independent of the spacecraft orientation. Not only was instrument coalignment a big driver in design, so was instrument jitter, or how much the instruments shake due to the gimbals and other moving items on the spacecraft. The Lucy gimbals shake the instruments less than 5 microradians; or to put it another way, the angle formed by the top and bottom of a quarter when viewed from 3 miles away.

The gimbals are a heritage design, tracing their lineage through the GOES weather satellites and several missions that are orbiting Mars. The smooth output motion of the gimbal is due to its high gear reduction zero-backlash harmonic drive, and a very fast feedback control loop of the motor. A high bandwidth control loop within the motor electronics ensures the gimbal is moving in the way the spacecraft has commanded, while an outer loop on the spacecraft monitors the gimbal positions and adjusts gimbal commands as necessary to maintain the desired pointing.

5 Guidance, Navigation, and Control Subsystem

Lucy's Guidance, Navigation and Control (GNC) subsystem is responsible for determining the attitude and pointing of the spacecraft and IPP, controlling the attitude of the spacecraft and instrument pointing platform orientation, and controlling the firing of the thrusters to change the trajectory of the spacecraft. Lucy is a 3-axis attitude stabilized spacecraft, meaning the GNC subsystem controls how all three axes of the spacecraft are oriented in inertial space.

For attitude determination (AD) Lucy employs gyroscopes, star trackers, and sun sensors. The gyros are used to measure the relative rotation of the spacecraft. Even if the spacecraft doesn't know its exact orientation in space it can use the gyro rate measurement to maintain control of the spacecraft and move about to find the sun or to find stars with the star tracker. To determine the spacecraft absolute orientation in inertial space, Lucy uses a star tracker with two cameras. The star tracker cameras take pictures of deep space, compare the stars it sees in the images to known stars in a catalog, and provides the spacecraft with its attitude orientation in inertial space. The final component used in determining the spacecraft attitude are the 4 head sun sensors which detect where the sun is relative to the spacecraft and are used to point the spacecraft solar arrays at the sun if the star trackers are not able to see stars.

For attitude control (AC) Lucy uses a variety of actuator components. For nominal control, Lucy uses a suite of three reaction wheels, which use the principle of "equal and opposite action" to point the spacecraft. Applying torque to the reaction wheels produces an equal and opposite torque on the spacecraft body, turning it in a desired direction. Lucy uses three orthogonal reaction wheels for control, though they are not co-aligned with the spacecraft body axes. A fourth, spare wheel is not co-aligned with any of the other three, allowing it to be used for three axis control in case of failure of one of the main wheels. The reaction wheels allow the spacecraft to pitch over and track the Trojans at rates up to 1 degree per second during encounters.

For attitude control during spacecraft emergencies and for changing the trajectory of the spacecraft, GNC controls the three types of thrusters on Lucy. The eight ACS (Attitude Control System) thrusters arranged such that a set of four thrusters can be fired to produce a pure torque about a principal axis of the spacecraft while producing very small residual force on the spacecraft. The six TCM (Trajectory Correction Maneuver) thrusters are used to impart small to medium forces on the spacecraft to change its trajectory and the single Main Engine is used for large trajectory corrections of the spacecraft, velocity changes greater than 50 meters per second.

During the flyby encounters with the Trojan Asteroids, and the rehearsal flyby of the Main Belt Asteroids Dinkinesh and Donaldjohanson, the GNC subsystem uses the AD and AC to coarsely point the spacecraft at the asteroids, pitching about the Lucy Y axis as it passes by. GNC also controls the pointing of the Instrument Pointing Platform (IPP) for high accuracy pointing of the science instruments, to perform off-points and scans with the instruments, and, for the binary encounters, to rapidly retarget between the two asteroids.

Due to the large distances from Earth that the Trojan encounters occur, about 5.7 AU, the knowledge of where the Trojans are relative to the spacecraft will not be well known as the spacecraft approaches the targets. Even using the L'LORRI instrument in the months leading up to the encounters to image and improve the location knowledge, there can still leave hundreds of kilometers of location uncertainty during the encounter. With the Trojans being about 15-100 km in diameter, this uncertainty could cause the science instruments to completely miss the target at the ideal opportunity to image and collect data. For this reason, Lucy employs a Terminal Tracking System (TTS) to autonomously image the Trojans in the final hours of approach and to update the onboard knowledge of the target locations without intervention from the ground operators.

The Terminal Tracking System is comprised of a Terminal Tracking Camera (TTCam) to image the Trojans in the final hours and minutes of close approach, as well as on-board software to update the knowledge location of the targets. The redundant TTCams' field of view are wide enough to image the entirety of the largest of the Trojans at their closest approach. The TTCam pixel resolution of 74 micro-radians is fine enough to allow it to resolve the smallest of the Trojans within the final hour of its closest approach.

Lucy uses a combination of Centroiding, State Estimation and Target Reference Generation software to update an onboard model of where the target will be relative to the spacecraft at any given time.

The first part of the TTS software, Centroiding, finds the target's bright pixels in the image, determines the center of brightness (COB) of the brightly lit area of the target, and passes that information, along with an estimate of where the TTCam was inertially pointing at the time of the image, along to the next step in the TTS software, the Encounter State Estimator (ESE). The Encounter State Estimator, and the Binary Encounter State Estimator used for the binary encounters, uses the Centroiding measurement of the Trojan location as inputs to its Kalman Filter, which mathematically converges on the unknown flyby parameters. These flyby parameters, the close approach distance, time of closest approach, and out of plane location of the target, are used by the final piece of software in TTS, the Target Reference Generator (TRG).

The TRG uses the ESE converged flyby parameters to provide a more accurate estimate of the location of both the Center of Mass (COM) of the Trojan at any point in time during the encounter as well as an estimate for where the Center of Brightness will be at any point in time. Either of these estimates, COM and COB, can be used to point the IPP and science instruments at the targets at any time during the close flyby. In general, COM is used when the full target is illuminated or when trying to image multiple portions of the target. Center of Mass or Center of Brightness pointing modes can be used by the science team depending on the exact observation. For wider angle observations, it is often best to use center of mass observation, while for high resolution narrow angle observations, it is better to target the bright surface.

While the TTS method of imaging and updating an encounter target's flyby states has been used on other flyby missions, such as Stardust, Lucy will be the first mission to fly by a binary system and not only update the flyby conditions but also autonomously estimate the orbital parameters of the binary system. As Lucy approaches the Patroclus and Menoetius near equal mass binary the two targets will initially both be in the field of view of TTCam and the science instruments. However, during the hours and minutes leading up to close approach, the angular separation between the two Trojans will increase, necessitating that the IPP move back and forth between them to obtain the desired science. The special binary version of ESE, BESE, uses images taken during the hours leading up to the close approach to update its onboard model of the binary system orbit parameters, feeding this to TRG in addition to its flyby estimates.

6 Main Structure

The Lucy spacecraft is Lockheed Martin's seventh generation composite interplanetary spacecraft and consists of a 26-inch diameter core cylinder with an integral launch vehicle adapter, aft and middle decks that interface to gusset panels tied to the core, and a forward deck supported by struts from the mid-deck. The aft and mid-decks provide radial shear load paths into the cylinder, while gussets serve to transfer vertical loads into the core cylinder. In addition to the primary structure, there is a variety of secondary structures on the spacecraft such as component mounting brackets and the solar array hinge fittings.

The Lucy spacecraft main structure is primarily constructed from lightweight composite materials. The structure weighs a mere 135 kg and supports the entire spacecraft wet mass of 1500 kg (3307 lbs.). The efficient spacecraft structure accounts for an amazingly small (9%) fraction of the spacecraft mass.

6.1 Propulsion Subsystem

The spacecraft propulsion system consists of the following key components, all of which are readily available within industry, with the exception of the fuel and oxidizer tanks, which are developed and built custom for Lucy:

- Two fuel tanks
- One oxidizer tank
- One helium pressurant tank
- One bi-propellant main engine assembly
- Six monopropellant trajectory correction maneuver thrusters
- Eight monopropellant attitude control thrusters
- One mechanical pressure regulator
- Numerous latch valves, pyro valves, service valves, and filters

The propulsion system stores the fuel and oxidizer needed for all propulsive events over the life of the mission. The Lucy mission plan defined five deep space maneuvers utilizing the bi-propellant main engine, with a longest burn up to 2600 seconds. The propulsion system provides (rotational) attitude control with small thruster forces and (translational) velocity forces to accomplish trajectory correction.

6.2 Thermal Subsystem

The thermal subsystem maintains the Lucy spacecraft and its components within temperature specifications in the harsh environment of space across an extreme range of environments. A combination of temp sensors, heaters, multi-layer insulation blankets, special coatings, heat spreaders and isolators are leveraged on Lucy. The maximum flight system heater power utilization is projected to be less than 192 Watts.

6.3 Electrical Harness

The total length of the Lucy spacecraft electrical harness is a little over 2 miles of wire (average length of wire of about 15 to 20 feet). The biggest section of the harness bundle is about 4-inches in diameter at the splice area near the C&DH. One of the biggest challenges was optimizing harness routing to minimize 60 kgs (131 lbs.) wire mass.

The spacecraft electrical harness has 364 connectors and over 1300 splices, with 12,847 electrical connections. There are 8 field joints and 40 field joint connectors. The aluminum foil overwrap weighs about 13 pounds.

7 Computation, Fault Protection, & Communications

7.1 Command & Data Handling Subsystem

The Lucy spacecraft contains two fully redundant flight computers, or Command & Data Handling (C&DH) units. The C&DH is responsible for all on-board computing logic, component communication, and data storage. The C&DH and related flight software control all aspects of spacecraft operation, relying on a raditioned hardened RAD750 PowerPC processor.

Different than terrestrial computers, space-rated processors must tolerate an abundant spectrum of harmful high energy particles regularly found in the space environment. The central processor is complemented with dozens of radiation hard ASICs and FPGAs to provide reliable computing for the dozen year mission.

7.2 Flight Software & Fault Protection

The C&DH houses flight software that is responsible for all aspects of spacecraft control. Onboard sequencing with high-level language constructs achieve a programmable spacecraft without reloading flight software. The spacecraft flight software contains about 350 K lines of code logic (with more than 1M total lines of code). The combination of target tracking and operational sequence programmability enables multiple encounters over mission life.

The flight software also includes logic to monitor, detect and respond to any on-board anomaly without ground controller intervention. The spacecraft is designed to operate autonomously (with no mission operations controller intervention) for up to 30 days and thus requires a complex system to detect and respond to any issue that may arise.

7.3 Telecom Subsystem

Lucy's X-band telecommunication system is responsible for transmitting spacecraft telemetry and receiving commands from the Mission Operations Center via the Deep Space Network. Lucy uses a 25 Watt tube amplifier and a 2 meter diameter high gain antenna to transmit data to Earth with data rates ranging from 2 kbps to 352 kbps. Lucy encounter science data return rates range from 10 kbps to 50 kbps and requires weeks of near continuous communications to receive the total data set. The spacecraft also contains a low gain antenna and medium gain antenna which provide communication coverage over the wide range of Earth distances and Sun, Earth, and spacecraft geometry. One way light time (the speed of light) can be up to 55 minutes at far Earth ranges during Lucy's mission.

8 Conclusion

Like all planetary spacecraft, Lucy is designed to perform the scientific mission while minimizing mass and providing all necessary power and control resources to obtain the great science results. Through extensive use of qualified components combined into subsystems and a configuration supporting Lucy's unique Trojan encounters at outer-planet sun ranges, the Lucy spacecraft meets its requirements within the cost requirements of a Discovery-class mission. Key contributors to Lucy success builds upon prior successes of using solar arrays beyond 2 AU, GN&C algorithms to autonomously execute asteroid tracking during encounters, and in-flight programmability to accomplish science goals.

Acknowledgements The Lucy spacecraft is only possible due to the dedicated work of the hundreds of people and many orgnaziations involved in designing and building the spacecraft under the leadership of Hal Levison, Principal Investigator.

Declarations

Competing Interests The author affirms that there is no conflict of interest and that the subject of this manuscript did not involve research on human nor animals.

References

Blume WH (2005) Deep impact mission design. Space Sci Rev 117:23–42. https://doi.org/10.1007/s11214-005-3386-4

Bolton SJ, Lunine J, Stevenson D, et al (2017) The Juno mission. Space Sci Rev 213:5–37. https://doi.org/10.1007/s11214-017-0429-6

Jakosky BM, Lin RP, Grebowsky JM, et al (2015) The Mars atmosphere and volatile evolution (MAVEN) mission. Space Sci Rev 195:3–48. https://doi.org/10.1007/s11214-015-0139-x

Levison HF, Marchi S, Noll KS, et al (2025) NASA's Lucy Mission to trojan asteroids: unraveling the history of the outer Solar System. Space Sci Rev 221. https://doi.org/10.1007/s11214-025-01173-7

Saunders R, Arvidson R, Badhwar G, et al (2004) 2001 Mars Odyssey mission summary. Space Sci Rev 110:1–36. https://doi.org/10.1023/B:SPAC.0000021006.84299.18

Smith PH, et al (2008) Landing site characterization experiments, mission overviews, and expected science. J Geophys Res 113:E00A18. https://doi.org/10.1029/2008JE003083

Space Science Reviews (2024) 220:47
https://doi.org/10.1007/s11214-024-01082-1

Mission Design and Concept of Operations for the Lucy Mission

Catherine Olkin[1] · Michael Vincent[2] · Coralie Adam[3] · Kevin Berry[4] · Jacob Englander[5] · Sheila Gray[6] · Hal Levison[2] · Julien Salmon[2] · John Spencer[2] · Dale Stanbridge[3] · Brian Sutter[6]

Received: 5 July 2023 / Accepted: 27 May 2024 / Published online: 11 June 2024

Abstract
The Lucy mission is NASA's 13th Discovery-class mission and the first mission to the Trojan asteroids. The spacecraft conducts flybys of 8 Trojan asteroids over the course of 12 years. A series of 3 Earth Gravity Assists are used to increase the aphelion of the spacecraft's orbit and to target the final Trojan asteroid flyby. Over the course of 2 years the spacecraft conducts 4 flybys in the L4 swarm to explore 6 Trojan asteroids, which includes two small satellites. Near the end of the mission, Lucy flies past the near-equal size binary, Patroclus-Menoetius, in the L5 swarm. The concept of operations for the Trojan flybys invokes a standard timeline for spacecraft operations to allow a science sequence that is tailored to each Trojan asteroid. The concept of operations enables efficiency of observations and resiliency in the observing sequence to robustly meet the Lucy science requirements.

Keywords Asteroids · Spacecraft · Trajectory design · Space mission

1 Baseline Mission Design

1.1 Overview

The Lucy mission design was developed to provide close flyby observational opportunities of a diverse population of Jupiter Trojan asteroids. A 6-year resonant orbit, essentially a Jupiter Trojan cycler, with apoapsis ranges slightly above 5 AU, and periapsis ranges of slightly less than 1 AU, was adopted as the baseline skeleton profile from which the targeting of specific Trojans could advance.

✉ C. Olkin

1 Muon Space, Mountain View, CA, USA

2 Southwest Research Institute, Boulder, CO, USA

3 KinetX, Simi Valley, CA, USA

4 NASA Goddard Spaceflight Center, Greenbelt, MD, USA

5 Johns Hopkins University Applied Physics Laboratory, Laurel, MD, USA

6 Lockheed Martin, Littleton, CO, USA

Using the resonant orbit constraint, an initial set of targets chosen for their spectral and mass diversity was reduced to flybys of (3548) Eurybates, (21900) Orus in the L4 swarm, and (614) Patroclus/Menoetius in the L5 swarm with Earth flybys in December of 2024 and 2030. A search for additional targets that were in close proximity to the nominal trajectory was undertaken resulting in the inclusion of (15094) Polymele and (11351) Leucus to the L4 observations. In a similar manner, opportunities to observe a main belt asteroids to exercise the flight system and ground system prior to the L4 encounters were evaluated and resulted in the inclusion of (52246) Donaldjohanson and (152830) Dinkinesh.

Having designed the heliocentric Trojan encounter orbits, the task of attaching that orbit with a December 2024 EGA to a 2021 launch opportunity as required by the Discovery program Announcement of Opportunity commenced. Following a significant optimization undertaking, the design of a series of orbits that enabled both a 21-day launch period and a single heliocentric Trojan encounter profile was developed. This strategy placed the Lucy spacecraft onto the identical heliocentric trajectory following the 2nd Earth gravity assist regardless of launch date. This capability allows the spacecraft, operations, and science to focus their efforts on a sequence of very specific encounter characteristics with minimal variability.

This general mission architecture has essentially remained unchanged since selection (Fig. 1), allowing the Lucy spacecraft, instrumentation, and operations to advance in a linear fashion through the development cycles with minimal redesign due to mission changes.

The first encounter is a flight test of the system with the main belt asteroid Dinkinesh on November 1, 2023; the second encounter is with Donaldjohanson in April 2025. The early timing of these two encounters gives the operations team time to fully address any unexpected behavior prior to the first L4 Trojan encounter.

The Eurybates encounter on August 12 of 2027 has been enhanced by the discovery of a moon, Queta, orbiting the large Trojan (Noll et al. 2020; Brown et al. 2021). About 1 month later, a flyby of Polymele and its recently discovered satellite will be performed on September 15, 2027. Flybys of Leucus on April 18, 2028 and Orus on November 11, 2028 will conclude the L4 Trojan encounters.

Small deterministic Deep Space Maneuvers (DSM) and smaller Trajectory Correction Maneuvers (TCM) will be performed to encounter each object at the proper lighting and distance requirements. Lucy will flyby Earth on December 26, 2030 and use it to adjust the trajectory to encounter the near equal size binary pair of Patroclus/Menoetius on March 3, 2033. The mission is completed after transmission of the data following the Patroclus/Menoetius encounter, however Lucy will remain on its stable Trojan cycler orbit and return to the L4 and L5 swarms repeatedly over the next few million years.

1.2 Launch

Lucy launched on an Atlas V 401 with a Centaur upper stage, provided by United Launch Alliance (ULA). The launch mission design team was tightly integrated among GSFC, KinetX, Lockheed-Martin, the Launch Services Program (LSP) at KSC, and ULA. Lucy was designed with a primary launch period in 2021 and a backup in 2022. The primary launch period consisted of 23 daily launch windows, each approximately 60 minutes long, beginning on October 16th, 2021. The 2022 launch period was constructed by removing Earth gravity assist (EGA) 1 from the mission design and launching on any of 17 days surrounding the date corresponding to EGA1 in the nominal design.

Because Lucy visits ten small bodies in one mission, small changes in the launch mission design drive changes to the entire mission. However, it was not possible to formulate

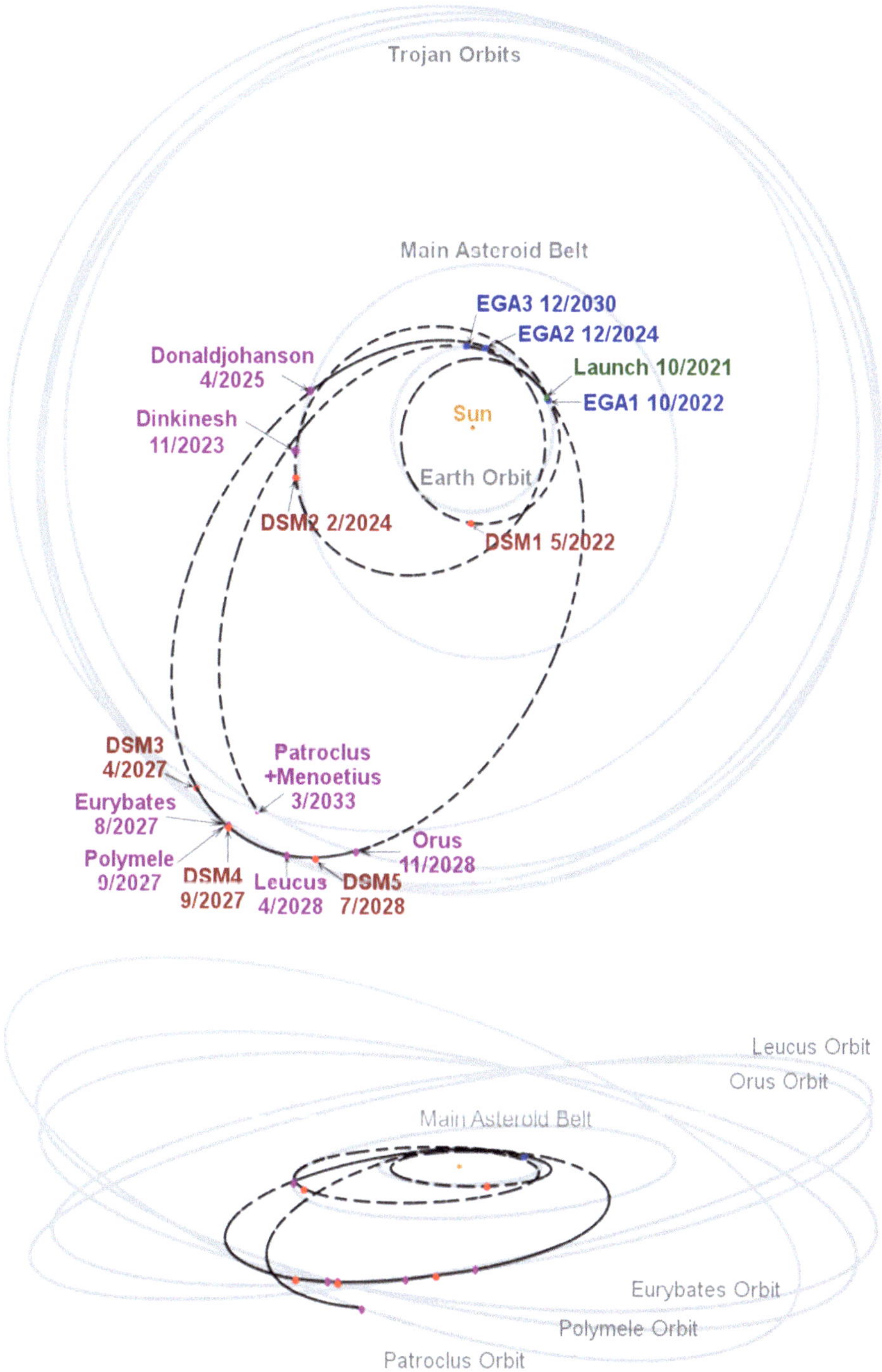

Fig. 1 The Lucy trajectory (black dashed line) and the orbits of Earth, the Main Belt Asteroid Donaldjohanson and the Lucy Trojan targets (gray solid lines) as seen from two different perspectives, a, and b

the entire mission design from the launch pad to Patroclus-Menoetius (PM) as a single optimization problem because there is no optimization software package that can properly model launch, atmospheric flight, and interplanetary flight. The Lucy team therefore chose to use the Evolutionary Mission Trajectory Generator (EMTG; Englander and Ellison 2020) to optimize the interplanetary trajectory and ULAs Spartan software to model the launch, with both tools modeling the hyperbolic injection maneuver.

The trajectory optimization was done in three stages. In the first stage, before ULA was selected as the launch partner, the mission was optimized entirely in EMTG and the launch vehicles ascent to the parking orbit was not modeled. Instead, the parking orbit was approximated as a circle with 28.5 degree inclination and passing over KSC. The injection burn was modeled as an impulse. EMTG was used to optimize the timing of this impulse, the right ascension of the ascending node of the parking orbit, and then the entire Lucy mission all the way to PM ten years later. This approach was sufficient to generate an initial set of launch targets twice the hyperbolic excess energy (C3), right ascension of the launch asymptote (RLA) and declination of the launch asymptote (DLA) for a single opportunity in each of 21 days, and to provide an optimized ΔV budget that changed by less than 1% in later iterations.

Once ULA joined the team, a second iteration was done to a higher level of fidelity. ULA generated optimal ascent trajectories to meet the launch targets for each of the 21 days, and generated Trajectory Interface Point (TIP) states for each launch opportunity. TIP takes place 20 minutes after the injection burn is completed and includes the small delta-v caused by the separation mechanism. GSFC then re-optimized the entire mission-to-go using EMTG. Since the original estimate of TIP time by GSFC in the first iteration was less accurate than the high-fidelity modeling of TIP by ULA, this resulted in an outbound trajectory that was sub-optimal. The TIP epoch mismatch of 20 minutes resulted in designs with 5 m/s of deterministic v at TCM-1, 30 days after launch.

The design process changed in the third iteration, both to remove the deterministic TCM-1 and to accommodate a new constraint that Lucy may not operate the main propulsion system closer than 0.89 AU to the Sun. The third iteration was similar to the second except that instead of re-optimizing the mission-to-go from the ULA-provided TIP state, GSFC used the parking orbit design from ULAs launch optimization as an initial condition for a larger optimization problem that included a finite burn model of the injection and then the rest of the mission all the way to PM. At this time an additional two days were added to the launch period, resulting in 23 opportunities. EMTG varied the initial epoch and duration of the injection maneuver but the direction was fixed along the velocity vector. New launch targets were harvested from this design and passed back to ULA. ULA performed a final round of launch trajectory optimization and provided new TIP epochs and states, which this time were within 17 seconds of the EMTG solution. A final round of EMTG optimization was performed using the TIPs as an initial condition, resulting in a reduction of TCM-1 to 1 m/s at the center of each daily window. The final targets are shown in Fig. 2.

ULA then generated optimized launch trajectories to the final target set for every five minutes within a two hour window centered on each days optimal time. This resulted in 575 TIP states and epochs, each of which was used to optimize an interplanetary trajectory. ULA also provided 7x7 Injection Covariance Matrices (ICMs) for each TIP. The Lucy navigation team at KinetX then rendered the entire post-launch trajectory and performed a statistical analysis in the MIRAGE-PIRATE (Knittel et al. 2019) toolchain using the TIP states and ICMs as an initial condition and the EMTG trajectory to provide targets.

Lucy launched on October 16th, 2021, at the first opportunity in the available window. The delivery was incredibly accurate, achieving the targeted C3 to within 0.003 km^2/s^2, RLA

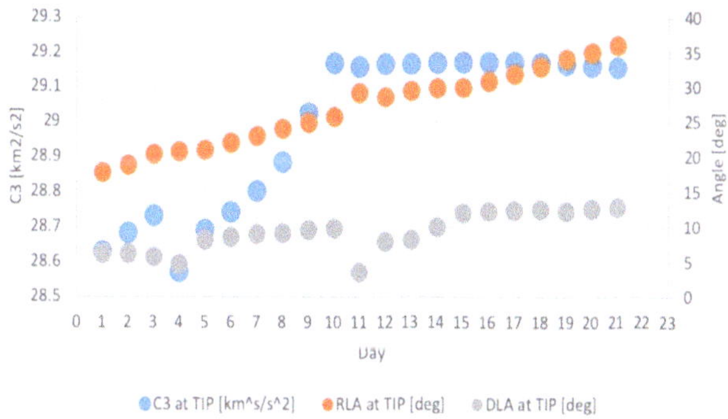

Fig. 2 C3, RLA, and DLA across 23 launch opportunities from October 16, 2021 to November 7, 2021

within 0.002 deg, and DLA within 0.003 deg. As a result, we were able to cancel the first two trajectory correction maneuvers that had been scheduled for November and December of 2021.

1.3 Earth Gravity Assist 1

The pre-launch trajectory optimization found that the optimal date for the first Earth Gravity Assist (EGA1) was one year after launch. Since Lucy launched on the first day of the launch window, October 16, 2021, EGA1 occurred on October 16, 2022. The post launch cruise trajectory placed Lucy into a heliocentric orbit with an initially increasing solar range, reaching a maximum of 1.16 AU before dropping to a minimum of 0.835 AU. The purpose of EGA 1 was to increase the energy of the orbit by raising the semi-major axis, increasing the orbit period to roughly two years with a maximum solar range of 2.3 AU and a minimum solar range of 0.86 AU. If Lucy had been unable to launch within our primary launch period, then the backup launch opportunity one year later would have eliminated the need for EGA1 entirely, at the cost of spacecraft mass. Figure 3 shows the EGA-1 flyby trajectory relative to the Earth.

A maneuver schedule for the launch – EGA1 timeframe was designed before launch, consisting of one deterministic deep space maneuver (DSM) and six statistical trajectory correction maneuvers (TCM's), to correct for errors in launch injection, orbit determination and trajectory prediction, and maneuver execution. The term 'deterministic' is reserved for maneuvers required in the reference trajectory assuming no errors in orbit determination, trajectory prediction or maneuver execution, hence DSM's are deterministic maneuvers. The term TCM is typically reserved for maneuvers which correct for the statistical errors. However, to avoid confusion for operations, DSM's are also given a TCM number so that propulsive maneuvers can be tracked in a continuous fashion.

The Lucy spacecraft has two propulsive maneuver modes, hydrazine trajectory correction maneuver (TCM) thrusters for maneuvers under 50 m/s, and a bi-propellant main engine (ME) for maneuvers over 50 m/s. Nominally, only one deterministic deep space maneuver (DSM) was required for the Launch to EGA-1 phase which was planned a little over six months after launch, with a delta-v of about 6.5 m/s in the pre-launch optimized trajectory. Hence DSM-1 was executed using the TCM thrusters.

TCM's 1 and 2 were planned to correct for launch injection errors at 30 and 60 days after launch. TCM-2a was added post-launch to test the propulstion system. DSM-1, was required to keep Lucy on the optimized trajectory and to target the updated EGA-1 aimpoint based on a trajectory reoptimization. TCM-4 was added to correct for any potential DSM-1 maneuver execution error. Finally, TCM's 5 and 6 were added at 30 and 10 days prior to

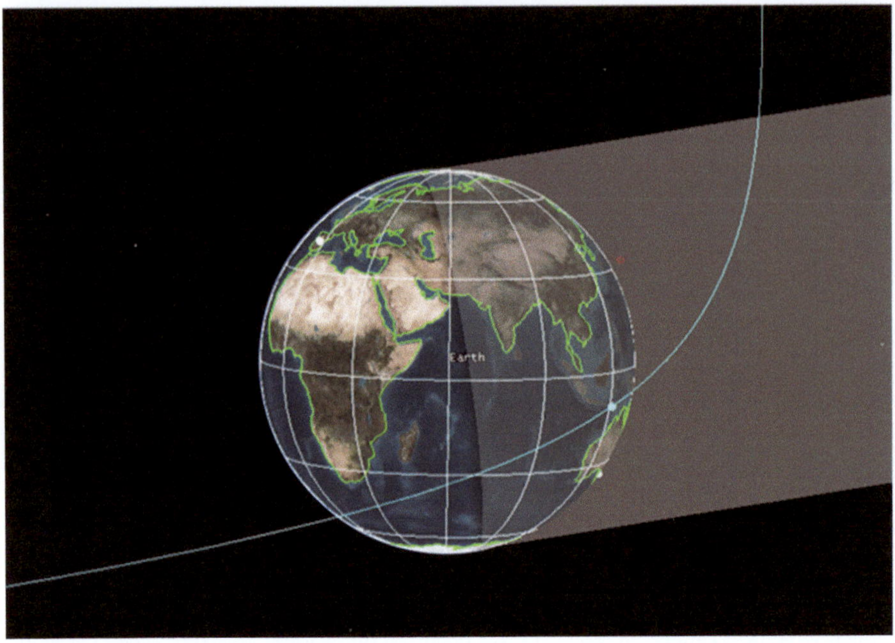

Fig. 3 The Lucy trajectory during EGA1. This for the opening of the launch period

the encounter to target the EGA encounter target in the B-plane and correct for OD and trajectory prediction errors. A potential collision avoidance maneuver one day prior to the encounter (TCM-6a) was added should the Conjunction Assessment Risk Analysis (CARA) team determine that the probability of collision (Pc) during the Earth encounter exceed 1 in 10,000. Finally, TCM's 7 and 8 were added to correct for errors in the Earth flyby modeling. The maneuver schedule is given in Table 1.

To ensure that at no point in the launch to EGA-1 cruise phase the probability of Earth impact or atmospheric entry is less than 1% should no additional maneuvers be performed, a progressive targeting strategy was employed walking in the perigee altitude. This was done by first optimizing the trajectory, then performing a statistical maneuver (Monte Carlo) analysis incorporating predicted orbit determination (OD) errors from an OD covariance analysis, and maneuver execution errors provided by the GNC (Guidance, Navigation and Control) team, to determine the impact probability for each subsequent maneuver. If it was found that the impact probability was less than 1%, a new trajectory optimization was performed with a higher perigee altitude constraint, and the Monte Carlo analysis was performed again. This iteration continued until the overall mission delta-v was minimized while satisfying the 1% constraint. Figure 4 shows the as-flown B-plane targeting history from launch to EGA_1.

Once TCM-4a was executed, the reference trajectory used for the maneuver design, based on OD016, was held fixed and no further reoptimizations were adopted, although they continued to be performed and tracked to determine whether TCM's 5 and 6 would need to be executed. There was some drift in the trajectory compared to the reference trajectory during the approach, but the trajectory optimization showed that these small errors could be absorbed in post-EGA-1 maneuvers.

Figure 5 shows the approach B-plane history following TCM-4a. The red X indicates the target. Some drift was observed on the approach, increasing the periapsis distance by around

Table 1 Lucy Launch to EGA-1 Maneuver Schedule

Maneuver	Date	Delta-V (m/s)	Purpose
Launch	Oct 16, 2021	N/A	Explore the Jupiter Trojans!
TCM-1	Nov 15, 2021	Cancelled	Correct for launch injection error
TCM-2	Dec 15, 2021	Cancelled	Correct for launch injection error
TCM-2a	Mar 2, 2022	1.252	Propulsion system test
DSM-1 / TCM-3	June 7, 2022	4.204	Target 80,915 km EGA altitude 15.8% radial EGA aimpoint bias
TCM-4	June 21, 2022	1.531	Target 2,187 km EGA altitude 13.0% radial EGA aimpoint bias
TCM-4a	Aug 3, 2022	0.415	Target 1,861 km EGA altitude Un-biased EGA aimpoint
TCM-5	Sep 16, 2022	Cancelled	Target 350 km altitude EGA-1 targeting
TCM-6	Oct 6, 2022	Cancelled	EGA-1 targeting
TCM-6a	Oct 15, 2022	Cancelled	Collision avoidance maneuver
TCM-7	EGA + 10 days	Cancelled	Correct for EGA error
TCM-8	EGA + 30 days	Cancelled	Correct for EGA error

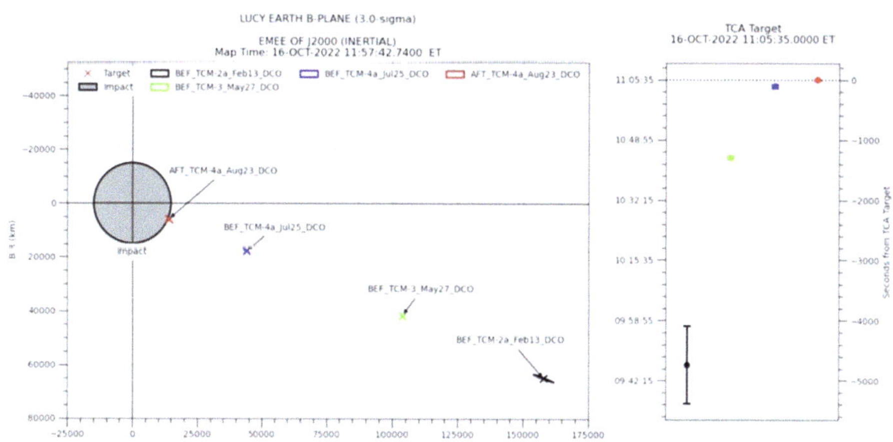

Fig. 4 EGA B-plane Aimpoint History

10 km, but this was deemed a positive occurrence by the GNC team as it reduced aerodynamic torques on Lucy's very large solar arrays. Since there was a negligible delta-v penalty and a benefit to the spacecraft, TCM's 5 and 6 were cancelled. This eliminated the need to perform TCM's 5 and 6, also retiring the risk associated with executing propulsive maneuvers close to the flyby during a period of high spacecraft activity. The final, reconstructed B-plane intercept is shown in Fig. 5,

A final collision avoidance maneuver was prepared and ready to be executed 12 hours prior to perigee to adjust the time of perigee passage should the need arise. The Lucy project coordinated with the CARA team at GSFC to determine if such a maneuver was needed to

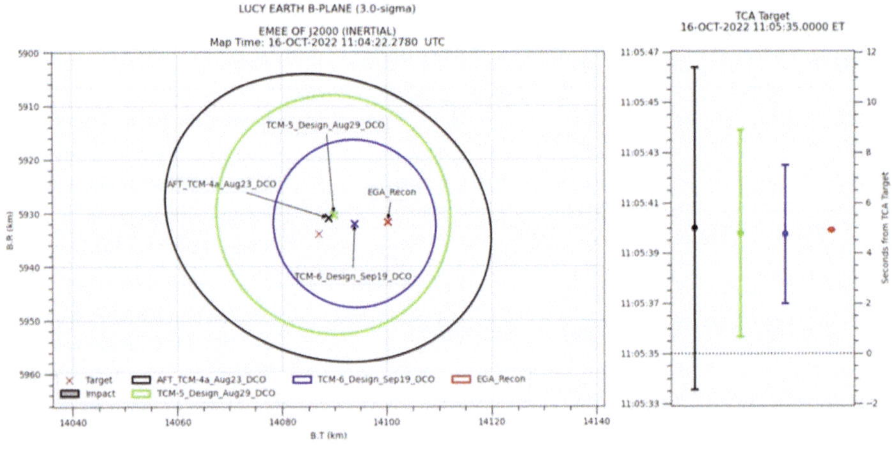

Fig. 5 Lucy B-plane History After TCM-4a

Table 2 EGA-1 B-plane Accuracy (EME2000)

Trajectory	B.R (km)	B.T (km)	Periapsis time (UTC)
Reference Trajectory	5933.946	14,086.887	16-OCT-2022 11:04:26.564
EGA Reconstruction	5931.616	14,100.149	16-OCT-2022 11:04:30.750
Difference	−2.33	13.262	4.186 s

avoid a possible collision with any other objects in the space catalog. The navigation team provided the CARA team trajectories and associated uncertainties, beginning 7 days prior to launch. The CARA team had three categories of risk: low risk with $P_c < 1e\text{-}7$, medium risk with $1e\text{-}4 > P_c > 1e\text{-}7$, and finally high risk if $P_c > 1e\text{-}4$. In the event that it had been necessary to maneuver, two "pre-canned" maneuvers were prepared in advance to change the time of perigee, arriving either two or four seconds earlier, which were also assessed by the CARA team. If necessary, the lowest risk trajectory would have been chosen 24 hours before perigee and a diversion maneuver would have been executed 12 hours later. Lucy was fortunate in that no objects were within the high or medium risk categories, so a decision was made not to execute TCM-6a.

EGA1 provided an opportunity for unique in-flight instrument calibration activities using the Earth and Moon on approach and departure. The alignment with the Earth, Moon, and Sun was favorable on trajectories that launch in the beginning of the launch period, but not in the later portion of the launch period, so it was fortuitous that Lucy able to launch on the first day.

The last propulsive maneuver executed was 54 days prior to the EGA. In this time period, there were daily momentum wheel desaturation maneuvers and a relatively large solar radiation pressure perturbation due to the large size of the solar arrays, which was challenging to model. Table 2 shows the reference trajectory B-plane intercept used for targeting and final reconstructed B-plane from post-EGA orbit determination.

EGA1 also provided an opportunity for unique in-flight instrument calibration activities using the Earth and Moon on approach and departure (see chapter by Spencer et al.).

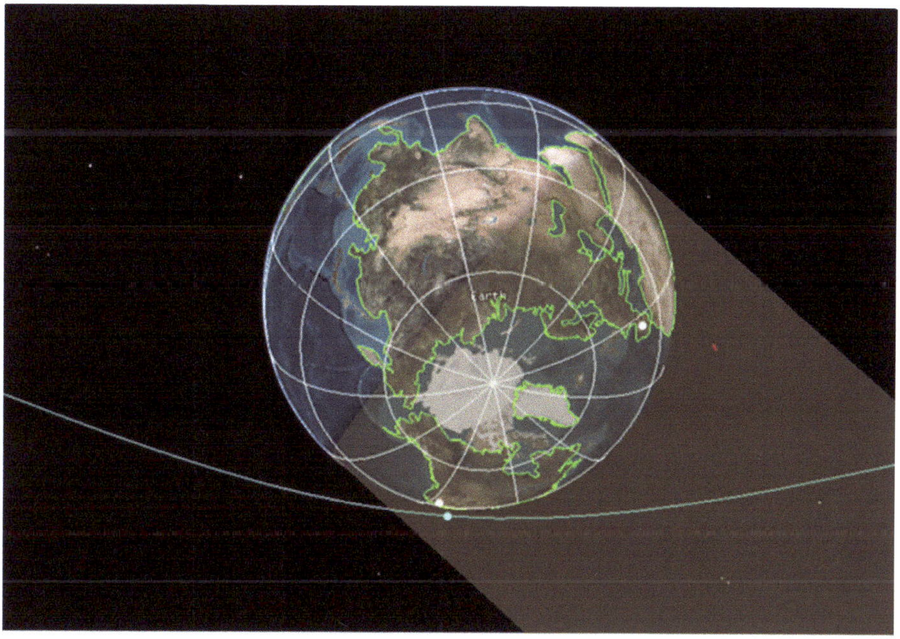

Fig. 6 The Lucy trajectory during EGA2 for launch at the opening of the launch period

1.4 Earth Gravity Assist 2

The second Earth Gravity Assist (EGA2; Fig. 6) occurs on December 13, 2024, which is approximately two years after EGA1. Roughly half way in between EGA1 and EGA2, the second Deep Space Maneuver (DSM-2) is performed to set up the proper Earth flyby conditions at EGA2 to encounter the main belt asteroid Donaldjohanson. DSM-2 will be performed on the bipropellant main engine and is the largest maneuver of the mission with a ΔV magnitude between 760 m/s and 912 m/s (depending on launch date). Due to the size of DSM-2 and the long duration until EGA2, the perigee target will not need to be biased. The EGA2 B-Plane uncertainty at the time of DSM-2 is so large that the probability of Earth impact or atmospheric entry is always smaller than 1%. The subsequent TCMs occurring 30 days and 7 days before EGA2 will bias the perigee target as-needed to ensure that the spacecraft is never on an Earth intercept trajectory.

EGA2 has a nominal perigee altitude ranging from 350 km at the open of the launch period to 580 km at the close of the launch period, with an eclipse duration of about 20 minutes. This gravity assist further increases the semimajor axis of Lucys heliocentric orbit, increasing the orbit period to roughly six years. Similar to EGA1, the Navigation team will coordinate with CARA to assess the need of a collision avoidance maneuver and execute it if necessary.

1.5 Dinkinesh and Donaldjohanson Encounter

The first two encounter for Lucy are both flight tests before encountering the Trojan asteroids. There are no science requirements to accomplish at either of these encounters, although scientific data will be collected. This rehearsal will reduce risks for the Trojan flybys by exercising the system in flight. The closest approach distance to Dinkinesh is 425 $\pm$ 6 km and for Donaldjohanson it is nominally 1000 km to have a similar angular rate at the flyby for our most pressing case, Polymele.

On April 20, 2025, the Lucy spacecraft will flyby the Main Belt Asteroid (52246) Donaldjohanson (DJ) on the way out to the L4 swarm of Trojan asteroids.

As part of the rehearsals, the on-board terminal tracking will be taking data with the Terminal Tracking Camera (T2Cam) and the state estimation will be computing the asteroid target location. The priority is to establish state estimation is working properly and it is not required to enable terminal tracking (using the estimated state to point the Instrument Pointing Platform).

The rehearsal will also exercise the navigation TCM process, and the ability to estimate planetary ephemerides, and that of the spacecraft relative to the flyby targets, using radiometric data types and optical navigation images using the LLORRI instruments supplemented by images from T2Cam. Radiometric data types will be collected as well as optical navigation images using the L'LORRI instrument. There will be two TCMs to target the delivery of the spacecraft to the aim point, one 30 days before closest approach and one 7 days before closest approach. The knowledge update process will also be exercised in the rehearsal where the post-TCM spacecraft and the flyby target ephemerides are sent to the spacecraft 4 days prior to the encounter.

Even though there are no science requirements tied to the encounters, both asteroids are interesting scientific targets. Dinkinesh will be the smallest main belt asteroid to be visited by a spacecraft, extending the detailed knowledge of asteroid surface properties to a smaller size regime.

Donaldjohanson is a member of a young collisional family (Nesvorný et al. 2015) and will be the first collisional family member visited by spacecraft. Science data will be collected during approach to DJ. On approach the phase angle is 15° and, therefore, on departure if the instruments were pointing at DJ they would be pointing near the Sun. For the health and safety of the instruments, the Instrument Pointing Platform will point away from DJ to a location that ensures it will not point within 40 deg of the Sun.

1.6 L4 Trojan Encounters

The Lucy mission will conduct 4 flybys of 6 Trojan asteroids in the L4 swarm. For each of these encounters the closest approach distance was chosen to accomplish the Level 1 science objectives (Olkin et al. 2021b). The nominal closest approach distance is 1000 km from the center of the Trojan asteroid on the sunward side (Table 3). This distance is close enough to meet our mass determination requirement of 25% for Eurybates, Leucus and Orus, but not Polymele, the smallest of our required targets (Levison et al. 2021). The closest approach distance will be re-evaluated prior to building the final encounter sequence and could be revised. The Polymele encounter has a nominal closest approach distance of 434 km. All the Trojan encounters meet all science requirements for closest approach distances that are about 3σ from the aim point.

The spacecraft approaches each of the Trojan asteroids in the L4 swarm from a solar phase angle between 82° to 126° (see Table 3). The spacecraft will pass on the sunward side of the Trojan asteroid, therefore, the phase angles will decrease leading up to closest approach. The decreasing phase angle corresponds to increasing illuminated surface area for shapes that are not pathological (spherical, oblate, top shaped).

The velocity in Table 3 is relative to the Trojan asteroids. At these velocities and closest approach distances, the spacecraft sweeps through phase angles from 75° to 0° in $\sim$3 minutes.

Table 3 Encounter Circumstances

Encounter	Date	Ph. ang. deg	C/A dist km	Velocity km/s	Earth range AU	Sun range AU
Eurybates	08/12/2027	81	1000	5.7	5.6	5.7
Polymele	09/15/2027	82	434	6.0	6.2	5.7
Leucus	04/18/2028	104	1000	5.9	4.9	5.7
Orus	11/11/2028	126	1000	7.1	6.2	5.3

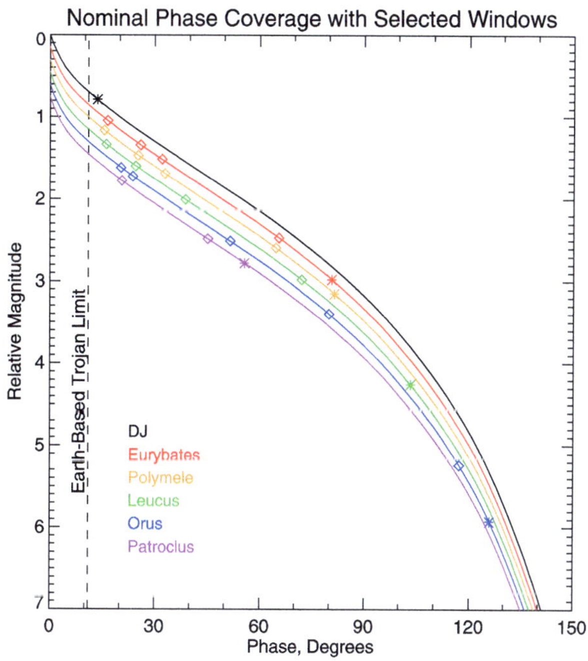

Fig. 7 The relative magnitude as a function of phase angle for each of the Trojan asteroids and the Lucy Main Belt Asteroid flyby target. Each of the asteroids is offset for readability. The maximum phase angle observable from Earth is indicated by a dashed line. The phase angle on approach to the flyby is denoted by a *. Lucy will nominally observe the Trojan asteroids at the phase angles indicated by the diamonds to build up knowledge of the phase curve of the Trojans before encounters

The closest approach distance for Queta, the small moon of Eurybates, is uncertain at this time because of the uncertainty in its orbital parameters (see chapter by Noll et al.).

From Earth, the Trojan asteroids cannot be observed at a phase angle of more than 11°, therefore, observations of the Trojan asteroids at moderate phase angles are planned well in advance of the Trojan encounters to inform estimates of expected brightness and exposure times. The strawman plan for photometric observations of our targets is shown in Fig. 7.

The Navigation team will use radiometric data types and optical navigation to deliver the spacecraft to the aim point. The B-plane delivery uncertainty in the closest approach distance will be no greater than 75 km (3σ) and the knowledge of the spacecraft location relative to the Trojan asteroid at 4 days before closest approach will be less than 100 seconds (3σ) in the time of flight direction, and 50 km (3σ) in each of the radial and cross track directions. For Polymele, the requirement on the delivery uncertainty to the aim point is reduced to 40 km to achieve the mass determination objective.

In order to deliver the spacecraft to the aim point, there are a series of TCMs. The first of these occurs at E-30 days (30 days before closest approach). The next occurs at E-7 days. Each of these TCMs has a data cut off 3 days before the maneuver execution. For the knowledge update at E-4 days, there is a data cutoff at E-5 days.

1.7 Earth Gravity Assist 3

The third Earth Gravity Assist (EGA3) occurs on December 27, 2030, which is approximately two years after the Orus encounter. However, the maneuver that targets EGA3 is the fifth Deep Space Maneuver (DSM-5), which occurs on July 23, 2028 (nearly 4 months before the Orus encounter). At roughly 350 m/s, DSM-5 targets the Orus encounter and sets up the proper Earth flyby conditions at EGA3 to encounter the Patroclus/Menoetius binary system in March of 2033 at a specific time where both targets will be observable but not overlapping from the spacecraft's perspective. Unlike the previous EGAs, the nominal EGA3 perigee target results in a 1% probability of Earth impact or atmospheric entry, so biasing will not be necessary in the design of DSM5 or any of the subsequent TCMs.

EGA3 is similar to EGA1 and EGA2, but with a flyby altitude of 600 km and an eclipse duration of 11 minutes. This gravity assist will change the heliocentric orbit inclination by nearly 9 degrees from 1.83 to 10.75. The Navigation team will coordinate with CARA to assess the need of a collision avoidance maneuver and execute it if necessary.

The TCMs preceding and following EGA3 are also used to correct any errors in the timing of the Patroclus/Menoetius binary encounter. To provide the best possible observations of both Trojan asteroids, arrival at the system will be timed to minimize the angular displacement of the two bodies without any overlap during the approach. This improves the terminal tracking solution by increasing the time that the two targets are in the T2Cam field of view. Patroclus and Menoetius have an orbital period of about 4.28 days about their gravitational center, hence this geometry also repeats every 4.28 days. The selection of the current encounter date and time is a reflection of the current understanding of the ephemeris of the two objects. It should be noted that the uncertainty in this ephemeris propagated 15 years into the future means this date is merely a design point that will be determined with more confidence as time passes by.

1.8 L5 Trojan Encounter

The Lucy spacecraft will fly past the near equal size binary pair Patroclus and Menoetius (otherwise known as the Patroclus-Menoetius binary, PMB) in the L5 swarm of Trojan asteroids. The closest approach will occur in March 2033. The exact day and time for the flyby will be determined by the time of the Deep Space Manuever (DSM-5) that sets up the geometry for Lucy's third Earth Gravity Assist (Dec. 2030) which targets the PMB encounter. DSM5 occurs after the Leucus encounter and before the Orus encounter. This is roughly between April and November 2028. The timing of the PMB encounter is critical because there are constraints on the relative locations of Patroclus and Menoetius. As the spacecraft is approaching Patroclus and Menoetius, the two objects should not be aligned as seen by the spacecraft or illuminated terrain would be unobservable. Additionally, there is limited range of motion of the gimbal in the direction that points the instrument pointing platform in the direction out of plane of the encounter (the encounter plane is composed of the velocity vector of the spacecraft and the vector from the spacecraft to the Trojan asteroid at closest approach). The instrument pointing platform can point +9 deg to −15 deg out of plane, therefore, limiting the out of plane distance Patroclus can be from Menoetius in the out of plane direction.

The Lucy spacecraft will fly past Menoetius before Patroclus. The encounter aim point will be defined relative to Menoetius. Currently the aim point for the Menoetius encounter is planned to be 1075 km from the center of Menoetius and on the sunward side. This puts the closest approach distance to Patroclus at 1245 km for our nominal encounter on March 3, 2033. On approach to the PMB, the phase angle will be 56 deg and the encounter velocity is 8.8 km/s with a Earth range of 5.2 AU and a Sun range of 5.4 AU.

The B-plane uncertainties and knowledge requirements are the same as described for the L4 Trojan asteroid targets.

After prime mission, Lucy will continue flying through the L4 and L5 swarms.

2 Trojan Encounter Concept of Operations

2.1 Encounter Architecture

The Concept of Operations for the Trojan flybys share many of the same elements, but there are modifications for each encounter. This section will describe the common elements across encounters and later sections will highlight some of the modifications made for specific encounters.

2.1.1 Baseline Timeline

For each of the Trojan encounters there is a baseline of activities that are common across encounters. Figure 8 shows selected common elements near closest approach. The L'LORRI instrument is turned on at E-60 days; L'Ralph is powered on at E-12 days; L'TES and the Terminal Tracking cameras are powered on at E-6 days. The first observations as part of the Trojan flyby occur at E-60 days with optical navigation to support ECM1 (Encounter Correction Maneuver) at E-30 days. Science observations begin at E-12 days and continue in general to E+20 days. The sequence of science observations for individual targets will have many commonalities with others, but will also incorporate unique aspects that depends on target specific parameters. The final planned TCM to refine the encounter trajectory occurs at E-7 days with an opportunity at E-5 days for a contingency TCM if needed. At E-4 days, auto-recovery mode is enabled. If a fault occurs when autorecovery is enabled (from E-4 days to E+4 days), the fault protection software will try to resolve the fault and rejoin the science sequence. Also, at E-4 days, the Final Knowledge Update is uploaded to the spacecraft and includes the best known ephemeris of the Trojan asteroid target along with other configuration files.

A unique aspect of the Lucy mission is that near closest approach science observations are initiated based on range to the target rather than time. This is commanded using the Trojan Event File (TEF) starting at E-4 days. The TEF is a series of observation in virtual machine language (VML) blocks (Grasso and Lock 2008) and their parameters ordered by range to the target. The range to the target is determined initially from an on-board ephemeris. When the terminal tracking is enabled, the range to the target is determined from the state estimation of the terminal tracking. At a later time, designated in the command sequence, the position of the IPP is driven by the state estimation solution. The reason the Lucy mission uses range-based observations is to ensure observations are taken with the desired resolution to accomplish the science objectives. The state estimation will cause a fluctuation in the time a given range is achieved compared to the on-board ephemeris or even a previous estimate of the state. This fluctuation could, in principle, result in a command needing to be executed before the previous one has completed. This is problematic for observations with

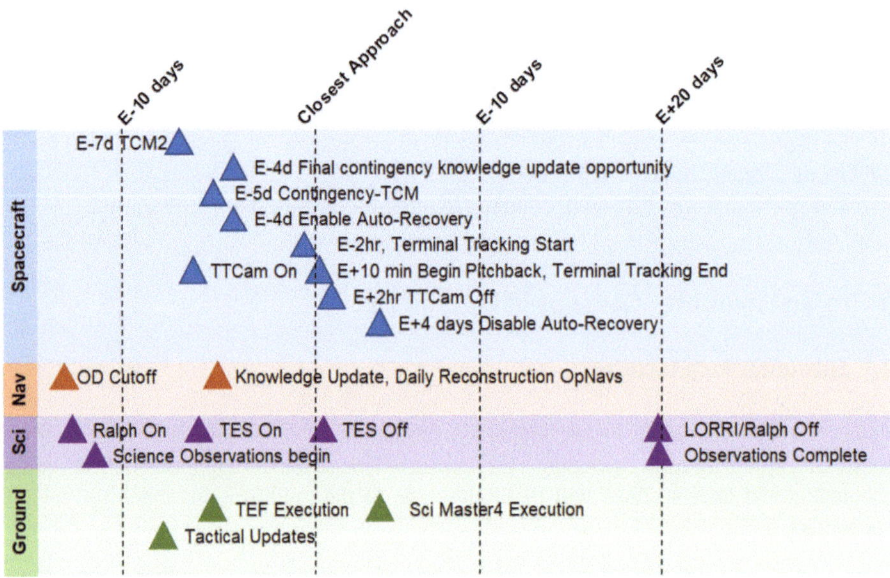

Fig. 8 Key activities during a Trojan asteroid flyby starting about 15 days before closest approach and continuing to 20 days after closest approach

the same instrument. For example, if a color scan with Ralph should execute given the current range estimate but the instrument is still completing a current observation, the pending color scan will be skipped. It is thus necessary to ensure that sequences can accommodate variations in the time-distance relationship and avoid collisions between consecutive instructions. This is ensured by scheduling sufficient time between subsequent observations/VML blocks of the same instrument.

2.1.2 Spacecraft Attitude During Encounters

The spacecraft and IPP attitude change over the course of the flyby to keep the instrument boresights pointed at the Trojan asteroids during the encounter. The concept of operations for the spacecraft and IPP attitude is dependent on the solar phase angle on approach. For encounters that approach from the sunward side of the Trojan asteroid where the solar phase angle is less than 90 degrees, Lucy is oriented in a "feet first" flyby which means that the spacecraft approaches the asteroid with the main engine ("feet") closer to the velocity vector than the IPP ("head"), see Olkin et al. (2021a). This is the encounter geometry for the Donaldjohanson, Eurybates, Polymele and PMB encounters. On approach, the high gain antenna (HGA) is pointed to the Earth with the spacecraft Y-axis orthogonal to the velocity vector. At E-2 hrs, the spacecraft attitude changes from pointing the HGA to Earth to aligning the spacecraft Z axis with the plane defined by two vectors:(1) the spacecraft to Trojan asteroid and (2) the Trojan relative velocity vector. To do this, the spacecraft rolls about the Z-axis. This rotation is small (less than 11°). The location of the Trojan asteroid is given by the on-board ephemeris at first. Once terminal tracking is enabled, the definition of this plane is determined by the state estimation from the terminal tracking. The time that state estimation is enabled is defined in the command load and will be based on simulations of the terminal tracking and state estimation performance based on the encounter geometry and best estimate shape models for the Trojan asteroid.

As the spacecraft approaches the Trojan asteroid, the spacecraft will rotate about the Y axis. This motion will take the solar arrays off the Sun and the HGA away from Earth. At E+10 minutes past closest approach, the spacecraft will perform a pitchback maneuver and rotate about the Y-axis to undo the rotation that took place during closest approach. This rotation will take about 35 minutes. During this time, the IPP will rotate to keep the Trojan in the instruments' FOVs and will return the spacecraft to a power positive state and communications with Earth through the HGA will be reinstated. For the Trojan encounters where the spacecraft is approaching from the unilluminated hemisphere (Dinkinesh, Leucus and Orus), the basic idea is the same except the spacecraft approaches with the IPP close to the velocity vector.

2.1.3 Science Planning and an Example Timeline

Science planning begins with the Level 1 science requirements (see Levison et al. 2021). To facilitate that planning, the science team has developed a series of short documents or "Measurement Techniques" that describe in detail attributes of the observations needed to accomplish the science objectives. An example illustrating the level of detail in the measurement technique is the range of illumination angles desired for observations designed to identify craters (discussed more in the next section). It is difficult to detect craters without shadows so the illumination angle of the observation is important. Measurement techniques are defined for the prime method for accomplishing the science as well as for backup observations including those that could be accomplished with an alternate instrument.

The Eurybates encounter will occur at a relative velocity of ~6 km/s and with a nominal close approach distance of 1000 km. Signal-to-noise considerations require observations to be done at a phase angle smaller than 82°. For the Eurybates encounter, the asymptotic phase angle is 81° on approach and 99° on departure. The phase angle will be 82° approximately 10 min after passage at closest approach. As such, most of the primary science observations will be done on the approach side for Eurybates.

Figure 9 shows a portion of the timeline of the encounter, with markings of where the instruments are being used. Instruments can be operated simultaneously, to the exception of MVIC and LEISA which are components of the RALPH instrument. Observations for the Level-1 science objectives begin around 12 days before closest approach (E-12 days), with L'LORRI observations designed to detect potential satellites and determine their orbits. The Satellite Orbit Determination Observations use a series of 19 image sets that start at E-11.8 days and ends at E-1 day. The goal is to sample the Szebehely sphere for a variety of orbital periods, and to get at least 3 data points per possible orbit. The cadence of the observations is adjusted as Lucy get closer to the target in order to adequately sample distant and close-in satellites.

Next, global coverage observations begin using L'LORRI, LEISA, and MVIC. These observations do not have a resolution requirement, but must be done at a cadence that ensures the entirety of the target's available sunlit surface is observed while aiming for the highest resolution possible. As Lucy approaches the Trojan asteroid, the phase angle varies little at long distances, and new portion of the target become visible solely through the asteroid's rotation. To ensure both complete coverage and redundancy, we perform 29 L'LORRI observations spaced by 1/27th of a rotation (0.325 hrs), as well as, 9 observations with MVIC and LEISA spaced by 1/7th of a rotation (1.25 hrs).

For some of the requirements that relate to observations of craters, there is an additional criterion that comes into play: the phase angle of the observation must be greater than 40° and smaller than 75°. This is further complicated by the uncertainty in the spacecraft's delivery, which change the time/range at which a particular phase angle configuration can be

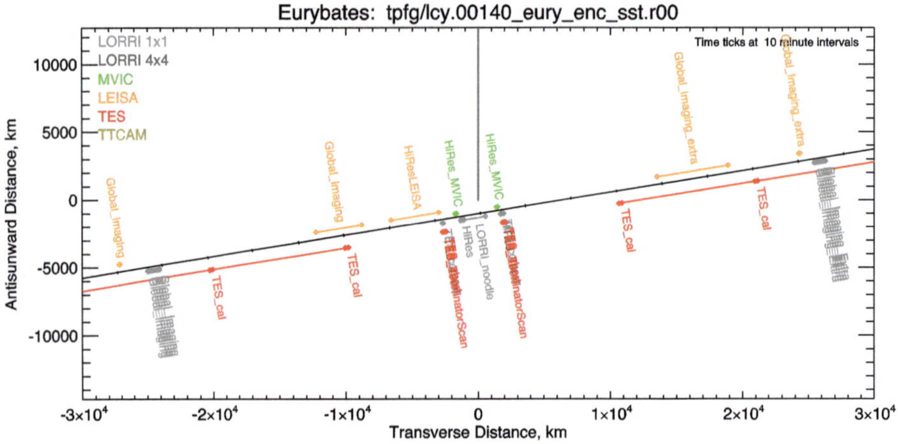

Fig. 9 A portion of the timeline of the Eurybates encounter

Table 4 Range and distance at which the desired phase angles occur for the 3σ low and high delivery at Eurybates. The first two columns are on approach and the last two columns are on departure

	75° phase angle	40° phase angle	40° phase angle	75° phase angle
3σ low case	8825 km at E-12m15s	1410 km at E-03m00s	1080 km at E+01m35s	2275 km at E+06m00s
3σ low	10258 km at E-29m25s	1639 km at E-03m39s	1255 km at E+01m50s	1644 km at E+07m00s

achieved. Table 4 shows the time and ranges at which the 40° and 75° phase angles happen for the 3σ low and high delivery (which correspond to a close approach distance of 925 km and 1075 km, respectively). The main consequence is that the opportunities to place observations that will give adequate data for all possible deliveries is reduced.

When planning science observations, we have adopted the stance to try and achieve our Level 1 Science Requirements with as few observations as possible. This leaves space for redundancy by giving more opportunities to place backup observations, and it also will allow us to implement additional observations to accomplish additional science. For example, an MVIC scan at 1900 km gives enough coverage and resolution to satisfy Level 1 Science Requirement 5 (panchromatic observation of 500 km^2 at 100 m resolution), 10 (color observation of 700 km^2 at 1500 m resolution) and 11 (color observation of 150 km^2 at 600 m resolution). These can be backed up by an additional MVIC scan at +1250 km. In general, we try to maximize the time between a Primary and Backup observation to build in resiliency.

Most of the other Panchromatic Level-1 Science Requirement are accomplished through the observations performed for global coverage, or via T2CAM imaging. The exception is the highest-resolution requirement of 10 km^2 at 29 m resolution (the latter is then expected to be improved to 14 m via deconvolution of successive L'LORRI exposures). This is implemented by a series of L'LORRI exposures (at a 1 second cadence) with very short exposure times, approximately 10 ms (to mitigates smear), as Lucy passes through closest approach.

For LEISA observations at Eurybates the combined areal coverage/resolution requirement is achieved by triggering a LEISA observation on approach at 6760 km. For time efficiency, this observation is divided into two separate scans. One covers the shorter wavelength with a shorter integration time and the other covers the long wavelength portion of the detector with a longer integration time. It takes about 10 min to complete the combination of these two scans. This observation ends just before we have to start moving the IPP to perform a terminator scan with L'TES (see next paragraph). Because of the lengthy duration of these scans, it is not possible to place a satisfactory backup on the departure side, as the scan could not be completed before the phase angle reaches 82 deg and the signal to noise degrades below the requirement. Instead, the last scan for global coverage at −12,500 km is a backup, leaving over 6.5 min between the prime and backup.

The L'TES instrument will take scans every 2 s starting 24 hours before closest approach (only interrupted by brief calibration operations approximately every 30 min). The only special pointing for L'TES will be to perform measurement of the target's unilluminated areas. This is implemented by having the IPP slowly drift across the target through the terminator. In order to measure the temperature of the Trojan asteroid's unilluminated surface at least 50% of the detector's FoV must be on the target without a view of the illuminated terrain. The relative sizes of the L'TES field of view and that of the target's unilluminated area is a combination of the phase angle (lower phase angle means larger portion of the target being illuminated) and Lucy's range to the target (closer range means the smaller the size of the TES FoV projected on the surface). At Eurybates, the configuration is such that the Primary and Backup terminator scans must be implemented on the approach and departure side, at 2290 km and 1575 km respectively.

A tool to simulate the encounters was developed to demonstrate that the area and resolution requirements of Lucy's Level 1 science objectives were met. This tool allows us to estimate the coverage and resolution we can expect to come out of a given science sequence. Figures 10 to 13 shows resolution maps for combined Panchromatic observation (L'LORRI, T2CAM and MVIC-Pan channel), MVIC-Color channels, LEISA, and L'TES. On each map,

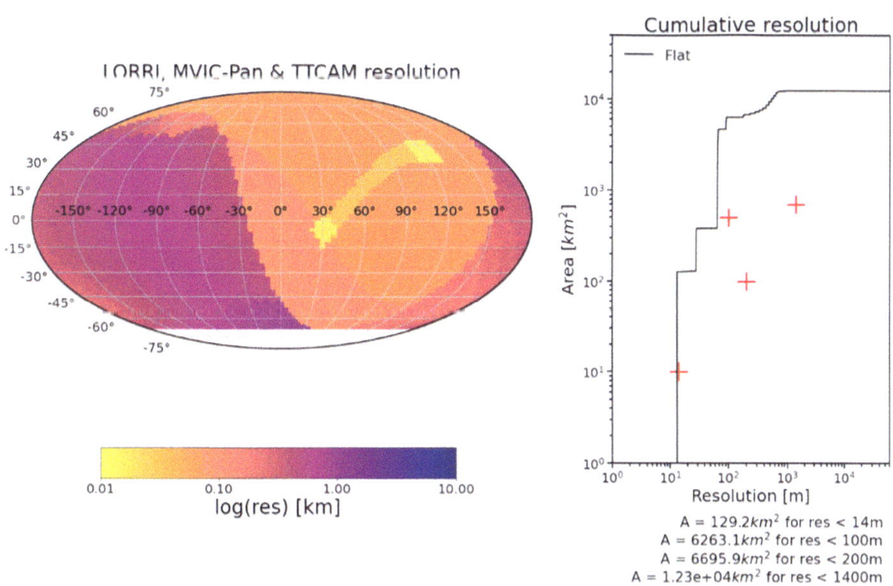

Fig. 10 Resolution map for all panchromatic observations using L'LORRI, T2CAM and MVIC-Pan. The map shows improved resolution where deconvolution of consecutive L'LORRI exposures is expected

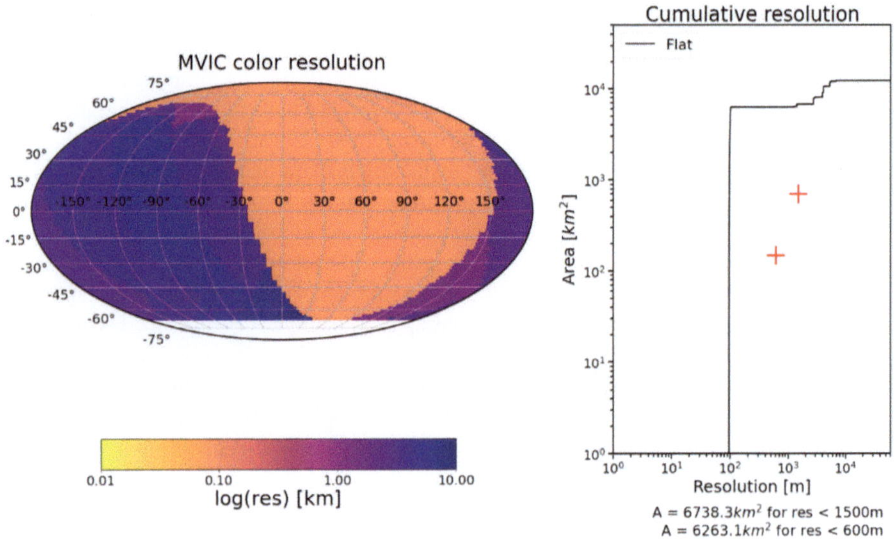

Fig. 11 Resolution map for all MVIC-Color observations

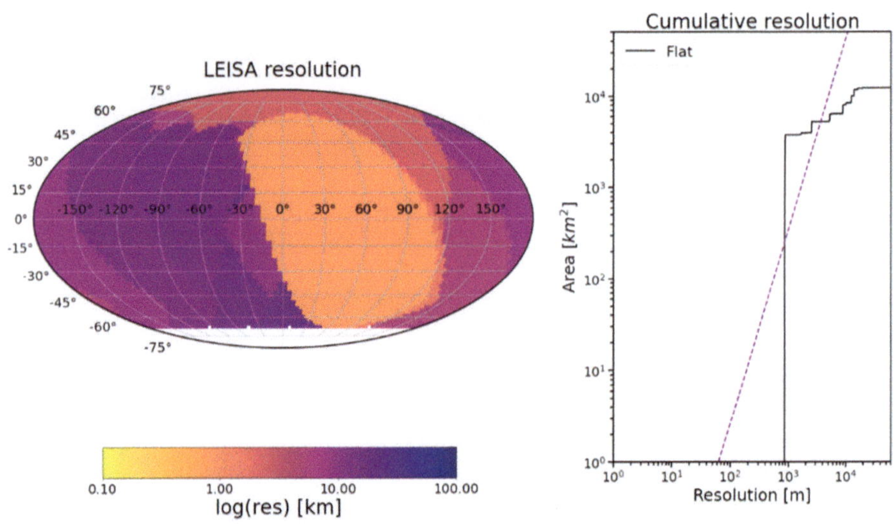

Fig. 12 Resolution map for all LEISA observations

the right side shows the cumulative resolution with science requirement of resolution/coverage showed as red crosses (for LEISA, the red dashed line shows the power-law that is the areal/resolution requirement for LEISA). The example sequence is estimated to give us sufficient coverage and adequate resolution to meet all Level 1 Science requirements.

Additionally, this tool can be used to perform a Monte-Carlo analysis of each science sequence. The Monte-Carlo analysis allows variation in Lucy's delivery to the encounter aim point, the on-board knowledge of the position of the target, and various properties of the target (shape, size). Using results of the state estimation convergence from the terminal

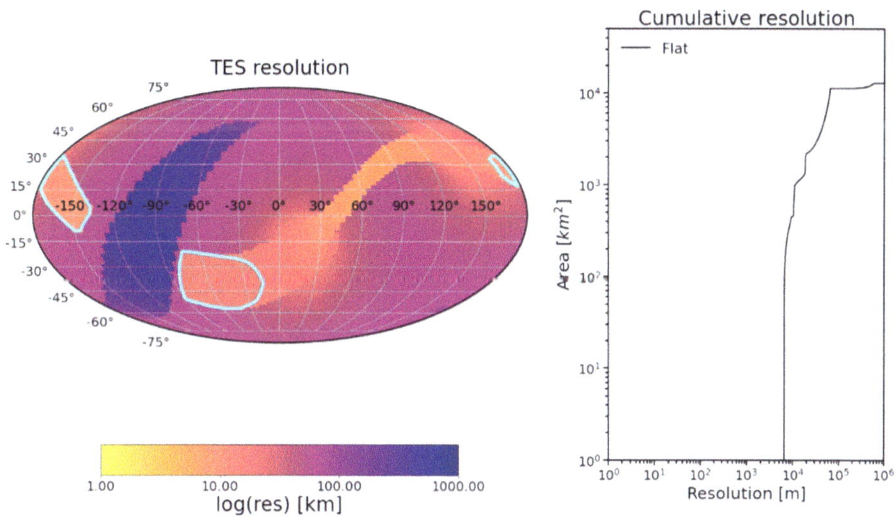

Fig. 13 Resolution map for all L'TES observations. The blue contour shows unilluminated areas observed

tracking, the Monte-Carlo analysis can investigate the resilience of a science sequence to the variation in the range estimates over time.

2.1.4 Resiliency in Encounter Planning

Spacecraft flybys are a time critical operation. If an opportunity is missed, it cannot be picked up at a later time. For this reason, the concept of operations includes a number if design elements that make the encounter resilient and time efficient. When the science timeline for observations is constructed, the highest priority observations are the ones that accomplish the Level 1 science objectives (Olkin et al. 2021b). The next priority is backup observations for the Threshold science objectives. The back up observations are preferably placed more than 5 minutes from the prime observations for an objective. This time spacing is important because in at least some failure modes, the fault protection system can recover to join the science sequence in about 5 minutes.

In addition to backup observations, we consider additional observations that can address the objective with another instrument if possible. This provides an ability to address an objective, perhaps in a degraded manner, if one of the instruments has a problem during the encounter. The L'LORRI and L'Ralph-MVIC instruments can provide backups for each other, however, with the different resolutions and fields of view of the instruments, the backup observations with an alternate instrument may be degraded.

The Lucy mission has incorporated tactical updates, to mitigate Trojan surprises, that modify select instrument activities no later than E-4 days, without modifying the command sequence files, sequences or block libraries or overall block/sequence execution timing. These updates will be designed in a manner that does not require the Flight Team to re-build tested and verified sequences (i.e., the update cannot change sequences or block timing). The intent is to prevent activity duration increase that could lead to adjacent activity overlap and conflicts. But it does not preclude activity duration reduction. For L'LORRI, the tactical update is exposure level adjustment to go up or down with a maximum adjustment factor of 2 via a global variable. For instance, a 100 ms exposure observation could be updated to either 50 ms or 200 ms. For Ralph MVIC, the tactical update is a TDI row adjustment, a default readout of 32 TDI rows could be changed to 16 or 64 depending on target brightness but would not change the scan rate of the Ralph mirror and the corresponding observation duration.

3 Conclusions

The flyby mission architecture allows 8 Trojan asteroids to be visited by one spacecraft. This is important for the first exploration of the Trojan asteroids and allows Lucy to observe the diversity of the Trojan asteroids (Levison et al. 2021). The use of a trajectory with a periapse near Earth and an apoapse near the Trojan swarms provided an opportunity to visit both the L4 and L5 Trojan swarms. Additionally, the mission design allowed for a rehearsal of the flyby with the main belt asteroids Dinkinesh and Donaldjohanson which reduces risk for the Trojan encounters.

The concept of operations was streamlined across Trojan encounters as much as feasible to help the operation team and to reduce risk by having a repeatable pattern of activities. The mission team's goal was to provide a platform that would allow flexibility in observation planning in the future. This was accomplished by having the system allow for observations by any instrument with only minimal restricted times. To be efficient during the flyby, Lucy includes a terminal tracking system that estimates the location of the Trojan target. This collapses the relatively large (in angular terms) uncertainty ellipse that would need to be covered without the terminal tracking capability.

Funding The *Lucy* mission is funded through the NASA Discovery program.

Declarations

Competing Interests The authors declare no competing interests.

References

Brown ME, Levison HF, Noll KS et al (2021) The orbit and density of the Jupiter Trojan satellite system Eurybates-Queta. Planet Sci J 2:170. https://doi.org/10.3847/PSJ/ac07b0

Englander J, Ellison D (2020) Evolutionary Mission Trajectory Generator. www.github.com/JacobEnglander/emtg. Accessed 8 July 2021

Grasso CA, Lock PD (2008) VML sequencing: growing capabilities over multiple missions. In: AIAA SpaceOps conference. https://doi.org/10.2514/6.2008-3295

Knittel J, Stanbridge D, Williams K (2019) Automated navigation analysis for the Lucy mission. In: Horneman KR et al (eds) Astrodynamics 1999. Advances in the astronautical sciences, vol 171. Univelt, p 2859

Levison HF, Olkin CB, Noll KS et al (2021) Lucy Mission to the Trojan asteroids: science goals. Planet Sci J 2:171. https://doi.org/10.3847/PSJ/abf840

Nesvorný D, Brož M, Carruba V et al (2015) Identification and dynamical properties of asteroid families. In: Michel P, DeMeo F, Bottke W (eds) Asteroids IV. University of Arizona Press, Tucson, pp 297–321

Noll KS, Brown ME, Weaver HA et al (2020) Detection of a satellite of the Trojan asteroid (3548) Eurybates—a Lucy Mission target. Planet Sci J 1:44. https://doi.org/10.3847/PSJ/abac54. arXiv:2008.01858

Olkin CB, Levison HF, Vincent M et al (2021a) Redundancy in the science implementation of NASA's Lucy Mission to the Trojan asteroids. In: IEEE Aerospace Conference. https://doi.org/10.1109/AERO50100.2021.9438406

Olkin CB, Levison HF, Vincent M, Noll KS, Andrews J et al (2021b) Lucy Mission to the Trojan asteroids: instrumentation and concept of operations. Planet Sci J 2:172. https://doi.org/10.3847/PSJ/abf83f

Space Science Reviews (2025) 221:91
https://doi.org/10.1007/s11214-025-01208-z

Managing the Lucy Mission

Donya Douglas-Bradshaw[1] · Arlin Bartels[1] · John Andrews[2] · Vince Elliott[1] · Chris Carson[1] · Jim Rogers[1]

Received: 4 August 2023 / Accepted: 6 August 2025 / Published online: 1 October 2025

Abstract
Lucy is NASA's 13th Discovery-class mission. It will be the first mission to encounter seven Trojan asteroids in a series of five flybys. Lucy is led by Dr. Hal Levison of the Southwest Research Institute. As the Principal Investigator, Dr. Levison is accountable to NASA for the success of the investigation, with full responsibility for its scientific integrity and execution within committed cost and schedule. Day to day management of the Lucy project is delegated to Goddard Space Flight Center. During Lucy's development, the project experienced a once-in-a-lifetime pandemic that threatened the team's ability to meet the planetary launch window. Despite this challenge, Lucy launched on time and under budget. This article discusses key factors that enabled project success.

Keywords Project management · NASA · Discovery mission · Announcement of opportunity · PI led mission · NPR 7123.1 · NPR 7120.5 · Lucy · GSFC · Goddard · Trojan · Jupiter · Covid-19 · Coronavirus · Lockheed Martin · SwRI · Arizona State University · Johns Hopkins University Applied Physics Lab · LORRI · L'Ralph · TES · Spacecraft · Instrument · Pandemic · Launch · KSC

1 Introduction

Lucy, a NASA Discovery Class mission, was selected via the 2014 Discovery Announcement of Opportunity and successfully launched at the opening of its planetary launched window on October 16, 2021. The Lucy mission name honors the influence that the *Australopithecus afarensis* hominin fossil named "Lucy" has had in advancing our understanding of the history of our species. Its name embodies the mission goal to similarly advance the understanding of the formation and evolution of our Solar System. The Jupiter Trojan Asteroids, which lead and follow Jupiter in its orbit by approximately 60 degrees°, have never been visited by a spacecraft. Lucy will investigate seven primitive asteroids near the

✉ D. Douglas-Bradshaw
donya.douglas-bradshaw@nasa.gov

1 National Aeronautics and Space Administration (NASA), Goddard Space Flight Center (GSFC), Code 434, Greenbelt, MD 20771, USA

2 Department of Space Studies, Southwest Research Institute (SwRI), 1050 Walnut St., Suite 400, Boulder CO, 80302, USA

Jupiter L_4 and L_5 Lagrange points where planetesimals from the outer planetary system have been preserved since early in Solar System history. This rich diversity of targets is enabled by fortuitous orbital alignments exploited by Lucy's unique trajectory.

2 Management Approach

The aim of project management is to deliver a successful outcome that meets the needs of stakeholders while balancing constraints such as time, cost, and performance. Lucy's management team developed an executable plan based on proven management practices. The plan involved clearly defined goals and objectives, unambiguous lines of authority, effective communication, results-driven design, a high performing team, project planning and control that kept the team on schedule and within cost, and a rigorous risk management strategy. These elements are discussed in the following sections.

2.1 Mission Implementation Team

Dr. Hal Levison, the Lucy PI, assembled a world-class team of institutions and highly experienced personnel to partner on this exciting mission of the first reconnaissance to Trojan bodies. The mission implementation team included team members from the Southwest Research Institute (SwRI), NASA Goddard Space Flight Center (GSFC), and Lockheed Martin (LM), all with demonstrated track records implementing PI-led missions. They had demonstrated experience executing within cost and schedule constraints while returning high value science to NASA and the planetary and space science communities. The team was further strengthened with the addition of John's Hopkins University Applied Physics Laboratory (APL), Arizona State University (ASU), and GSFC as heritage instrument providers; all three recognized providers of planetary science instrumentation. KinetX, the lead for navigating to Pluto on New Horizons and Bennu on OSIRIS-Rex, rounded out the team.

The Lucy project development organization is shown in Fig. 1. GSFC is responsible for project management, mission system engineering, safety and mission assurance, and mission operations. LM provided the spacecraft bus, spacecraft integration, and performs mission operations. GSFC led the navigation and mission design with support from KinetX. The Science Operations Center is provided by SwRI. APL provided the Lucy High-Resolution Visible Imager (L'LORRI), ASU provided the Lucy Thermal Emissions Spectrometer (L'TES), and GSFC provided the Lucy panchromatic and color visible imager and Infra-Red (IR) spectroscopic mapper panchromatic and color visible imager and IR spectroscopic mapper (L'Ralph).

The most important factor in Lucy's success was the team itself. The team was fully staffed with high performers, a strong bond, and a sense of camaraderie that permeated every aspect of the team's dynamic. All team members were fully dedicated and committed to both the mission and to each other. The high performing nature of the team enabled the project to absorb changes in key leadership positions with little to no impact and to overcome a myriad of technical challenges. They worked hard and celebrated their successes, demonstrating they were truly "Lucy Strong".

2.2 Goals and Objectives

On any mission, clearly defined goals and objectives are paramount to ensuring that everyone involved has an understanding of what they are working towards. The team was deeply

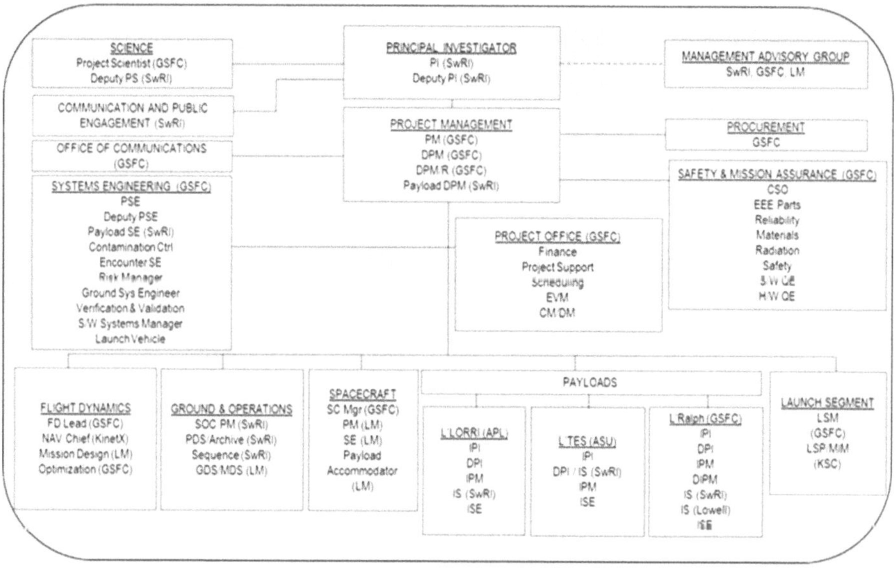

Fig. 1 Lucy Project Organization Chart

aware of the mission's critical importance to NASA and the planetary community on a wide range of topics stemming from advanced understanding of the origin and evolution of the solar system. The objective of the Lucy mission is to provide crucial inputs to four of the ten Priority Questions for planetary science as expressed in the 2013 Decadal Study:

1. Determine the compositional differences between the various classes of Trojans, measure the size-frequency distribution (SFD) of the craters at these targets, search for satellites and rings around the targets, and measure the physical and compositional differences between the two components of a near-equal mass binary to answer what were the initial stages, conditions and processes of Solar System formation.
2. Determine the compositional makeup, geological features, physical surface properties (i.e., regolith characteristics like thermal inertia), and bulk density of Trojans and compare surface properties of a C-type family member to freshly exposed material on a D-type asteroid to answer how the giant planets and their satellites accrete, and is there evidence that they migrated to new orbital positions.
3. Determine the compositional differences between the various classes of Trojans to answer what governed the accretion and what roles did bombardment by large projectiles play.
4. Determine the compositional differences of known classes of Trojans to answer what were the sources of primordial organic matter.

Lucy's science objectives are achieved using flight-heritage, highly capable instruments. The mission Level-1 requirements and success criteria are defined in the Planetary Missions Program Plan – Program Level Requirements Appendix for Lucy, and controlled within the Marshall's Space Flight Center's Planetary Missions Program Office. The Lucy level-2 requirements are defined in the Mission Requirements Document (Lucy-MGMT-REQ-0001) and controlled within GSFC's Technical Data Management System.

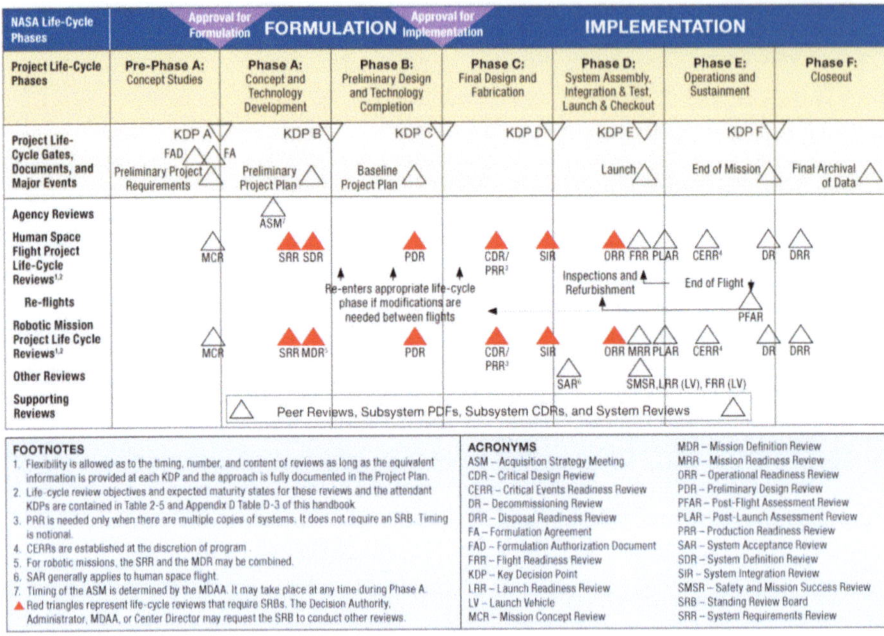

Fig. 2 NASA Space Flight Project Life Cycle from NPR 7120.5E

2.3 Lines of Authority and Governance Structure

The Lucy project management processes and plans are aligned with the guidance provided by NASA Policy Directive (NPD) 1000.1, NASA Strategic Plan. NASA manages projects based on life cycles, Key Decision Points (KDPs), and evolving programmatic and technical products during each life-cycle phase, as described in NASA Procedural Requirements, NPR 7120.5, NASA Space Flight Program and Project Management Requirements and NPR 7123.1, NASA Systems Engineering Processes and Requirements.

Project authority is delegated from the Associate Administrator for the Science Mission Directorate through the Planetary Science Division (PSD), to the Planetary Mission Program Office (PMPO) at Marshall Space Flight Center (MSFC). The Science Mission Directorate Program Management Council governed the mission. The Lucy PI was responsible for formulation and maintains responsibility for implementing the mission. During Phases B/C/D, Dr. Levison delegated day-to-day decision-making authority to GSFC's Lucy project manager (PM). During Phases E/F, Dr. Levison delegates day-to-day decision-making authority to GSFC's Space Science Mission Operations (SSMO) PM. The SSMO project assumed responsibility for management of the Lucy mission in Phases E/F following the successful Post Launch Assessment Review.

Figure 3 shows how the Lucy project office relates to the PMPO, PSD, GSFC Center Management, and NASA Headquarters Management, while Fig. 4 shows the governance structure for the Lucy project.

2.3.1 Planetary Missions Program Management

The PMPO provides the capability to integrate the programmatic resource requirements for all projects in the Discovery Program. The Discovery program manager is accountable

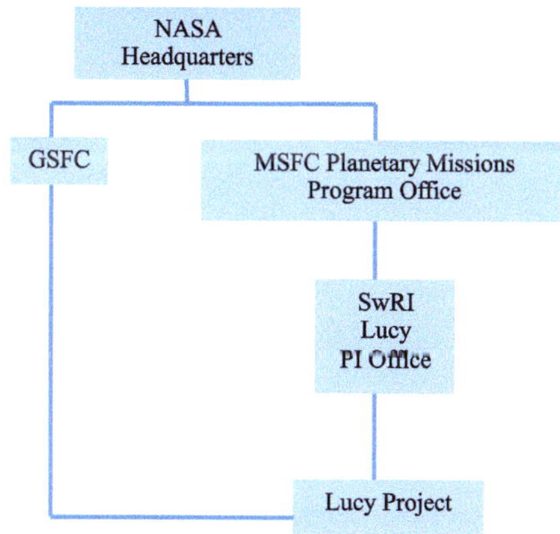

Fig. 3 Project Office Relationship to GSFC Management and NASA Headquarters

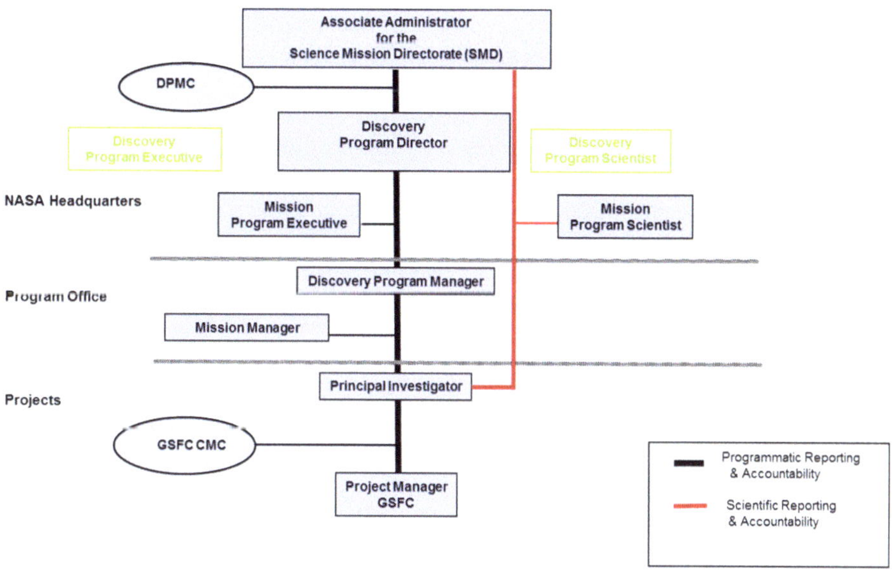

Fig. 4 Lucy Governance Model

to PSD Director for technical and programmatic progress of planetary missions, including Lucy. All Lucy project resources are budgeted and accounted for through PMPO from PSD.

2.3.2 Center Management

The Lucy development project office was located at GSFC and SSMO for mission operations is also located at GSFC. GSFC's Center Director reports to NASA's Administrator (through NASA's Associate Administrator). GSFC's Center Director is accountable for overall success of the Lucy project, via their acceptance of the delegated responsibility for managing project implementation and as the institutional authority for SSMO.

2.3.3 Principal Investigator

The Lucy mission is led by the PI. The PI has sole responsibility and accountability to the PMPO and PSD for the successful execution of the Lucy mission within cost and schedule. The PI is supported at SwRI by the deputy PI (DPI), the deputy project scientist, and the PI Support Office.

The PI served as the final decision authority on all matters related to maintaining the scientific integrity of the mission, staying within the PI-managed mission cost cap, and ensuring on-time delivery of the spacecraft to meet the fixed planetary launch window. The PI delegated day-to-day management and development decision-making authority to the PM within these guidelines and delegates day-to-day operations to the operations PM. The PMPO at MSFC holds the contract that funds the PI, his support office, and the science team.

2.4 Communication – Internal and External

Lines of authority help establish clear communication and decision-making, allowing for effective and efficient project execution. Good communication is essential for any project, but particularly for a NASA project where multiple teams and stakeholders may be involved. For the Lucy project, clear and consistent communication helped ensure everyone was informed and on track and connected partners regardless of home organization.

Lucy used partner institutions' best practices to facilitate efficient and open communication internal to the team. Management tag up meetings were held on a weekly basis with all mission elements and partner institutions. In addition, the leadership team, including the PI, Deputy-PI, PM, Deputy-PM, and Project Scientist met on a weekly basis to facilitate communication among science, engineering, and business personnel. The team generated weekly status reports for distribution across the project. The leadership team also established multiple working groups that met weekly to work the design. In addition to standing meetings, the Lucy project held in person multi-day monthly status reviews (MSR) that alternated between the partner institutions.

Lucy's communication plan also included routine status reporting to various stakeholders to ensure its project execution strategy managed stakeholder expectations. However, once Lucy entered the implementation phase, the team also implemented "First Stories" to quickly inform all levels of management at each of the partner institutions, as well as, members of the Standing Review Board of emerging technical issues that could substantially impact schedule. First Stories clearly conveyed a high level summary of the issue, the basis of decisions, progress, and critical milestones or decision points. Prior to any major review or key decision point, the development team briefed the independent review teams on current project issues and major risks. The open and honest communication fostered an air of trust.

2.5 Results-Driven Design

Lucy utilized a results-driven approach to the overall mission design, with an emphasis on successfully achieving the science objectives. The Lucy spacecraft and instruments were based on designs with extensive, successful flight history. There was no new technology development necessary for the mission. The majority of the hardware needed only minor design modifications to accommodate unique Lucy mission requirements. While some elements required a new design or were operating in different environments, none required technology development. It is worth noting that while the mission did not require technology development, there were lessons learned associated with manufacturability of the larger solar arrays.

2.6 Project Planning and Control

"Schedule is king" was an often-used phrase to emphasize the importance of prioritizing schedule adherence over other factors. Lucy's constrained 21-day launch period induced a focus on schedule performance – this is an important distinction for planetary missions as compared to Earth-orbital missions. Since delaying launch would have had extreme mission impacts, a concerted effort was directed towards maintaining the schedule pace. This was achieved through active and balanced management of technical scope, schedule, and cost, i.e. the triple constraints, and healthy levels of cost reserves and funded schedule margin. Time became a precious commodity ("time is money, but no amount of money will help when you're out of time"). Phase C/D had been well planned at healthy resource levels corresponding with previous analogous missions. The constrained launch period, by extension, constrained the development duration. Paying for risk mitigation activities was addressed as early as possible with an eye towards controlling scope creep.

Rigorous cost reserve management equated to lifecycle cost (LCC) control for the nearly $1 billion mission. The baseline was struck at selection (KDP-B). Every adjustment to the LCC baseline was justified, recorded, tracked, and reported monthly and at every lifecycle review. This provided the Lucy PI/PM with continuous awareness of available cost reserves. Strategically, the Lucy project utilized any and all available resources to get to the launch site on time with a qualified spacecraft and payload complement.

2.6.1 Schedule Management and Cost Control

The focus of schedule management was to develop and maintain a centralized Microsoft Project schedule database or Lucy Integrated Master Schedule (IMS). The IMS captured the time-phased required work scope of the Lucy project and was organized by the work breakdown structure (WBS). The IMS chronologically choreographed all of the milestones, major events, agency-level and project-level reviews, and logically related them to each other to identify the various critical path(s). It also tracked the total schedule slack at any given time throughout the project's lifecycle. In addition, the IMS aligned task schedule dates with the time-phased budgets that existed in the various budgeting systems deployed on the project. The IMS was the system of record for schedule data for the project. The Lucy schedule management plan met or exceeded the requirements contained in NPR 7120.5E and any risks to meeting project schedules were communicated and tracked using the Lucy risk management process. KDP and launch readiness milestones were controlled by the PMPO at MSFC. Schedule risks, issues, and concerns were communicated to the PMPO, GSFC management, and NASA Headquarters program executive using pre-established status reporting process.

The Lucy IMS was fully constructed for every major lifecycle review. Schedule evaluation of the rolled up project schedule was performed through analysis and was reviewed biweekly providing visibility into the planning and scheduling milestones to help ensure project objectives were met. Due to the size of the Lucy flight system in comparison to the payloads and ground system development efforts, it was necessary to perform routine monthly assessments in a de-integrated form, all while verifying the overarching critical path and other interfaces through analysis. In this way, true schedule performance could be readily addressed with hardware providers or ground systems further off the critical path.

Improved insight was gained using this approach to schedule control. The project implemented and maintained a schedule management system capable of facilitating the development, maintenance, analysis and control of the IMS.

All flight hardware and ground elements developed their schedules in coordination with the Lucy lead scheduler. Schedules from all hardware and ground elements were delivered monthly. Each partner institution was responsible for maintenance of their schedules and provided status for the schedule monthly at the PI Management Reviews (PIMRs). In addition, the Lucy lead scheduler performed monthly analysis (to include critical path, milestone burndown, observation of trends, etc.) of the Lucy master schedule and of the inputs from the partner institutions, reports to the project, and provided feedback to the institutions.

The Lucy lead scheduler developed performance metrics and reports to support all project reviews (e.g., MSRs, etc.) and meetings. Routine schedule assessments such as Top 3 Critical Paths, 12-Month "Rolling Wave"/Critical Milestone Schedule, Schedule Margin Tracking, and Goddard Schedule Assessment Tool Performance Metrics were reported on a monthly cadence. These assessments created a crosshatch of validation to schedule health and have been very effective on planetary missions where adherence to schedule is imperative to mission success. All schedule data was produced directly from the project master schedule and provided to project organizations via the project web-based Technical Data Management System. The IMS was baselined with project approval at KDP-C (Confirmation Review). Schedule data was combined with cost data for earned value management (EVM) reporting.

Through the 58 months of development, from mission selection to launch, Lucy met all of its major milestones and though challenges arose, the schedule replanning effort to mitigate disruptions was superior. Technical challenges are commonplace and expected; mitigating them in the COVID-19 environment was new. COVID-19 added an additional wrinkle to the mitigation efforts coupled with hardware and technical challenges alike. Mitigation creativity allowed for addressing development challenges in the new pandemic climate. Employing new or variations of existing mitigations such as early engagement of the long-lead acquisition process, smaller but separate floor and oversight teams, and the acquisition of the Defense Priorities & Allocations System rating to enable Lucy contractors to continue working during the domestic shutdowns allowed Lucy to stay on track.

2.6.2 Reserve Release Approach

On Lucy, the release of reserves was under the authority of and subject to review by the PM through PI-delegated authority (within certain pre-defined restrictions). Decisions that impacted schedule or requirements were addressed via the Configuration Control Board process. The process of change control, resource allocation, and release of reserves was essential to the success of Lucy. The PI was kept informed of the reserve releases through regular, weekly, and monthly project reporting. The Configuration Control Board process was used to control the mission baseline and track any technical changes. Technical changes that did not impact the mission baseline were managed by the Project Systems Engineer (PSE).

Projects carry three allotments of reserves or margins; Technical, Schedule, and Cost. Overall mission level technical resource reserves (mass, power, data rate, etc.) were held by the Lucy PSE. The Spacecraft Manager and Deputy Project Manager for Payloads (DPM-P/L) were allocated portions of the technical resource margins by the Lucy PSE, allocated as needed to support their respective technical teams. In the event more technical resources were required, the Lucy PSE further engaged, secured concurrence of the PM, and was able to approve release of mission-level technical reserves.

On a timeframe appropriate for the development phase, all technical resource current-best-estimates were updated so that the technical resources and technical reserves could be reallocated to reflect the progress of work and maturity of the design.

Overall mission-level schedule reserve or margin was held by the Lucy PM. The PM would allocate schedule margin for the three development teams (instrument, spacecraft, and ground system) as they developed their schedules. The DPM-P/L and Spacecraft Manager would request use of schedule margin from the PM to support their respective technical development efforts. In the event more schedule margin was needed, the Lucy PM was responsible for authorizing the release of these reserves.

The project schedule contained funded and unfunded (schedule slack) schedule margin. The project complied with GSFC requirements (GPR 7120.7) for minimum funded schedule margin (FSM) throughout development. FSM was incorporated into each project element's cost and schedule estimates. Cost margin needed for utilizing unfunded schedule slack were also held by the PM. Schedule slack is defined as the unfunded duration between the nominal instrument, spacecraft or spacecraft subsystem, or element completion date and the delivery date to the next higher level of system integration. FSM is defined similarly but for the funded duration between the completion date and the delivery date to the next higher level of system integration. Both are measured in work days.

At the outset of development, Lucy carried 125 work days of FSM at the mission level against a policy requirement of 102 work days. This level of margin was allowed to slowly erode over time at a rate dependent upon the phase with an outcome of zero by launch. Lucy schedules were updated monthly at PIMRs and MSRs during Phase C, and more frequently during integration and test through launch. During this process, overall schedule accomplishment was evaluated per the baseline schedule, and FSM held by the development teams could be readjusted by the PM. Through Phase C/D and the tracking of the baseline schedule, mission-level FSM fell below the policy requirement only twice, in March 2020 due to the initial blow of COVID-19 and again in July 2020 due to the impact of a propulsion tank weld failure. Within one to two months of each episode, FSM levels were restored to be in compliance primarily due to the use of reordering/replanning the future schedule. This capability to adjust the future was enabled by two mission characteristics: (1) the experience and expertise of the spacecraft prime contractor, LM, and (2) the project strategy of procuring known hardware and materials early in Phase B to mitigate notorious supply chain challenges often experienced but commonly overlooked in planning.

Overall mission-level cost reserves were held by the Lucy PM for issues that might arise during development and operations. At mission selection, in January 2017, the project carried $122 million or 32 percent in cost reserves to conduct development. Cost reserve management utilized a process that documented and traced every entry from reserve request or cost threat, to lien, to encumbrance, and finally to baseline that is fully documented through the configuration management process. Through launch, more than 450 requests had been reviewed and evaluated for reserve release and over $128 million had been encumbered (cost reserves were augmented for the 2019 Government Shutdown and COVID-19 impacts). Cost reserves were requested formally by the respective WBS lead to the PM. If accepted, a lien was held against the project reserve budget. If the issue triggering the reserve request became a manifest, i.e., became a requirement to implement; the lien transitioned to an encumbrance and the project reserve budget was decremented by the lien amount. Only the Lucy PI and/or PM were able to approve the use of project cost reserves. Cost reserve management was so essential to Lucy programmatics that a dedicated member of the resources team was assigned the role of Cost Reserve Coordinator to assist requests through this process and bi-weekly Threats meetings were held without fail with project leadership to discuss and disposition requests.

2.6.3 Approach to Earned Value Management

Lucy's approach to implementing EVM met the requirements imposed upon the project as a result of NPR 7120.5E, NASA's Federal Acquisition Regulation, and as directed by the Discovery 2014 Announcement of Opportunity (AO). No waivers were requested for the Lucy EVM construct.

The LM spacecraft contract exceeded the $100 million threshold, requiring certified EVM System (EVMS) reporting of which LM already possessed. The L'LORRI (APL) and L'TES (ASU) instrument developments took place within the SwRI Payload Management contract as subcontracted efforts. The contract value did not exceed $100 million, but was more than the $20 million threshold requiring EVM reporting from SwRI. SwRI had experience in EVM from previous hardware developments. The L'Ralph instrument was built at GSFC and exceeded the $20 million threshold for EVM reporting. Lucy's EVM approach was consistent with the best practices for implementing EVM at GSFC and was detailed in the Lucy EVM Plan released in July 2018. Since GSFC did not have a certified EVMS, Lucy implemented the guidelines of EVM as defined by Electronic Industries Alliance-748, EVM System, to the best of its ability. EVM training was made available to Lucy personnel prior to implementing the Lucy EVMS that was established for Phase C/D beginning in January 2019.

Lucy delivered earned value reports covering the full project – all WBS elements, including KinetX and level of effort management activities for which there were no separate earned value reporting requirements (because the value of these items fell below the dollar threshold for EVM requirements). Monthly schedule and cost reporting were utilized within the project to reconstruct the EVM performance for those efforts without EVM reporting requirements.

Over the 32 months of Phase C/D development and Earned Value monitoring, Lucy never fell below 0.90 in either the cumulative Schedule Performance Index or Cost Performance Index culminating in an on-time launch (1.0 Schedule Performance Index) and cost performance greater than 0.95 Cost Performance Index (despite the impact of COVID-19). With the initial pandemic impact of substance appearing in March 2020, the project began to baseline COVID-19 mitigation efforts by June 2020, only three months later, thereby neutralizing the prospect of long-term cost overruns and schedule delays.

2.7 Risk Management

All projects carry risk, and Lucy was no exception. The Lucy project managed risk through a continuous surveillance approach. The implemented Lucy Continuous Risk Management Plan operated consistently with the "B" risk classification, as defined in NPR 8705.4 Risk Classification for NASA Payloads. Lucy's Risk Management Board (RMB), comprised of senior Lucy personnel, met regularly to facilitate and lead the Continuous Risk Management process. The RMB, chaired by the Lucy PM, met at least monthly to review, assess, and update mission risks and their mitigation plans. All Lucy staff, including the instrument, flight, ground, and science teams were responsible for identifying risk and commensurate mitigation schemes; this data was then presented to the RMB for review. Risks that were deemed yellow or red were managed at the RMB level, and risks that fell below that threshold were managed by the WBS element leads. The spacecraft, instrument, and ground system teams also implemented risk management and reported on all risks on a monthly basis for evaluation and potential escalation to the project RMB, or closed as mitigated.

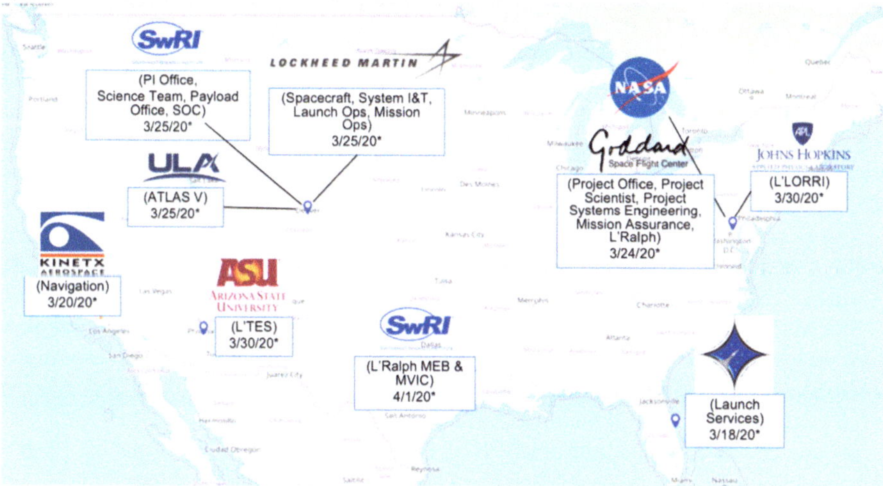

*Date when states issued stay-at-home orders or Centers transitioned to mandatory telework.

Fig. 5 Initiation of COVID-19 Protocols Across All the Partner Institutions

Only the Lucy PI and PM were authorized to expend project-held resources for mitigation, assign responsibility for risk mitigation plans, and approve risk closure and/or acceptance. The PM reported the highest risk items to GSFC's Center Management Council and NASA HQ as part of the MSR.

3 Leading Through a Once-in-a-Lifetime Pandemic

The nature of NASA projects often means that plans can change quickly. The project manager must be able to respond to these changes and adjust the project plan as needed. This became more evident during the pandemic.

In March 2020, five months before the planned start of spacecraft level assembly, test, launch, and operations (ATLO), the President declared a nationwide state of emergency as a result of the spread of the highly infectious novel coronavirus disease (COVID-19). By April 1, NASA transitioned to mandatory telework, and most states had issued mandatory stay-at-home orders. The Lucy spacecraft provider, as well as their supply chain, implemented changes to curb the spread of the virus. In some instances, these changes included closing facilities, affecting hardware development for varying lengths of time. Travel restrictions sharply reduced the amount of travel for typical integration support and minimized the number of personnel allowed on-site. The three Lucy instruments, propulsion subsystems, and solar arrays experienced significant schedule erosion jeopardizing the start of ATLO and ultimately our ability to meet the October 2021 launch readiness date. The team was challenged to maintain the health and safety of the Lucy team, remain within the Agency Management Agreement, and meet the planetary launch window.

First things, first. Lucy established project-wide COVID-19 protocols to keep the team safe: 1) Onsite support was reduced to essential personnel providing touch labor; 2) Teams were segregated, to the extent possible; 3) project travel was restricted; 4) the project provided protective equipment, and 5) institutions performed enhanced cleaning and disinfecting.

The project utilized a proactive "what-if" approach to assess "risk exposure", identify trade space, and to develop work around plans for a number of scenarios, including work stoppages due to COVID, late hardware delivery, supply chain issues, inability to travel, etc. The team scrubbed the instruments and spacecraft test programs for optimization and to identify areas where the project could assume a higher risk posture to save schedule. For example, "testing as you fly" and "flying as you test" were critical considerations. New delivery dates were established for the instruments. The new dates included contingency, as well as schedule slack. The project provided additional oversight via daily tag-ups and weekly schedule reviews to give increased confidence that the instruments would meet the new dates.

The project aggressively spent reserves to mitigate COVID-19 impacts and meet the planetary launch window: 1) augmenting teleworking capabilities, including virtual inspections; 2) bringing on surge support for the spacecraft and instruments; 3) identifying opportunities to buy back schedule; and, 4) building instrument mass simulators.

In June 2020 the project implemented its Feasible ATLO Shortened Timeline (FAST) Plan, which:

- Accommodated late delivery of key hardware, restored the critical path funded schedule margin, and maintained the ATLO start date and Kennedy Space Center ship date.
- Maximized team resiliency to COVID-19, through both segregation of personnel, and utilization of weekend and alternate shift work where feasible.
- Utilized tools such as Zoom and Microsoft Teams to enable off-site personnel to monitor integration activities as well as test telemetry "over the shoulder" of onsite personnel.

Part way through ATLO, the project conducted a strengths, weaknesses, opportunities, and threats analysis to review the FAST plan. The project identified a total of twenty-three weaknesses/threats and twenty opportunities to mitigate schedule risks and further increase team resiliency for the remainder of ATLO and launch processing.

We could not lose site of the "human factor". The project's success thus far, had hinged on the social/relationship equity built up over the years. They continued to make a concerted effort to maintain the sense of team spirit with new creativity, including:

- Virtual parties/gatherings, a team peer award program, "accomplishments of note", established peer "buddies" for support, virtual bingo, social distanced gatherings, etc.
- Project leadership traveling to work sites to show support to personnel required to work onsite.
- Adding more frequent team wide meetings to facilitate communication.

It worked! The project did not have to compromise the ground test campaign. The spacecraft shipped to the launch site and launch operations proceeded as planned. Lucy launched on time and under budget during a pandemic; compressing 9.5 months of work into 7 months, while keeping the team safe!

4 Lessons Learned

Lucy was successful in navigating COVID and launching on time for a number of reasons:

- They maintained a "people first, mission always" mindset.
- They balanced time, cost, and performance, with risk as the discriminator.
- They had healthy cost reserves, which were used to buy additional support and time.

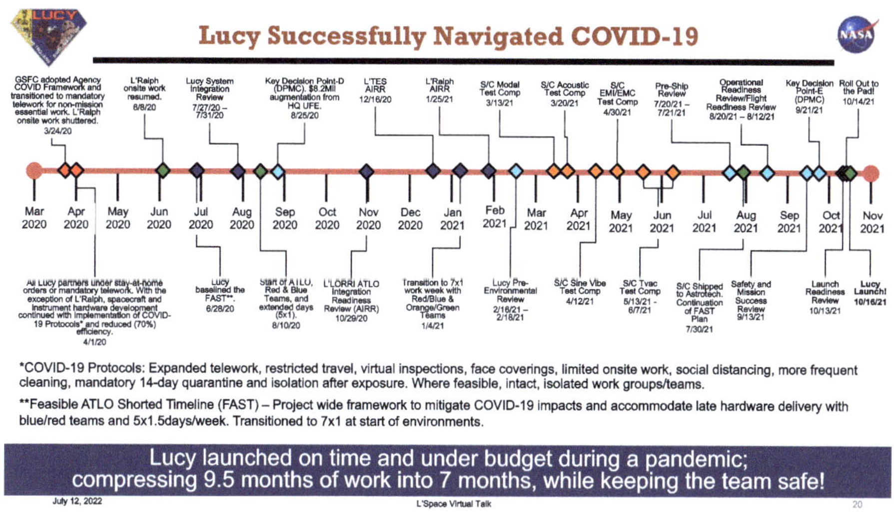

Fig. 6 Lucy Successfully Navigated COVID-19

- Risk management was continuous. They developed mitigation strategies with offramps.
- They embodied the characteristics of family... openness, grace, appreciation, interest in each other and their lives. They stepped in to help each other, filling in the gaps as needed.
- They worked hard and celebrated every success, big or small, together.
- Schedule ***was*** king. Every decision took the potential schedule impact into account.
- They were single-minded in their vision – to get to launch and have a successful mission.

Although the team was extremely successful, FAST came at a price – the project pushed hard for 10.5 months under very taxing circumstances. As time progressed, team brittleness because apparent. This would not have been sustainable under normal circumstances.

Glossary

APL	John's Hopkins University Applied Physics Laboratory
ASU	Arizona State University
ATLO	Assembly, Launch, Test, and Operations
DPI	Deputy Principal Investigator
DPM	Deputy Project Manager
EVM	Earned Value Management
EVMS	Earned Value Management System
FAST	Feasible ATLO Shortened Timeline
FSM	Funded Schedule Margin
GPR	Goddard Procedural Requirement
GSFC	Goddard Space Flight Center
IMS	Integrated Master Schedule
IR	Infrared
KDP	Key Decision Point
LCC	Lifecycle Cost
L'LORRI	Lucy LOng Range Reconnaissance Imager
MSFC	Marshall Space Flight Center
MSR	Monthly Status Review

NASA	National Aeronautics and Space Administration
NPD	NASA Policy Directive
NPR	NASA Procedural Requirement
PI	Principal Investigator
PM	Project Manager
PMPO	Planetary Mission Program Office
PSD	Planetary Science Division
PSE	Project Systems Engineer
RMB	Risk Management Board
SSMO	Space Science Mission Operations
SwRI	Southwest Research Institute
L'TES	Lucy Thermal Emissions Spectrometer
WBS	Work Breakdown Structure

Acknowledgements The authors would like to thank Mr. Michael Donnelly, Mr. John Loiacono, Mr. Robert Jenkins, Mr. Gary Davis, Mr. Mike Amato, and Mr. Charles Baker, and Mr. Matt Cox for their leadership and contributions to the Lucy mission.

Declarations

Competing Interests The author affirms that there is no conflict of interest and that the subject of this manuscript did not involve research on human nor animals.

Space Science Reviews (2023) 219:86
https://doi.org/10.1007/s11214-023-01030-5

The Terminal Tracking Camera System on the NASA *Lucy* Trojan Asteroid Discovery Mission

J.F. Bell III[1] · Y. Zhao[1] · E. Cisneros[1] · M. Beasley[2] · C. Olkin[2] · M.A. Caplinger[3] · M.A. Ravine[3] · J.A. Schaffner[3] · M.J. Clark[3] · J. Shamah[3] · P. Faiks[4] · S. Mottola[5] · C. Adam[6] · E. Lessac-Chenen[6] · B.J. Bos[7]

Received: 17 June 2023 / Accepted: 21 November 2023 / Published online: 13 December 2023

Abstract
The Terminal Tracking Camera system (TTCam) on the NASA *Lucy* Trojan asteroid Discovery mission consists of a pair of block redundant cameras and their associated electronics that are mounted on the spacecraft's Instrument Pointing Platform and co-boresighted with the rest of the mission's science payload instruments. The primary function of the TTCams is as a navigation system designed to provide an autonomous onboard late pre-encounter update of the location of each asteroid flyby target relative to the spacecraft. However, once the terminal tracking function is complete, the TTCam system will also provide 11.0°×8.2° field of view broadband (425-675 nm) images during the close approach phase of each asteroid flyby that will be used for scientific analyses like shape modeling and assessment of each target's geology and topography. This paper provides an overview of the TTCam cameras and electronics, the science-focused requirements that the system is designed to meet, pointers to pre-flight calibration and in-flight calibration details for the cameras, as well as a high-level summary of the kinds of science that these images will enable for the mission.

Keywords NASA *Lucy* mission · Trojan asteroids · Spacecraft instrumentation · Deep space cameras

1 Introduction

The *Lucy* mission was chosen for flight in 2017 as part of NASA's Discovery Program of low-cost, competitively selected deep space robotic science missions. The mission was suc-

✉ J.F. Bell III
Jim.Bell@asu.edu

[1] School of Earth and Space Exploration, Arizona State University, Tempe AZ, USA

[2] Southwest Research Institute, Boulder, CO, USA

[3] Malin Space Science Systems, Inc., San Diego CA, USA

[4] Lockhed Martin Space, Littleton, CO, USA

[5] German Aerospace Center (DLR), Berlin, Germany

[6] KinetX Inc., Tempe, AZ, USA

[7] NASA Goddard Spaceflight Center, Greenbelt, MD, USA

cessfully launched on October 16, 2021 and conducted its first Earth Gravity Assist (EGA) flyby on October 16, 2022. After a flyby of main belt asteroids (152830) Dinkinesh in 2023 and (52246) Donaldjohanson in 2025, *Lucy* will conduct five flyby encounters of eight members of Jupiter's Trojan asteroid population between 2027 and 2033. The primary goals of the *Lucy* mission include assessing the diversity of the Trojan asteroids, determining their surface composition, assessing their geology, determining their bulk properties (including mass and density), and searching for satellites and rings. Additional details about the science goals of the *Lucy* mission can be found in Levison et al. (2021).

To achieve these goals, the *Lucy* spacecraft carries a primary science payload consisting of a multispectral camera, infrared imaging spectrometer, high-resolution panchromatic imager, and a thermal infrared spectrometer. Additionally, two spacecraft subsystems also act as secondary science payloads: the spacecraft's communications system will be used to determine the mass of each Trojan target via Doppler tracking, and a wider-field panchromatic camera system will be used to perform autonomous terminal tracking of the Trojan targets during the close approach subphase of each flyby. These Terminal Tracking Cameras (TTCams) are part of the guidance, navigation, and control subsystem of the Lockheed Martin spacecraft's Flight System (Cox et al. 2024), and are designed to help provide updated knowledge of the target locations for the spacecraft's science instruments mounted on the two-axis articulated Instrument Pointing Platform (IPP). However, the TTCams will also provide extensive imaging coverage over a wide range of phase angles immediately before, during, and after each flyby (after the system's terminal tracking function is complete), thus providing unique constraints on the shape and volume of each target as well as additional coverage of each target for geologic characterization and mapping. In turn, estimates of both the mass and volume will lead to an estimate of the density of each target. Additional details about the *Lucy* mission's science goals, payload, IPP, and mission operations concept can be found in Levison et al. (2021) and Olkin et al. (2021).

This paper describes the requirements, design, testing, and expected performance characteristics of the *Lucy* TTCam imaging system, with a focus on the instrument characteristics and performance that will enable the additional science results described above. Additional details about the specific timing and algorithms that will be used by the TTCam system for terminal tracking of each target are provided in Good et al. (2022), and additional details about the pre-flight and in-flight calibration of the TTCam system are provided in Zhao et al. (2024).

2 *Lucy* Terminal Tracking Camera Requirements

The driving science-relevant requirements for TTCam (Table 1) are set by the camera's primary function as a navigation system designed to provide an autonomous onboard late pre-encounter update of the location of each target relative to the spacecraft. By definition and "to do no harm" to the system's ability to achieve its primary target tracking function for the mission, Requirements 1 through 4 were defined as being fully adequate for TTCam images to be able to meet their secondary science requirements related to shape modeling/volume determination (e.g., Mottola et al. 2024) and additional geologic characterization (e.g., Marchi et al. 2023). Requirements 5 and 6 enable higher-fidelity images for science analysis as well as finer temporal (phase angle) sampling in the few minutes around each closest approach. Neither of these science-focused requirements will impact the ability of the TTCam system to meet its primary navigation and tracking function. Additional requirements and specifications for the TTCam system that are primarily related to the system's tracking function are described in Good et al. (2022).

Table 1 Science-Related Requirements for the *Lucy* Mission Terminal Tracking Camera System

1	Provide sufficient sensitivity to allow an image with a Signal to Noise Ratio (SNR) of ≥ 10 for a target with a radiance factor (I/F) of 0.0020 at a solar distance of 5.7 AU and for an integration time of ≤ 400 msec.
2	Have a Field of View (FOV) sufficiently large to completely encompass each target throughout each flyby.
3	Have an Instantaneous Field of View (IFOV) sufficiently small to resolve a target with visible dimensions of 52×45 km at 51,400 km distance and at a solar phase angle of 125° to greater than 4 pixels wide and 10 total pixels of area.
4	Acquire images of a V magnitude 6 star with SNR > 5 for a 10 second exposure.
5	Flat field images at SNR > 100 shall be generated during pre-flight ground calibration.
6	The cadence of image acquisition during the closest approach phase of each flyby (once the terminal tracking function has completed) shall enable $\leq 10°$ sampling of solar phase angles.

Requirements 1 though 4 drive the choice of detector and its Quantum Efficiency (QE), the detector size and pixel pitch, the $f/\#$ of the optics, and the aperture size. For example, the driving case for Requirement 2 (Field of View: FOV) is the flyby of the (617) Patroclus/Menoetius Trojan binary pair in 2033. The approximately equal-sized asteroids each have an estimated projected size on-sky of $\sim122\times113$ km. Thus, for a nominal close approach flyby distance of ~1000 km, the TTCam FOV required to encompass each of these objects would need to be at least $\sim10.8°\times8.1°$, including a 25% to 45% margin to accommodate errors on the estimated diameter and spacecraft pointing. Further, given the rectangular aspect ratio of typical digital imaging sensors, the orientation of the camera on the spacecraft was chosen such that the wider axis is along the direction of the largest expected pointing errors. As another example, to obtain images at least 4 pixels wide of a 52×45 km object at 51,400 km distance (simulating an asteroid being tracked on approach) would require the optics to have an Instantaneous Field of View (IFOV) of $\leq\sim253\times219$ µrad/pixel, with 219 µrad being the driving case minimum pixel scale for Requirement 3. Finally, Requirements 1 and 4 also provide specific constraints on the QE, read noise, gain, full well, and transmissivity of the optics that would be required to meet the specified extended source or point source SNR levels.

Requirement 5 (flatfields for science) was designed to be met by increasing the number of flatfield images acquired for each camera during pre-flight calibration to increase the SNR of co-added flatfield images via simple photon counting (square root of the number of images combined) statistics. Requirement 6 (phase angle sampling) is designed to enable excellent shape modeling of each target, and will be met by increasing the cadence of imaging of each target during the few minutes around closest approach, starting after the primary target tracking function of the system has completed. The stressing case for this requirement is the 2027 flyby of (15094) Polymele, where an imaging cadence of ~13 seconds between subsequent images will be required to enable solar phase angle sampling in increments of $\sim10°$.

3 Target Tracking and Optical Navigation Support Objectives

As described in Good et al. (2022), the TTCam's primary role in the *Lucy* mission is to provide center-of-body measurements that allow the encounter solution to converge on the range to the target asteroid. *Lucy*'s science observations during the close approach phase of each encounter are range-triggered and an accurate solution is required to achieve *Lucy* science objectives.

In addition, the *Lucy* Navigation team will use TTCam images taken within the minutes surrounding each encounter close-approach for trajectory reconstruction. This was added to the baseline optical navigation (OpNav) plan during development, in order to meet the Trojan closest approach radial distance accuracy requirement. Additionally, the OpNav team will characterize the in-flight optical distortion and boresight alignment of the TTCams relative to the L'LORRI boresight (see Sect. 7.2 and Zhao et al. 2024).

4 Design Considerations and Heritage

In addition to meeting the specific performance requirements identified in Table 1, other important objectives in the selection of a specific implementation of the TTCam imaging system were to minimize size, mass, and power requirements as well as cost, and to maximize system heritage acquired from previous deep space missions. Given those constraints, Lockheed Martin Flight System engineers and managers chose to procure an imaging system developed by Malin Space Science Systems, Inc. (MSSS; San Diego, CA) that is based on their ECAM line of science-grade polychromatic cameras (MSSS 2013; Ravine et al. 2016). The CMOS sensor, camera acquisition and storage/processing electronics, and camera flight software and logic chosen for TTCam have high heritage from the TAGCAMS imaging system flown successfully on the NASA OSIRIS-REx mission (Bos et al. 2018, 2020) and the Junocam imaging system flown successfully on the NASA Juno mission to Jupiter (Hansen et al. 2017). The optics design ultimately selected for the TTCam system is a unique prescription designed to specifically meet the *Lucy* mission requirements, but is based on many previous MSSS-developed deep space refractive optical designs.

5 Instrument Description

5.1 CMOS Sensor and Camera Head Electronics

Figure 1 shows the basic block diagram for the two fully block redundant TTCam camera heads and their associated Digital Video Recorder (DVR) electronics cards (each known as a "slice"). The heart of each TTCam is a 2752×2004 pixel monochrome CMOS imaging sensor from Aptina/ON Semiconductor (Model MT9P031). The sensor has 2.2 μm pixels and is operated at 10 MHz clock rate in Electronic Rolling Shutter mode, resulting in a ~680 msec full-frame readout time from the top to bottom row of each image (ON Semi 2017). On-chip electronics convert amplified photon-induced voltages to digital signals with a 12-bit dynamic range (Data Numbers or DN from 0-4095) and a selectable analog gain from 1× to 3× for this implementation. The QE of the detector ranges from ~55% to ~35% over the 425 to 675 nm bandpass of the system, with a peak of ~62% near 470 nm (ON Semi 2017).

Due to a limit on the image size that can be transferred into the spacecraft's terminal tracking algorithm, only the pixels in the 2592×1944 active (light sensitive) image area are retrieved and used for imaging during terminal tracking mode observations (Fig. 2). For calibration mode images not taken during asteroid tracking, additional rows of masked pixels outside of the active area are downlinked to provide 2592×2000 pixel images that enable assessment of the bias plus dark current background signal levels over time and temperature.

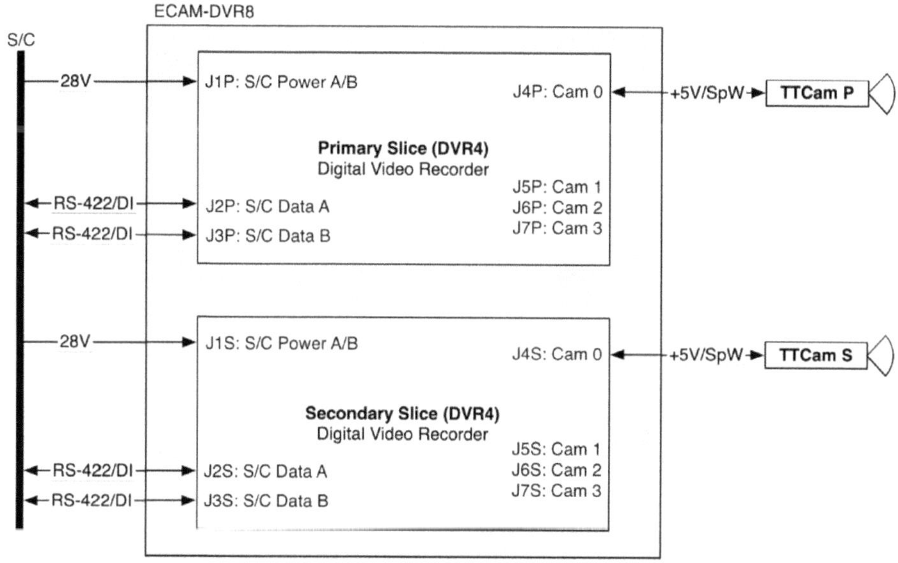

Fig. 1 TTCam electronics block diagram. Additional external harnessing for thermal control is not shown

These calibration mode images provide a way to validate a model (Zhao et al. 2024) for the removal of these effects in the calibration of the 2592×1944 terminal tracking mode images lacking masked pixels.

Based on the *OSIRIS-REx* TAGCAMS experience using this same sensor (see Bos et al. 2020 Fig. 6 and its associated text), only the masked pixels in rows 1966 to 1989 and columns 16 to 2607 in Fig. 2 are used for the assessment of background signal level in calibration mode images that contain masked pixels. For asteroid tracking and encounter images downlinked without masked pixels, a modeled background is used to remove that background signal from the raw data (see Sect. 6.1 below and Zhao et al. 2024).

The TTCam sensors are controlled by dedicated electronics in each camera head assembly. The camera head electronics are built around a rad-hard FPGA that performs all logic functions, including CMOS sensor power control, clock generation, and command and Spacewire interfaces. In addition, the FPGA can perform a limited amount of pixel processing, including 12-bit to 8-bit companding (see Sect. 5.3), prior to transfer of imaging data to the DVR.

5.2 Digital Video Recorder (DVR)

The TTCam DVR receives +28 V of nominal bus voltage and receives commands from the *Lucy* spacecraft, provides switched +5 V of power to the CMOS sensors and camera head electronics, and captures image data from the camera heads over a SpaceWire interface. Camera-specific flight software and logic developed by MSSS is loaded into each slice of the DVR and runs on a Microblaze soft processor within a Xilinx FPGA. The DVR controls the storage of images in flash memory (up to ∼1600 uncompressed 8-bit full-sensor images can be stored per camera) and the transmission of image data as well as health and status telemetry back to the spacecraft computer via an RS-422 asynchronous serial interface. The DVR also runs internal fault management (including single-event-upset mitigation) and can

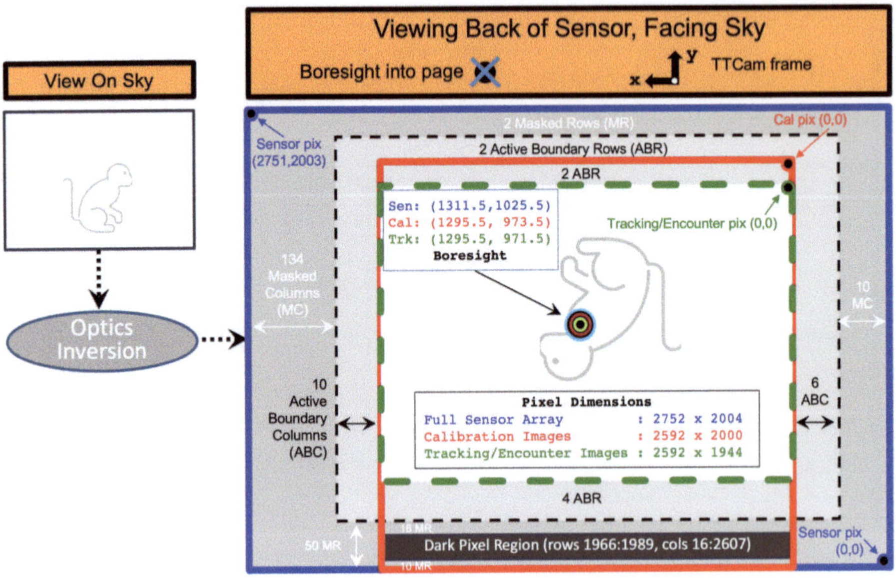

Fig. 2 Cartoon diagram of the full TTCam1 CMOS detector array as it is oriented in the TTCam coordinate frame, showing the full sensor coordinate system (blue outline), and the pixel areas and TTCam coordinate system for downlinked images in Calibration Mode (red outline) and in Tracking/Encounter Mode (green outline). Pixel area dimensions are not to scale. The sensor's active image area is in white, the active (unmasked) boundary pixel area is in light gray, the masked pixel area is in dark gray, and the masked pixel area used for bias and dark current assessment is in black. An image of a simian is shown as it appears on the sky and as it is projected onto the sensor array by the instrument optics. All indexing is 0-based center-of-pixel. The vertical (y) axis of the downlinked images runs anti-parallel to the sensor array coordinates. The sensor geometry is the same for TTCam2, but the camera and images are rotated 180° about the optical axis relative to TTCam1

Fig. 3 CAD model of the three main components of the TTCam camera head and cutaway view of the integrated system

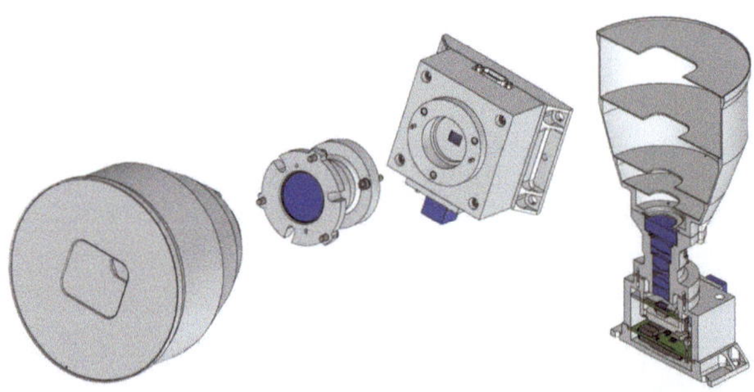

perform optional onboard lossless (Huffman first-difference) or lossy (JPEG) compression. The DVR contains two independent slices, one per camera head, in a single aluminum housing. Each slice contains redundant data connectors to enable it to interface with either the primary or backup *Lucy* flight computer.

5.3 Optomechanical Design

Figure 3 shows a CAD model representation of the three main components of each TTCam camera head: electronics box, optics assembly, and sunshade. The electronics box (which

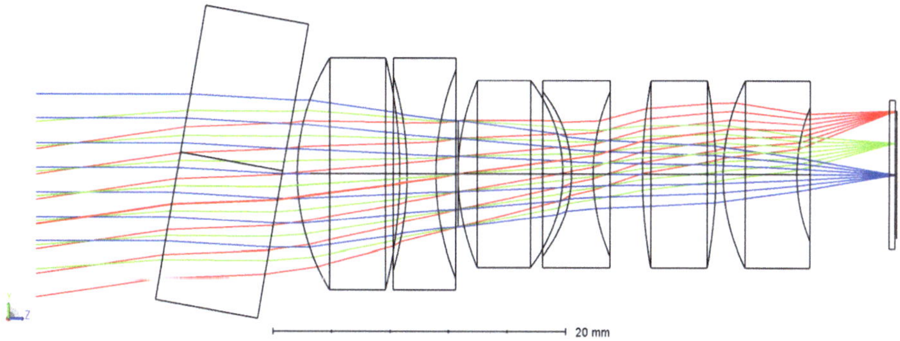

Fig. 4 TTCam optics ray trace. The front window is intentionally tilted by ~5° to help mitigate stray light

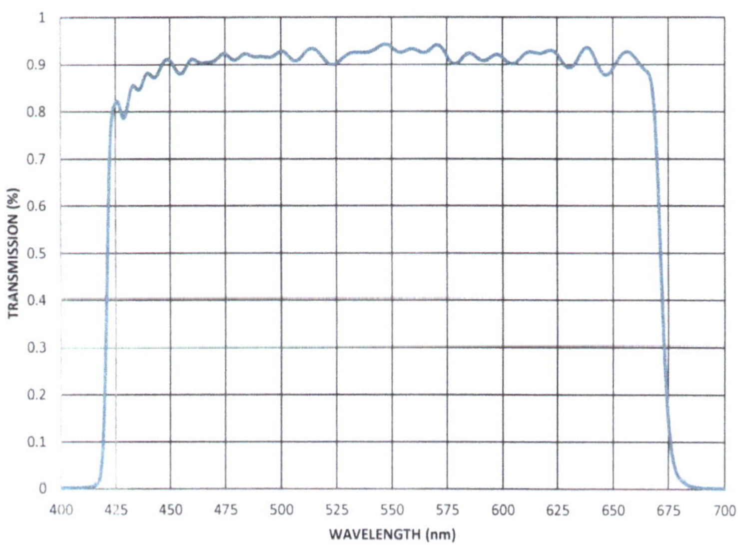

Fig. 5 Overall transmission of the TTCam optics + bandpass filter as a function of wavelength

houses the CMOS sensor and its FPGA electronics) uses the same mechanical design as that used on the OSIRIS-REx TAGCAMS (Bos et al. 2018).

MSSS subcontracted to Collins Aerospace (Carlsbad, CA) for the TTCam lens design, fabrication, and component-level testing. The optics (Fig. 4) consist of a seven-element refractive design, including an intentionally-tilted front window to help mitigate stray light. In addition, a 425-675 nm bandpass filter (Fig. 5) is part of the tilted front window, to enable broad-band imaging in the wavelength region where the sensor QE is highest. The TTCam optics are designed to be athermal over the expected operating temperature range of the cameras (−40 °C to +55 °C). The optics alone were required to meet a 30% Modulation Transfer Function (MTF) requirement at the Nyquist frequency (227 line pairs/mm), and the overall system (optics + filter + CMOS sensor) was required to have an MTF greater than 10% at Nyquist. Despite the relatively wide field of view, distortion in the lens design is expected to be very small, only ~0.12% in the corners of the field. Stray light in the system is also expected to be extremely small, with the strongest ghost images predicted to have an

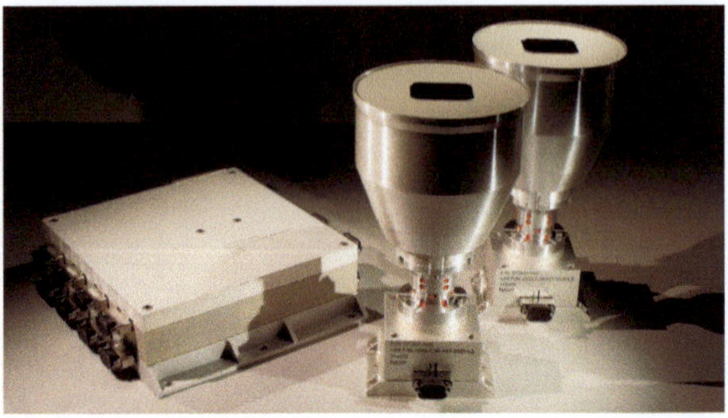

Fig. 6 Photo of the *Lucy* TTCam flight Digital Video Recorder (left) and the two flight camera head electronic boxes and sunshades (from left to right: serial numbers 194503 – the Primary camera – and 194504 – the Secondary camera). For scale, the DVR is 19×16×7 cm in size and each camera head electronics box is 8×8 cm square.

intensity of 0.5% or less of the primary image. These very small distortion and stray light levels have been validated in flight (Zhao et al. 2024).

5.4 Onboard Image Processing

Onboard processing of raw TTCam images is performed both within the DVR flight logic prior to transferring the images to the *Lucy* spacecraft, and then within the spacecraft flight software itself, after image transfer but prior to eventual downlink. First, the raw 12-bit DN values in each pixel are compressed to 8-bits, nominally using an onboard square-root lookup table designed to avoid Poisson (shot) noise being encoded or downlinked in the telemetry (a process known as "companding"; e.g., Malin et al. 2013; Bell et al. 2017). For some cruise image tests of different companding schemes, the 12-bit to 8-bit compression is achieved either using a divide-by-16 linear scaling, or by clipping the 12-bit images down to just their least-significant 4 bits. Then, the 8-bit companded but otherwise uncompressed images are transferred to the spacecraft computer, where they are further compressed using the lossless PPMd algorithm (Prediction by Partial Matching, variant d; e.g., Jumar et al. 2018) to further decrease the average number of bits per pixel to be downlinked.

5.5 Integration and Accommodation

Figure 6 shows the TTCam flight DVR and camera heads (Primary = DVR1 = serial number 194503; Secondary = DVR2 = serial number 194504). Additionally, two ~76 cm harnesses fabricated by Glenair (Glendale, CA) connect each slice of the DVR to its corresponding camera head. The camera head electronics box, optics module, and sunshade were integrated at MSSS in the winter and spring of 2020, and a series of ambient and thermal vacuum condition Verification and Validation (V&V) and calibration tests were performed at MSSS in June and July, 2020 (see Sect. 6 and Zhao et al. 2024). The flight DVR and flight camera heads were delivered to the Lockheed Martin *Lucy* mission Assembly, Test, and Launch Operations (ATLO) team in July, 2020, and were installed on the spacecraft's Instrument Pointing Platform shortly thereafter (Fig. 7). For harness and connector accommodation, TTCam1 and TTCam2 are rotated by 180° about the optical axis relative to each other, and thus the entire sky to sensor to downlinked image coordinate frame relationship shown in Fig. 2 is repeated for TTCam2, but rotated 180° relative to TTCam1. As described in more detail in Olkin et al. (2021) and Good et al. (2022), both the Primary and Secondary TTCams are co-boresighted with the rest of the *Lucy* instruments, and all instrument FOVs are completely contained within the TTCam FOVs.

Fig. 7 Photo of the Lucy TTCam1 (T1) and TTCam2 (T2) flight cameras (with red "Remove Before Flight" covers) installed on the spacecraft's Instrument Pointing Platform. The large cylinder to the upper left of T1 is the L'LORRI telescope (L), and the smaller black cylinder below T2 is the telescope for the L'TES infrared spectrometer (TES) [see Olkin et al. (2021) and Good et al. (2022) for additional instrument and spacecraft details]. Photo taken in Aug. 2021 at Lockheed's pre-launch integration facility at Cape Canaveral, Florida, and used with permission of Lockheed

6 Expected Instrument Performance

6.1 Preflight Predictions

During the design phase of the TTCam development process, a model was generated to predict the SNR that could be achieved for Trojan asteroid targets imaged by the TTCam system. Known/verifiable parameters in the model included the QE, read noise, gain, and dark current rate of the sensor, the transmission of the optics, and the flyby speed at each encounter (to factor in potential image smear). Assumed or variable characteristics of the model included the heliocentric distance of the observations, and the nature of the surface phase function of the target asteroids. Table 2 shows an example of the resulting SNR predictions for TTCam observations of each of the *Lucy* flyby targets for assumed Lambertian ("Albedo" columns) and Lommel-Seeliger ("LS" columns) surface phase functions. These predictions, since validated from Earth and Moon observations in flight (see Sect. 7.2 and Zhao et al. 2024), provide confidence that the actual flight cameras would be able to meet the SNR requirements specified for the system (Table 1).

6.2 Preflight Characterization Measurements

After subsystem integration and initial functional testing, the two flight and one flight spare (serial number 194501) TTCams went through an ambient and thermal vacuum V&V and calibration testing program at MSSS that included assessments of CMOS sensor and electronics gain, read noise, linearity/full well, and dark current rate, as well as system radiometry, bandpass performance, flatfield behavior, and assessments of MTF and FOV/distortion. Detailed stray light measurements and modeling were also conducted during ATLO (see

Table 2 Predicted SNR from *Lucy* TTCam observations of each of the Trojan flyby targets

Body – Trojan Swarm	Average Diameter (km)	Solar Phase (°)	Encounter Velocity (km/s)	Stability (°/s)	Closest Approach (km)	Sun Range (AU)	Dia./FOV @C/A	IFOV (m/pix)	Exp. Time (ms)	Albedo Model	Albedo SNR	LS Model	LS SNR
Eurybates – L_4	46.4	81	5.8	0.09	1000	5.7	0.32	74.1	20.0	0.050	47.4	0.077	59.3
Polymele – L_4	15.1	81	6.0	0.27	400	5.7	0.26	29.6	20.0	0.090	64.2	0.077	59.3
Leucus – L_4	65	104	5.9	0.09	1000	5.6	0.45	74.1	20.0	0.080	61.5	0.087	64.2
Orus – L_4	31	126	7.1	0.09	1000	5.3	0.22	74.1	20.0	0.080	65.1	0.086	67.5
Patroclus/ Menoetius – L_5	132/ 122	56	8.8	0.09	1244/ 1074	5.4	0.79/ 0.74	92.2/ 79.6	20.0	0.047	50.5/ 48.5	0.058	54.1/ 56.3

Sect. 6.3) to take into account the potential scattered light effects of other payload elements and subsystems on the IPP and spacecraft. Details and results from many of these pre-flight tests and analyses are described in Zhao et al. (2024), and Table 3 summarizes the resulting derived or measured properties and characteristics of the TTCam system.

6.3 Assembly, Test, and Launch Operations (ATLO) Testing

Testing in ATLO was primarily driven by the terminal tracking needs of the mission. The minimum focal distance of the TTCams (~200 m) precluded meaningful imaging in the ATLO test environment, so pre-made images were loaded onto the DVR to be sent to the spacecraft during testing. See Good et al. (2022) for a discussion of the image simulation and testing.

Additional scattered light testing was done during ATLO as part of a larger, mission level scattered light test. Earlier simulations had indicated that sunlight might be reflected from the L'LORRI baffles in certain orientations of the spacecraft on approach or departure from a target. Testing was done with a setup designed to simulate the various potential Trojan encounter geometries (Fig. 8). Actual scattered light levels were much lower than predicted by the earlier modeling, and thus scattered light is predicted to have no impact to either science or terminal tracking functionality. This prediction has been validated from dedicated cruise scattered light characterization observations (see Sect. 7.2 and Zhao et al. 2024).

7 Mission Operations and Data Products

7.1 Operational Considerations

Because the flyby velocities are so high (~6-9 km/sec) and the encounters occur so far from Earth (42 to 47 light-minutes), the primary function of the TTCam imaging system is to provide an important late (shortly before each flyby) update of the location of each target asteroid relative to the *Lucy* spacecraft (Cox et al. 2024). TTCam's navigation and tracking function will thus enable more accurate pointing of the much higher-resolution and smaller-FOV instruments on the *Lucy* science payload. Since the system will have completed that job at least several minutes before each close approach but will continue collecting images throughout the entire encounter, TTCam images will also be available to help fulfill some of the science objectives of the *Lucy* mission as well.

Operationally, however, using TTCam during the close approach phase of each flyby will involve a change in data collection cadence. Specifically, TTCam images used for the Flight

Table 3 *Lucy* Mission Terminal Tracking Camera (TTCam) Specifications

Optics	*Description*
Focal length	29.7 mm (all-refractive; fixed focus)
Focal ratio	*f*/2.95
Depth of Field	≈200 m to ∞ (hyperfocal: 133.7 m; near-focus: 66.8 m)
entrance pupil aperture	9.9 mm
aperture area	77 mm^2
exit pupil diameter	7.1 mm
t/#	< 3.36
pixel solid angle	5.5 × 10^{-9} sr
throughput ($A_0 \bullet \Omega$)	4.22 × 10^{-7} mm^2 sr
IFOV	74.1 µrad/pixel (74.1 m/pixel @ 1000 km)
FOV	11.0° × 8.2° (13.7° diagonal) (193×143 km @ 1000 km)
Filters	1
Filter bandpass	Panchromatic: 425-675 nm
Effective Wavelength	Avg. Red Trojan: 537.8 nm (FWHM: 246.7 nm)
(Filter × QE × Sun × Trojan)	Avg. Less-Red Trojan: 534.1 nm (FWHM: 245.2 nm)
Optics Transmission	> 80%
Optics Distortion	0.12% average in corners of field of view
MTF	0.19 (optics+filter+sensor) at Nyquist = 227 l.p./mm
Detector & Electronics	*Description*
Sensor	CMOS: ON Semi Model MT9P031; Electronic Rolling Shutter mode
Clock Rate	10 MHz in 8-bit mode
Exposure Durations	~0.0 to 30.7675 sec; commanded in units of 0.1 msec
Electronic Rolling Shutter	Min. exp.: 121 or 60.4 µsec (8-bit or 12-bit); Frame time ~680 msec
Array size	2592 × 1944 active pixels (2752 × 2004 total)
Preferred Dark Pixels (*cf.* Fig. 2)	rows 1966-1989, cols 16-2607 (region 4 of Fig. 6 in Bos *et al.*, 2020)
Pixel size	2.2 µm (square pixels)
Gain, Read Noise, Full Well	1.8 e^-/DN; 7 e^-; 5400 e^-
Quantum Efficiency	> 50% at 550 nm
Digitization	12 bits/pixel; single gain, no offset states
DVR to Camera data interface	SpaceWire, 100 Mbit/sec
DVR-S/C Cmd/Housekeeping interface	RS-422, 57.6 Kbit/sec Asynchronous Serial
DVR-S/C video telemetry interface	RS-422, 8.33 Mbit/sec Synchronous Serial
Memory	8 GB non-volatile RAM (≈1600 8-bit images) per camera
Camera Head Power	1.5 W standby and 2.0 W imaging, per camera head
DVR Power	6.0 W standby and 8.0 W imaging, per camera board
Image Compression Options	*Description*
Uncompressed	12-bit data (0-4095 DN); No compression
Lossless	In DVR: First-difference, Huffman; In spacecraft: PPMd
Lossy	In DVR: JPEG, commandable compression quality (1–100)
Companding	In DVR: 12-bit to 8-bit square-root encoding (lookup table)
Physical	*Description*
Data Volume: full raw image	2752×2004×12 bits = 8.27 MBytes/image
Data Volume: 8-bit companded	2752×2004×8 bits = 5.52 MBytes/image
Data Vol: 8-bit then lossless comp.	2.76 MBytes/image, assuming 2:1 lossless compression
Dimensions: CMOS Sensor	5.7 × 4.3 mm (7.1 mm diagonal)
Dimensions: Camera Head	8 × 8 × 22 cm (each camera)
Dimensions: DVR	19 × 16 × 7 cm
Mass: Camera Head	0.53 kg, CBE (each camera)
Mass: DVR	1.90 kg, CBE

System's navigation and tracking function will be acquired at a cadence of approximately 15 seconds per image, starting several hours before closest approach. Those tracking and encounter images will be rapidly transferred from the camera head to the DVR and into

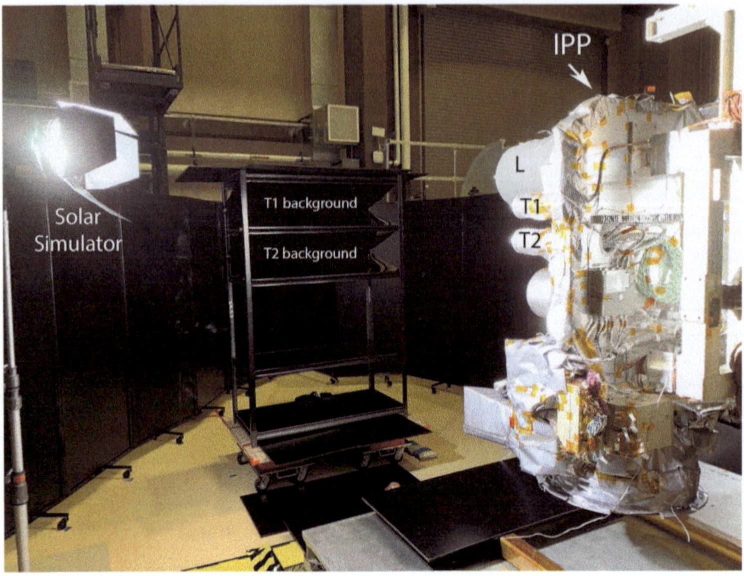

Fig. 8 Photograph of the setup used for TTCam stray light image testing during ATLO on May 6, 2021 at the Lockheed Spacecraft facility in Littleton, CO, and used with permission of Lockheed. The solar simulator lamp was oriented at different positions to simulate the viewing geometries of the Instrument Pointing Platform (IPP) during the various encounters. The fields of view of the Primary TTCam (T1) and Secondary TTCam (T2) were filled by dark background targets. Models had predicted that sunlight scattering off the L'LORRI telescope baffle (L) might produce contaminating background signature in the TTCams that could interfere with their tracking and navigation function. The actual tests showed extremely low scattered light levels in all simulated geometries, however

the spacecraft computer in 8-bit companded mode without the masked dark pixels. These smaller images are required because of a limit on the image size that can be transferred into the spacecraft's terminal tracking algorithm. Once the tracking function is complete, however, (typically around 2 minutes before close approach), the cadence of TTCam imaging can be increased to provide the required phase angle sampling of each target (Requirement 6 in Table 1). These "science phase" TTCam images will also not include the masked pixel area of the CMOS array, because the images will still be getting acquired as part of the earlier navigation and tracking sequence.

Separately, as part of periodically executed in-flight calibration campaigns begun shortly after launch and expected to continue before and after the Trojan flybys (see Sect. 7.2), we are acquiring nearly full-sensor calibration images of star fields and the Earth and Moon that do include the masked pixel area. These calibration images are being used to validate the pre-flight radiometric calibration predictions for the TTCams, to monitor their performance over time, and to adjust the dark current models and/or bad pixel maps for the cameras, if needed (Zhao et al. 2024).

Many more details about the operational use of TTCam and the other *Lucy* payload elements can be found in Olkin et al. (2021) and Good et al. (2022).

7.2 In-Flight Calibration Observations and Plans

Since launch, a series of in-flight observational tests have been initiated to test and validate the performance of the TTCams to the fidelity necessary to perform the required asteroid tracking, navigation reconstruction, and science tasks. The results from these calibration activities are being used to refine the calibration pipeline before reaching the first Trojan asteroid, as well as to provide valuable performance metrics for planning each encounter (Zhao et al. 2024).

The first TTCam functional checkout was performed just 20 days after launch, and ensured that the instruments were in good health. Star field exposures varying from 0.0 to 30.0 seconds provided a first opportunity to measure the detector bias, read noise, and dark current when temperatures were near their highest (near 1 AU). Additional star field and stellar cluster imaging was conducted 120 days and 7 months after launch, to validate the geometric calibration of the cameras and to test various image companding modes that could be used throughout the remainder of the mission's long cruise phase. Detector and optics performance will continue to be measured periodically during flight, and especially important will be observations planned during a series of 4 AU deep space calibration activities, when the thermal conditions are expected to more closely match the Trojan encounter temperature profiles.

A special set of TTCam observations was acquired during the first Earth Gravity Assist (EGA1) flyby one year after launch, to validate the radiometric calibration of the TTCams primarily based on imaging of the Moon. TTCam observations during EGA1 provided a successful demonstration of IPP pointing and TTCam extended source image quality (Fig. 9), as well as the required validation of the instrument's calibration pipeline and radiometric calibration coefficients, which yielded derived radiance factor (I/F, an approximation of surface reflectivity; Hapke 1993) values of high and low albedo lunar terrains that are within $\sim$15% or less of their expected modeled I/F values (Zhao et al. 2024). More details about the *Lucy* EGA1 observations and results can be found in Spencer et al. (2023).

Among the more critical engineering in-flight activities has been a series of IPP calibration alignment sequences conducted during the launch + 120 days and EGA1 flyby. The purpose of these activities has been to determine the relative alignments between the TTCams and the L'LORRI and L'TES fields of view, as well as to fine-tune the alignment knowledge between the IPP and the spacecraft's star trackers. To perform this activity using star fields, a preliminary optical distortion calibration was required to effectively solve for the instrument boresights in each image. The observations conducted 120 days after launch, for example, consisted of slewing the IPP gimbal angles through 8 target fields (6 on the inner gimbal, and 2 on the outer gimbal.) These target fields were selected to maximize the number of stars captured in the TTCam and L'LORRI FOVs, within operational constraints. The images were processed through the KinetX camera calibration suite (Pelgrift et al. 2018) to estimate the distortion model parameters and focal length, such that the residuals are zero-mean with a standard deviation in both image dimensions on the order of the expected uncertainty in the center-finding algorithm, and showing no systematic structure related to the stars' positions within the FOV. In-flight assessment of geometric distortion has met pre-flight design and testing expectations, and initial results from early cruise are provided in Zhao et al. (2024).

Such observations also establish the TTCam coordinate system to spacecraft coordinate system pointing matrix and also provide the initial post-launch calibration of the optical distortion of the cameras over the usable FOV to better than 1 pixel at the edge of the FOV ($<0.05\%$ of the focal length). The post-launch value of the camera focal length is also estimated as part of the solution for the distortion model parameters. By performing these observations near 1 AU and then again at 4 AU, we will also quantitatively characterize the sensitivity of the alignment and distortion calibrations to camera operating temperature and begin to establish if this calibration changes as the cameras age.

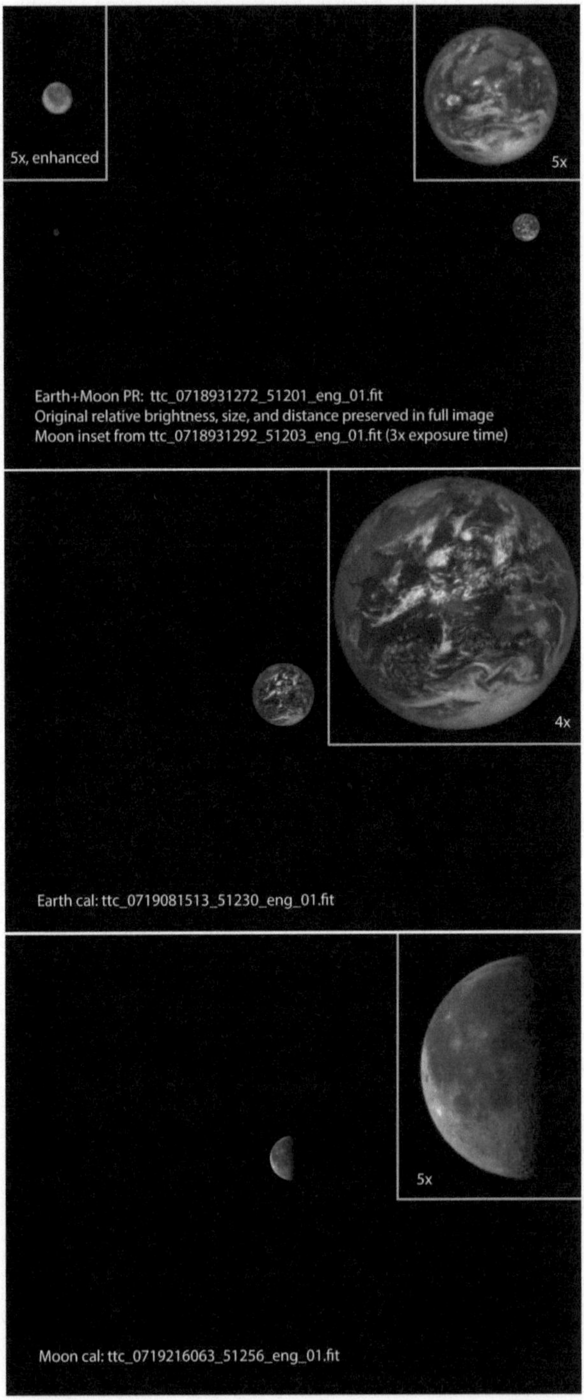

Fig. 9 Examples of TTCam images of the Earth and Moon acquired as part of the EGA1 imaging tests on Oct. 16, 2022. Insets show enlarged views

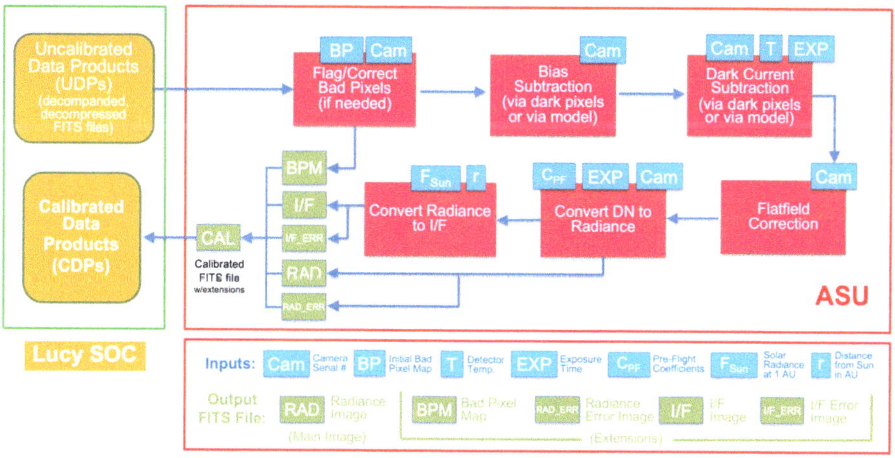

Fig. 10 Flow chart describing the TTCam ground data calibration pipeline. Inputs (in blue) include the camera ID, initial bad pixel map, CMOS detector temperature, exposure time, pre-flight radiometric conversion coefficients, the solar radiance at 1 AU convolved over the TTCam bandpass, and the heliocentric distance of the target object. See Zhao et al. (2024) for detailed descriptions of the pipeline outputs (in green)

Stray and scattered light characterization activities are also being performed during cruise, the first of which occurred almost 18 months after launch at 1.7 AU, and a second set in colder thermal conditions will be acquired once the spacecraft reaches 4 AU. These tests, as well as pre-flight scattered light tests (Sect. 6.3), provide images that provide a qualitative assessment of the expected magnitude of stray and scattered light in TTCam images at the most challenging expected encounter geometries for the *Lucy* asteroid target encounters. The initial 1.7 AU tests confirmed the expectation based on the optics design (Sect. 5.4) that scattered light will not be a significant contributor to the signal in TTCam tracking or encounter images acquired at the expected few to tens of millisecond exposure times (Zhao et al. 2024).

7.3 Ground Data Processing and Calibration

The TTCam calibration pipeline, which is part of the *Lucy* mission's Science Operations Center (SOC), takes in uncalibrated data products (UDPs) in Flexible Image Transport System (FITS) format and turns them into calibrated data products (CDPs), also as FITS files, and including extensions. Figure 10 shows a basic flowchart of the calibration pipeline. The pipeline takes in, as input, a list of UDPs, and uses UDP header information to find the relevant calibration files and parameters needed for each flight model. The pipeline outputs the CDP as a single FITS file, with the main FITS image containing the derived radiance of the scene, plus four FITS extensions that contain: (1) a bad pixel map; (2) a radiance error image; (3) an I/F image; and (4) an I/F error image. Details on the inputs, algorithms, and outputs of this calibration pipeline are provided by Zhao et al. (2024).

7.4 Expected Science Analyses and Data Products

Although primarily a spacecraft subsystem, because of its relatively wide angle FOV, large pixel count, and high-cadence capabilities, TTCam offers an invaluable complement to the mission's narrow-angle imagers for achieving *Lucy*'s science goals. Especially during the

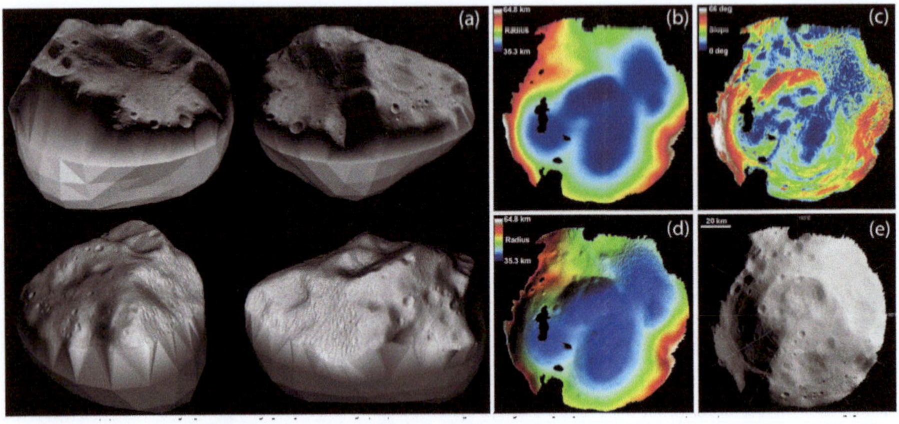

Fig. 11 (a) Textured shape model of asteroid (21) Lutetia obtained partly from Rosetta OSIRIS images comparable to what we expect from the Lucy TTCam images (Preusker et al. 2012). Also shown are other example derived products from OSIRIS images, including (b) a color-coded Digital Terrain Model; (c) a slope map; (d) a hill-shaded relief map; and (d) an orthoimage mosaic. All panels in Lambert Azimuthal Equal-area projection and based on Preusker et al. (2012)

few minutes immediately before and after the respective closest encounters with each target, TTCam will be able to image the entire sunlit part of each body within a single frame and capture the rapidly changing viewing geometry during the flyby. As described below, these observations will help address a number of specific *Lucy* mission science goals. In addition to the systematically processed calibrated images, TTCam data will also be released in the form of higher-level products such as controlled and uncontrolled mosaics, ortho-rectified maps, local Digital Terrain Models (DTMs), thematic maps, rendered views, and fly-over movies.

7.4.1 Trojan Asteroid Shape Modeling and Volume Estimation

TTCam flyby images of each Trojan target body from different perspectives, together with the data from the other *Lucy* imagers, will enable the generation of shape models and DTMs by means of methods like stereophotogrammetry and photoclinometry (e.g., Capanna et al. 2014; Mottola et al. 2024) as well as limb and terminator tracking. For flybys during which the spin state of the object (spin rate and axis orientation) prevents the observation of the whole target, the shape model will be complemented with a ground-based model obtained through lightcurve and stellar occultation observations, as done for example by Preusker et al. (2012) for the *Rosetta* mission (21) Lutetia flyby (Sierks et al. 2011).

The shape models derived from TTCam images and other data will provide the basis for the determination of the size, volume, and refinement of the rotation properties of the targets (e.g., Fig. 11). The volume determination, in conjunction with the mass estimation from the radio-science experiment, will enable the determination of the object's bulk density. The presence of a misalignment between the figure principal axis and the spin axis would give hints about the internal mass distribution. Furthermore, by providing information about the local illumination and viewing conditions on the surface, the shape model will enable photometric studies to be performed, to reveal the physical properties of the regolith.

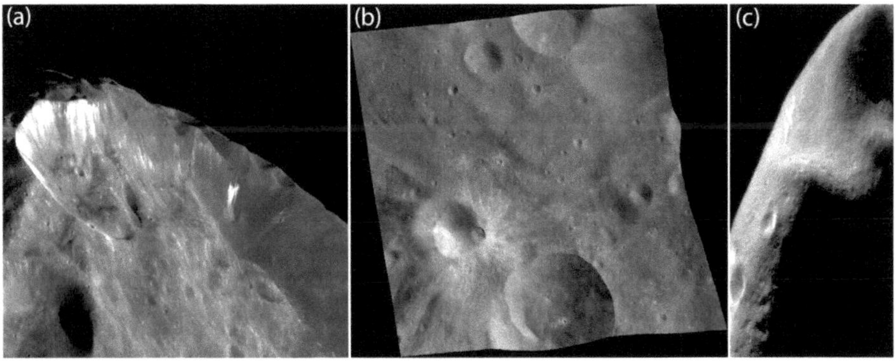

Fig. 12 Example images of small bodies all at the typical expected image scale (~70 m/pix) of TTCam Trojan close approach images. (a) Impact cratering surface gradation processes on Saturn's moon Phoebe, from the Cassini mission (JPL Photojournal image PIA06075); (b); Impact cratering, subsurface exhumation, and surface degradation processes on part of main belt asteroid (4) Vesta, from the Dawn mission (Dawn image F2_372565839); and (c) Impact and potential tectonic processes on the near-Earth asteroid (433) Eros, from the NEAR-Shoemaker mission (NEAR image M0126540610F4)

7.4.2 Trojan Asteroid Geology and Topography

As described by Olkin et al. (2021) and Marchi et al. (2023), an important additional scientific use of TTCam images of the *Lucy* target asteroids will be to help characterize their geology and geologic history. Given the field of view and pixel scale of the TTCams, and the planned flyby distances, none of the Trojan targets are expected to over-fill the TTCam field of view. Thus, the TTCam image data set for each flyby will provide "whole disk" views of the entirety of the sunlit surface of each target body, at pixel scales around 30 to 90 m/pixel, depending on the specific flyby distances.

Even though TTCam images will have much lower resolution than those from the other *Lucy* science payload imagers, they will provide superior coverage of each Trojan target, and will thus directly address a number of specific *Lucy* science goals (Levison et al. 2021). For example, as shown in Fig. 12, imaging at typical TTCam pixel scales can allow a global assessment of the morphology, size distribution, and areal distribution of impact craters on small bodies. In addition, TTCam images could provide evidence for other albedo features and/or geologic structures or units that could be potentially indicative of erosional, tectonic, or perhaps even cryovolcanic processes on Trojan asteroids. Finally, the multiple emission angle views of each target provided by TTCam imaging during the flyby will enable the creation of stereoscopic images and DTMs for a significant fraction of the sunlit surface.

8 Summary

This paper summarizes the requirements, design, expected in-flight performance, in-flight calibration plans, and expected science potential of the NASA *Lucy* Trojan asteroid Discovery mission's Terminal Tracking Camera (TTCam) imaging system. The details of the optics, electronics, and the calibration and performance information provided here is primarily focused on the eventual scientific use of TTCam images to contribute to the mission's overall science goals (Levison et al. 2021). Additional details about the use of the TTCam system for their primary *Lucy* mission function – as a navigation system designed to provide an autonomous onboard late pre-encounter update of the location of each asteroid target relative to the spacecraft – can be found in Olkin et al. (2021) and Good et al. (2022), and the technical details on the pre-flight and in-flight calibration of the cameras can be found in Zhao et al. (2024).

Acknowledgements We are extremely grateful to the members of the *Lucy* mission team, MSSS staff, NASA support staff, and their many subcontractors and vendors who kept the design, development, fabrication, and testing of the TTCam system on track despite the numerous hardships and risks associated with the 2020-2021 global CoVID-19 pandemic. We thank Danielle Smilovsky and Oketa Basha (ASU) for support on the development and testing of the TTCam instrument calibration pipeline, and Jeanne Ladewig at Lockheed Martin for her many years of support of TTCam design, development, and integration.

Funding This work was funded by grants and contracts from the NASA *Lucy* Discovery mission.

Declarations

Competing Interests The authors have no competing or financial interests to declare that are relevant to the content of this article.

References

Bell JF III, Godber A, McNair S, Caplinger MA, Maki JN, Lemmon MT, Van Beek J, Malin MC, Wellington D, Kinch KM, Madsen MB, Hardgrove C, Ravine MA, Jensen E, Harker D, Anderson RB, Herkenhoff KE, Morris RV, Cisneros E, Deen RG (2017) The Mars Science Laboratory Curiosity rover Mastcam instruments: Preflight and in-flight calibration, validation, and data archiving 4:396–452. https://doi.org/10.1002/2016EA000219

Bos BJ, Ravine MA, Caplinger M, Schaffner JA, Ladewig JV, Olds RD, Norman CD, Huish D, Hughes M, Anderson SK, Lorenz DA, May A, Jackman CD, Nelson D, Moreau M, Kubitschek D, Getzandanner K, Gordon KE, Eberhardt A, Lauretta DS (2018) Touch And Go Camera System (TAGCAMS) for the OSIRIS-REx Asteroid Sample Return mission. Space Sci Rev 214:37. https://doi.org/10.1007/s11214-017-0465-2

Bos BJ, Nelson DS, Pelgrift JY, Liounis AJ, Doelling D, Norman CD, Olds RD, May CW, Witherspoon R, Church E, Huish D, Adam CD, Sahr E, Kidd J, Drozd K, Owen WM, Moreau MC, Seals LT, Butt J, LeDuc D, Alkiek K, Chevres-Fernandez LR, Gordon KE, Khlopenkov K, Haney C, Bhatt R, Hikes J, Wolfram A, Caplinger M, Ravine MA, Getzandanner K, Bartels A, Lauretta DS (2020) In-flight calibration and performance of the OSIRIS-REx Touch And Go Camera System (TAGCAMS). Space Sci Rev 216:71. https://doi.org/10.1007/s11214-020-00682-x

Capanna C, Jorda L, Gesquière G, Gaskell RW, Groussin O, Gutiérrez P, Hviid SF, Lamy PL, Preusker F, Rodionov S, Scholten F, Vibert D (2014) Multi-resolution stereophotoclinometry by deformation, a new 3D shape reconstruction method applied to ROSETTA/OSIRIS images. In: AGU fall meeting abstracts. P41C-3941

Cox M et al (2024) The Lucy spacecraft. Space Sci Rev 220

Good PG, Faiks P, Pisano WJ (2022) Terminal Tracking for the Lucy Trojan Asteroid Mission. In: 44th Annual AAS Guidance, Navigation and Control Conference. AAS 22-131. https://ntrs.nasa.gov/citations/20220001851

Hansen CJ, Caplinger MA, Ingersoll A, Ravine MA, Jensen E, Bolton S, Orton G (2017) Junocam: Juno's outreach camera. Space Sci Rev 213:475–506. https://doi.org/10.1007/s11214-014-0079-x

Hapke B (1993) Theory of reflectance and emittance spectroscopy. Cambridge University Press, New York

Jumar R, Maaß H, Hagenmeyer V (2018) Comparison of lossless compression schemes for high rate electrical grid time series for smart grid monitoring and analysis. Comput Electr Eng 71:465–476. https://doi.org/10.1016/j.compeleceng.2018.07.008

Levison HF, Olkin CB, Noll KS, Marchi S, Bell JF III, Bierhaus E, Binzel R, Bottke W, Britt D, Brown M, Buie M, Christensen P, Emery J, Grundy W, Hamilton VE, Howett C, Mottola S, Pätzold M, Reuter D, Spencer J, Statler TS, Stern SA, Sunshine J, Weaver H, Wong I (2021) Lucy mission to the Trojan asteroids: science goals. Planet Sci J 2:171. https://doi.org/10.3847/PSJ/abf840

Malin Space Science Systems (MSSS), Inc. (2013) ECAM Imaging System. http://www.msss.com/brochures/ecam.pdf

Malin MC, Edgett K, Jensen E, Lipkaman L (2013) Mars Science Laboratory Project Software Interface Specification (SIS): Mast Camera (Mastcam), Mars Hand Lens Imager (MAHLI), and Mars Descent Imager (MARDI) Experiment Data Record (EDR), Reduced Data Record (RDR), and PDS Data Products. Version 1.2, JPL D-75410, Online at http://pds-imaging.jpl.nasa.gov/data/msl/MSLMST_0001/DOCUMENT/MSL_MMM_EDR_RDR_DPSIS.PDF

Marchi S, Bell JF III, Bierhaus B, Spencer J (2023) Surface geology of Jupiter's Trojan asteroids. Space Sci Rev 219:44. https://doi.org/10.1007/s11214-023-00985-9

Mottola S, Britt D, Brown ME, Buie MW, Noll K, Pätzold M (2024) Shapes, rotations, photometric and internal properties of Jupiter Trojans. Space Sci Rev 220

Olkin CB, Levison H, Vincent M, Andrews J, Gray S, Good P, Noll K, Marchi S, Christensen P, Reuter D, Weaver H, Pätzold M, Bell JF III, Hamilton V, Dello Russo N, Simon A, Beasley M, Grundy W, Howett C, Spencer J, Ravine M, Caplinger M (2021) Lucy Mission to the Trojan Asteroids: Concept of Operations and Instrumentation. Planet Sci J 2:172. https://doi.org/10.3847/PSJ/abf83f

ON Semiconductor (2017) MT9P0311/2.5-Inch 5 Mp CMOS Digital Image Sensor. Data Sheet online at https://www.onsemi.com/pdf/datasheet/mt9p031-d.pdf

Pelgrift JY, Sahr EM, Nelson DS, Jackman CD, Benhacine L, Bos BJ et al (2018) In-flight calibration of the OSIRIS-REx optical navigation imagers. In: 1st annual RPI workshop on image-based modeling and navigation for space applications. Troy, New York

Preusker F, Scholten F, Knollenberg J, Kührt E, Matz K-D, Mottola S, Roatsch T, Thomas N (2012) The northern hemisphere of asteroid (21) Lutetia—topography and orthoimages from Rosetta OSIRIS NAC image data. Planet Space Sci 66:54–63. https://doi.org/10.1016/j.pss.2012.01.008

Ravine MA, Schaffner JA, Caplinger MA (2016) ECAM: a modular spaceflight imaging system – first flight deliveries. In: 3rd international workshop on instrumentation for planetary. #4106

Sierks H, Lamy P, Barbieri C, Koschny D, Rickman H, Rodrigo R, A'Hearn MF, Angrilli F, Barucci MA, Bertaux J-L, Bertini I, Besse S, Carry B, Cremonese G, Da Deppo V, Davidsson B, Debei S, De Cecco M, De Leon J, Ferri F, Fornasier S, Fulle M, Hviid SF, Gaskell RW, Groussin O, Gutierrez P, Ip W, Jorda L, Kaasalainen M, Keller HU, Knollenberg J, Kramm R, Kührt E, Küppers M, Lara L, Lazzarin M, Leyrat C, Lopez Moreno JJ, Magrin S, Marchi S, Marzari F, Massironi M, Michalik H, Moissl R, Naletto G, Preusker F, Sabau L, Sabolo W, Scholten F, Snodgrass C, Thomas N, Tubiana C, Vernazza P, Vincent J-B, Wenzel K-P, Andert T, Pätzold M, Weiss BP (2011) Images of asteroid 21 lutetia: a remnant planetesimal from the early solar system. Science 334:487–490. https://doi.org/10.1126/science.1207325

Spencer J et al (2023) The first Lucy Earth flyby (EGA1). Space Sci Rev 219. https://doi.org/10.1007/s11214-023-01034-1

Zhao Y, Bell JF III et al (2024) Pre-flight and in-flight calibration and performance of the Terminal Tracking Cameras (TTCams) on the NASA Lucy Mission. Space Sci Rev 220

Space Science Reviews (2023) 219:82
https://doi.org/10.1007/s11214-023-01028-z

The Lucy Long Range Reconnaissance Imager (L'LORRI)

H.A. Weaver[1] · J.P. Wilson[1] · S.J. Conard[1] · J.D. Adams[1] · S. Begley[1] · J. Burgum[1] · E.H. Darlington[1] · N. Dello Russo[1] · R. Hacala[1] · S. London[1] · M.F. Morgan[1] · G. Murphy[1] · T. Nelson[1] · A. Shah[1] · J.R. Spencer[2] · H. Taylor[1] · T. Boehmer[1] · L. Burke[1] · C. Drabenstadt[1] · C. Henry[1] · S. Ling[1] · C. Porter[1] · J. Yin[1]

Received: 27 July 2023 / Accepted: 20 November 2023 / Published online: 30 November 2023

Abstract
NASA's *Lucy* mission spacecraft was launched on 16 October 2021 and will perform the initial in situ investigation of the Jovian Trojan asteroids (Levison et al. 2021, 2024). The Lucy LOng Range Reconnaissance Imager (L'LORRI) is a panchromatic visible light (420–795 nm, 50% QE points), narrow-angle (field of view = 0.29°), high spatial resolution (1.0″ pixel^{-1}) imager used on the *Lucy* mission for both science observations and optical navigation. L'LORRI is designed to provide maps of the sunlit portions of the Trojan surfaces to a resolution of ~10 m (after deconvolution), which will enable crater counting to constrain the surface ages. L'LORRI's high sensitivity and large dynamic range permits imaging of the low albedo Trojans at moderately large phase angles (down to I/F values of ~0.0014 with SNR ≈ 30 using an exposure time of 100 ms), as well as providing early acquisitions of the Trojans during the approach phase, searches for Trojan activity that are ~10× better than can be obtained from Earth, and deep searches for potential Trojan satellites down to $V \approx 20.4$ at spatial resolutions far surpassing that available from Earth. This paper describes the L'LORRI instrument design and the requirements that drove the design. We present results from L'LORRI's ground calibration campaign, summarize the L'LORRI in-flight calibration plan, and describe typical L'LORRI operations scenarios during the Trojan flybys. We also present an analysis of in-flight data taken during the first year of *Lucy* operations, which show that most aspects of L'LORRI's performance are nominal (i.e., as predicted), but the telescope's point spread function is slightly degraded relative to pre-flight predictions. Nevertheless, L'LORRI is still expected to fulfill all of its scientific objectives, which should revolutionize our view of the Jovian Trojans.

Keywords Trojan asteroids · Asteroids · Kuiper belt objects · Instruments · Cameras

1 Introduction

The *Lucy* mission is a NASA Discovery program that was selected in 2016 to perform the first in situ exploration of the Jovian Trojan asteroids (Levison et al. 2021, 2024). These Trojans are thought to have been emplaced in the Lagrange L4 and L5 regions early in the solar system's history during the radial migration of the giant planets (cf. Nesvorný et al. 2013) and most have been in dynamically stable orbits for the past 4.5 Gyr. As such, the

Extended author information available on the last page of the article

Jovian Trojans are fossil remnants that may provide a unique tracer of the solar system's early dynamical, physical, and chemical evolution.

The Lucy LOng Range Reconnaissance Imager (L'LORRI) is a panchromatic visible light camera that will provide the highest resolution images of the surfaces of the bodies targeted for flybys during the *Lucy* mission. These images will be used to perform detailed geological mapping of the surfaces of the *Lucy* targets, which will enable crater counting to constrain the ages of these surfaces. L'LORRI will also be used to perform sensitive searches for potential satellites of the targeted bodies and any rings or dust associated with possible current or past surface activity. In addition to its use as a scientific imager, L'LORRI also serves as the primary optical navigation camera for *Lucy*, allowing early acquisition of the flyby targets and systematic monitoring during the approach phase to enable making accurate trajectory correction maneuvers to the flyby aim points.

In this paper, we describe L'LORRI's objectives, its design requirements that enable achieving those objectives, and the ground calibration results used to characterize its performance prior to launch. We also discuss how L'LORRI is used operationally during the *Lucy* mission, we outline the in-flight calibration plan that will be used to measure and trend L'LORRI's performance over the mission's lifetime, and we provide some results from L'LORRI observations taken during the first year of in-flight operations.

L'LORRI is the successor to the LORRI instrument on the *New Horizons* mission (Cheng et al. 2008; Weaver et al. 2020), which has continued to operate successfully in space for more than 17 years without any degradation in performance. L'LORRI essentially performs the same functions on the *Lucy* mission that LORRI has been performing on the *New Horizons* mission. However, there are some differences between L'LORRI and LORRI, which we describe here.

2 L'LORRI Objectives and Their Implementation

2.1 Science Objectives

The *Lucy* mission's scientific objectives and the spacecraft and instrument requirements necessary to achieve those objectives are summarized in NASA's Program Level Requirements Appendix (PLRA) for Lucy. The PLRA specifications relevant to L'LORRI for each Trojan flyby are:

- To support global imaging for shape models and the identification of geological units, obtain images spaced by 1/25 to 1/13 of a rotation over a full rotation period during the approach phase, and obtain images of the entire sunlit surface at a series of phase angles between 15° and 25° during the close encounter phase.
- To support stereo imaging for the creation of elevation models, obtain images of an area $\geq$100 km^2 with sub-spacecraft resolution of $\leq$200 m at two emission angles.
- To enable the investigation of landform degradation at different latitudes, obtain images that cover an area $\geq$500 km^2, or 50% of the sunlit area, and regions from the equator to $\geq$60°, with characteristic resolution $\leq$100 m.
- To investigate the impact crater size distribution, obtain images that cover $\geq$700 km^2, or 80% of the sunlit surface, that can resolve craters $\geq$7 km across (i.e., cover the large end of the distribution), and obtain images that cover $\geq$10 km^2 that are capable of resolving craters $\geq$70 m in diameter (i.e., cover the small end of the distribution).
- Search the region of stable satellite orbits to detect any bodies $\geq$2 km in diameter, assuming a geometric albedo $\geq$0.04.

Each of the above PLRA specifications are flowed down to more refined requirements in the Lucy project "Mission Requirements Document" (MRD), which lays out the high-level requirements for the spacecraft subsystems and instruments. The MRD requirements in turn drive even more detailed requirements for each instrument, which must be implemented by the individual instrument teams and verified by the *Lucy* project.

To summarize, L'LORRI's main high-level objectives are: (1) obtain high-resolution images of the Trojan surfaces, (2) perform searches for satellites and dust around the Trojans, and (3) obtain the optical navigation images required to support spacecraft trajectory corrections. These are essentially identical to the objectives of *New Horizons* LORRI, except the targets were Pluto and its satellites rather than the Jovian Trojans. LORRI successfully achieved all of its objectives during the Pluto flyby, and L'LORRI's requirements have been tailored to ensure that its objectives can be met during the *Lucy* flybys.

2.2 Requirements

As an optical imaging system, L'LORRI is required to obtain high-resolution, panchromatic images under low light conditions. There are no requirements to perform multispectral imaging, which is handled by another instrument on the *Lucy* payload (Reuter et al. 2023). The driving requirements for the instrument involve resolution (point spread function, or PSF), signal-to-noise ratio (SNR), and environments. The key design requirements for L'LORRI are summarized in Table 1. Except where noted, all of these requirements were verified during the ground calibration campaign conducted at APL, with specific results discussed later in this paper.

The original LORRI instrument was designed to return high SNR imagery from Pluto at 33 au from the Sun, whereas L'LORRI's science targets have heliocentric distances in the range 5.2-5.7 au. Although the sunlight level at the Trojans is 30-40× larger than at Pluto, *Lucy's* Trojan targets have average albedos that are ∼15× smaller than Pluto's average albedo, so the surface brightness levels during the Trojan flybys might only be ∼2-3× larger than the Pluto surface brightness levels encountered during the *New Horizons* flyby. L'LORRI must be able to measure the brightest regions on the Trojans (I/F∼0.4) without saturating and also support imaging of the darker Trojan surfaces at relatively high solar phase angles (down to I/F ∼ 0.0014). L'LORRI can achieve these objectives with a combination of high (≥3000) instantaneous dynamic range and the use of two or more exposure times.

L'LORRI is a catadioptric telescope based on a Ritchey-Chrétien design, with hyperbolic-shaped primary and secondary mirrors made of silicon carbide, and a 3-element field-flattening lens assembly mounted in the hole of the primary mirror (Fig. 1). The photodetector consists of a frame transfer CCD detector with associated readout electronics. Table 2 provides a summary of the key characteristics of L'LORRI. The mass value (11.24 kg) includes the OTA (6.96 kg), the FPU (1.01 kg), the DPU (1.77 kg), all harnesses (0.72 kg), and all thermal blankets (0.78 kg). The power value (9.92 W) includes operational power for the FPU (2.11 W), operational power for the DPU (4.48 W), and the OTA operational heater power (3.33 W), but excludes the power drawn from the spacecraft-controlled decontamination heater to maintain the OTA at its specified value (∼2.3 W).

L'LORRI is mounted on an instrument pointing platform (IPP) located outside the *Lucy* spacecraft (Fig. 2 and Fig. 3) and is exposed to more challenging environments than the original LORRI, which was mounted inside the *New Horizons* spacecraft. Unlike LORRI, L'LORRI is subjected to a highly variable thermal environment. The silicon carbide (SiC) optical telescope assembly (OTA) structure was supposed to mitigate any issues associated

Table 1 Main L'LORRI instrument level requirements

ID	Requirement
LL-INST-3	The L'LORRI bandpass shall cover at a minimum 500 nm to 850 nm with a sensitivity over the entire range that is at least half the peak sensitivity within this range[1]
LL-INST-5	L'LORRI shall have a PSF with FWHM $\leq$ 15 µrad over the entire CCD[2]
LL-INST-7	The full FOV shall be 5.0 mrad $\pm10\%$, square
LL-INST-9	The instantaneous FOV (IFOV) shall be 4.94 µrad $\pm10\%$
LL-INST-11	L'LORRI shall have SNR ≥ 7 for a point source with $V = 15.8$ in a single 10 s exposure in 4×4 mode with the PSF centered on a pixel
LL-INST-13	L'LORRI shall have SNR ≥ 6 for a point source with $V = 15.2$ in a single 5 s exposure in 1×1 mode with the PSF centered on a pixel
LL-INST-15	L'LORRI shall have SNR ≥ 4 for a point source with $V = 19$ in a series of 10 co-added images, each with a 10 s exposure in 4×4 mode with the PSF centered on a pixel
LL-INST-17	L'LORRI shall have SNR ≥ 25 for a resolved target with I/F=0.0014 at a heliocentric distance of 5.7 au with a 100 ms exposure
LL-INST-19	L'LORRI shall have a point source transmittance (PST) for out-of-field stray light $\leq$0.02 at angles $\geq$ 1 deg
LL-INST-21	L'LORRI shall have a dynamic range $\geq$3000 in a single exposure
LL-INST-27	L'LORRI shall provide image data with an absolute radiometric accuracy of $\pm10\%$ after application of a calibration routine
LL-INST-29	L'LORRI shall provide image data with a relative radiometric accuracy of $\pm2\%$ from exposure to exposure and pixel to pixel
LL-INST-31	L'LORRI shall provide image data with a relative radiometric accuracy of $\pm2\%$ from pixel to pixel within a single image

These instrument-level requirements flow down from higher level scientific requirements specified in the *Lucy* Mission Requirements Document (MRD), whose requirements in turn flow down from NASA's Program Level Requirements Appendix (PLRA) for the *Lucy* mission.

[1] A waiver was obtained in this case because the bandpass was reduced by design to mitigate artifacts from optical ghosts.

[2] This requirement was verified for the required FOV but not for the full CCD.

with the predicted large range of temperatures during the Trojan flybys (−75 C to −125 C), but ground testing (see later discussion) showed that the optical performance degraded significantly as the temperature was lowered. A spacecraft-controlled contamination heater can be used to warm the L'LORRI OTA when necessary to improve the optical performance during science operations. L'LORRI is also required to withstand slightly higher radiation levels and have higher micrometeoroid impact survivability requirements compared to LORRI.

Although the launch vehicles for *New Horizons* (ATLAS V 551) and *Lucy* (ATLAS V 401) are similar, the launch load requirements for L'LORRI were set significantly higher than for LORRI, which meant that the L'LORRI OTA required a stronger mechanical structure. The exposure to external environments did relax requirements on accommodation of the CCD, as maintaining it at a cold temperature no longer required a radiator. During the approach to the Trojans, L'LORRI is required to image the Trojan surfaces at SNR $\geq$ 25 in a single image with an exposure time of 100 ms. The resolution requirement is to have a full-width half-maximum (FWHM) point spread function of $\leq$15 µrad to obtain imagery of the Trojans sufficient to meet the science objectives after accounting for jitter from the spacecraft. Although most scientific imaging benefits from having the tightest possible PSF (i.e., the smallest possible FWHM), optical navigation objectives generally

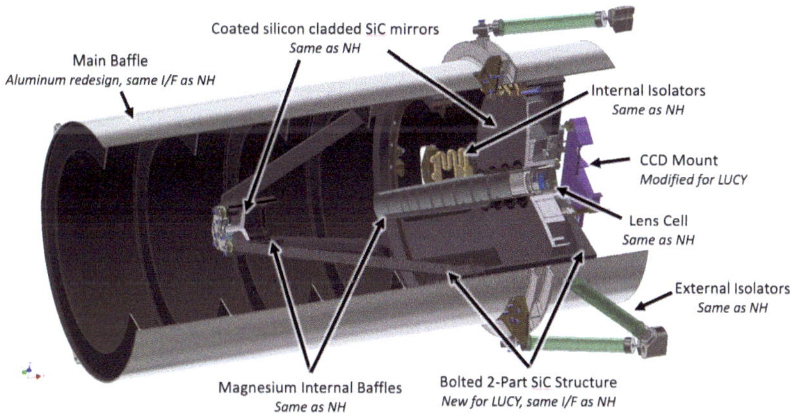

Fig. 1 CAD drawing of L'LORRI showing its mechanical design and differences with *New Horizons* LORRI. Both telescopes have a Ritchey-Chrétien optical design and a 3-element field-flattening lens assembly mounted in the hole of the primary mirror. L'LORRI's main baffle and vanes were constructed from a single block of aluminum, while NH-LORRI has a baffle composed of a ceramic composite with aluminum vanes installed separately. L'LORRI's optical telescope assembly (OTA) structure is comprised of two separate silicon carbide pieces that are bolted together, whereas they were fused together for NH-LORRI. Both telescopes use essentially identical CCDs, but NH-LORRI's requires an attached radiator for cooling while L'LORRI's does not. Both telescopes have identical mechanical isolation systems (internal and external) to mitigate potential vibrational coupling to external environments that could degrade the optical performance. Here "I/F" is short for "interface"

benefit from having a broader PSF so that stellar images are not under sampled. L'LORRI is designed to meet imaging requirements at cold temperatures, with the OTA near −95 C. At room temperature the performance is only required to be stable to allow for trending of performance under ambient laboratory conditions.

L'LORRI has a 4×4 pixel binning mode, for which its limiting magnitude requirement is $V \approx 19$. The in-flight throughput measurements discussed later suggest that a single L'LORRI 4×4 image with an exposure time of 64.9 s can achieve SNR ≈ 5 on a star having $V \approx 21$ if there are few nearby stars (i.e., the background field is relatively sparse) and there is no solar scattered light.

2.3 Planned Observations

Each Trojan flyby is comprised of several phases (see Olkin et al. 2024 for a more extensive discussion of *Lucy* operations), all of which require L'LORRI activities:[1] Pre-Acquisition (E−300d to E−60d), Acquisition (E−60d to E−30d), Approach (E−30d to E-4d), Close Approach (E−4d to E+4d), and Departure (E+4d to E+20d). Even before pre-acquisition, during the so-called "cruise" phase, L'LORRI will observe the targeted Trojans roughly every year to investigate their phase behavior under geometrical conditions unavailable from Earth-based observatories, which will inform some details of L'LORRI's measurements (e.g., the selection of exposure times) during the later flybys.

During the pre-acquisition phase, the spacecraft subsystems and L'LORRI will be configured to enable the subsequent phases (e.g., set the appropriate thermal environment for L'LORRI). L'LORRI will perform systematic optical navigation ("OpNav") observations

[1] "E" is short for "Encounter" and "d" is short for "days"; thus, E−60d is short for 60 days before encounter.

Table 2 Summary of L'LORRI key parameters

Item	Description
Optical Telescope Assembly (OTA)	L3H Ritchey-Chrétien optical design with 3-element field flattener lens assembly Silicon Carbide (SiC) structure, SiC mirrors coated with high reflectance dielectric 20.8 cm primary mirror diameter, $\sim$11% central obscuration Focal length = 263.446 cm OTA in-flight operating temperature is approximately −95 C No moving parts
Focal Plane Characteristics	Teledyne-e2v 47-20 frame transfer Charge Coupled Device (CCD) detector CCD frame transfer time $\approx$12 ms 1024 × 1024 optically active pixels, 13 µm square pixels AR-coated, backside-thinned, backside-illuminated CCD 1 × 1 and 4 × 4 (rebinned) output formats Field of View (FOV) = 505.190 milliradians = 0.28946 deg (square) Pixel FOV (IFOV, 1 × 1 format) = 4.9335 µrad = 1.0176 arcsec (square) Correlated Double Sampling (CDS) with 12-bit Analog-to-Digital Converter (ADC) Full well $\approx$80,000 e (linear range) Anti-blooming technology to mitigate the effects of saturated targets in the FOV Electronics noise $\approx$19 e Dynamic range $\approx$3500 (single image) Gain: 21.1 e DN^{-1} (1 × 1), 20.0 e DN^{-1} (4 × 4) Dark current $\leq$0.040 e s^{-1} $pixel^{-1}$ (1 × 1 at operating temperature of −80 C) Available exposure times: 0 ms to 64,900 ms at 1 ms spacings 1 Hz maximum frame rate (minimum time between consecutive images is 1 s)
Mass and Power	Total Mass = 11.24 kg; Power = 9.92 W
Wavelength Range	Panchromatic (no filters) with $\sim$63% peak QE 420–795 nm at 50% of peak QE, 380–860 nm at 10% of peak QE
Photometric Accuracy	$\sim$2% (1σ) absolute for solar-type spectral energy distribution (SED) after in-flight calibration $\leq$10% (1σ) absolute for non-solar-type SEDs after in-flight calibration $\leq$1% (1σ) relative for signal-to-noise ratio (SNR) $\geq$100

during the Acquisition and Approach phases at a cadence specified by the OpNav team to enable accurate spacecraft targeting to the designed aim point (i.e., miss distance from the Trojan). The OpNav imaging will be complemented by additional science imaging for better characterization of the Trojan's behavior (e.g., measure light curves for body shape determinations), perform observations of known satellites (e.g., Queta for Eurybates and the Patroclus-Menoetius binary), search for new satellites, and search for dust and other signs of Trojan activity.

The most scientifically interesting L'LORRI imaging will occur during a several hour period surrounding closest approach to the Trojan. At that time, the instrument pointing platform (IPP) will be commanded to point the L'LORRI boresight at the sub-spacecraft location on the Trojan. L'LORRI itself will be commanded into a special "noodle" mode in which the exposure time will alternate between short ($\sim$15 ms) and long ($\sim$100 ms)

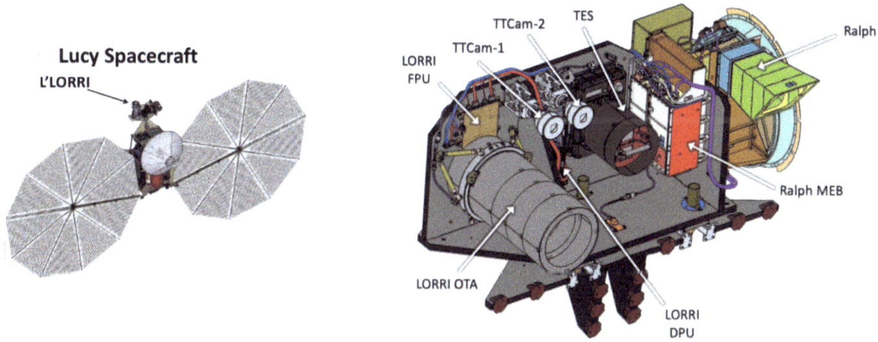

Fig. 2 Left: Drawing of the *Lucy* spacecraft with its two large solar arrays, each of which is ~7 m in diameter. L'LORRI and the other instruments are mounted on the instrument pointing platform (IPP), which stands off from the spacecraft and can be rotated in two orthogonal directions by two different gimbal mechanisms. Right: Magnified view of the IPP showing the locations of L'LORRI and the other instruments. "FPU" and "DPU" are the L'LORRI Focal Plane Unit and Data Processing Units, respectively. "MEB" is the L'Ralph electronics module

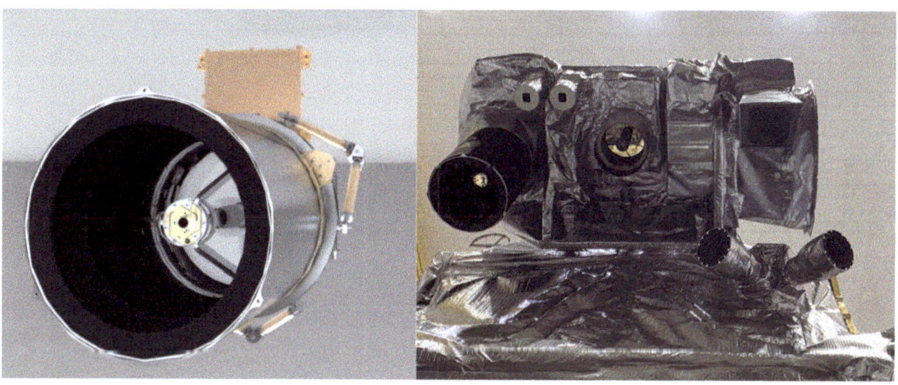

Fig. 3 Left: CAD drawing of L'LORRI, not including its thermal blankets. The focal plane unit (FPU) is the brown box located behind the optical telescope assembly (OTA). Right: Photo showing all the *Lucy* mission instruments covered in thermal blankets and mounted on the instrument pointing platform (IPP), which itself is mounted on top of the spacecraft, shortly before the entire spacecraft was encapsulated within the launch vehicle fairing in September 2021. In addition to the components identified in Fig. 2, two orthogonally pointed star trackers are mounted on the spacecraft just below L'Ralph

exposure times every second to expand the dynamic range of the imaging and capture with high SNR both high and low reflectance regions on the Trojan surface, while minimizing image smear caused by Trojan surface rotation and relative motion during each exposure.

In addition to the scientific imaging during each encounter, L'LORRI will be exercised multiple times throughout the cruise phase to verify nominal functionality and perform calibration observations. The cruise phase data will enable systematic monitoring and trending of L'LORRI's performance over the full mission duration. This includes detailed in-flight characterization of L'LORRI's point spread function (PSF) during observations of star clusters, investigation of optical ghosts by observing bright stars, investigating solar scattered light in different observational geometries (especially under conditions mimicking the Trojan flyby geometry), and measuring the absolute sensitivity by observing a solar-type absolute calibration standard star. Near the time when the first Earth Gravity Assist (EGA1) was performed in October 2022, we executed extensive observations of the Moon's surface, which tested L'LORRI's ability to measure craters with sub-PSF resolution using deconvolution techniques, similar to what is planned for measuring the Trojan crater size distributions during those flybys (Robbins et al. 2023a,b).

Subject to Lucy mission budget and operational constraints, various "cruise science" L'LORRI observations will be conducted that will enhance the mission's scientific return beyond its baseline activities, which are focused on the initial spacecraft investigations of the Jovian Trojan population.

3 Instrument Description

3.1 Overview

The L'LORRI instrument development was led by The Johns Hopkins University Applied Physics Laboratory (JHU-APL, or APL for short) with many of the original LORRI staff reprising their roles on L'LORRI after almost two decades. Integration and test (I&T) efforts began in the spring of 2020 and coincided with the outbreak of the COVID-19 pandemic, which presented significant logistical and safety challenges to the team throughout the entire I&T campaign, as well as after delivery of the instrument to the spacecraft. Despite these challenges, L'LORRI was delivered to the *Lucy* spacecraft in late-October 2020, it exhibited nominal performance during the subsequent ground testing, and it was launched into space on the *Lucy* spacecraft in October 2021.

L'LORRI has three sub-assemblies in close proximity connected by electrical harnesses. These are the Optical Telescope Assembly (OTA), which was manufactured by L3-Harris SSG (L3H), the focal plane unit (FPU), and the data processing unit (DPU). L'LORRI has two DPUs for redundancy, and either DPU can be used with either of two spacecraft command and data handling (CDH) computers (cross-strapping). The baseline plan is to use only one of the DPUs for the entire mission, with the second DPU only being powered for periodic functional testing (cold redundancy). All components are mounted to the instrument pointing platform (IPP) on the spacecraft (Fig. 2). Unlike LORRI, L'LORRI does not have an aperture door because the contamination risk was judged by the *Lucy* project to be smaller than the risk of a door opening failure.

L'LORRI is electronically shuttered and has no moving parts. The OTA focuses light onto a Charge Coupled Device (CCD; model 47-20 supplied by Teledyne-e2v), which is part of the FPU. The CCD is a 1024 × 1024 pixel (optically active region), back-thinned, backside-illuminated device that incorporates anti-blooming technology to eliminate bleeding of the electrons along columns when bright targets saturate (i.e., when the signal in a pixel exceeds the full-well capacity of ~80,000 electrons). The FPU contains a single electronics board connected to the CCD.

For the highest resolution observations, all optically active pixels are read out from the CCD ("1×1" format) with a square IFOV (individual pixel field-of-view) of $1.0''$ pixel^{-1} and a square FOV of $1024'' \times 1024''$. The pixels can be also be re-binned by a factor of 4 in each direction (i.e., column and row directions) during CCD readout ("4×4" format), which reduces the data volume by a factor of 16, increases the SNR because the readout noise is added only once for 16 pixels, and results in an IFOV of ($4.0''$ pixel^{-1}) but with the same FOV. During readout in either format, the analog signals are processed using correlated double-sampling (CDS) and converted to 12-bit data numbers (DNs) using an analog-to-digital converter (ADC; only 12 bits of a 14-bit device are used), yielding a valid DN integer range from 0 to 4095. The digitized image is then transferred to flash memory in the DPU (see below).

Whenever L'LORRI is active, the CCD is exposed to whatever scene is in the FOV. The clocking of the CCD includes a "frame scrub", followed by exposure to the scene for the commanded integration time, followed by a "frame transfer" in which the CCD rows are sequentially transferred from the optically active area to the image storage region, followed by a readout of the image storage region to the downstream electronics. Owing to the finite transfer time, photons from the source continue to be integrated during the fast image shift, thereby producing the so-called transfer smear. However, as long as the source is not saturated, transfer smear can be effectively and deterministically removed during post-processing at the cost of some additional photon noise. Unlike the case for LORRI, the L'LORRI scrub and transfer times are exactly equal (11.7762 ms), which considerably simplifies the algorithm used to remove the smear transfer signal in the calibration pipeline.

L'LORRI does not have any color filters. Instead, L'LORRI maximizes sensitivity by providing panchromatic imaging over a wide bandpass: 420–795 nm at the 50% throughput points. Image exposure times range from 0 ms to 64,900 ms in 1 ms increments. Images can be taken at a maximum rate of once per second, or at any commanded cadence longer than that.

A flexible autoexposure mode is available whenever the scene being imaged has unknown intensities, but autoexposure can only be used for exposure times $\leq$900 ms. Although we plan to check the autoexposure capability during in-flight L'LORRI functional tests, we don't currently have plans to use this mode in-flight, except possibly during observations to characterize solar scattered light. When making in-flight L'LORRI observations of a target with unknown, or poorly characterized, brightness, we will usually opt to use two different *manual* exposure times that span the dynamic range of interest, rather than relying on autoexposure mode.

A "trigger" mode is also available that enables L'LORRI to determine autonomously when a target has entered the FOV, via analysis of a 32-bin histogram of the image, and then run an onboard macro routine that saves images to the flash memory for a specified cadence and duration after the trigger condition has been satisfied. We do not anticipate using trigger mode during any of the Trojan flybys, but we successfully tested it during L'LORRI observations of the Moon near the time of the first Earth gravity assist (EGA-1). Trigger mode functionality is also routinely tested in-flight using the L'LORRI internal lamps.

Each DPU has 64 Gb of flash memory, which can store over 4400 1×1 format images, which is much larger than the expected number of images collected during any single Trojan flyby ($\sim$2000 1×1 images). Although there is room in flash memory to store over 70 thousand 4×4 format images, the total number of images stored in either format is limited by flight software design to a maximum of 11,264, which is the number of possible entries in the onboard image allocation table. The images stored in the DPU's flash memory can be transferred by command to the spacecraft's command and data handling (CDH) computer, which losslessly compresses them to its solid state memory as part of the transfer process. The spacecraft can send selected images to its telecommunications hardware for downlink to the antennas of NASA's deep space network (DSN). The DPU handles all L'LORRI communications with the spacecraft.

The spacecraft controls several survival heaters situated throughout the instrument, as well as a decontamination heater, which is used to warm the optics during all science activities. The L'LORRI DPU controls several operational heaters that are used to maintain

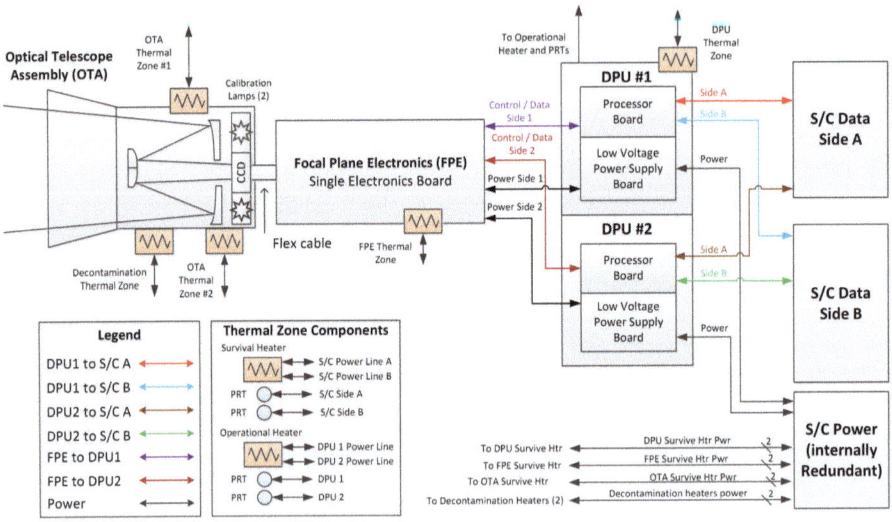

Fig. 4 L'LORRI block diagram showing the major components of the instrument and its interfaces to the *Lucy* spacecraft

instrument temperatures within acceptable ranges during operations. Figure 4 is a block diagram of L'LORRI showing the main instrument components.

3.2 Design Requirements and Trades

The stringent optical, thermal, and structural requirements for the original *New Horizons* LORRI OTA presented many design challenges, with maintenance of high resolution without the need for focus adjustments as the primary driver. This led to the use of Silicon Carbide (SiC) for the structure and reflective optics, which, in principle, would allow for operation over the wide thermal range expected during the different *Lucy* Trojan flybys. Thus, the L'LORRI OTA uses the same basic design as LORRI with a few exceptions. The higher design launch loads for *Lucy* required additional mass in the OTA structure for more strength. The primary baffle material changed from a composite to aluminum, which is cheaper and can be manufactured quicker. The bandpass was optimized to reduce ghosting, while still maintaining sufficient signal to meet performance requirements. LORRI was mounted inside the *New Horizons* spacecraft, and a radiator was needed to cool the LORRI CCD down to its specified operating temperature (approximately −80 C). But this radiator was deleted from the L'LORRI design because L'LORRI is mounted to the Instrument Pointing Platform (IPP), which is open to cold space allowing passive cooling of the CCD to its desired operating temperature (also near −80 C).

The L'LORRI electronics required significant redesign as requirements in general for electronics have changed significantly over the last two decades. New controls and quality requirements as well as additional radiation requirements, including Enhanced Low Dose Rate Sensitivity (ELDRS) testing, was needed. The thermal blankets around the OTA were changed to a less reflective material to reduce reflections into other instruments. Several of the major trades for L'LORRI centered on reducing mass, which had strict requirements during the design phase. This included changing the material of the DPU housing from aluminum to magnesium for mass reduction.

3.3 Optical Design

The L'LORRI OTA is a Ritchey-Chrétien design with an emphasis on high system throughput to measure Trojan dark surfaces at high phase angles, search for faint satellites near the main bodies, and assist with optical navigation. Since the *Lucy* science objectives for L'LORRI were so similar to the LORRI objectives on *New Horizons*, L'LORRI adopted essentially the same optical design as LORRI. During the last two decades, there has been an evolution in the techniques L3H uses to manufacture and process SiC optics, and most of the L'LORRI OTA components benefitted from those innovations. However, the spare primary mirror (M1) from the *New Horizons* LORRI program became the M1 for L'LORRI because an error was discovered during the final processing of the newly manufactured L'LORRI M1 mirror and schedule pressure precluded waiting for a new M1 to be produced and tested before the *Lucy* launch. The unintended variation of focus with temperature discovered during the ground testing of L'LORRI (discussed in more detail later) may be related to the use of the original LORRI M1 in the L'LORRI OTA, but there is no direct evidence to support that hypothesis.

The complete L'LORRI OTA design was evaluated with a Structural, Thermal, and Optical Performance (STOP) analysis. This included time-dependent analysis during both the approach and departure phases associated with the Trojan flybys. Fortunately, the OTA thermal environment changes insignificantly during the few hours centered on closest approach owing to the OTA's high thermal inertia. Stray light was modeled in FRED, a commercial optical analysis software package.

3.4 Thermal Design

The L'LORRI thermal environment posed a design challenge relative to the heritage case. LORRI was mounted within the interior of the *New Horizons* spacecraft, and the OTA temperature has been passively maintained near −80 C throughout the entire mission. In contrast, the L'LORRI is mounted outside the *Lucy* spacecraft with predicted OTA temperatures ranging from −120 C to −75 C for different Trojan flybys, assuming there is no active thermal control. The design challenge was to minimize defocus due to thermal gradients arising in the L'LORRI OTA metering structure.

The main OTA structure was manufactured using SiC 55a to produce a nearly athermal design that enables passive operation. The high conductivity of SiC acts to minimize thermal gradients, and SiC's low coefficient of thermal expansion minimizes the thermal strain impact of such gradients. Invar 36, a good match to SiC 55a over the temperature range of interest, was chosen for the metallic inserts that allow bolting together of the OTA assembly to improve its strength. All invar inserts, as well as the secondary mirror foot, were epoxy bonded to the SiC.

The OTA is mounted inside the telescope baffle tube, which is made of aluminum. The baffle tube provides a relatively uniform cold sink along the length of the telescope that helps to reduce longitudinal thermal gradients. The telescope is mounted to the spacecraft via long G-10 isolators, which isolate the OTA conductively from the spacecraft deck. The entire OTA (except for the entrance aperture) is covered with multilayer insulation (MLI), which minimizes radiative coupling between the OTA and space. The secondary mirror is mounted to a flexured Invar 36 plate, which in turn mounts to the end of the SiC metering structure. The secondary and primary mirror magnesium baffles are flexure mounted to the OTA structure (SiC) and the lens cell (Invar 36), respectively. The CCD mount plate is attached to the OTA structure by titanium flexures, which serve the dual purpose of mitigating thermal strain and thermally isolating the CCD from the structure. The G-10 mounts for the overall OTA have titanium post flexures on either end. Additionally, the OTA itself is mounted to the aluminum baffle tube via titanium isolators, which act to mitigate thermal strain and vibrations and to provide conductive isolation.

The baseline mission design ran the decontamination heaters with setpoints of −10 C for the first six months of the mission to promote early desorption of water vapor and at −40 C for the rest of the mission (except during science or optical navigation operations when the OTA is kept cooler) to mitigate against potential condensation of volatiles. The CCD must be maintained at a temperature colder than −70 C while acquiring science data to achieve acceptable levels of dark current. This requirement is met passively from cooling by the OTA, which balances the heat transfer from the FPU that generally runs at a warmer temperature.

The L'LORRI in-flight temperatures were predicted via a finite difference thermal model that included all conductive and radiative heat transfer. A thermal balance test of the instrument was performed that validated the thermal model, and a separate thermal balance test was performed at the spacecraft level to further validate the model.

The STOP analysis for the different encounter geometries demonstrated that a passive design produced minimal gradients in the OTA metering structure. Optical performance with the expected gradients was included in the final performance analysis and was within the performance specification on resolution. However, ground testing showed that the OTA focus varied significantly with temperature and slightly violated the resolution requirement at the coldest temperatures. Thus, we decided to re-purpose the spacecraft-controlled decontamination heater to force the OTA to a temperature near −95 C for all science observations. This new strategy should produce similar optical performance for all L'LORRI imaging, including all of the *Lucy* mission flybys.

3.5 Mechanical Design

The structural design objectives of the OTA were to minimize mass while maintaining performance over operational temperatures and allowing for the stiffest design that will survive the launch environment. The requirements were a maximum mass of 7.3 kg, a minimum resonant frequency of 60 Hz, and survival under launch-induced vibration and stresses. The inherently high stiffness-to-weight ratio of SiC (∼4.5 times that of aluminum) allowed the fabrication of a low-mass structure with a light-weighted primary mirror of open back, hub-mounted design to minimize weight (identical to the original LORRI).

The main baffle tube, including its vanes, was fabricated from a single block of aluminum. The smaller internal baffles were fabricated from a light-weight magnesium alloy (ZK60A). Vibration isolation was required to survive the launch environment. Titanium isolators were incorporated to mount the structure to the main baffle tube at three points approximately at its center of gravity location. Another important structural design consideration was to minimize any potential for mount-induced distortion of the optics. Intimately connected to this design consideration is the requirement for the OTA to mount to a surface with only moderate mounting coplanarity, namely, a spacecraft honeycomb panel. To avoid degradation of optical quality, a three-point mount was adopted, with the bases of each of the three mount locations on the OTA outfitted with a ball joint that can be loosened and re-tightened, if necessary. At the interface between the main baffle tube and the OTA inner assembly, the vibration isolators also act to mitigate any mount-induced strains. The flexurized mount plate at the secondary mirror serves the dual purpose of controlling thermally induced distortions, as well as mount-induced distortions; the CCD flexure mounts also serve this dual purpose. Due to its mass, the primary mirror was not flexure-mounted, as the low resultant frequency and dynamic responses would have increased the risk to the OTA under vibration. Instead, the primary mirror is hub-mounted, with a post and a foot bolted via three invar inserts to a mount plate. The mount plate is in turn bolted to the structure. Because the lens cell is made of Invar and mounts to the structure, close to the primary mirror, the mount plate helps to separate any induced thermal strain in the structure from being transferred to the primary mirror.

3.6 Instrument Integration, Focus, and Alignment

When the OTA was assembled at L3H, a convex spherical reflector was centered at the focus using interferometry. Metrology, combined with knowledge of the shim size used to connect the reflector to the carrier plate, determined the location of the focus relative to the interface location on the carrier plate. L3H also provided a reference mirror on the back of the secondary mirror mount, such that the telescope line-of-sight was parallel to the reference mirror's normal. Two optical reference flats at right angles to the line-of-sight were mounted to the LLORRI metering structure for use in alignment monitoring after mounting to the spacecraft.

The depth of focus for L'LORRI at the detector, with mechanical tolerances on the CCD, allowed initial shim sizes to be selected for system focusing at the L3H-provided focus location. A 300-mm aperture, f/5 off-axis parabolic collimator was used to project a point-like image into L'LORRI. This image was produced by a laser unequal path interferometer (LUPI) at the collimator focus. A series of exposures were made at several points in the L'LORRI field by moving within the field using a fold mirror. At each location, the spot was centered on a detector element by viewing the live image from the CCD through ground support electronics, and balancing the wings of the image symmetrically about the center detector element. This was repeated for several focus adjustments of the collimator, with slight measured changes from nominal to allow for deterministic shimming of L'LORRI. Data from these exposures were examined to find the best L'LORRI focus versus the collimator adjustment. New shims were then installed to move the plane of the CCD onto the plane of best focus. Once at nominal focus, the CCD was centered on the optical axis by use of a theodolite viewing both the reference mirror on the back of the secondary mount and the primary mirror. The theodolite was auto-collimated on the reference mirror, and the azimuth and elevation recorded. The theodolite then viewed the four corners of the CCD off the primary mirror, and these azimuth and elevation values were recorded and averaged. The CCD location was shifted in the plane of best focus such that the average of the four CCD corners was within tolerance of the normal to the reference mirror.

After the best focus was found in ambient conditions, L'LORRI was installed in the APL optical calibration facility (OCF) for a focus check at flight-like thermal and vacuum conditions. This was done by viewing a collimated beam, which projected a small point-like image into L'LORRI, and stepping the spot over the field by tilting L'LORRI on a gimbal platform.

Prior to delivery to the spacecraft, L'LORRI's line-of-sight was measured relative to the optical reference flat mounted to the back of the secondary mirror support. Additionally, the roll angle was measured by viewing the CCD corners, and referencing to the orthogonal alignment mirrors. These data, combined with measurements performed referencing the flats to the spacecraft coordinate system, showed that L'LORRI's line-of-sight was within mission requirements. Tracking of the L'LORRI alignment references through the spacecraft environmental test program did not show any significant movement relative to the spacecraft coordinate system.

Table 3 L'LORRI Contamination Requirements

Component	Requirement
External Surfaces	400 A/2
Internal Surfaces	250 A/2
CCD	350 A/2

These requirements refer to beginning-of-life values for the *Lucy* mission (i.e., valid at launch). Requirements refer either to Particulate or Non-Volatile Residue (molecular) contamination.

3.7 Contamination Control

The L'LORRI telescope assembly remained under a nitrogen purge during all phases of integration and test until launch, except for limited times when images were taken or when put under vacuum. The internal cleanliness requirements for the L'LORRI OTA per the L'LORRI specification are given in Table 3. The outgassing of the entire instrument was verified to be below 5×10^{-12} g cm^{-2} s^{-1} during thermal vacuum testing at the highest survival temperatures allowed for the instrument.

3.8 Electronics

As shown in Fig. 4, the L'LORRI electronics consist of the DPU and FPU. The DPU contains two printed circuit cards: the low-voltage power supply (LVPS), and the processor board (PB). The DPU is the primary interface between the spacecraft and the FPU, which is directly connected to and controls the CCD. There are two DPUs to provide redundancy; either DPU can be used with either of two redundant spacecraft CDH computers. Only one DPU side is active at a time; the other is a cold spare.

3.8.1 FPU

The L'LORRI FPU is required to read out a complete image in one second, with the charge level in each pixel represented by a 12-bit binary word. The FPU noise is required to be <40 electrons per pixel, well above the CCD read noise, which is calculated to be about 10 electrons at the readout time of about 0.7 μs per pixel. In general, exposures of 20 to 200 ms are typical for L'LORRI science images, although bias images are taken with the minimum commanded exposure time of 0 ms, the maximum exposure time is 64.9 s, and optical navigation images typically use 5 s and 10 s exposure images.

The FPU also includes two small incandescent bulbs mounted on the same sub-board as the CCD. These lamps can illuminate the CCD via light scattered off the back end of the OTA structure (Fig. 5), so that testing can be performed even when imaging is not possible through the optics (e.g., functional testing). These lamps were from the same batch of lamps as the original *New Horizons* LORRI, which produced output light levels that have been stable to ∼1% for the entire (ongoing) mission. The illumination pattern produced by the lamps provides a means of testing whether there has been any change in the relative pixel response over the entire CCD, including whether there has been any change in any foreign object debris (FOD) on the CCD (see Sect. 4.2.7).

The L'LORRI CCD uses a three-phase architecture for charge transfer in the image, memory zone, and serial registers. For optimum performance, these phases must be clocked

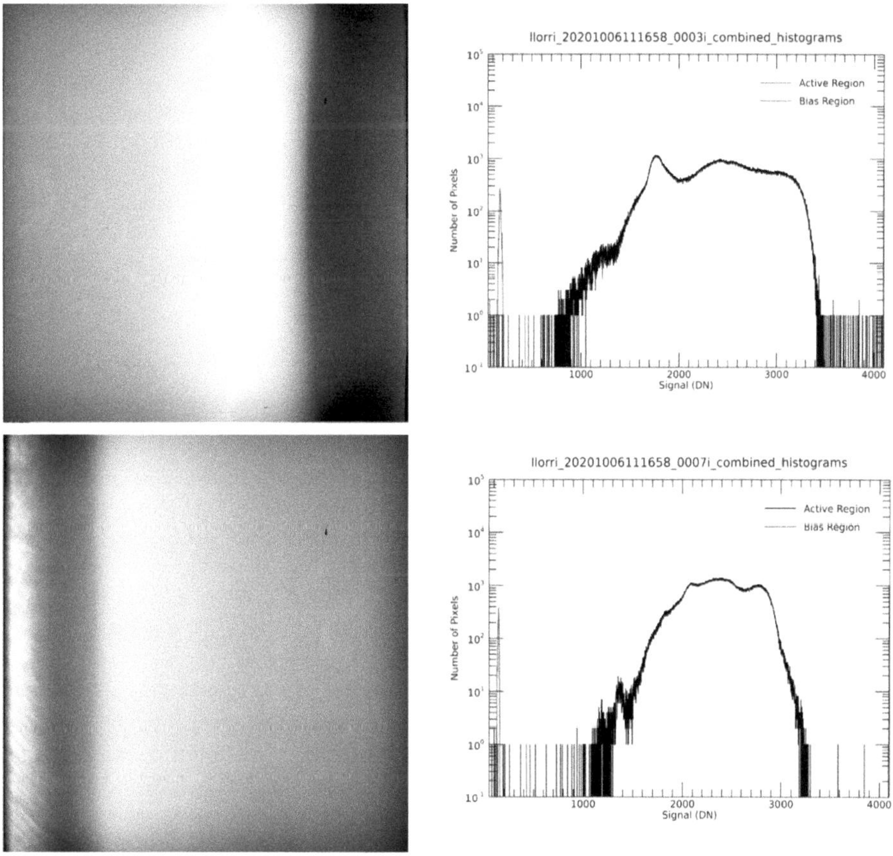

Fig. 5 Top left: An image from internal lamp 1. Top right: Histogram of the image from lamp 1. Bottom left. An image from internal lamp 2. Bottom right: Histogram of the image from lamp 2. Both images are displayed on a linear scale spanning 1000 DN to 3500 DN. Both lamps were set to a level of 100, and the image exposure time was 100 ms

with specific waveforms and voltage levels. These phase lines are highly capacitive, particularly for the image and memory areas of the chip, and they experience mutual capacitive coupling. L'LORRI uses drivers that are designed to drive high capacitance loads at the required voltage levels from logic level inputs. They are switching, not linear, devices so that low and high voltage levels are obtained by a suitable choice of supply voltages, and transition rates must be adjusted at the output. This is done with series resistance, which adds to the internal switch resistance of the drivers, forming a simple time constant with the capacitance of the CCD phase.

The CCD requires 29 V bias for the output field effect transistor. A charge pump with pre- and post-regulation was used to generate this voltage. The L'LORRI FPU uses an integrated circuit that performs correlated double sampling, signal amplification, and analog to digital conversion to 12 bits (after dropping the two least significant bits of the 14-bit ADC) at maximum rates of 6 MHz, comfortably above the pixel readout rates, which are ~1.5 MHz. The integrated double sampling circuit is potentially susceptible to latch-up from ionizing radiation and heavy particle impact in space, therefore the L'LORRI FPU incorporates latch-up protection circuitry that was tested in proton irradiation facilities.

3.8.2 DPUs

The L'LORRI processor board (PB) controls the instrument via interfaces to the LVPS board. The PB communicates to the spacecraft using an RS-422 link, which receives commands and transmits engineering data. The PB uses an FPGA processor and runs FORTH code. There are dual, redundant interfaces on the spacecraft, which provide command, data handling, and telemetry functions. One interface is active and one is a back-up; L'LORRI provides both interfaces (low voltage differential signal, or LVDS, for high-speed image data and RS-422 for low-speed command and telemetry) to both sides of the spacecraft. Secondary functions of the L'LORRI PB include the ability to:

- Store and transmit the image and header information
- Calculate a 32-bin histogram
- Generate test patterns without an FPU present
- Command the FPU mode and exposure times

The PB can generate test pattern images without an FPU present. The first pattern consists of a horizontal ramp, and the second pattern consists of a vertical ramp. The imager interface FPGA can also receive data from the spacecraft once a second, and those data are used to set the FPU mode and exposure time. The instrument engineering data are transmitted by the spacecraft within a separate data stream from the science data, and the engineering data must be associated with individual images in ground processing. However, the critical header information is encoded into the images themselves, at the cost of 56 pixels in the first row of each image. The FPGA also calculates a 32-bin histogram of the FPU image data currently being transmitted. This histogram is then made available to the instrument for future exposure time calculations (e.g., for AUTO exposure mode). The FPGA collects the FPU status and temperature data, making it available to the instrument for thermal control purposes. The LVPS provides power at the voltage required by the other boards within the DPU and by the FPU. The input voltage from the spacecraft is 24-34 V. The LVPS board provides switching to control power on/off to the FPU and the instrument operational heaters.

The 64 Gb of flash memory resides on the DPU PB. When images are taken, they are written sequentially to flash memory, where they can be retrieved non-sequentially by spacecraft command using either the observation identification number (OBSID), which is set by a L'LORRI command, or by memory index number, which is managed by the L'LORRI flight software. L'LORRI images are typically transferred from flash memory to the spacecraft CDH memory within a second, but the CDH flight software also losslessly compresses each image and formats it for downlink as part of the transfer process, which typically takes $\sim$23 s for each image. Additional image transfers are not allowed until the previous processing has been completed. The flash memory can be erased by L'LORRI command, typically after verification that the previously stored data have been successfully downlinked to Earth. Note that the flash memory in one DPU cannot be accessed by the other DPU. That is, the two flash storage devices, like the DPUs themselves, are completely independent of each other.

3.8.3 Flight Software

The L'LORRI processor shares a common design with that of many instruments that APL has flown. This common design extends to the software. The common flight software (CFS) provides packet telemetry and command handling services. Besides handling L'LORRI-specific packets, the CFS automatically generates a variety of standard packets, including housekeeping and status, command echo, memory dump, and alarm packets. Similarly, besides handling L'LORRI-specific commands, the CFS also handles standard commands for memory loads and memory dump requests. The CFS also provides storage and execution of command sequences.

The CFS has timekeeping, voltage and current monitoring, and memory management services, and a standard boot program. The L'LORRI-specific flight software controls heaters, collects voltages, currents, and temperatures from the LVPS board, and manages the FPU. In the FPU, the software controls the exposure time, either by manual command or automatically based on the hardware-provided image histogram, generates an image header, and enables routing of the image to the flash memory and then to the spacecraft. The software also controls the FPU's test patterns and internal lamps.

4 Ground Calibration Results

L3H delivered the flight model (FM) L'LORRI OTA to APL on 2020 April 14. L'LORRI underwent extensive ground testing at APL starting in May 2020 and ending with shipment to the *Lucy* mission spacecraft facility at Lockheed-Martin (LM) in Littleton, Colorado in late-October 2020. In this section, we describe the results from the ground testing at APL, which served to characterize L'LORRI and verify that it met its requirements. The testing at LM verified that L'LORRI's performance on the spacecraft was generally consistent with the results from the ground testing at APL, so we won't explicitly discuss the LM test results here.

The L'LORRI ground tests at APL are summarized in Table 4. Some of the early tests, including the pre-environmental calibration, were performed with the engineering model (EM) FPU and EM DPU because the flight models (FMs) were not yet available. However, the entire FM system was used during environmental testing in September 2020, and during the post-environmental calibration conducted in October 2020. No changes were made to L'LORRI after the post-environmental calibration.

The L'LORRI optical performance tests at APL were generally conducted either in a cleanroom, or in an adjacent optical calibration facility (OCF). L'LORRI in-flight conditions were simulated in the OCF (Fig. 6), and measurements in that facility during the post-environmental calibration were the most critical ones for verifying conformance with L'LORRI's requirements. Thus, we will mainly focus here on the results from the post-environmental calibration, except for results on optical ghosts, scattered light, and CCD saturation from measurements made in the APL cleanroom under ambient laboratory conditions.

For the OCF tests, L'LORRI was mounted on a rotation stage inside a thermal vacuum (TVAC) chamber. For PSF and focus measurements, L'LORRI was pointed at the output of a collimator mounted just outside a large (40.6 cm diameter) TVAC chamber window. The collimator has an off-axis, confocal Cassegrain design with a focal length of 1866 mm, an exit pupil of 304 mm (L'LORRI's entrance pupil is 208 mm), and a fold mirror that can scan across a $1.1^\circ \times 2.4^\circ$ field (L'LORRI's field of view is 0.29°). The pinhole in the collimator could be precisely moved along the optical axis to produce focus curves, that is, PSF measurements as a function of focus location along the L'LORRI optical axis. The collimator provided both broad bandpass (panchromatic) light, as well as light from a spectrograph to measure L'LORRI's spectral response (i.e., sensitivity as a function of wavelength).

L'LORRI could also be rotated to view a calibrated integrating sphere through a second 40.6 cm diameter window for flat-field and linearity measurements. Two different lamps

Table 4 L'LORRI ground tests

Test ID	Timeframe (2020)	Comments
Special Early Tests	May	Measure scattered light and characterize optical ghosts. Also characterize amplifier undershoot/overshoot. All of this testing used the FM OTA but with an EM FPU and an EM DPU. All measurements made in APL cleanroom under ambient conditions with red (632 nm) laser illumination from a LUPI.
Preliminary Focus	Late-May and June	Preliminary focus characterization using the FM OTA, EM FPU, and EM DPU. Tests conducted in both the APL cleanroom under ambient conditions and in the OCF under flight-like conditions.
Pre-Environmental Calibration	June 27–July 8	Calibration measurements in flight-like environments with the FM FPU and EM DPU.
Thermal Cycle/Balance	September 5-17	Mostly performed to verify thermal models but included the best measurements of bias and readnoise, including their variations with temperature.
Post-Environmental Calibration	October 1-7	Calibration measurements in flight-like environments with the FM FPU and the FM DPU.
ATLO TVAC	May 2021	L'LORRI testing while integrated on the *Lucy* spacecraft.

EM refers to an Engineering Model, and FM refers to the Flight Model. OTA stands for the Optical Telescope Assembly. LUPI is a Laser Unequal Path Interferometer. OCF refers to the APL Optical Calibration Facility. ATLO stands for Assembly, Test, and Launch Operations. TVAC refers to testing in a Thermal Vacuum chamber under flight-like conditions. All tests were conducted in 2020, except for ATLO TVAC.

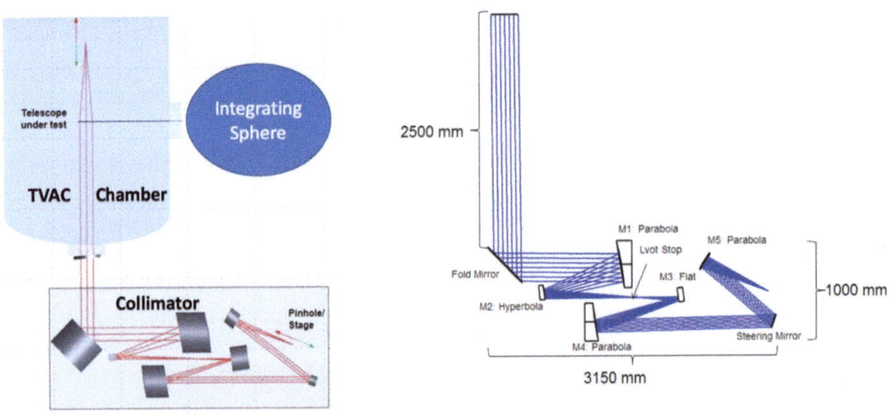

Fig. 6 Left: Drawing of the APL optical calibration facility (OCF), where the main L'LORRI ground calibration was conducted. L'LORRI was placed in the thermal vacuum (TVAC) chamber. Point Spread Function (PSF) and focus measurements were conducted by illuminating L'LORRI through a large window in the TVAC chamber using the collimator located just outside the TVAC chamber. L'LORRI could also be rotated to view an integrating sphere through a second window for flat-field and linearity measurements. Right: Ray trace showing the collimator optical path

were used to illuminate the integrating sphere: a plasma discharge lamp provided a reasonably good approximation to the solar spectrum for most of the visible light portion of the spectrum, while a quartz halogen lamp was used to mimic the solar spectrum at the longer

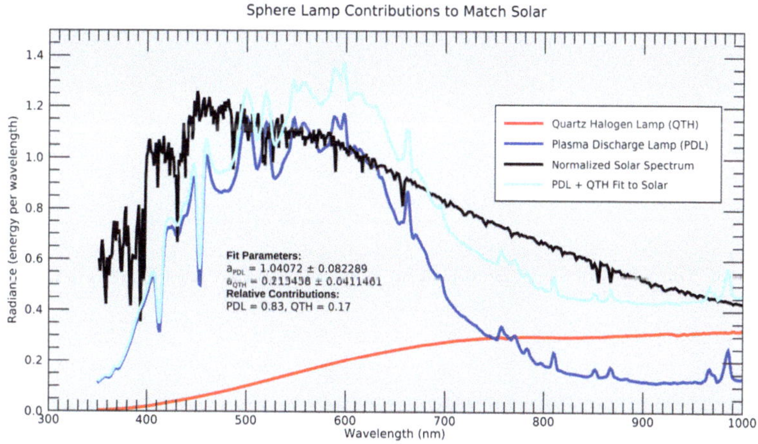

Fig. 7 The solar spectrum is compared to the spectra of the two lamps used to illuminate the integrating sphere during the L'LORRI calibration measurements. The linear combination of the two lamp spectra that provides the best fit to the solar spectrum is also displayed. To produce the best fit, the plasma discharge lamp has a weighting factor of 0.83 and the quartz halogen lamp has a weighting factor of 0.17, which are the weighting factors used to produce the final L'LORRI flat field

wavelengths. The relative spectral distributions of each lamp, and the linear combination that best matches the solar spectrum, are displayed in Fig. 7.

4.1 Optical Performance

We first discuss the optical performance, which includes:

- Focus variation with temperature
- Point spread function (PSF) quality and encircled energy (EE)
- Scattered light
- Ghosts

4.1.1 Focus Variation

Contrary to its design specification, the L'LORRI OTA is *not* athermal, as demonstrated by testing at both L3H and APL. Figure 8 shows that the L'LORRI focus location changes by up to 750 μm over the full range of operating temperatures (i.e., from +20 C to −125 C). As previously discussed, the change in focus with temperature motivated us to use the spacecraft-controlled decontamination heater to force the L'LORRI OTA temperature to be near −85 C for all science observations. As discussed in Sect. 5, preliminary in-flight testing suggests that an even colder OTA temperature (near −95 C) produces better optical performance. In any case, up to ∼2.3 W of extra power is required to heat the OTA to the desired temperature during in-flight operations, which is available from a spacecraft-controlled decontamination heater.

The location of the L'LORRI CCD relative to the best focus position has a relatively large uncertainty of ±100 μm. Based on the post-environmental calibration data, our best estimates for the L'LORRI focus locations as a function of OTA temperature are depicted in Fig. 9. At the originally targeted OTA temperature of −85 C, the OTA best focus may lie up to ∼400 μm in front of the CCD, or the best focus may be only ∼200 μm in front of

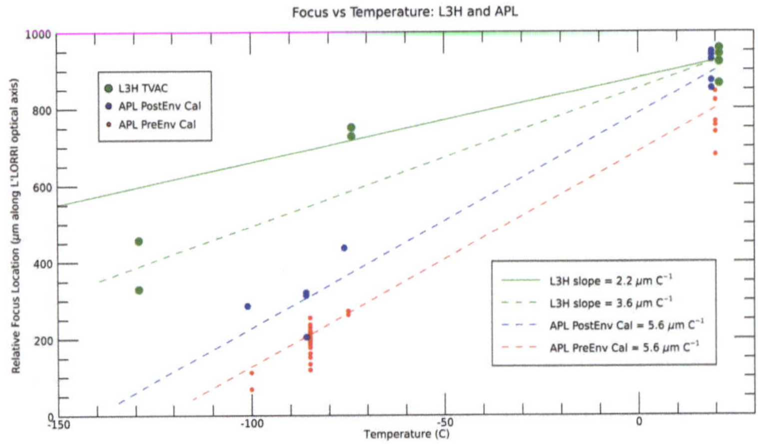

Fig. 8 Plot showing focus versus temperature results for the on-axis location for both L3H and APL measurements. The uncertainty in the focus location appears to be about ±100 μm for both sets of data. The APL data seem to show a steeper change in focus location with temperature, but both data sets show a large change in focus location as the OTA is cooled to the temperature expected for the *Lucy* mission science observations (probably close to −95 C)

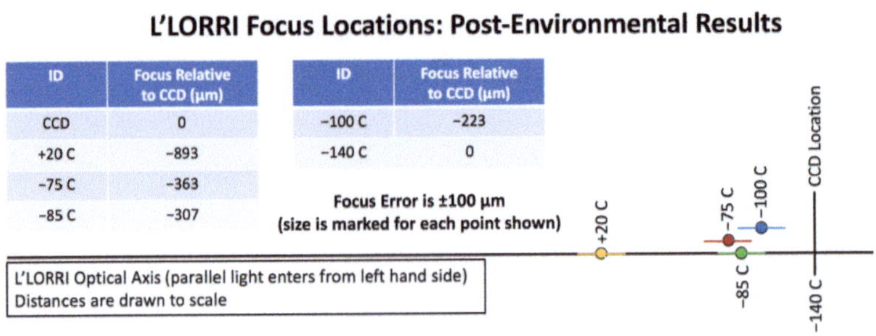

Fig. 9 Figure showing the estimated L'LORRI focus locations relative to the CCD location for various temperatures based on the post-environmental ground calibration measurements. The uncertainties in the focus locations are ±100 μm. For the *Lucy* mission science and optical navigation observations, the OTA will be thermostatically controlled to a temperature near −95 C, with the final thermal set point determined after more in-flight tests are conducted

the CCD. Fortunately, the focus curve for the L'LORRI OTA is relatively flat, and the PSF quality only degrades modestly when moving off best focus (see Sect. 4.1.2).

4.1.2 PSF Quality and Encircled Energy (EE)

During testing at APL, we performed focus runs comprised of the following activities:

- Rotate the collimator beam in altitude and azimuth to position it at 13 locations across the L'LORRI FOV
- At each FOV location, drive the collimator focus stage through 17 different positions (i.e., motions of the image along the L'LORRI optical axis), approximately ±8 locations about the optimal focus position, using more closely spaced positions near nominal focus to enable better fitting of the focus curve

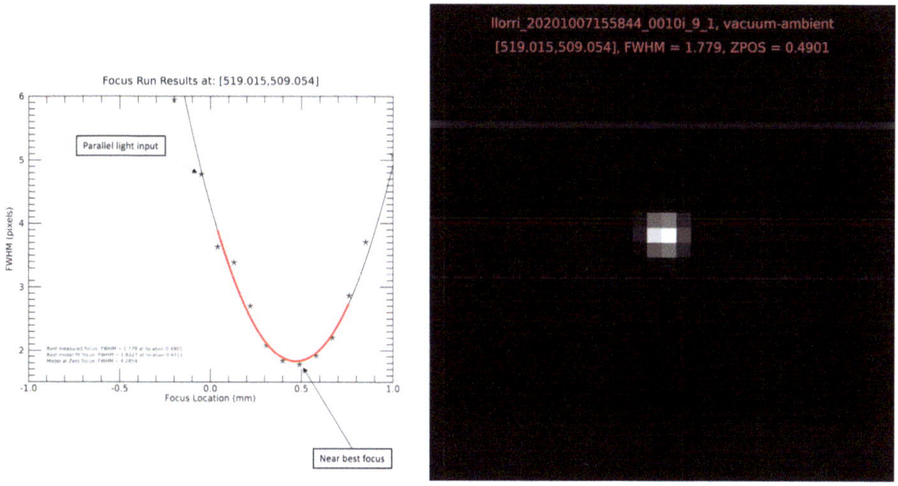

Fig. 10 Left: Focus curve for the OTA at room temperature (+20 C) near the center of the CCD. The curve is relatively sharp with the PSF FWHM depending strongly on the focus location. Right: PSF near the best focus location, which is relatively symmetrical with a FWHM of 1.8 pixels. The display uses a linear stretch grayscale ranging from the minimum to maximum values of the PSF

- Take 3 images at 1 second cadence at each piston location using an exposure time of 10 ms

L'LORRI has ~1 arcsec pixels, which imposes severe constraints on the quality and stability of the collimator image. The APL OCF facility was commissioned only shortly before the start of the L'LORRI ground calibration campaign, and the L'LORRI team had to troubleshoot multiple issues that arose during the course of this testing. In particular, the team struggled to control environmental factors that affected collimator image stability (e.g., not allowing anyone in the collimator room during optical testing, turning off vacuum pumps, turning off thermal control fans, etc.; image stability was affected by both the collimator stability and the stability of L'LORRI's mounting fixture within the OCF thermal vacuum chamber). Despite using extremely short L'LORRI exposure times, the collimator image motion was typically at least 1-2 L'LORRI pixels during each image. Thus, the PSFs measured during our testing at APL are generally degraded relative to their true quality. The image quality was judged by fitting an elliptical Gaussian to the PSF and quoting the geometrical mean of the major and minor axis diameters as the "FWHM". L'LORRI has a requirement that the PSF must have FWHM $\leq$3 pixels over most of the CCD. The "encircled energy" (EE), defined as the ratio of the signal contained within the peak pixel relative to the signal integrated over the entire PSF, is another measure of image quality. L'LORRI must have EE $\geq$ 0.04 (i.e., EE $\geq$ 4%) to meet its most challenging SNR requirement.

The L'LORRI PSF has a sharp focus curve and a tight PSF (FWHM $\leq$ 2 pixels) (best fit Gaussian) at room temperature (Fig. 10). The PSF is roughly circularly symmetric, and the EE is ~20% over much of the field at best focus (Fig. 11). However, EE decreases to ~5% for the out-of-focus PSF at the location where the CCD is mounted when parallel light is sent into the L'LORRI OTA under room temperature conditions.

The L'LORRI PSF monotonically degrades in quality as the temperature is lowered, mainly a result of increasing astigmatism with decreasing temperature. This behavior was neither expected (i.e., not consistent with the specification), nor has it been successfully

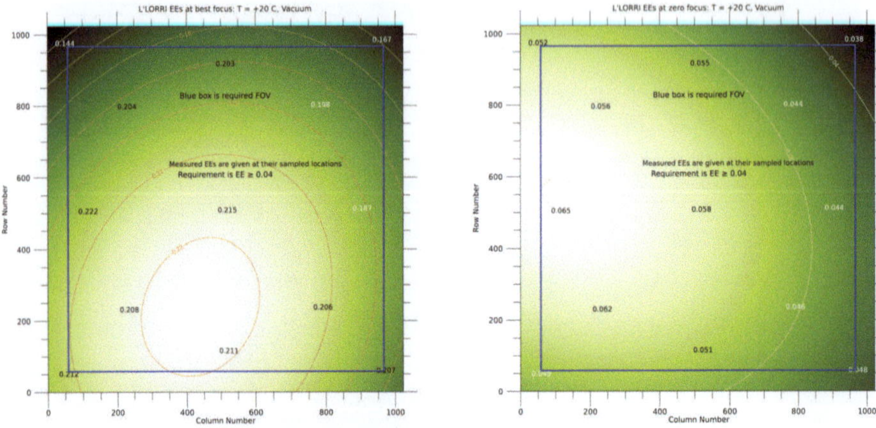

Fig. 11 Left: EE values across the CCD at best focus when the OTA is at room temperature (+20 C). The color background image is created by fitting a 2-dimensional polynomial of degree 2 to the measured EE values (at the positions indicated on the image) and is displayed using a linear stretch ranging from the minimum to maximum EE values. Smaller EE values are darker green, and larger EE values are lighter green. The EE values are ~20% across most of the CCD. Right: The same but with parallel light input to the OTA, which is focused ~900 μm in front of the CCD (see Fig. 9). The EE values are ~5% across most of the CCD

modeled by L3H. Given the change in focus with temperature and the degradation in image quality at lower temperatures, we decided to use the spacecraft-controlled decontamination heater to fix the OTA temperature at the predicted OTA temperature for the "warmest" Trojan flyby (−85 C) for *all* L'LORRI science and optical navigation observations, rather than allow the OTA temperature to vary over the large range predicted for the different Trojan flybys (−85 C to −120 C). We also tried to put the CCD at the best focus location for an OTA temperature of −85 C, but that was not completely successful (see Sect. 4.1.1) owing to discrepancies between focus data taken in the APL cleanroom (which we used to shim the CCD location) and data taken in the APL OCF.

Despite these difficulties, L'LORRI still meets its *Lucy* mission requirement over the required FOV when the OTA temperature is −85 C. Figure 12 shows the focus curve at −85 C, which is much flatter than the curve at room temperature, and therefore more forgiving in focus location. The FWHM is expected to be in the range 2.5-2.7 pixels over the FOV (Fig. 13), and the EE should be ~9% over the FOV (Fig. 14). The in-focus PSF at −85 C is more symmetrical than the out-of-focus PSF, but the FWHMs are comparable (Fig. 15).

We note that the PSFs from the ground calibration were degraded by image instability, and we expected the in-flight performance to be better, assuming spacecraft jitter didn't significantly degrade the in-flight PSF. Although the initial in-flight testing showed that spacecraft jitter didn't compromise the PSF for exposure times shorter than about 10 s, the PSF was slightly degraded relative to our expectations when the OTA was near −85 C, causing us to consider operating the OTA at a colder temperature during L'LORRI science observations. See Sect. 5 for further discussion.

4.1.3 Scattered Light

The scattered light properties of L'LORRI were quantified by obtaining point source transmittance (PST) measurements in the APL cleanroom. This involved measuring L'LORRI's

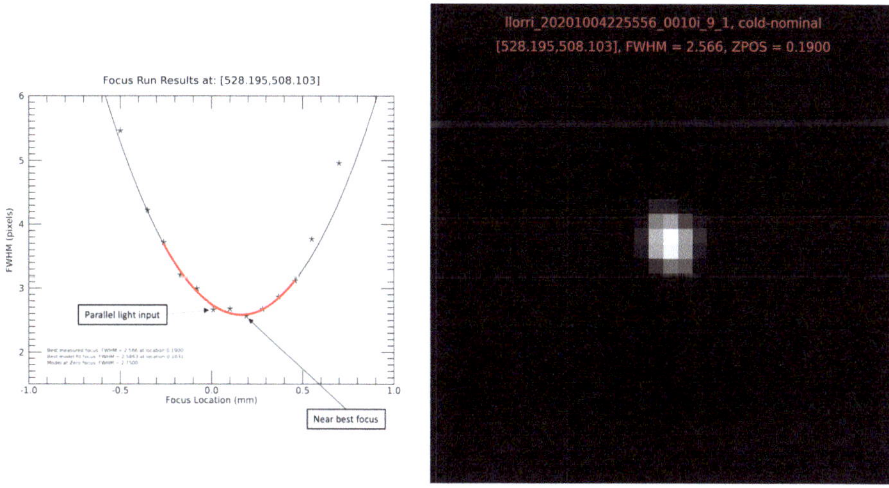

Fig. 12 Left: Focus curve for the OTA at −85 C near the center of the CCD. The curve is relatively flat with the PSF FWHM changing only moderately with the focus location. Right: PSF near the best focus location, which has FWHM ≈ 2.6 pixels. The display uses a linear stretch grayscale ranging from the minimum to maximum values of the PSF

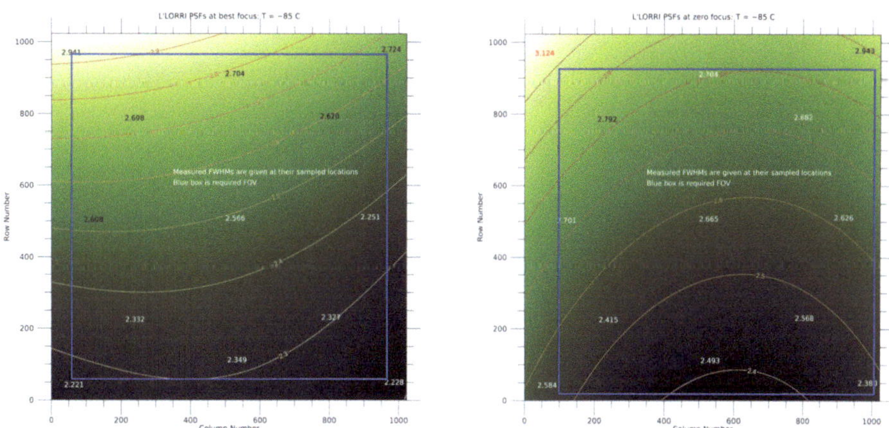

Fig. 13 Left:FWHM values across the CCD at best focus when the OTA is at −85 C, which we targeted as the OTA temperature for all *Lucy* mission observations prior to launch. The color background image is created by fitting a 2-dimensional polynomial of degree 2 to the measured FWHM values (at the positions indicated on the image) and is displayed using a linear stretch ranging from the minimum to maximum FWHM values. Smaller FWHM values are darker green, and larger FWHM values are lighter green. The FWHM values are ∼2.2-2.7 across the CCD. Right: The same but with parallel light input to the OTA, which is focused ∼300 μm in front of the CCD (see Fig. 9). The FWHM values are ∼2.4-2.9 across the CCD

response to point source illumination as a function of offset angle from the boresight (Fig. 16). The L'LORRI PST appears to be similar to the LORRI PST, which is expected given that both instruments have essentially identical optical designs. We originally hoped that the aluminum baffle used for L'LORRI, rather than the composite baffle used for LORRI, might reduce the scattered light for L'LORRI. But both baffles have aluminum vanes coated with Z306 black paint, and scattering from those surfaces is apparently dom-

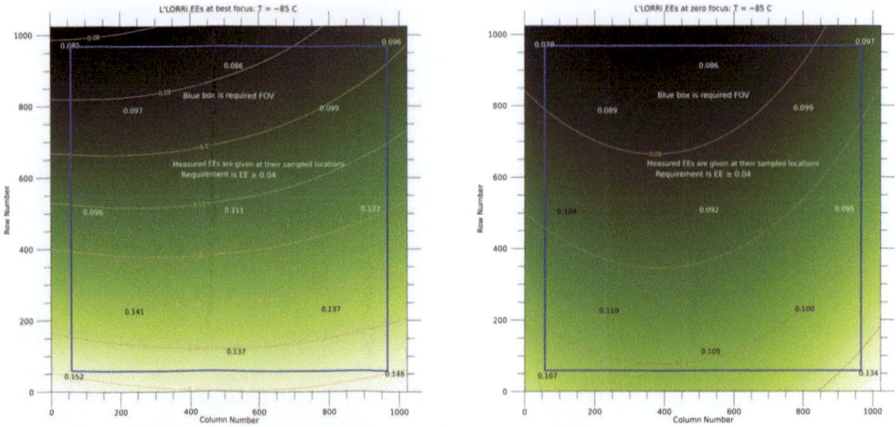

Fig. 14 Left:EE values across the CCD at best focus when the OTA is at −85 C, the pre-launch targeted OTA temperature for all *Lucy* mission observations. The color background image is created by fitting a 2-dimensional polynomial of degree 2 to the measured EE values (at the positions indicated on the image) and is displayed using a linear stretch ranging from the minimum to maximum EE values. Smaller EE values are darker green, and larger EE values are lighter green. The EE values are ∼11% near the center of the CCD. Right: The same but with parallel light input to the OTA, which is focused ∼300 μm in front of the CCD (see Fig. 9). The EE values are ∼9% near the center of the CCD

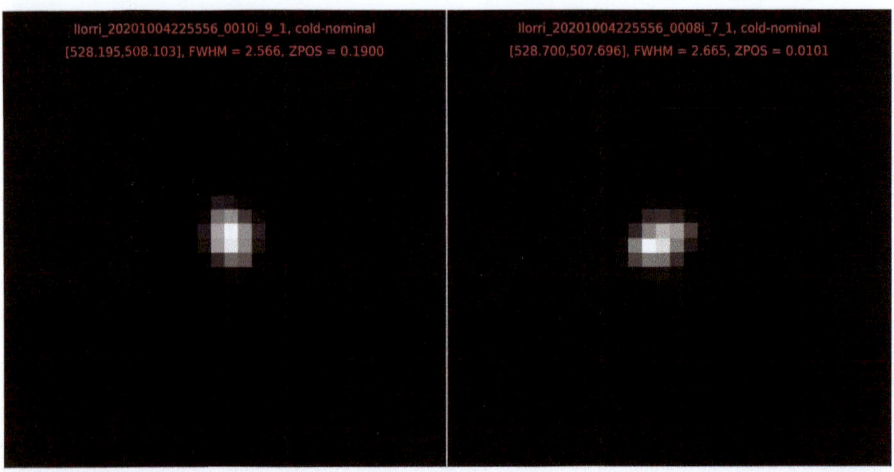

Fig. 15 Comparison of the PSF near the center of the CCD at −85 C for best focus (left) and for parallel light input to the OTA when the CCD is ∼300 μm behind the best focus. The best focus case is more symmetrical, but both PSFs have similar FWHMs. Both images are displayed using a linear grayscale ranging from the smallest to largest values in the PSF

inant in producing the measured PST values. The Ritchey-Chrétien optical design used for both instruments, with its central obscuration, limits how well scattered light can be suppressed. Nevertheless, L'LORRI meets its scattered light requirement with significant margin.

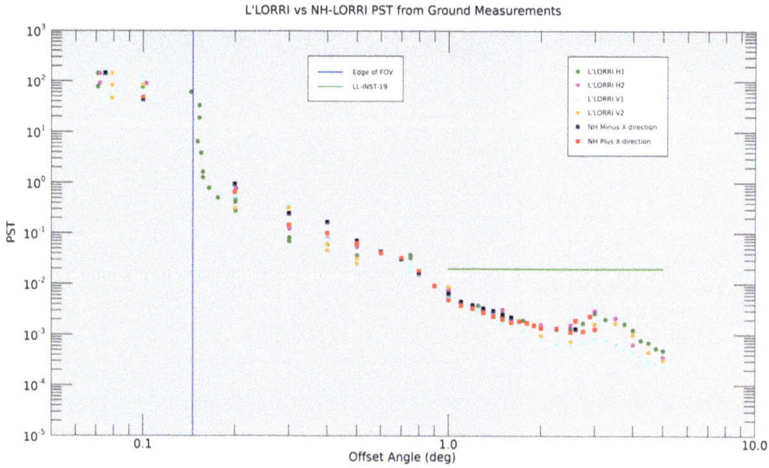

Fig. 16 The point source transmittance (PST) is plotted for both LORRI and L'LORRI. In both cases, the input illumination was provided by red (632 nm) laser light. Four different scan directions were obtained for L'LORRI, and two scan directions were obtained for LORRI. The clump of points plotted between offset angles of 0.07° and 0.08° actually refer to an offset angle of 0° but are separated in the plot for better visibility. The PST curves for LORRI and L'LORRI are similar, as expected since they have essentially identical optical designs. L'LORRI easily meets its scattered light performance requirement (LL-INST-19)

4.1.4 Ghosts

The L'LORRI optical design includes an assembly with three lenses mounted inside the hole in the primary mirror (M1). The purpose of the lens assembly is to flatten the image plane, which would otherwise be significantly curved across the full extent of the CCD. Unfortunately, however, the lenses produce ghosts from multiple reflections within the assembly, which manifest as pupil images on the CCD (see Fig. 17 for an example). If the lenses are labeled by L1, L2, and L3 in order of increasing distance from the OTA secondary mirror (M2), and the two surfaces of each lens are labeled by S1 and S2 (with S1 being the surface closer to M2), the four ghosts marked in Fig. 17 are produced by internal reflections from the surfaces listed in Table 5.

The ghost with the highest surface brightness is A, which is almost twice as bright as B. For wavelengths near the middle of L'LORRI's bandpass, the integral of the ghost signal in A is roughly 4×10^{-4} times the integral over the PSF of the illumination source. The surface brightness in A is roughly 3×10^{-6} times the peak signal from the illumination source. Thus, for example, a $V = 6.2$ star located just outside the L'LORRI FOV can produce a ghost whose surface brightness is approximately equal to the peak brightness of a $V = 20$ star in the L'LORRI FOV. The presence of optical ghosts produced by nearby bright objects should be considered when planning L'LORRI observations, but no ghosts are produced for objects more than 0.37° from the L'LORRI boresight. Ghosts of similar morphology and brightness were also observed for *New Horizons* LORRI (Cheng et al. 2010).

4.2 Detector Performance

4.2.1 Bias

As is typical for CCDs, an electronic bias offset is applied to the CCD signal to prevent negative values at the analog-to-digital chip (ADC) that converts electrons to data numbers

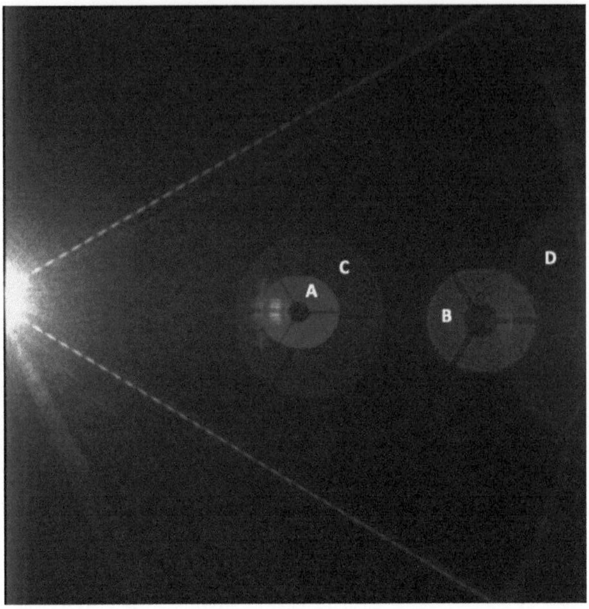

Fig. 17 Image showing ghosts produced by a bright point source located just outside the L'LORRI field of view (FOV). The four brightest ghosts are produced by multiple reflections within the field flattening lens assembly, as described in Table 5. For this image, the input light illumination was produced by a diode laser with a wavelength of 635 nm. The image is displayed using an ASINH stretch ranging from −10 DN to 3835 DN

Table 5 L'LORRI optical ghosts

Ghost ID	Surface Reflections Producing the Ghosts
A	L3S2 to L3S1
B	L3S1 to L2S2
C	L3S1 to L2S1
D	L3S2 to L1S1

Ghost IDs refer to the labels marked in Fig. 17. The nomenclature for the internal surface reflections that produce the ghosts is discussed in the body of the paper.

(DNs). In the L'LORRI calibration pipeline, the removal of this bias is a two step process: first, a global bias level is removed based on the average signal level for pixels sampled from the optically inactive region of the CCD, and then the pixel-to-pixel bias variation is removed using a so-called *superbias* image, which has a median level of 0 DN. Figure 18 shows the superbias image for 1 × 1 format, and Fig. 19 shows the superbias image for 4 × 4 format. In both cases, the bias levels in the lowest rows differ significantly from the bias levels in the rest of the image because there is a surge in current during the initial image-to-storage transfer that temporarily shifts the bias level.

For 1 × 1 format images, four columns in the inactive region are captured, which means a total of 4096 pixels (1024 × 4) are used to compute a robust average of the global bias level. For 4 × 4 format images, two columns in the inactive region are captured, which means a total of 512 rebinned pixels (256 × 2) are used to compute a robust average of the global bias level. However, there is a systematic offset between the bias level in the active and inactive regions of the CCD (Fig. 20), which is likely associated with different capacitive effects in the active and shielded regions of the CCD. The offset is 3.2 DN for 1 × 1 images and 5.1 DN for 4 × 4 images. The L'LORRI calibration pipeline accounts for this offset.

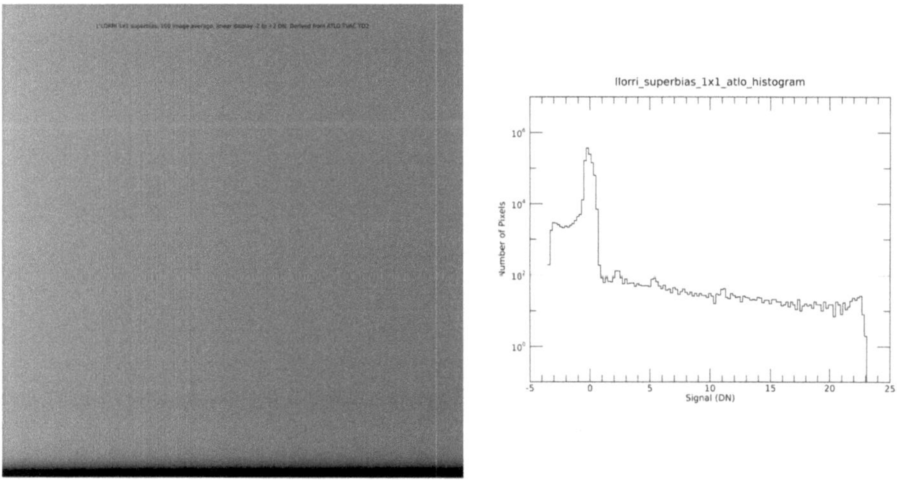

Fig. 18 Left: Mean of 100 1×1 bias images ($t_{exp} = 0$ ms), which is called a *superbias* image. The serial output amplifier is closest to the pixel at the bottom left of the image. Right: Histogram of the superbias image. The positive tail is associated with the lowest rows of the image, and the negative tail is associated with the rows just above them

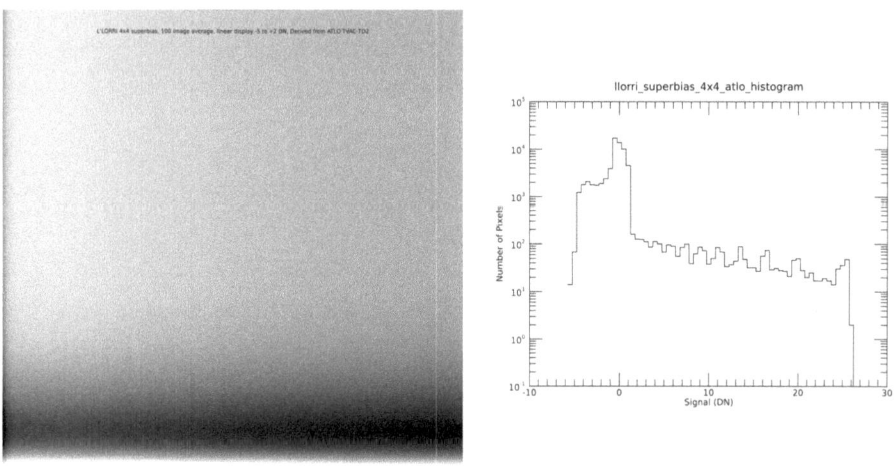

Fig. 19 Left: Mean of 100 4×4 bias images ($t_{exp} = 0$ ms), which is called a *superbias* image. Right: Histogram of the superbias image. The positive tail is associated with the lowest rows of the image, and the negative tail is associated with the rows just above them

For LORRI images, the bias level was a linear function of the temperature of the FPU board, with the bias level increasing by $\sim$1.2 DN for every 1 C change in the board temperature. In contrast, for L'LORRI images the bias level is essentially constant over the entire predicted temperature range of the FPU board during *Lucy* operations (Fig. 21).

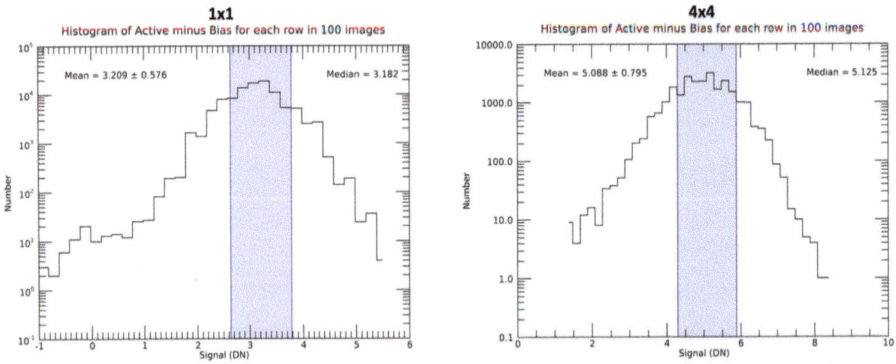

Fig. 20 Histograms showing the offset in bias level between the active and inactive regions of the L'LORRI CCD: the robust mean for the bias region was subtracted from the robust mean for the active region for each row in each of 100 1×1 images and 100 4×4 images. The exposure time was 0 ms for all the images. The shaded regions show $\pm 1\sigma$ about the mean differences. The mean of the active region is 3.2 DN larger than the mean of the bias region for 1×1 images and 5.1 DN larger for 4×4 images. See the text for further discussion

Fig. 21 L'LORRI CCD bias level as a function of FPU board temperature, obtained during thermal cycle measurements at APL. The FPU board temperature is predicted to lie in the range −29 C to +4 C for *Lucy* observations, where the bias level is essentially constant. The CCD temperature was relatively high (+5 C) for these measurements, which affects the global bias level but should not affect the variation with FPU temperature

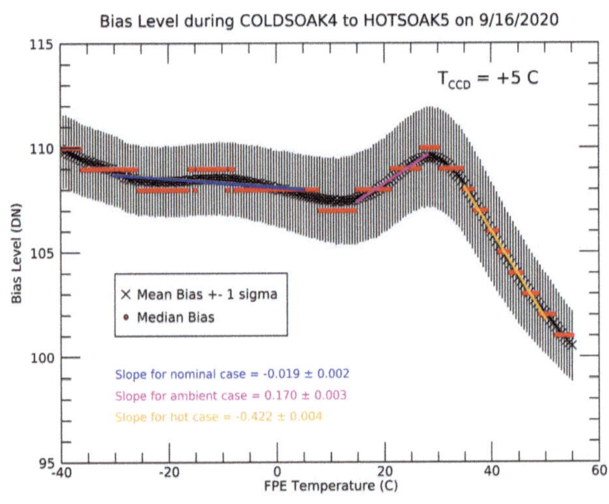

4.2.2 Electronics Noise

The difference of two bias images ($t_{exp} = 0$ ms) is analyzed to produce L'LORRI noise histograms. Following the technique outlined by Janesick (2001), we randomly sample 25×25 pixel subarrays (7×7 pixel subarrays for 4×4 format images) of the difference image for 50,000 randomly selected regions across the entire CCD to obtain a histogram of noise values, which includes the CCD read noise and any other noise introduced by the downstream electronics. Noise pickup from a thermal control rack compromised the measurements taken during the pre- and post-environmental testing, but the environment was much quieter during thermal cycle/balance testing, which provided an opportunity to measure the true L'LORRI system electronics noise. An example is shown in Fig. 22, which has both a noise histogram and a plot showing the variation of the noise with CCD temperature. For the temperatures relevant for the *Lucy* mission, the L'LORRI noise is well below the re-

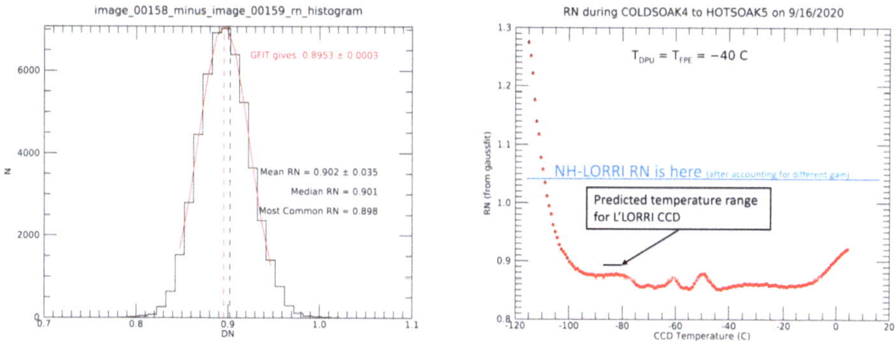

Fig. 22 Left: L'LORRI noise histogram at a CCD temperature of -83 C, which is within the range predicted for in-flight conditions. The noise is $\sim$0.9 DN, which is well below the requirement of 1.9 DN. Right: Plot showing the variation of noise with CCD temperature. The temperature range predicted for in-flight conditions is indicated, as is the noise value measured for *New Horizons* LORRI

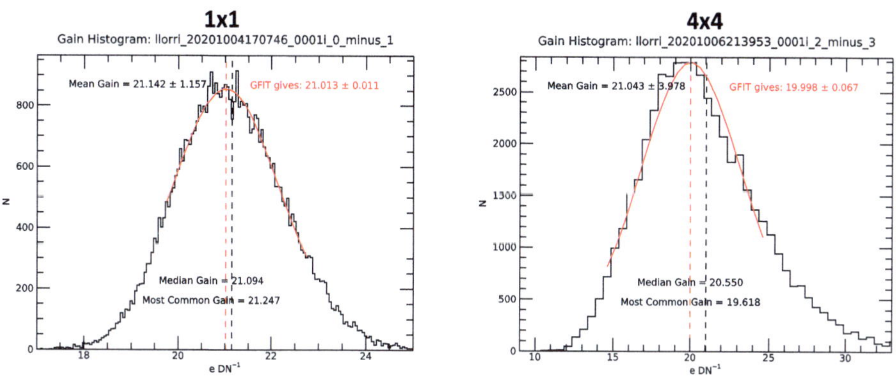

Fig. 23 Gain histograms for 1×1 format (left) and 4×4 format (right). The gain for 1×1 images is $\sim$21.0 e DN^{-1}, and the gain for 4×4 images is $\sim$20.0 e DN^{-1}

quirement (0.9 DN measured vs 1.9 DN requirement) and is also significantly smaller than the value measured for *New Horizons* LORRI.

4.2.3 Gain

The electronics gain, which is the scale factor that enables conversion from electrons detected to DN, is slightly different for 1×1 versus 4×4 images. Again, we follow the technique outlined by Janesick (2001) to produce a gain histogram for each of the two CCD formats using two well-exposed images with essentially constant illumination over the entire CCD. We use differences of two flat field images taken from the post-environmental calibration conducted in October 2020. Examples of gain histograms are shown in Fig. 23. The gain for 1×1 images ($\sim$21.1 e DN^{-1}) is slightly smaller than the gain for 4×4 images ($\sim$20.0 e DN^{-1}).

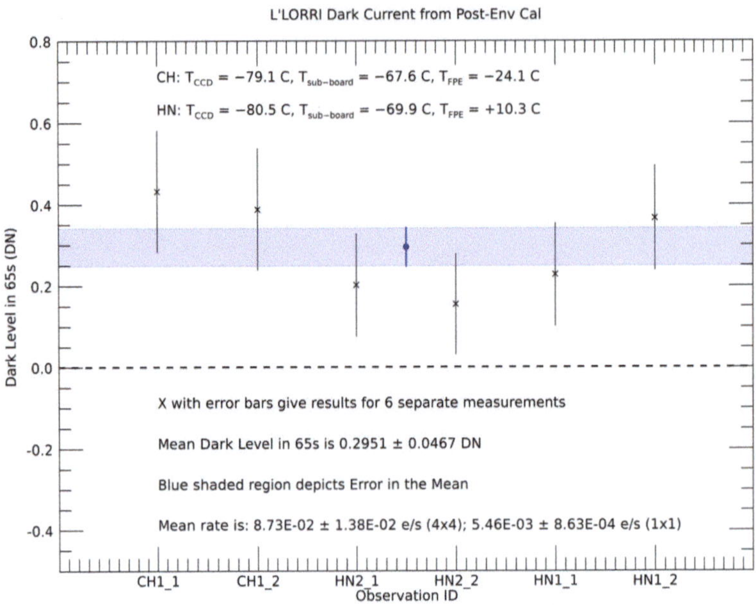

Fig. 24 Dark current measurements from the post-environmental calibration at APL are plotted for two different thermal environments for the electronics, but with the CCD near −80 C in all cases. The measured dark current is extremely low, comparable to that measured in-flight for *New Horizons* LORRI. See the text for further discussion

4.2.4 Dark Current

At the predicted in-flight CCD temperatures, the dark current is expected to be essentially negligible. Nevertheless, we attempted to measure the dark current under flight-like conditions during the post-environmental calibration at APL. By taking three consecutive pairs of 0 ms and 64,900 ms images in 4×4 format, and then subtracting the shorter exposure image from the longer exposure image, we were able to measure the dark current accumulation rate even when the CCD was cold. We performed measurements with L'LORRI in two different thermal environments: one with cold electronics and hot OTA ("CH" case) and one with hot electronics and nominal temperature OTA ("HN" case). In both cases, the CCD temperature was approximately −80 C. Measurements were taken with both DPUs in the HN case ("HN1" for DPU1 and "HN2" for DPU2). Only pixels in the optically-inactive region of the CCD were used to avoid contamination from any residual background light. The results are displayed in Fig. 24 and demonstrate that the dark current from the L'LORRI CCD is extremely small, comparable to that measured for *New Horizons* LORRI.

4.2.5 Linearity and Dynamic Range

Linearity and dynamic range were characterized by pointing L'LORRI at an integrating sphere and taking a series of images with different exposure times while the lamp intensity was held constant. By definition, the dynamic range refers to the maximum net signal level (after subtraction of the bias) in the linear response portion of a single exposure divided by the electronics noise (0.9 DN). The results presented here, which demonstrate that L'LORRI's response is linear to within $\sim 1\%$ and the dynamic range for a single exposure is ≥ 4200, are independent of which L'LORRI DPU side was used for the measurements.

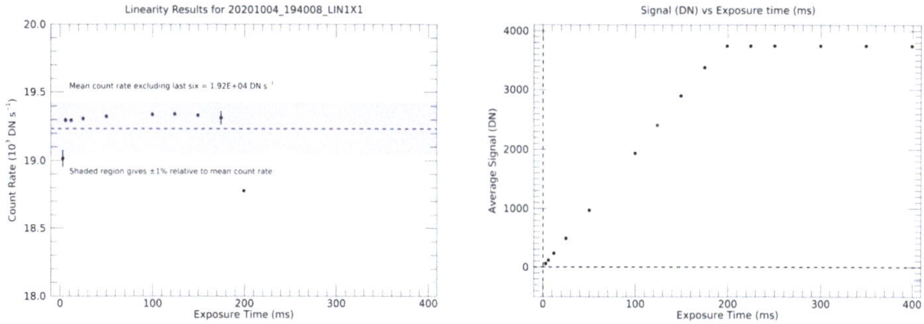

Fig. 25 The plot on the left shows that the measured count rate for 1 × 1 images is within 1% of the mean value for all exposure times (the blue shaded region), except for the saturated ones. The plot on the right shows that the response remains linear up to at least a net signal level of 3400 DN after bias subtraction

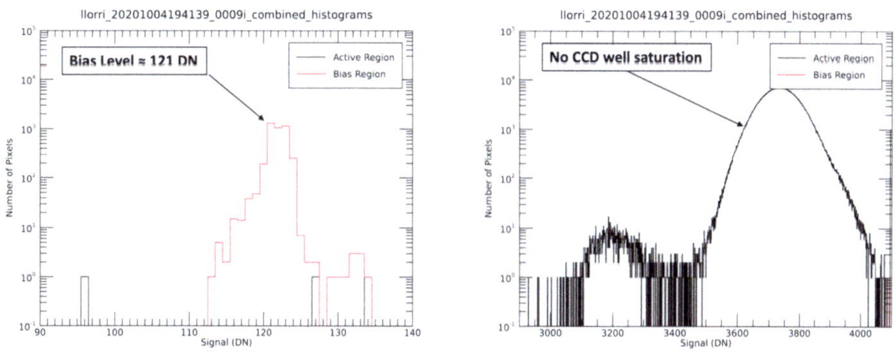

Fig. 26 The plot on the left shows a histogram of the bias region of the CCD for the last non-saturated point in Fig. 25. The plot on the right shows the high end of the same histogram and shows no evidence of CCD charge well saturation out to at least a *raw* signal level of ~4000 DN. Since the bias level is ~121 DN, the net signal for linear response extends to at least ~3880 DN. Given that the electronics noise is ~0.9 DN, the dynamic range is ≥4310, which is ~44% larger than the requirement (3000)

4.2.6 Flats

Flat-fielding refers to the process of removing the pixel-to-pixel sensitivity variations in the image, which can be induced by either optical or CCD non-uniformity effects. An image obtained by illuminating the L'LORRI aperture uniformly with light is called a "flat." During ground calibration testing, flats were obtained by using a calibrated integrating sphere to provide uniform illumination across the full L'LORRI aperture. Two different light sources were used, as described earlier, but they produced flats that differed by ≤1% for the vast majority of pixels. Flats were obtained in multiple thermal environments meant to mimic in-flight conditions, but they varied by ≤0.5% for the vast majority of pixels. Flats taken with the two different L'LORRI DPUs were essentially identical at the 0.1% level. The absolute intensity of the input illumination was measured using a calibrated photodiode.

The flat-field reference file used in the L'LORRI calibration pipeline was produced by averaging 100 flats taken when the OTA was at −85 C during the post-environmental calibration using weighted averages of flats from the two different light sources. These average images were debiased and desmeared, and then normalized to an average value of 1

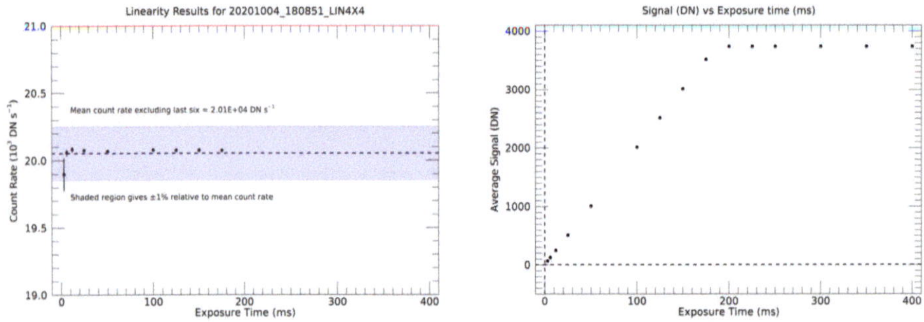

Fig. 27 The plot on the left shows that the measured count rate for 4 × 4 images is within 1% of the mean value for all exposure times (the blue shaded region), except for the saturated ones. The plot on the right shows that the response remains linear up to at least a net signal level of 3500 DN after bias subtraction

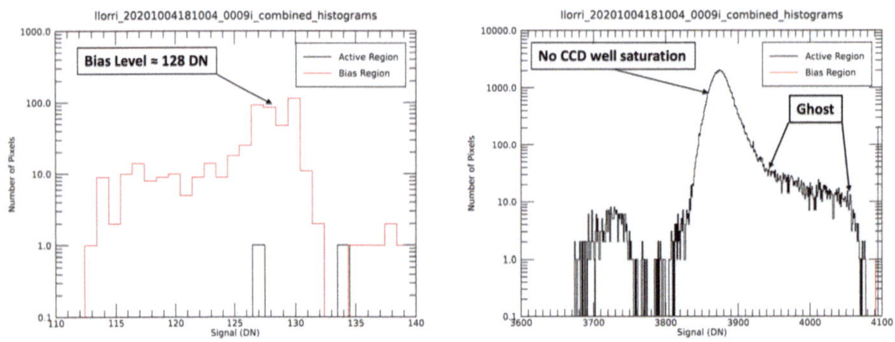

Fig. 28 The plot on the left shows a histogram of the bias region of the CCD for the last non-saturated point in Fig. 27. The plot on the right shows the high end of the same histogram and shows no evidence of CCD charge well saturation out to at least a *raw* signal level of ∼4060 DN. Since the bias level is ∼128 DN, the net signal for linear response extends to at least ∼3800 DN. Given that the electronics noise is ∼0.9 DN, the dynamic range is ≥4220, which is ∼40% larger than the requirement (3000)

(Fig. 29). The raw weighted average included a ghost of the field-flattening assembly near the center of the CCD (the integrating sphere overfilled the L'LORRI aperture) and scattered light near the bottom and right edges of the CCD. We tried to remove both of these artifacts, but their residuals are visible in the final flat, although only at the ∼1% level, or less.

If S (in units of DN) is an image of a target that has already been debiased and desmeared, and if FF is the reference flat-field image, then the flat-fielded (i.e., photometrically-corrected) target image (C; units are DN) is given by C = S/FF. The effects of the CCD particles (Sect. 4.2.7) are almost completely removed by flat-fielding, except near the center of the most prominent particle located in the upper right of the CCD. That particle is approximately 17 pixels high by 7 pixels wide, and two of its pixels have normalized intensities of ∼0.15, which means their effects are difficult to remove. Excluding row 0 and column 0, only 52 pixels in the L'LORRI flat have normalized intensities ≤0.8, and 45 of those are in the single large particle. A histogram of the normalized flat is shown in Fig. 30.

Fig. 29 L'LORRI flat from a weighted average of 100 flats taken with two different lamps after subtraction of the central ghost and scattered light near the bottom and right edges. The image is displayed using a linear intensity scale ranging from 0.9 to 1.1. The prominent black area in the upper right is from a particle sitting on the CCD. Several smaller particles can also be seen across the image, but flat-fielding essentially removes all traces of them in the calibrated image. All the particles have been fixed in location on the CCD since September 2020

4.2.7 Artifacts

As is the case with all instruments, L'LORRI has artifacts that must be considered when interpreting its images. Here we discuss two of L'LORRI's known artifacts: foreign object debris (FOD) on the CCD and L'LORRI's behavior when imaging bright sources that saturate the CCD full-well. Two other L'LORRI artifacts were discussed previously: optical ghosts were covered in Sect. 4.1.4, and the bias offset between the active and non-active regions of the CCD were covered in Sect. 4.2.1.

L'LORRI's CCD does not have a protective cover. Once the CCD is installed behind the OTA, dust and other particles (generally called FOD) can fall unimpeded onto the surface of the exposed CCD. Although we took great care to keep the area around the CCD clean, L'LORRI's environmental testing produced multiple opportunities for contamination events.

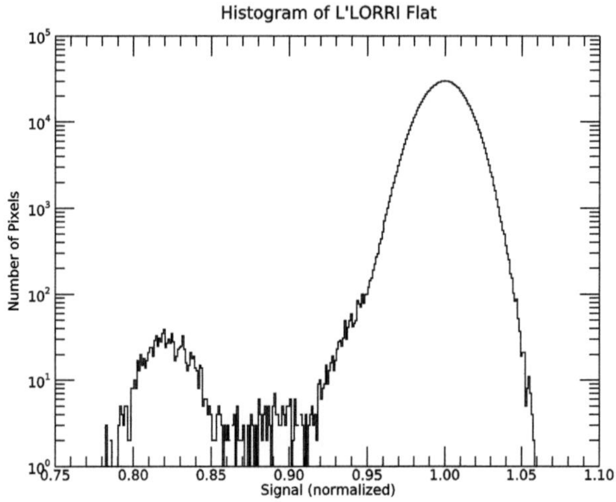

Fig. 30 The figure shows a histogram of the L'LORRI normalized flat in Fig. 29. The vast majority of pixels lie within 10% of the histogram peak. The second histogram peak centered near 0.83 is comprised almost exclusively from pixels in row 1 of the CCD

During the instrument-level vibration testing at APL, when the OTA was mounted vertically with the CCD facing upwards, a rather large particle was shaken loose during the thrust-direction vibe and fell down onto the CCD surface. This particle was displaced sideways by hundreds of CCD pixels during the subsequent vibe in one perpendicular direction, but has since remained in its current location, including after another vibration perpendicular to the other two axes. The particle also did not move during the spacecraft-level vibration during ATLO. Perhaps this particle was displaced from the L'LORRI lens assembly; we had previously noticed FOD within the lens assembly but could not remove it after the OTA was assembled. In any case, this large particle has now been in the same location for over a year, which suggests it is now a permanent feature.

Figure 31 shows a magnified view of the L'LORRI flat field centered on the large FOD. The particle has dimensions of approximately 17 pixels (0.221 mm) by 7 pixels (0.091 mm) with 45 pixels having transmittances ≤ 0.80. Two of the pixels have transmittances of ~ 0.15 and are not removed by the flat-fielding process used to produce calibrated L'LORRI images. Thus, we have marked these two pixels as "bad" in the quality factor extension of the calibrated FITS image. There is other FOD on the CCD (see Fig. 29), but their effects are minimal because the flat-fielding process essentially removes them in the calibrated image.

Saturated images of point sources can also produce artifacts in L'LORRI images. These were characterized most extensively for the EM FPU on 2020 May 29 under ambient laboratory conditions in the APL cleanroom during illumination of the OTA by a LUPI using red laser light. Point source saturation effects for L'LORRI were also measured during the pre-environmental calibration test with the FM FPU, but the degree of saturation was much less than explored during the earlier testing. No point source saturation tests were conducted during the post-environmental calibration.

The *New Horizons* LORRI images of bright stars in 4×4 format have "black tails" (when the image is displayed using a grayscale ranging from black for the lowest DN values and white for the high DN values) along the CCD row direction, which is due to amplifier undershoot/overshoot in response to a rapidly changing analog input signal. This effect has essentially been eliminated in L'LORRI images, but another artifact has been introduced for high levels of saturation. As shown in Fig. 32 and Fig. 33, the bias level in the rows above the row containing the saturated target is depressed by up to ~ 30 DN when the input light is ~ 100 times the saturation level. This produces a dark area in the image that spans entire

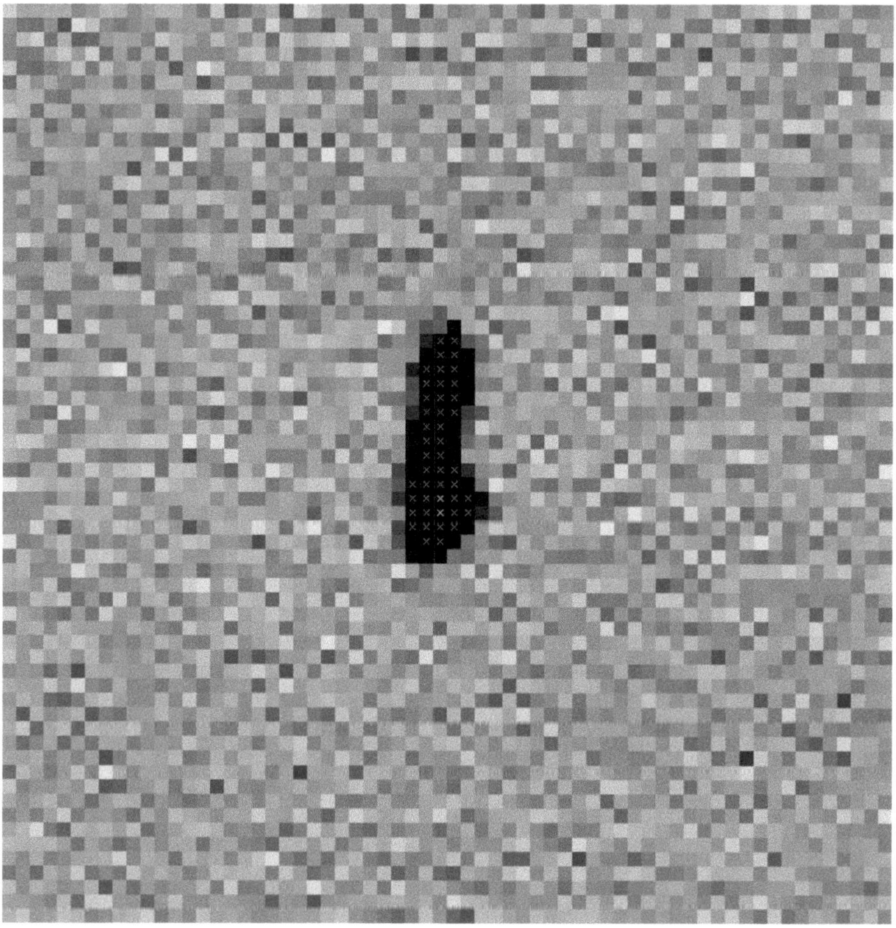

Fig. 31 Magnified view (64 × 64 pixels) of the L'LORRI flat field image (see Fig. 29) centered on the largest particle (also called "foreign object debris," or FOD) lying on the CCD. Red X symbols mark where the normalized transmittance is ≤0.80; the two pixels where the transmittance is ∼0.15 are marked with yellow X symbols. This particle is approximately 17 pixels (0.221 mm) by 7 pixels (0.091 mm), and there are a total of 45 pixels with transmittances ≤0.80

CCD rows (Fig. 32). Row profiles in the region of the saturated pixel also show different bias levels on either side of the saturated pixels (i.e., left-to-right asymmetry; see Fig. 33).

When individual pixels are saturated in 4 × 4 format, very high amplitude analog signals are presented to the ADC possibly causing its internal protection diodes to activate and start discharging the reference capacitors. Recovery back to the nominal bias level is estimated to take ∼6 ms (based on the relevant circuit's RC time constants), which corresponds to ∼8 rows worth of pixel readout time, which roughly corresponds to the size of the "dark rows" artifact. No similar effect has ever been observed in *New Horizons* LORRI images, nor has been it observed in L'LORRI 1 × 1 images. Since all the close flyby imaging of *Lucy's* targets will be taken in 1 × 1 format, we do not expect this dark rows artifact to be an issue for the L'LORRI images taken near the time of closest approach. Perhaps, however, this artifact

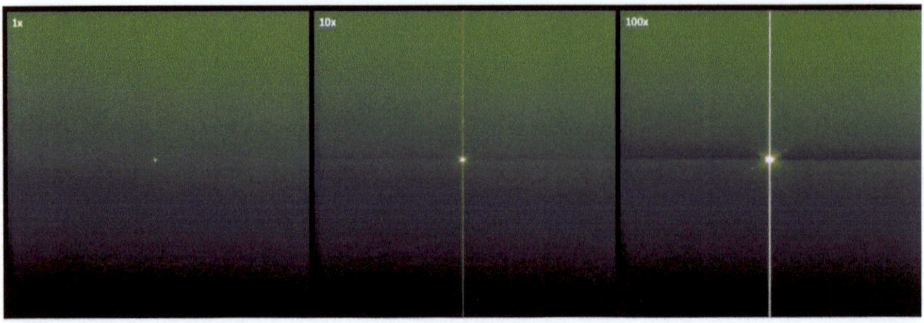

Fig. 32 Three different raw 4 × 4 L'LORRI images with different levels of pixel saturation are shown: 1, 10, and 100 times saturated from left to right. The number of saturated pixels in the images are 1, 5, and 13 respectively. The exposure time was 50 ms in each case, and the light intensity was increased by using different levels of neutral density filters. All images are displayed using an ASINH intensity stretch ranging from 290 DN to 4095 DN and a "cubehelix" color scale for better visibility of different intensity levels. The vertical streak is associated with the CCD frame transfer (*not* CCD charge bleeding) and becomes stronger as the input light becomes brighter. As the input light level becomes stronger, a dark region extending across entire CCD rows becomes more prominent. All the images were taken in a cleanroom under ambient laboratory conditions and have an intensity ramp from the bottom to the top of the CCD owing to dark current accumulated during the readout from bottom to top (there is more time to integrate dark signal for the pixels in the higher rows). See the text for further discussion

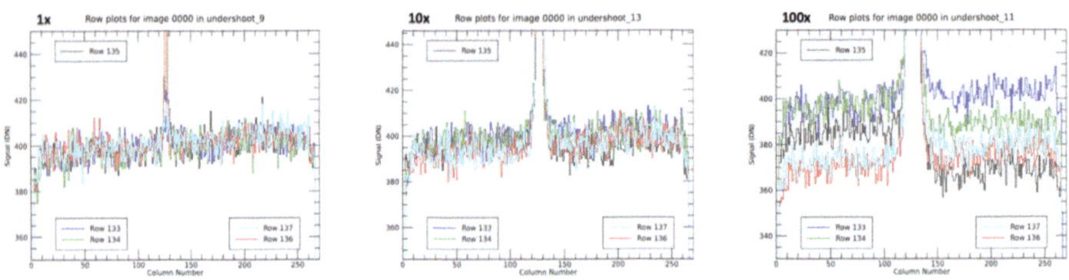

Fig. 33 Row profiles across the images in Fig. 32 are displayed. In each plot, the intensity profile for the row containing the pixel with the brightest light, as well as profiles for ±2 rows about that peak row, are displayed. For the image with one saturated pixel (1× saturation), all the row profiles lie on top of each other. For the 10× saturated image, there is an offset of ∼5 DN between the right hand and left hand sides (i.e., relative to the peak column) of the peak row (row 135), but there are no offsets for the other rows. For the 100× saturated image, there is an offset of ∼15 DN between the right hand and left hand sides of the peak row, there is a left-right offset of ∼5 DN for one row below the peak row (row 134), and the lower rows have systematically larger signal than the upper rows

might be seen if there are bright stars in the background field during deep observations taken to search for potential satellites near *Lucy* targets.

4.3 System Throughput

For sensitivity considerations, we consider two types of targets: "point" and "diffuse". A point source is one that is not spatially resolved by L'LORRI (e.g., "star-like"), as will be the case for distant observations of the *Lucy* mission targets (e.g., for initial OpNav target acquisition) and searches for unresolved satellites. A "diffuse" source is one that is spatially resolved by L'LORRI, as will be the case during near-encounter observations of the surfaces of the *Lucy* mission targets.

For a photon counting optical system like L'LORRI, the signal detected (S in electrons) in an image pixel with exposure time t (in seconds) can be expressed as:

$$S = t\,\Omega \int_\lambda I\,A_{\mathrm{eff}}\,d\lambda = t\,\Omega\,A \int_\lambda I\ \mathrm{QE}\,d\lambda \text{ (diffuse target)} \tag{1}$$

$$S = t\ \mathrm{EE} \int_\lambda F\,A_{\mathrm{eff}}\,d\lambda = t\ \mathrm{EE}\ A \int_\lambda F\ \mathrm{QE}\ d\lambda \text{ (point target)} \tag{2}$$

And the noise ("N" in electrons) can be expressed as (for both diffuse and point targets):

$$N = \sqrt{S + SL + FT + (I_d t) + \mathrm{RN}^2} \tag{3}$$

Where:

Ω = pixel IFOV (sr)
A = unobscured input aperture area (cm^2) of the OTA
A_{eff} = System effective area = $A * \mathrm{QE}$ (cm^2)
QE = system quantum efficiency (dimensionless; including all losses)
EE = ensquared energy = fraction of point source total signal contained in peak pixel (dimensionless)
I = diffuse target's radiance (photons cm^{-2} s^{-1} sr^{-1} nm^{-1})
F = point target's flux, or irradiance (photons cm^{-2} s^{-1} nm^{-1})
SL = signal produced by solar scattered light (electrons)
FT = signal produced by the CCD frame transfer process (electrons)
I_d = CCD dark current (electrons s^{-1} $pixel^{-1}$)
RN = electronics noise (electrons; includes the CCD read noise)

The integrals are over all wavelengths.

The L'LORRI signal level depends on two key parameters: the total system quantum efficiency (QE) and (in the point source case) the amount of energy from a point source concentrated in a single pixel (EE). We already discussed EE in Sect. 4.1.2. For point sources with SNR ≥ 5 in a single pixel, the SNR can typically be improved by a factor of ~ 2 by summing the signal over several pixels (aperture photometry) or by employing PSF fitting photometry. These techniques can be employed when the SNR is only marginally above the requirement for a single pixel.

The system quantum efficiency (QE) is given by:

$$\begin{aligned} \mathrm{QE} = A_{\mathrm{eff}}/A &= L_{\mathrm{obscure}}\ T_{\mathrm{M1}}\ T_{\mathrm{M2}}\ T_{\mathrm{L1}}\ T_{\mathrm{L2}}\ T_{\mathrm{L3}}\ \mathrm{QE}_{\mathrm{CCD}} \\ &= L_{\mathrm{obscure}}\ T_{\mathrm{optics}}\ \mathrm{QE}_{\mathrm{CCD}} \end{aligned} \tag{4}$$

Where:

L_{obscure} = loss factor associated with obscuration by the secondary mirror and OTA spider
T_{x} = transmittance of the individual optical element "X"
M1 = OTA primary mirror
M2 = OTA secondary mirror
L1 = first lens in the field flattener lens assembly
L2 = second lens in the field flatter lens assembly
L3 = third lens in the field flatter lens assembly
T_{optics} = total transmittance of all optical elements

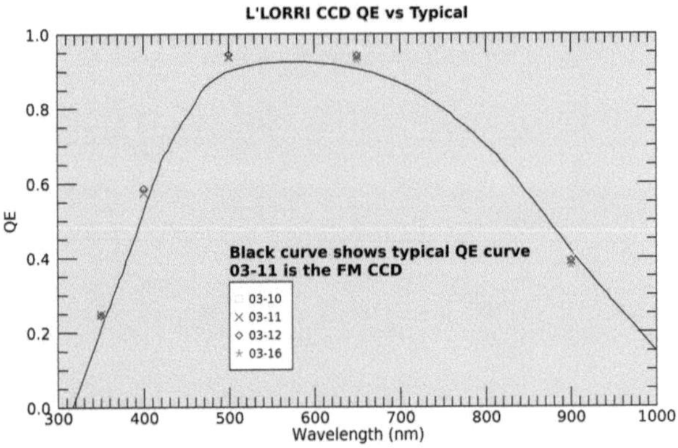

Fig. 34 QE_{CCD} curves were measured by Teledyne-e2v for each of the devices procured for L'LORRI. In all cases, the L'LORRI values lie above the "typical" QE curve over most of the optical wavelength range. Device 03-11 was selected as the L'LORRI flight CCD

QE_{CCD} = Quantum Efficiency of the CCD

Note that $A = \pi(20.8)^2/4 = 339.8\ \text{cm}^2$ (a L'LORRI requirement is that the M1 mirror must have a clear aperture ≥20.8 cm) and $L_{obscure} \geq 0.88$ (another L'LORRI requirement).

Although not explicitly stated in the equation above, all quantities in the QE equation are a function of wavelength, except A and $L_{obscure}$. L'LORRI is a panchromatic instrument, which means that its output signal is proportional to the integral of the above factors over all wavelengths.

We refer to the"pivot" wavelength (λ_{pivot}), which is one way of characterizing the "effective" wavelength for a broadband optical instrument. The pivot wavelength is defined as:

$$\lambda_{pivot} = \sqrt{\frac{\int QE * \lambda\, d\lambda}{\int QE/\lambda\, d\lambda}} \tag{5}$$

The pivot wavelength for L'LORRI is calculated to be 603.0 nm, whereas *New Horizons* LORRI had a pivot wavelength of 607.6 nm.

L'LORRI is designed to have the same QE as LORRI, or better. In setting the transmittance requirements for the optical elements, APL specified to L3H that the transmittances for the L'LORRI optical elements must be greater than, or equal to, the measured LORRI values. L3H used the same optical coating prescriptions for the M1 and M2 mirrors that LORRI used, but the original coating vendor is no longer in business. Nevertheless, the new optical coating vendor (Infinite Optics Incorporated, IOI) achieved transmittances consistent with the L'LORRI specifications. As previously noted, the spare LORRI M1, after precision cleaning at L3H, became the flight M1 for L'LORRI. The transmittance data from IOI for the L'LORRI lens assembly was essentially identical (to within 2%) to the LORRI values.

L'LORRI uses the same model CCD (CCD47-20), with the same specifications (backside-thinned, backside-illuminated, with "Midband AR" coating and anti-blooming) and the same vendor (formerly e2v, now Teledyne-e2v), as used on LORRI. The CCD quantum efficiency (QE_{CCD}) is determined by the AR coating and the intrinsic response of the silicon substrate. The vendor provided QE_{CCD} measurements for each CCD delivered to the L'LORRI program, and the curves lie above the "typical" values for all the CCDs (Fig. 34).

Figure 35 shows the LORRI system QE as a function of wavelength, as determined from the absolute calibration measurements of the solar-type reference star HD 37962. The LORRI system QE is approximately 50% over much of the visible wavelength range (e.g., 480–700 nm). LORRI is a panchromatic instrument, which means that its output signal is

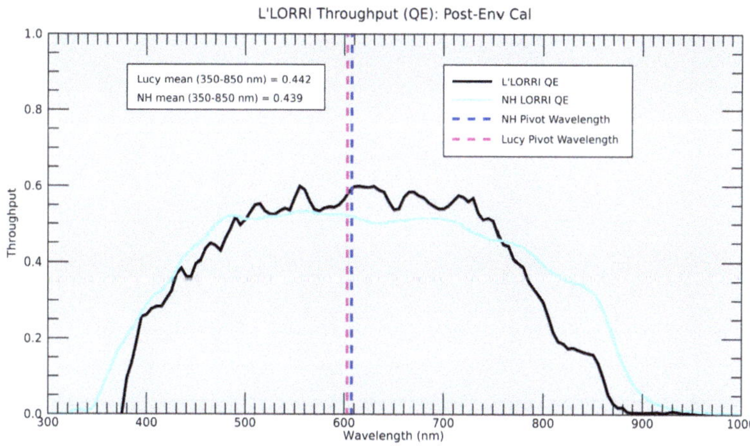

Fig. 35 The system quantum efficiency (QE) is plotted as a function of wavelength for both L'LORRI and LORRI. The mean QE over the wavelength range 350–850 nm is ~44% for both L'LORRI and LORRI. The locations of the L'LORRI and LORRI pivot wavelengths are shown, as are the standard visible photometric bands in the Johnson-Landolt system. L'LORRI's pivot wavelength is located between the standard *V* and *R* band wavelengths

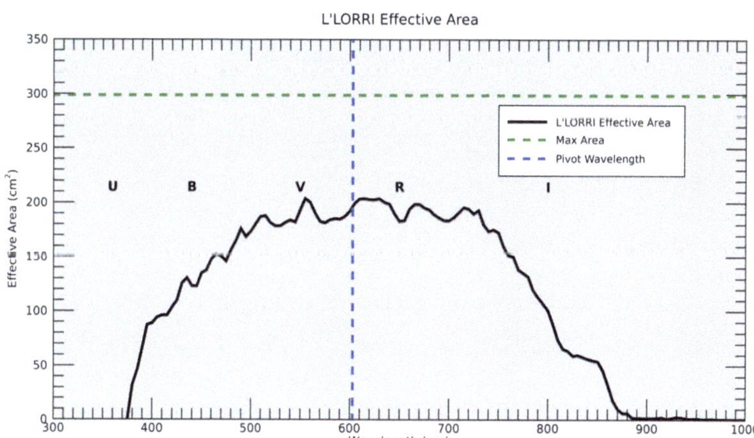

Fig. 36 LORRI's effective area is plotted as a function of wavelength. Given the estimated L'LORRI obscuration of ~12%, the theoretical maximum effective area is given by the green dashed line. The locations of the L'LORRI pivot wavelength and the standard visible photometric bands are also shown

proportional to the integral over all wavelengths of the product of the QE and the target's SED.

The commonly used quantity "effective area" (A_{eff}) of the optical system is just the area of L'LORRI's input aperture ($A = \pi * 20.8^2/4 = 339.8\ cm^2$) multiplied by the system QE. Given the estimated L'LORRI obscuration of ~12% (i.e., $L_{obscure} = 0.88$), the largest possible effective area for L'LORRI is ~300 cm^2. Figure 36 shows the L'LORRI effective area as a function of wavelength.

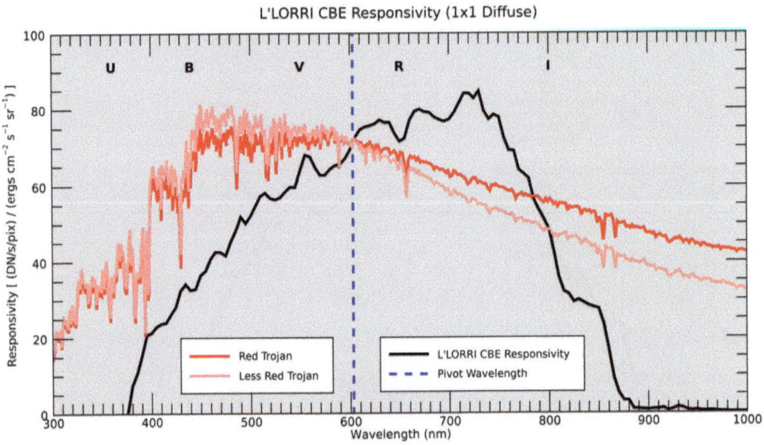

Fig. 37 L'LORRI's absolute responsivity is plotted as a function of wavelength. The in-flight measurements of an absolute calibration standard star indicate that these values should be multiplied by 1.139. Normalized spectra of the two types of Trojan spectra ("Red" and "Less Red") are overlain for comparison. The locations of the L'LORRI pivot wavelength and the standard visible photometric bands are also shown

The L'LORRI absolute responsivity curve, which enables the conversion from engineering units to physical units, can be derived from the QE curve using the following equation:

$$R_\lambda = A_{eff} * \Omega * \lambda/gain/hc \tag{6}$$

where:

R_λ is the responsivity ([DN s^{-1} $pixel^{-1}$] / [ergs cm^{-2} s^{-1} sr^{-1}])
A_{eff} is the effective area as defined above (cm^2)
Ω is the solid angle of a single pixel (sr)
λ is the wavelength of interest (Å)
gain is the CCD gain (e DN^{-1})
hc is the product of Planck's constant and the speed of light (1.986×10^{-8} ergs-Å)

Figure 37 shows the L'LORRI responsivity curve for 1×1 format based on results from the ground calibration program. The responsivity curve for 4×4 format can be obtained by multiplying the 1×1 curve by 16.4. The figure also compares the L'LORRI responsivity curve to the spectral energy distributions (SEDs) of the two types of Trojans being targeted by the *Lucy* mission ("Red" and "Less Red", or "Gray"). The L'LORRI response to the Trojans is the integral of the product of the Trojan's SED and the L'LORRI responsivity.

We used the ground calibration mainly to characterize the *relative* sensitivity as a function of wavelength. The absolute sensitivity based on the ground calibration was expected to be accurate to ~10-20%. Our best estimate for L'LORRI's absolute sensitivity will be derived from data taken during in-flight observations of an absolute calibration standard star. The irradiance (or flux) of the calibration stars we plan to use for L'LORRI are known to an absolute accuracy of ~1% from observations made by the *Hubble Space Telescope*. As discussed later, the first in-flight observation of an absolute calibration standard showed that L'LORRI's absolute response is ~14% larger than the nominal value from the ground calibration.

For a target whose spectral energy distribution (SED) is similar to that of the standard star, the L'LORRI absolute calibration should be accurate to a similar accuracy. However, for *Lucy* targets with SEDs significantly different from the SED of the calibration star, we estimate that the L'LORRI absolute calibration in those cases should be accurate to ~10-20% (i.e., similar to the accuracy of the ground calibration).

A L'LORRI exposure time calculator (ETC), which uses the responsivity curve discussed above, already exists. The ETC has recently been updated to include the results from the in-flight observation of an absolute calibration standard star. This ETC is used for both planning future observations (e.g., estimating SNR for planned observations) and for analyzing completed observations (e.g., for converting L'LORRI signals from engineering units into absolutely calibrated scientific units). The ETC includes noise from: the electronics, shot noise from the target itself, shot noise associated with CCD frame transfer process (i.e., frame transfer smear), shot noise from the CCD dark current (generally negligible), and shot noise from solar scattered light when the solar elongation angle (SEA) is $\leq 90°$. The ETC also permits selection of the fraction of the total signal from a point source that falls within the peak pixel (i.e., EE).

The L'LORRI calibration pipeline provides photometrically corrected images, but the intensities are still in engineering units (DN). Pixel intensities can be converted to scientific units using photometry keywords provided in the image data file. In Table 6, we provide photometry keywords for objects with three different SEDs: solar-like, similar to the average "red" Trojan spectrum, and similar to the average "gray" Trojan spectrum. We provide here two examples showing how to convert from engineering units to physical units: one for a diffuse target and one for a point (i.e., unresolved) target.

Consider a diffuse target whose spectrum is similar to that of red Trojan globally averaged SED. In this case, the RTROJANR photometry keyword in the header of the calibrated image should be used to convert from the observed count rate in a pixel to a radiance value at L'LORRI's pivot wavelength:

$$\mathrm{I} = \mathrm{S}/\mathrm{t_{exp}}/\mathrm{RTROJANR}\ \text{(diffuse target)} \tag{7}$$

where:

I is the diffuse target radiance (ergs $\mathrm{cm^{-2}\ s^{-1}\ Å^{-1}\ sr^{-1}}$) at λ_{pivot}
S is the measured signal in a pixel (DN)
t_{exp} is the exposure time (s)
RTROJANR is the L'LORRI diffuse photometry keyword for targets with an SED similar to a red Trojan

Since the solar flux ($F_\odot$) at a heliocentric distance of 1 AU at the L'LORRI pivot wavelength is 176 ergs $\mathrm{cm^{-2}\ s^{-1}\ Å^{-1}}$, the value for the radiance can be converted to I/F (where $\pi F = F_\odot$), which is a standard photometric quantity used in planetary science, using:

$$\mathrm{I/F} = \pi \mathrm{I} r^2/F_\odot \tag{8}$$

$$\rightarrow \mathrm{I/F} = (\mathrm{S}/\mathrm{t_{exp}}/\mathrm{RTROJANR}) * \pi r^2/F_\odot \tag{9}$$

where "r" is the target's heliocentric distance in au.

For unresolved targets (e.g., planetary targets observed at large ranges), the absolutely calibrated flux (also called the "irradiance") at the L'LORRI pivot wavelength can be determined using the point source photometry keywords. For a target with an SED similar to that of gray Trojan, the observed count rate integrated over the LORRI PSF can be related to the flux (not to be confused with "F" in "I/F") at the L'LORRI pivot wavelength by:

$$\mathrm{F} = \mathrm{S_{total}}/\mathrm{t_{exp}}/\mathrm{PTROJANG}\ \text{(point target)} \tag{10}$$

Table 6 L'LORRI photometry keywords

Keyword	Value (1 × 1)	Value (4 × 4)
RSOLAR	2.713×10^5	4.455×10^6
RTROJANR	2.784×10^5	4.571×10^6
RTROJANG	2.712×10^5	4.453×10^6
PSOLAR	1.101×10^{16}	1.130×10^{16}
PTROJANR	1.130×10^{16}	1.160×10^{16}
PTROJANG	1.101×10^{16}	1.130×10^{16}

The keywords starting with "R" are diffuse target sensitivity keywords and their values have units of (DN s^{-1} $pixel^{-1}$) / (ergs cm^{-2} s^{-1} $Å^{-1}$ sr^{-1}). The keywords starting with "P" are point target sensitivity keywords and their values have units of (DN s^{-1}) / (ergs cm^{-2} s^{-1} $Å^{-1}$). For point targets, the signal refers to values integrated over the entire instrumental PSF.

where:

F is the point target flux, or irradiance (ergs cm^{-2} s^{-1} $Å^{-1}$)
S_{total} is the total signal from the target integrated over the PSF (DN)
t_{exp} is the exposure time (s)
PTROJANG is the L'LORRI point source photometry keyword for targets with an SED similar to a gray Trojan

When observing stars, it is more common to convert the absolute flux to a magnitude in a standard photometric system. For convenience to users, we provide a prescription for converting L'LORRI signal rates to standard *V* magnitudes in the Johnson photometric system:

$$V = -2.5 \log(S/t_{exp}) + ZPT + CC - AC \tag{11}$$

where V is the magnitude in the standard Johnson *V* band (i.e., specifies the target's flux at 5500 Å), S (DN) is the measured signal in the selected photometric aperture, t_{exp} (s) is the exposure time, ZPT is the photometric zero point (18.93 for 1 × 1 and 18.96 for 4 × 4), CC is a color correction term (as specified in Table 7), and AC is an aperture correction term to convert from the flux collected in a specified synthetic aperture to the total flux integrated over the L'LORRI PSF.

The CC terms listed in Table 7 were calculated using SEDs for the listed blackbody temperatures, which roughly correspond to the listed stellar spectral types. We used the two generic types of Trojan spectra ("red" and "gray"), as adopted by the *Lucy* mission science team, to estimate CC for those cases. For typical L'LORRI observations of point sources, the SNR is optimized by integrating over a circular aperture with a radius of 5 pixels (1 × 1 format) or 3 pixels (4 × 4 format), in which case AC is either 0.10 or 0.00, respectively.

For example, the *V* magnitude for an A-type star observed by L'LORRI is given by:

$$V_{star} = -2.5 \log(S_{total}/t_{exp}) + ZPT - 0.036 \tag{12}$$

where V_{star} is the star's magnitude in the standard Johnson *V* band, S_{total} is the total signal integrated over the L'LORRI PSF (DN), t_{exp} is the exposure time (s), ZPT is the photometric zero point (18.93 for 1 × 1 and 18.96 for 4 × 4), and the color correction term is −0.036.

Table 7 Approximate L'LORRI color corrections

T_{bb}	Spectral Type	Color Correction (CC)
30,000	O (≥30,000)	−0.071
10,000	B (10,000–30,000)	−0.071
7500	A (7500–10,000)	−0.036
6000	F (6000–7500)	+0.027
5200	G (5200–6000)	+0.093
5800	G2V (5800)	+0.040
–	Sun	+0.00
3700	K (3700–5200)	+0.39
2400	M (2400–3700)	+1.2
–	Red Trojan	+0.028
–	Gray Trojan	−0.00

T_{bb} is the blackbody temperature that describes the spectral energy distribution (SED) used to calculate the L'LORRI signal rate for a target assuming the same V magnitude for all listed temperatures. The Spectral Type refers to the stellar classification assigned to a star whose effective temperature is in the range listed in parentheses, with the starting value equal to T_{bb}, except we used the actual SED of the Sun for that case. The Color Correction (CC) is the value that must be added to the raw V to correct for the SED of the target (see equation (11)). We used the two generic types of Trojan spectra ("red" and "gray"), as adopted by the *Lucy* mission science team, to estimate CC for those cases. Since target SEDs are rarely matched exactly by a blackbody SED, CC only provides an approximate color correction for converting L'LORRI count rates (DN s^{-1}) to Johnson V magnitude. If the SED of the target is known, the L'LORRI ETC can be used to produce a more accurate CC value.

4.4 Optical Navigation and Satellite Searches

L'LORRI is the *Lucy* mission's primary optical navigation camera. For each encounter, L'LORRI imaging will begin two months in advance of each flyby. Subsequent systematic observations by L'LORRI will be used by the optical navigation team to determine the trajectory correction maneuvers (TCMs) needed to guide the *Lucy* spacecraft to its designed aim point relative to each target.

All of *Lucy's* targets should be detectable by L'LORRI two months in advance of closest approach in co-added 1 × 1 images, but a few of the targets are near the sensitivity limit. In those cases, it will be prudent to take 4 × 4 images as well to guarantee detection of the target.

The search for potential satellites near each *Lucy* target is another important L'LORRI objective. By co-adding multiple 4 × 4 images, L'LORRI should be able to detect any satellites larger than ~2 km in diameter in the orbital stability zone of each *Lucy* target. Further discussion of the *Lucy* satellite search program can be found elsewhere in this collection (Noll et al. 2023).

5 Initial in-Flight Results

After the launch of the *Lucy* spacecraft in October 2021, L'LORRI was used during a variety of activities in the following year (Table 8). The initial commissioning was conducted ~1 month after launch to verify L'LORRI's basic functionality. These tests were conducted while the L'LORRI hardware was at relatively high temperatures to enhance outgassing but

Table 8 Main L'LORRI inflight activities during the first year after launch

Dates (UTC)	Activity	Comments
2021 Nov 08	Initial commissioning	Nominal engineering performance, all operating modes verified, data transfers verified
2022 Feb 14	IPP calibration	Success, nominal instrument alignments
	Pointing stability	Nominal performance verified
	Limited straylight test	Nominal performance verified
	Throughput measurement	Nominal performance verified
	PSF with OTA at −85 °C	Slightly degraded relative to nominal prediction, but still meets requirement over most of the FOV
2022 May 15,16	Earth-Moon solar eclipse observations	Successful observations
	Full Functional Test	Nominal performance verified
	PSF with OTA at −95 °C	Successful, but very limited data
2022 Sep 26,27	DART impact observations	Successful campaign conducted
2022 Oct 16	Scattered light/Ghost test	Nominal performance verified
	Lunar crater imaging	PSF limits accuracy of crater counts
	PSF with OTA at −95 °C	Data suggest modest improvement relative to −85 °C, but more data needed
	Trigger mode test	Successful demo during lunar scan
	L'TES co-alignment check	Success

"Nominal performance" means the data lie within the predicted range. See the text for further discussion.

were still suitable for checking voltages, currents, and a suite of instrument functions. All L'LORRI engineering data and instrument functions were nominal during this testing. We verified that every L'LORRI operational mode worked as designed, the internal calibration lamps operated as expected, and all data were transferred properly from L'LORRI to the spacecraft and then from the spacecraft to the ground. The optical performance could not be checked at this time because the high OTA temperature (near 0 °C) meant that the telescope was badly out of focus, but stars could be seen and were used to verify that the spacecraft pointing accuracy was within its specified requirements.

The first opportunity to check L'LORRI's optical performance when the OTA was near its originally planned operating temperature (−85 °C) was in February 2022. The primary objective of those observations was to measure the relative boresight locations of L'LORRI, TTCam, and Ralph-MVIC and calibrate the instrument pointing platform (IPP) by pointing at multiple different locations using IPP motions; all of these tasks were performed successfully and the measurements indicated nominal performance.

The February 2022 plan was loaded with several other activities designed to characterize L'LORRI's performance. One set of images was taken at a solar elongation angle of 85°, and the solar scattered light level was only ∼60-65% of the value of the predicted value based on NH-LORRI data. We took images of the extremely bright star Bellatrix ($V = 1.64$) to characterize optical ghosts (nominal performance), the far-field PSF (achieved), and our ability to accurately remove frame transfer smear for the shortest exposure time of 1 ms (achieved). We obtained a series of images of the solar-type (G2V) absolute calibration star HD 37962, using both CCD formats and both DPUs, to measure the total system throughput (Fig. 38).

Table 9 Adopted magnitudes and fluxes

Object	B	V	R	F_B	F_V	F_R
Star-0	0.000	0.000	0.000	6.96×10^{-09}	3.66×10^{-09}	2.18×10^{-09}
Sun	−26.055	−26.771	−27.158	$1.84 \times 10^{+02}$	$1.87 \times 10^{+02}$	$1.59 \times 10^{+02}$
HD 37962	8.528	7.810	7.446	2.70×10^{-12}	2.75×10^{-12}	2.29×10^{-12}

"Star-0" refers to an unresolved target whose spectral energy distribution (SED) gives 0 mag in the Johnson-Kron-Cousins (BVR) stellar magnitude system. The adopted central wavelengths of the B, V, and R bands are 4400 Å, 5500 Å, and 6500 Å, respectively. The listed fluxes (F_B, F_V, F_R) are in ergs cm^{-2} s^{-1} Å^{-1} and are mean values in ±50 Å regions centered on these wavelengths. The values for the Sun and HD 37962 are derived from spectroscopic data archived in a calibration database maintained by the Space Telescope Science Institute (STScI). The Star-0 fluxes are calculated using spectral data of the calibration star Vega (α-Lyrae), assuming Vega has $V = 0.035$ in all wavelength bands. See the text for further discussion.

The analysis of these latter observations deserves more discussion owing to some subtle issues regarding the absolute calibration. We had previously assumed (Weaver et al. 2020) that HD 37962 had $V = 7.850$ and $B - V = 0.65$, but we recently noticed discrepancies with values reported from other sources. For example, DR10 (data release 10) from the AAVSO Photometric All-Sky Survey (APASS) assigns the following magnitudes to HD 37962: $V = 7.881$ (no error is given and is presumed to be near 0), $B = 8.662 \pm 0.039$, which are both fainter and redder ($B - V = 0.781 \pm 0.039$) than what we have been using. Gaia measurements and photometric relations found in Evans et al. (2018) can be used to derive $V = 7.829$ for HD 37962. Given these discrepancies, we decided to establish our own magnitude values for HD 37962 using the irradiance (or flux) measured by the *Hubble Space Telescope*, which has an absolute accuracy of ∼1%. The results are given in Table 9. L'LORRI's absolute sensitivity is ∼14% larger than the value determined from the ground calibration and ∼19% larger than the value measured for LORRI.

An extensive set of observations were performed near the star S Monocerotis ($V = 4.5$) in the open star cluster NGC 2264 to study the effects of spacecraft jitter and exposure time on the quality of the PSF (Fig. 39). The spacecraft pointing moves around at the level of ∼1″ even during inertial pointings, but that generally doesn't affect the PSF quality except possibly for exposure times longer than about 30 s (the longest available exposure time is 64.9 s).

The images of stars taken during the February 2022 observations were slightly more elongated than expected (Fig. 40). Although we knew from ground testing that the L'LORRI OTA had significant astigmatism and would be significantly out-of-focus when operated at cold temperatures (see the earlier discussion), we were expecting the in-flight PSF to be somewhat *better* than the ground PSFs owing to the various issues (mainly uncontrolled vibrations) that artificially broadened the PSFs measured during the ground calibration.

Figure 41 compares the FWHMs across the FOV for one of the ground tests versus the in-flight case, both taken with the OTA primary mirror near −85 C. Although the spatial pattern of the PSFs is different for the two cases (for reasons we don't yet understand), the FWHMs for the in-flight case tend to be slightly larger (by ∼0.1 pixel) than the values for the ground case. The requirement that the PSF must have FWHM ≤ 3 pixels is satisfied for both cases across most of the FOV, but the region where the requirement is met is slightly larger for the ground case. The temperature gradient across the OTA (represented by the M1−M2 temperature difference) appears to be slightly larger for the in-flight case (1.1 C for ground versus 2.7 C for in-flight), which might partially explain its slightly poorer performance, but we have no data to verify that. In any case, we were motivated by these initial in-flight

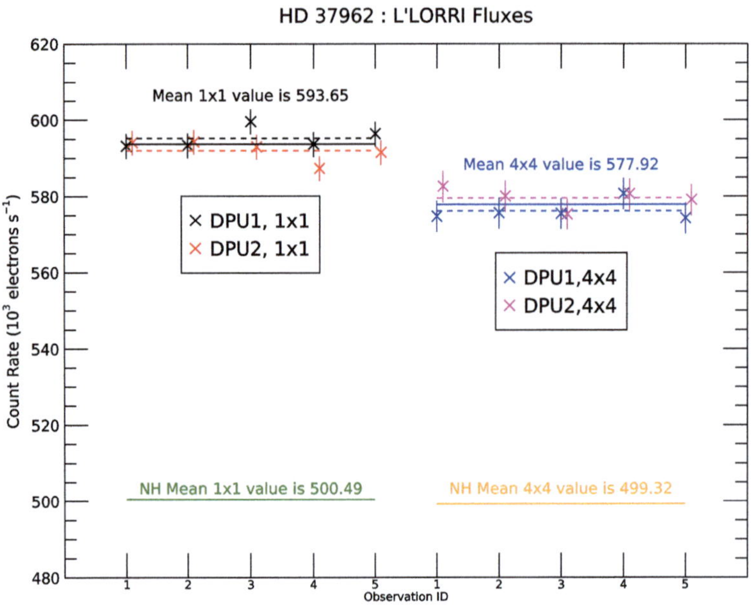

Fig. 38 L'LORRI observed the absolute calibration star HD 37962 ($V = 7.810$) in February 2022 to measure the total system throughput. The data from both the 1×1 and 4×4 images, and from both DPUs, are displayed. The two different DPUs give essentially identical results. The signal from the 1×1 images is $\sim$2.7% larger than the signal from the 4×4 images, but that is comparable to the uncertainty in converting the measured signal in particular synthetic photometry apertures to the total signal integrated over the PSF. The L'LORRI throughput is $\sim$19% larger than the throughput from LORRI

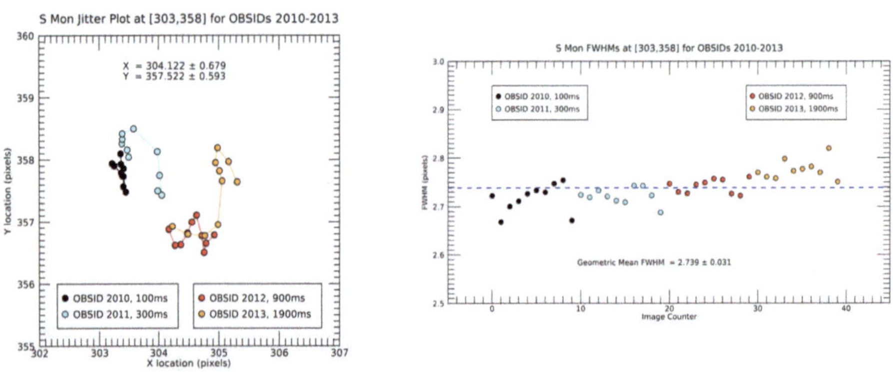

Fig. 39 The plot on left shows the apparent motion of a star during a series of 40 consecutive images taken in February 2022. The pointing was stable at the level of $\sim$1 L'LORRI pixel ($\sim$1.0″). The plot on the right shows the variation in the PSF FWHM for the 40 images. There is a slight increase in the FWHM with increasing exposure times, but the effect is small

measurements to investigate whether other thermal environments (i.e., colder or warmer OTA temperatures) might improve the PSF.

The observations in May 2022 provided an opportunity to obtain a limited set of data with the OTA at two additional temperatures: −75 C and −94 C. The images at −75 C were more

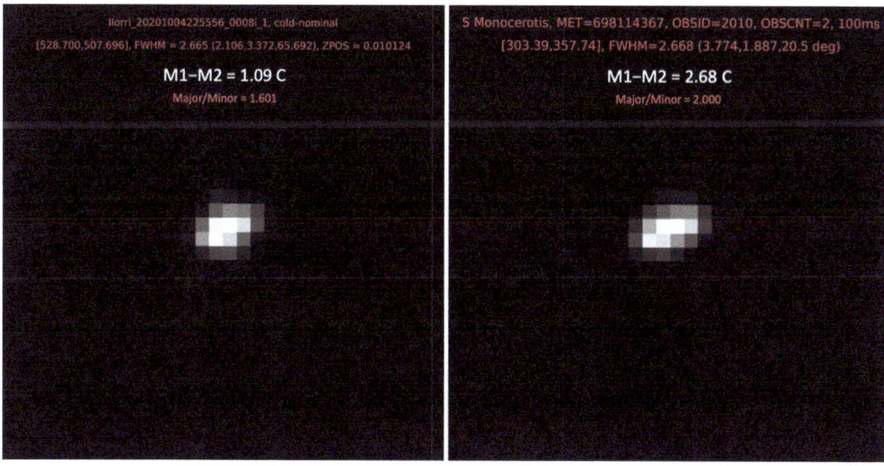

Fig. 40 The image to the left shows the L'LORRI PSF taken during ground testing with the OTA primary mirror at a temperature of −85 C. The image to the right shows the L'LORRI PSF taken during in-flight observations in February 2022 when the OTA primary temperature was −85 C. Although the two images have similar FWHMs (geometric mean of the best-fit elliptical gaussian is ∼2.7 pixels in both cases), the in-flight image is considerably more elongated (major/minor ellipse ratio of 2.0 vs 1.6). See the text for further discussion

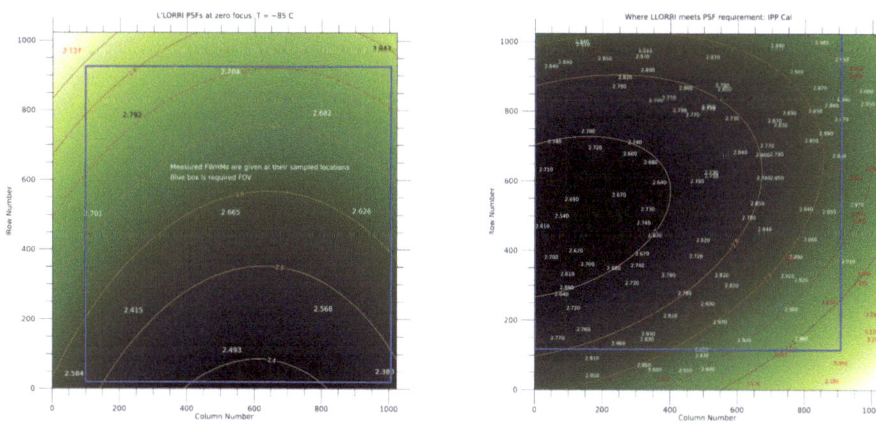

Fig. 41 The figure to the left plots the FWHM (geometric mean of the best-fit elliptical gaussian) of the L'LORRI PSF over the CCD for images taken during ground testing with the OTA primary mirror at a temperature of −85 C. The figure to the right plots the FWHM of the L'LORRI PSF over the CCD for images taken during in-flight observations in February 2022 when the OTA primary temperature was −85 C. The in-flight FWHMs are generally slightly larger than those obtained during ground testing. The blue boxes in each image show the size of the region where the PSF requirement must be satisfied, which is barely met by the ground case but not quite by the in-flight case. See the text for further discussion

symmetrical (rounder) than the others but were also much broader with FWHMs exceeding 3 pixels everywhere across the FOV because the CCD was far from the optimal focus at that temperature. However, the set of images taken at −94 C had PSFs that looked slightly sharper than those taken at −85 C, which prompted us to select an OTA thermal set point of −95 C for all of the fall 2022 activities, including the observations of NASA's Double

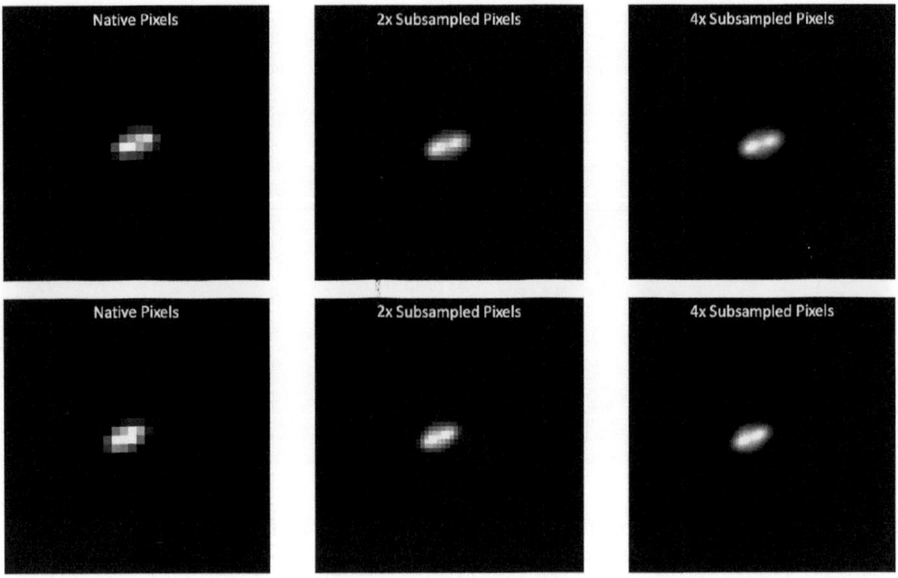

Fig. 42 Multiple images of a single isolated star have been combined to create Nyquist-sampled versions (Lauer 1999), which have then been displayed in three different formats: with native 1×1 pixels, with double-sampled pixels ("2x"), and with quadruple-sampled pixels ("4x"). The top row is derived from data taken in February 2022 when the L'LORRI OTA primary mirror was at −85 C; the bottom row is derived from data taken in October 2022 when the L'LORRI OTA primary mirror was at −95 C. Each image is centered on the pixel with maximum intensity, is displayed on a linear scale from 0 to the maximum value, and covers the same physical area on the CCD. The images at −95 C are more concentrated than those at −85 C suggesting that the resolution is slightly improved at the colder OTA temperature, but all images have a "double star" appearance because the OTA is out-of-focus and has significant astigmatism

Asteroid Redirection Test (DART) impact event and all the observations taken during EGA-1 in mid-October.

For EGA1 (whose observations are discussed in Spencer et al. 2024), we executed an extensive set of observations of the Moon, which provided an opportunity to measure lunar craters with L'LORRI and compare the results to ground truth collected by Moon-orbiting instruments. By taking a set of L'LORRI images of a relatively dense star field shortly after completing the lunar crater program, we created a properly (Nyquist) sampled PSF at −95 C, which could be used to deconvolve the lunar crater images and determine the spatial resolution that might ultimately be achievable during the flybys of the Jovian Trojans. The preliminary analysis of the lunar crater images is reported in an LPSC abstract (Robbins et al. 2023a), and the full analysis has recently been published in a journal (Robbins et al. 2023b).

Figure 42 shows a detailed comparison of the L'LORRI PSF for two OTA temperatures: −85 C and −95 C. In both cases, the PSFs (i.e., the response to a single unresolved target) were constructed from multiple dithered L'LORRI images using a technique described in detail by Lauer (1999) and previously applied successfully to LORRI images of Kerberos (Weaver et al. 2016), one of Pluto's small satellites. The PSF is clearly more compact for the colder temperature, and we are planning a more extensive set of observations with the OTA at −95 C in the spring of 2023 to determine if that thermal environment should be used for all future L'LORRI observations. On the other hand, even the PSF at −95 C has a "double star" appearance, which will limit the ability of deconvolution methods to squeeze higher resolution out of the L'LORRI images. We were originally expecting deconvolution to improve the resolution of L'LORRI images by approximately a factor of two, but the lunar crater data suggest that the improvement is likely to be closer to a factor of $\sqrt{2}$ (Robbins et al. 2023b).

Partly for calibration and partly for scientific purposes, *Lucy* conducted a comprehensive observing program of the Didymos (65803) near-Earth asteroid binary system, starting 12 hr before NASA's DART spacecraft impacted Dimorphos (the satellite of Didymos) and extending to 24 hr after the impact. All observations were performed using L'LORRI 4×4 format, and a total of 1549 images were collected documenting the impact event itself and the subsequent evolution of the debris cloud produced during the impact event. A full discussion of the results from the L'LORRI investigation of the DART impact event is provided in a separate publication (Weaver et al. 2023), but preliminary results were reported in Weaver et al. (2022).

6 Summary

L'LORRI is the highest resolution imager on the *Lucy* mission. Extensive ground testing of L'LORRI characterized its performance prior to launch, and additional in-flight characterization has been performed during the first year of *Lucy* operations. The total system throughput and CCD performance of L'LORRI have surpassed that provided by *New Horizons* LORRI, the heritage instrument that drove L'LORRI's design. L'LORRI's design goals were not achieved for focus and PSF behavior as a function of temperature; its PSF is considerably broader than LORRI's PSF and has a "double star" appearance that limits the ability to retrieve higher resolution via deconvolution techniques. Nevertheless, the spacecraft-controlled decontamination heater can be used to create a thermal environment that optimizes L'LORRI's optical performance and allows L'LORRI to achieve its scientific objectives. L'LORRI images of its targeted surfaces will reveal their geologies in unprecedented detail and will enable crater counting over several decades of size to constrain surface ages.

Information gleaned from L'LORRI images can be combined with color images and infrared spectral data from the L'Ralph instrument (Reuter et al. 2023) to investigate morphology-composition correlations, including searching for evidence of formerly buried material now on the surface. Morphological data from L'LORRI images can be combined with thermal inertia results derived from L'TES (Christensen et al. 2023) to determine if the surface porosity is associated with particular geological structures. The combination of TTCam (Bell et al. 2024) and L'LORRI imaging of the *Lucy* targets near the time of closest approach will provide important context for the data from both instruments.

L'LORRI will also serve as an excellent optical navigation camera to help guide the *Lucy* spacecraft to its designed aim points for each flyby encounter. During the approach to each *Lucy* target, L'LORRI will perform highly sensitive searches for any potential satellites or rings, as well as dust outflow from the targets that might be suggestive of cometary-like behavior.

The successful launch of *Lucy* on 2021 October 16 marked the beginning of a journey to determine what the Jovian Trojans can reveal about the early history of our solar system, and L'LORRI is expected to play a critical role in that investigation.

Glossary of Acronyms

ADC	Analog-to-Digital Converter
APL	Applied Physics Laboratory

AR	Anti-Reflection (coating)
ATLO	Assembly, Test, and Launch Operations
au	astronomical unit
CAD	Computer Aided Design
CCD	Charge Coupled Device
CDH	Command and Data Handling
CDS	Correlated Double Sampling
CFS	Common Fight Software
DN	Data Number
DPU	Data Processing Unit
EE	Encircled Energy
EGA	Earth Gravity Assist
FOD	Foreign Object Debris
FOV	Field of View
FPE	Focal Plane Electronics
FPGA	Field Programmable Gate Assembly
FPU	Focal Plane Unit
FWHM	Full Width Half Maximum
I/F	Intensity to (Solar) Flux ratio (reflectance)
IFOV	Instantaneous (single-pixel) Field of View
IPP	Instrument Pointing Platform
JHU	Johns Hopkins University
L3H SSG	L3Harris Sensors and Simulation Group (corporation)
L'LORRI	Lucy Long Range Reconnaissance Imager
LM	Lockheed-Martin (corporation)
LVDS	Low Voltage Differential Signal
LUPI	Laser Unequal Path Interferometer
NASA	National Aeronautics and Space Administration
OCF	Optical Calibration Facility
OpNav	Optical Navigation
OTA	Optical Telescope Assembly
PB	Processor Board
PLRA	Program Level Requirements Appendix
PSF	Point Spread Function
PST	Point Source Transmittance
QE	Quantum Efficience
SEA	Solar Elongation Angle
SNR	Signal-to-Noise Ratio
STOP	Structural, Thermal, and Optical Performance
TVAC	Thermal Vacuum
V	Visual Magnitude

Acknowledgements We thank Aaron Magner, Ray Sterner, Jake Strang, Brian Wolven, Lucas Wray, and Tmitri Zukowski for their critical support of the L'LORRI instrument I&T operations. We thank Dave Cunningham and John Stinchcomb for technical support during I&T, David Jones for leading the EMI-EMC testing, Chi Pham for radiation testing, Maria Spezio for L'LORRI flight software testing, Ryan Tillman for his materials expertise, Shirley Kha for electrical parts support, and Emory Toomey for thermal blanket support and fabrication. We thank the team at L3H (Dave Calabrese, Nangelie Ferrer, Ed Gilbertson, Jim Green, Evan Stryjewski, and Brian Zellers) for their efforts to produce the L'LORRI OTA. We thank the team at Teledyne-e2v (Ross Mackie, Paul Oliver, and Zoe Locke) for providing the L'LORRI CCDs on time. We thank Tod Lauer (NOIRLab) for creating the Nyquist-sampled PSFs. We thank Stefano Mottola for his comments on a draft of this paper and for his help in creating better L'LORRI flat fields. We gratefully acknowledge the heroic efforts of the *Lucy* personnel at the MOC and SOC who enabled the launch and successful operations of the mission on an extremely tight schedule in the middle of a global pandemic.

Funding Financial support for this work was provided by a contract from the Southwest Research Institute to The Johns Hopkins University Applied Physics Laboratory.

Declarations

Competing Interests The authors have no relevant financial or non financial interests to disclose.

References

Bell JF III, Zhao Y, Beasley M et al (2024) The terminal tracking camera system on the NASA Lucy Trojan Asteroid Discovery Mission. Space Sci Rev 219. https://doi.org/10.1007/s11214-023-01030-5

Cheng AF, Weaver HA, Conard SJ et al (2008) Long-Range Reconnaissance Imager on New Horizons. Space Sci Rev 140(1–4):189–215. https://doi.org/10.1007/s11214-007-9271-6. arXiv:0709.4278 [astro-ph]

Cheng AF, Conard SJ, Weaver HA et al (2010) Stray light performance of the long range reconnaissance imager (LORRI) on the New Horizons Mission. In: Space telescopes and instrumentation 2010: optical, infrared, and millimeter wave, p 77311A. https://doi.org/10.1117/12.859468

Christensen PR, Hamilton VE, Mehall GL et al (2023) The Lucy Thermal Emission Spectrometer (L'TES) Instrument. Space Sci Rev 219. https://doi.org/10.1007/s11214-023-01029-y

Evans DW, Riello M, De Angeli F et al (2018) Gaia Data Release 2. Photometric content and validation. Astron Astrophys 616:A4. https://doi.org/10.1051/0004-6361/201832756. arXiv:1804.09368 [astro-ph.IM]

Janesick JR (2001) Scientific charge-coupled devices. SPIE Press, Bellingham. https://doi.org/10.1117/3.374903

Lauer TR (1999) Combining undersampled dithered images. Proc Astron Soc Pacific 111(756):227–237. https://doi.org/10.1086/316319. arXiv:astro-ph/9810394 [astro-ph]

Levison HF, Olkin CB, Noll KS et al (2021) Lucy mission to the Trojan asteroids: science goals. Planet Sci J 2(5):171. https://doi.org/10.3847/PSJ/abf840

Levison HF, Marchi S, Noll KS et al (2024) Lucy science and mission overview. Space Sci Rev 220

Nesvorný D, Vokrouhlický D, Morbidelli A (2013) Capture of Trojans by jumping Jupiter. Astrophys J 768(1):45. https://doi.org/10.1088/0004-637X/768/1/45. arXiv:1303.2900 [astro-ph.EP]

Noll KS, Brown ME, Buie MW et al (2023) Trojan asteroid satellites, rings, and activity. Space Sci Rev 219:59. https://doi.org/10.1007/s11214-023-01001-w

Olkin CB, Vincent M, Adam C et al (2024) Mission design and concept of operations for the Lucy Mission. Space Sci Rev 220

Reuter DC, Simon AA, Lunsford A et al (2023) L'Ralph: a visible/infrared spectral imager for the Lucy Mission to the Trojans. Space Sci Rev 219:69. https://doi.org/10.1007/s11214-023-01009-2

Robbins SJ, Bierhaus E, Spencer JR et al (2023a) Testing the ability for deconvolution and Nyquist-sampling to allow detection of small impact craters: lunar proof-of-concept with Lucy's L'LORRI camera. In: 54th Lunar and Planetary Science Conference, LPI Contribution No 2806, id 2976

Robbins SJ, Bierhaus EB, Barnouin OS et al (2023b) Imaging Lunar craters with the Lucy Long Range Reconnaissance Imager (L'LORRI): a resolution test for NASA's Lucy Mission. Planet Sci J

Spencer JR, Bell JF III, Christensen PR et al (2024) The first Lucy Earth Flyby (EGA1). Space Sci Rev 220

Weaver HA, Buie MW, Buratti BJ et al (2016) The small satellites of Pluto as observed by New Horizons. Science 351(6279):aae0030. https://doi.org/10.1126/science.aae0030. arXiv:1604.05366 [astro-ph.EP]

Weaver HA, Cheng AF, Morgan F et al (2020) In-flight performance and calibration of the LOng Range Reconnaissance Imager (LORRI) for the New Horizons Mission. Proc Astron Soc Pacific 132(1009):035003. https://doi.org/10.1088/1538-3873/ab67ec. arXiv:2001.03524 [astro-ph.IM]
Weaver HA, Spencer JR, Marchi S et al (2022) Investigating the DART impact event with the Lucy LOng Range Reconnaissance Imager. In: AGU fall meeting abstracts, pp P54C–01
Weaver HA, Spencer JR, Marchi S et al (2023) Lucy observations of the DART impact event. Planet Sci J

Publisher's Note Springer Nature remains neutral with regard to jurisdictional claims in published maps and institutional affiliations.

Authors and Affiliations

H.A. Weaver[1] · J.P. Wilson[1] · S.J. Conard[1] · J.D. Adams[1] · S. Begley[1] · J. Burgum[1] · E.H. Darlington[1] · N. Dello Russo[1] · R. Hacala[1] · S. London[1] · M.F. Morgan[1] · G. Murphy[1] · T. Nelson[1] · A. Shah[1] · J.R. Spencer[2] · H. Taylor[1] · T. Boehmer[1] · L. Burke[1] · C. Drabenstadt[1] · C. Henry[1] · S. Ling[1] · C. Porter[1] · J. Yin[1]

✉ H.A. Weaver
hal.weaver@jhuapl.edu

J.P. Wilson
john.wilson@jhuapl.edu

S.J. Conard
steven.conard@jhuapl.edu

J.D. Adams
dewey.adams@jhuapl.edu

S. Begley
shawn.begley@jhuapl.edu

J. Burgum
james.burgum@jhuapl.edu

E.H. Darlington
hugo.darlington@jhuapl.edu

N. Dello Russo
neil.dello.russo@jhuapl.edu

R. Hacala
ryan.hacala@jhuapl.edu

S. London
shaughn.london@jhuapl.edu

M.F. Morgan
frank.morgan@jhuapl.edu

G. Murphy
graham.murphy@jhuapl.edu

T. Nelson
tyler.nelson@jhuapl.edu

A. Shah
amit.shah@jhuapl.edu

J.R. Spencer
spencer@boulder.swri.edu

H. Taylor
howard.taylor@jhuapl.edu

T. Boehmer
tyler.boehmer@jhuapl.edu

L. Burke
linda.burke@jhuapl.edu

C. Drabenstadt
christian.drabenstadt@jhuapl.edu

C. Henry
charles.henry@jhuapl.edu

S. Ling
sharon.ling@jhuapl.edu

C. Porter
christa.porter@jhuapl.edu

J. Yin
jie.yin@jhuapl.edu

[1] Space Exploration Sector, Johns Hopkins University Applied Physics Laboratory, 11100 Johns Hopkins Road, Laurel, 20723-6099, MD, USA

[2] Department of Space Studies, Southwest Research Institute, 1050 Walnut Street, Suite 300, Boulder, 80302, CO, USA

Space Science Reviews (2023) 219:69
https://doi.org/10.1007/s11214-023-01009-2

L'Ralph: A Visible/Infrared Spectral Imager for the Lucy Mission to the Trojans

D.C. Reuter[1] · A.A. Simon[1] · A. Lunsford[2] · H. Kaplan[1] · M. Garrison[1] · J. Simpson[1] · G. Casto[1] · Z. Dolch[1] · P. Finneran[3] · W. Grundy[4] · C. Howett[5] · P. Kim[1] · M. Loose[6] · T. Null[1] · F. Parong[1] · J. Rodriguez-ruiz[1] · P. Roming[7] · K. Smith[7] · P. Thompson[1] · B. Tokarcik[1] · T. Veach[7] · S. Wall[1] · J. Ward[8] · E. Weigle[9] · H. Levison[10]

Received: 27 June 2023 / Accepted: 29 September 2023 / Published online: 25 October 2023

Abstract

The Lucy Mission to the Trojan asteroids in Jupiter's orbit carries an instrument named L'Ralph, a visible/near infrared multi-spectral imager and a short wavelength infrared hyperspectral imager. It is one of the core instruments on Lucy, NASA's first mission to the Trojans. L'Ralph's primary purpose is to map the surface geology and composition of these objects, but it will also be used to search for possible tenuous exospheres. It is compact, low mass (32.3 kg), power efficient (24.5 W), and robust with high sensitivity and excellent imaging. These characteristics, and its high degree of redundancy, make L'Ralph ideally suited to this long-duration multi-flyby reconnaissance mission.

Keywords Trojan asteroids · Asteroids · Multi-spectral · Hyperspectral · Space mission

✉ D.C. Reuter
dennis.c.reuter@nasa.gov

1 NASA/GSFC, Code 693, Greenbelt, MD 20771, USA

2 American University, NASA/GSFC, Code 693, Greenbelt, MD 20771, USA

3 Jackson and Tull Inc., Beltsville, MD 20705, USA

4 Lowell Observatory, Flagstaff, AZ 86001 USA

5 Department of Physics, Clarendon Laboratory, Oxford University, Parks Rd., Oxford, OX1 3PU, UK

6 Markury Scientific, 518 Oakhampton St., Thousand Oaks, CA, 91361, USA

7 Southwest Research Institute, 6220 Culebra Rd., San Antonio, TX, 78228, USA

8 Newton-Engineering and Product Development, 5650 Rivertech Ct., Riverdale, MD, 20737, USA

9 Big Head Endian, LLC, 11785 181st RD, Burden, KS, 67019, USA

10 Space Sciences and Engineering Division, Southwest Research Institute (SwRI), 1050 Walnut St., Suite 400, Boulder, CO, 80302, USA

Abbreviations

A/D	Analog to Digital
ASIC	Applications Specific Integrated Circuit
AU	Astronomical Units
BHE	Big Head Endian
C&DH	Command and Data Handling
CCD	Charge Coupled Device
CDS	Correlated Double Sampling
DAC	Digital-to-Analog Converter
DN	Digital Number
DPSE	Differential Position Sensor Electronics
DPSS	Differential Position Sensor System
EGA	Earth Gravity Assist
FPE	Focal Plane Electronics
FOV	Field of View
FPGA	Field Programmable Gate Array
FSW	Flight SoftWare
FWHM	Full Width at Half Maximum
FWTM	Full Width at Tenth Maximum
GSFC	Goddard Space Flight Center
HK	Housekeeping
IFOV	Instantaneous Field of View
IPP	Instrument Pointing Platform
IR	Infrared
IRSM	Infrared Source Module
LEISA	Linear Etalon Imaging Spectral Array
LVF	Linear Variable Filter
LVPS	Low Voltage Power Supply
M4	Multi-Mission Mass Memory
MCE	Mechanism Control Electronics
MEB	Main Electronics Box
MLI	Multi-Layer Insulation
MVIC	Multi-spectral Visible Imaging Camera
NIR	Near Infrared
NIST	National Institute of Science and Technology
PSF	Point Spread Function
OVIRS	OSIRIS-REx Visible and Infrared Spectrometer
SIDECAR	System Image, Digitizing, Enhancing, Controlling, And Retrieving
SMM	Scan Mirror Mechanism
SMS	Scan Mirror System
SNR	Signal-to-Noise Ratio
SWIR	Shortwave Infrared
SwRI	Southwest Research Institute
TDA	Telescope Detector Assembly
TDI	Time Delay Integration
TIRS	Thermal Infrared Sensor
TVac	Thermal Vacuum
VIS/IR	visible/infrared

1 Introduction

Lucy, a multi-encounter mission to the Trojan asteroids in Jupiter's orbit, is the first mission to explore this class of objects. Launched on October 16, 2021, it will visit 8 Trojans over a 12-year period. The scientific rationale for Lucy and the overall mission planning are described in detail in Levison et al. 2021, and Olkin et al. 2021, and in several papers in this collection (Levison et al. 2024; Olkin et al. 2024). The Lucy mission is led by Principal Investigator Dr. Hal Levison of the Southwest Research Institute (SwRI) of Boulder, CO and is managed by SwRI and Goddard Space Flight Center in Greenbelt, MD. A core remote-sensing instrument on Lucy is L'Ralph, a visible/NIR camera and shortwave infrared spectral imager. The instrument's primary purpose is to measure surface characteristics, including geological processes, geomorphology, photometric properties, and surface composition. L'Ralph will also be used to measure the properties of comas (if they are present) and to obtain compositional measurements of rings and small satellites (again, if present). This paper describes L'Ralph and specifies its characteristics. L'Ralph is similar in concept to the Ralph instrument on the New Horizons mission (Reuter et al. 2008), however it has numerous science-based and operational improvements, and its performance is significantly better than New Horizons Ralph. Improvements of L'Ralph over NH Ralph include a much broader IR spectral range, the addition of onboard memory, and a scanning mirror that allows images to be taken without additional spacecraft motion. L'Ralph was provided by NASA's Goddard Space Flight Center (GSFC, Greenbelt, MD), however significant contributions were made by numerous entities throughout the US. L'Ralph also shares characteristics with the OSIRIS-REx OVIRS instrument (Reuter et al. 2018).

The L'Ralph instrument is mounted to the Instrument Pointing Platform (IPP) on the Lucy spacecraft (Levison et al. 2024, this collection). It has two elements, the Main Electronics Box (MEB), which provides the data and command interface between the instrument and the spacecraft and operates the instrument, and the Telescope Detector Assembly (TDA). The TDA consists of an imaging optical assembly, a scan mirror mechanism, the detectors, a passive thermal system for cooling the detectors as well as the assembly as a whole, and the calibration sources. The imaging assembly consists of a single telescope that feeds two sets of focal planes: 1) the Multi-spectral Visible Imaging Camera (MVIC), a visible to near-IR imager and 2) the Linear Etalon Imaging Spectral Array (LEISA), a short-wavelength IR spectral imager. To limit thermal background, the TDA is cooled to $<$ 195 K. The telescope uses an unobscurred, off-axis, three-mirror anastigmat design. The entire telescope assembly, including the three diamond-turned powered mirrors, is constructed from grain aligned 6061-T6 aluminum. The housing, excluding the attached covers but including the integral optical mounts, is fabricated from a single Al block. This combination of an all-Al structure and optics is lightweight, has uniform thermal expansion coefficients and is sufficiently thermally conductive that thermal gradients are not significant over the range of operating temperatures expected during the Trojan encounters (see Sect. 3.6 for the temperature ranges). These characteristics ensure that the optical performance of the instrument will meet requirements for all encounters. The highly baffled, 75-mm aperture, VIS/IR telescope provides ample sensitivity at all the encounters, while minimizing size and mass. The f/6 system's approximately 450-mm effective focal length offers a good compromise between photometric throughput and alignment stability. The mirrors are direct diamond turned off-axis aspheres, each with surface roughness below 6.0 nm RMS, fabricated by Coherent Inc. of Richmond, CA. In addition to an entrance baffle, stray light control is improved by using a field baffle at an intermediate focus between the secondary and tertiary mirrors, and by baffles near the exit pupil after the tertiary mirror. A dichroic beamsplitter transmits IR wavelengths longer than 0.95 μm to LEISA and reflects shorter wavelengths to MVIC.

MVIC is composed of 6 independent CCD arrays (channels) on a single substrate. Each of the CCDs has 5000 (cross-track) × 64 (along-track) optically active pixels and they operate in time delay integration (TDI) mode. The channels are panchromatic (350 – 950 nm), violet (375 – 480 nm), green (480 – 520 nm), orange (520-625 nm), phyllosillicate (625 – 750 nm), and near IR (750-950 nm). TDI operates by synchronizing the along-track transfer rate of each of the CCD's rows to the relative motion of the image across the detector's surface. In this way, each spot on the object being imaged transfers from one row to the next without being smeared and the integration time is increased by the number of rows used in the TDI observation, allowing high signal-to-noise measurements to be made even on fast scans. Thus, very large format images are obtained as the spacecraft scans the MVIC field of view (FOV) rapidly across the surface. The single pixel instantaneous field of view (IFOV) of an MVIC pixel is about 28.8 microradians and the cross-track width covered is about 144 milliradians (∼8.25°).

LEISA is a wedged filter infrared spectral imager that creates spectral maps in the compositionally important 0.95 – 3.95 μm shortwave infrared (SWIR) spectral region. It images a scene through a wedged filter (or Linear Variable Filter (LVF), Rosenberg et al. 1994) placed about 100 μm above a Mercury Cadmium Telluride (HgCdTe) detector array. The LVF is fabricated such that the wavelength varies along one dimension, the along-track scan direction. LEISA forms a spectral map by scanning the FOV across the surface in a push-frame fashion. The frame rate is synchronized to the rate of the scan, so that a frame is read out each time the image moves by the IFOV. The difference between a LEISA scan and a TDI scan is that in LEISA the row-to-row image motion builds up a spectrum while in TDI the motion increases the signal over a single spectral interval. The single pixel IFOV of a LEISA pixel is about 40 microradians and the maximum cross-track width covered is about 41 milliradians (∼2.35°)

2 L'Ralph Science Overview

The full scientific rationale for the Lucy mission to the Trojans is given in detail in Levison et al. (2021, 2024) and will not be repeated at length here. Briefly, the Trojan objects were scattered into the L4 and L5 Lagrange points in Jupiter's solar orbit early in the Solar System's development. In essence they are the "fossils" of planetary evolution existing in "archeological sites" of the Solar System that will be visited for the first time by the Lucy spacecraft. The objects to be explored (in order, (3548) Eurybates, (15,094) Polymele, (11,351) Leucus, (21,900) Orus, and the (617) Patroclus/Menoetius binary) represent a wide range of sizes, spectral reflectances, rotation rates, and possible sources. It should be noted that both Eurybates and Polymele have recently discovered moons (Noll et al. 2020; Buie et al. 2022) that will also be observed. The first exploration of the Trojans is both scientifically and publicly exciting. It will provide invaluable insights into the origin of, and ongoing processes in, the solar system.

The L'Ralph instrument will play a leading role in this exploration. It directly addresses seven of the Level 1, or primary, mission requirements (Levison et al. 2021) connected to the surface composition of the Trojans and how that is related to the surface topography and impact history. L'Ralph/MVIC shall obtain visible/NIR multi-spectral images at spatial resolutions and areal coverages that will help identify the shortwave spectral slope and the presence of species producing broadband spectral features (e.g., phyllosillicates and pyroxene) and their correlation with surface structures at sizes on the order of the expected cratering levels. L'Ralph/LEISA shall obtain hyperspectral maps at somewhat lower spatial resolution than MVIC, but still sufficient to identify correlations with geological features, including craters of the size expected to be found. These maps will cover the wavelength range of 0.95 to 3.95 μm (albeit with reduced sensitivity for $\lambda > 3.8$ μm) with a spectral full width at half maximum (FWHM) of 0.01 μm or less over the required 1 – 3.8 μm spectral range. This spectral range and resolution were chosen to achieve good sensitivity to the minerals, volatiles, and complex hydrocarbons that may be expected to be present (including, for example, H_2O (ice and mineral hydration feature), H_2S, CH_4, CH_3OH, SO_2 and NH_3), to characterize the CH_2 and CH_3 bands of aliphatic hydrocarbons, and to examine polycyclic aromatic hydrocarbons (PAHs). The LEISA spectral resolution elements are at least double sampled.

The targeted SNR for these observations is based on existing observations of outer Solar System objects. For example, organic molecules have been detected on the asteroid (24) Themis (Rivkin and Emory 2010, Campins et al. 2010), and on the satellites Iapetus and Phoebe (Dalle Ore et al. 2012; Cruikshank et al. 2008). In each case, the depths of the absorption bands are between 4% and 10% of the local continuum and cover 2 or more LEISA spectral resolution elements. Given this, spectra with SNR of 25 per spectral sample in the spectral range 3.2 to 3.6 microns will enable us to search for organics of the type and abundance present on these other bodies at greater than the 2-σ level in even the narrowest bands. A similar SNR in the 3-μm region will enable analyses that distinguish between H_2O ice and hydrated materials (e.g., Takir and Emery 2012) as well as determine phyllosilicate mineralogy, if phyllosilicates are detected (e.g., Takir et al. 2013). Spectra of Trojans themselves (Brown 2016) have also shown evidence of 4% or more absorption near 3.1 μm. Similarly, an MVIC SNR of 25 or greater will allow accurate retrievals of the spectral slope and broadband features corresponding to mineral absorption bands. In order to ensure sufficient margin to the SNR requirements, the albedo is assumed to be quite low (3% in the visible/NIR, 4% in the SWIR). This, combined with the fact that, to avoid illumination phase changes during an observation and to carry out multiple observations with all the Lucy instruments, the Level 1 observations are made either before or after closest approach when the solar phase angle ranges from a maximum of 82 degrees (Polymele) to a minimum of 54 degrees (Orus) and, finally, that there are limits placed on integration time to avoid spacecraft and IPP pointing drift smear, makes the SNR requirement a very strong driver on L'Ralph sensitivity.

L'Ralph's MVIC and LEISA reflectance maps will allow us to address many questions pertaining to the composition of the Trojans. Such as what is the spatial variability and scale size of surface features? What is the surface distribution of the various species? Are there areas of pure frost and mixed areas? Are there more complex species in selected regions of the surface? Is there a connection between geology and composition? Can we see the effects of space weathering in areas where collisions have exposed sub-surface material? Answers to these questions will significantly advance our understanding of the processes that occurred throughout the outer regions of our solar system during its formation and of the ongoing chemical and physical processes that are occurring there now. Figure 1 shows some Earth based spectra of Trojan objects and other small bodies. As is evident from this figure, there is a wealth of information to be gleaned from in-situ observations of this spectral region even from globally averaged spectra.

In addition to the Level 1 objectives, L'Ralph will address numerous Level 2 measurement goals. These include: MVIC and LEISA observations of Trojan moons if they are

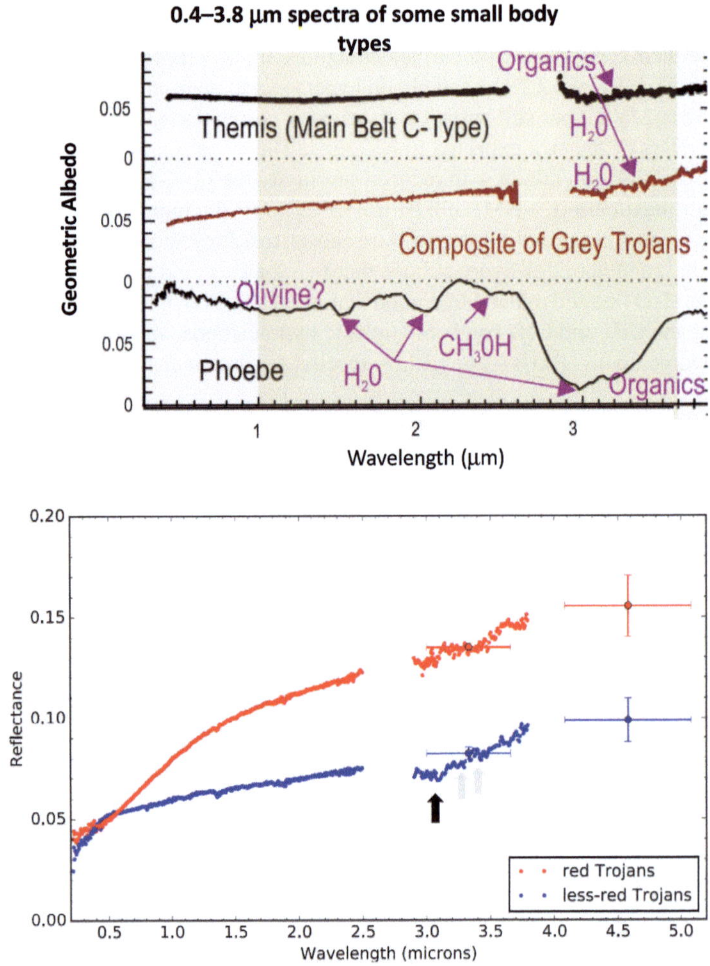

Fig. 1 Reflectance spectra of some small body objects (top) and Earth based spectra of red and less red Trojan object (bottom, Levison et al. 2021). As may be seen in these images, the hyperspectral measurements in the spectral range of 1 – 3.8 μm combined with multispectral measurements in the 0.4 to 0.9 μm regions are well suited to determining surface composition

found (two already have been), higher spatial resolution MVIC observations of some areas during the time near closest approach, higher spatial resolution LEISA observations over a limited spectral range during the time nearest closest approach, and MVIC and LEISA searches for a "coma" of volatile material in the time available before and/or after the Level 1 observations and closest approach. Table 1 summarizes the science objectives that determined the L'Ralph design, the measurement strategies that address these objectives and the derived instrument performance requirements. Note that the required spatial resolution for LEISA is dependent on the size of the Trojan and varies from 1 km for Polymele to 5 km for Patroclus. This is because the required resolution is set by the need to have a high probability of being able to measure a change in composition within resurfaced areas younger than 100 million years as predicted by cratering models.

Table 1 Science Objectives and Derived Instrument Requirements

Science Objective	Measurement Strategy	Derived Instrument Requirements			
		Spectral coverage	Resolution	Image Quality	Signal-to-noise
Determine the global distribution of compositional units.	Color images spaced by 1/6 to 1/3 of a rotation over a full target rotation.	No additional requirements	N/A	No additional requirements	No additional requirement
Search for compositional anomalies resulting from freshly exposed material.	Color images of an area $\geq$ 700 km^2 or 80% of the sunlit surface (whichever is less) capable of resolving features $\leq$ 1.5 km in size. Color images of an area $\geq$ 150 km^2 capable of resolving features $\leq$ 600 m in size	375 – 480 nm 480 – 520 nm 520 – 625 nm 625 – 750 nm 750 – 900 nm	85% of signal from 480-520 nm 91% of signal from 520-625 nm 91% of signal from 625-750 nm 87% of signal from 750-900 nm	Enclosed energy >54%; 50 mRad >83%; 76 mRad >91%; 168 mRad >98.2%; 242 mRad	SNR > 25 Scan 12 mrad target Distance = 5.7 AU I/F = 0.0014 Distance = 5.2 AU I/F = 0.005 1.2 s Maximum integration time No saturation on a scan for I/F a factor of 16 greater
Map the surface composition of the Trojan objects	Spectroscopic images shall detect features with relative spectral depth > 4% and spectral width of $\geq$ 70 nm in the spectral range 1.0–3.4 microns, when the object spans multiple spatial resolution elements. Spectroscopic images shall cover a spectral range 1.0–3.8 microns Spectroscopic images spaced by 1/6 to 1/3 of a rotation over a full target rotation	1.0 – 3.8 µm	$\Delta\lambda \leq$ 10 nm (Full Width Half Maximum) in 1.0 to 3.4 range µm	No additional requirement.	SNR > 25 for spectral features of FWHM of 70 nm in the range 1.0-3.4 um Scan 12 mrad target Distance = 5.7 AU I/F = 0.0019 Distance = 5.2 AU I/F = 0.0068 1.5 s Maximum integration time No saturation on a scan for I/F a factor of 13 greater
Search for compositional anomalies resulting from freshly exposed material	Spectroscopic images with a combination of spatial resolution (r) and areal coverage (A), where A is 25% of the object's area, that together satisfy the condition $r \leq 2(A/A_0)^\alpha$ km where $\alpha = 0.473$, $A_0 = 1470.6$ km^2.	No additional requirement	No additional requirement	Enclosed energy >55%; 136 mRad >82%; 204 mRad >91%; 312 mRad >98.2%; 1020 mRad	No additional requirement.

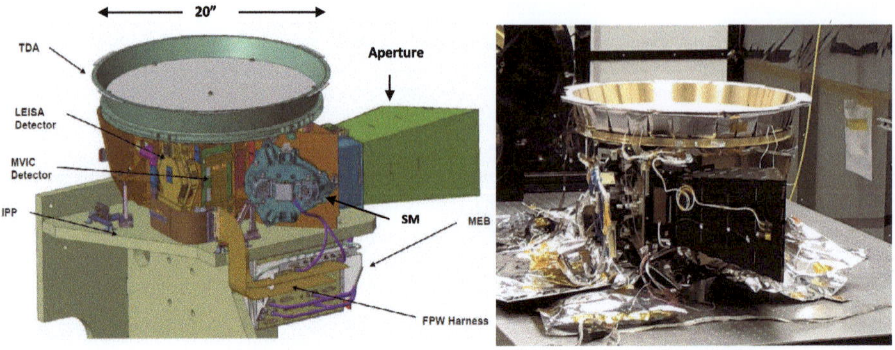

Fig. 2 Left) Model of the L'Ralph instrument mounted on the IPP. This shows both the TDA and the MEB with principal components labeled. (Right) Picture of the L'Ralph TDA, before the addition of most of the multi-layer insulation (MLI)

3 OPTO-Mechanical Design

Figure 2 shows a model of L'Ralph and a picture of the assembled instrument before addition of the final MLI. The major elements are labeled in the model. The mass of the instrument is 32.3 kg and the operational power is 24.5 Watts, both well within their allowable values (35.5 kg and 30 W, respectively). As shown in Fig. 2, L'Ralph has two main assemblies, the telescope detector assembly (TDA) and the Main Electronics Box (MEB). Both assemblies are mounted to the Instrument Pointing Platform (IPP). The TDA consists of the scan mirror assembly, the telescope optical elements, the baffling, the MVIC and LEISA focal planes, the two-stage passive radiator that cools the focal planes, the filament and blackbody calibration sources, and the solar calibrator. The TDA is mounted to the IPP by thermally isolating titanium flexures. Modeling done using results from instrument and spacecraft level thermal vacuum testing predict that during the Trojan encounters, the temperature of the TDA (not including the LEISA detector) will be between 160 and 200 K. The temperature of the electronics box, which is mounted directly to the IPP but thermally isolated from it, is predicted to be between 260 and 305 K. The low temperature of the TDA reduces the conductive and radiative thermal load on the focal planes. It also limits the background signal at the long wavelength end of LEISA. The inner stage of the externally mounted passive radiator cools the LEISA detector to < 115 K. The outer annulus maintains the MVIC CCDs at temperatures below 200 K. The 75 mm aperture, 450 mm focal length, f/6 optical system meets its spatial imaging requirements over the $8.25° \times 0.94°$ MVIC field of view and the $2.35° \times 3.37°$ LEISA FOV. The instrument parameters for L'Ralph are summarized in Table 2. Figure 3 shows a model of the TDA interior with a ray-trace diagram. Figure 4 shows a more detailed internal raytrace including the arrangement of scattering baffles and image stops.

3.1 The MVIC Focal Plane

The MVIC focal plane assembly, which was provided by SwRI, San Antonio TX, consists of a customized front side illuminated CCD array, provided by STA Inc., San Clemente, CA, mounted to a heat sink plate and placed directly behind a "butcher block" filter assembly. The array has six identical 5024-pixel TDI CCDs (5000×64 pixel photoactive area) on a single substrate. The extra 24 dark pixels (12 on each side of the 5000-pixel optically

Table 2 L'Ralph Instrument Parameters

Mass: 32.3 kg
Power: 24.5 Watt (maximum)
Telescope Aperture: 75 mm
Focal Length: 451.4 mm
f/#: 6
MVIC: Six Time Delay and Integrate (TDI) arrays
One 5024 × 64 Pixel Panchromatic TDI CCDs (350 – 950 nm)
Five 5024 × 64 Pixel Color TDI CCDs
Violet (375 – 480 nm)
Green (480 – 520 nm)
Orange (520 – 625 nm)
Phyllosilicate (625 – 750 nm)
Near-IR (750 – 900 nm)
Each CCD may be operated with 64, 32, 16, 8 or 4 TDI rows
13μm × 13μm pixels
Single pixel Field of View: 28.8 μrad × 28.8 μrad
TDI array FOV: 8.25° × 0.105°
Focal plane temperature: <200 K
Minimum TDI integration time (4 TDI, 7.2 msec rate): 27.8 msec
Maximum TDI integration time (64 TDI, 0.65 sec rate): 42 sec
LEISA: 2048 × 2048 element HgCdTe array operated in pushbroom mode
18μm × 18μm pixels
Single pixel Field of View: 40 μrad × 40 μrad
FOV: 2.35° × 3.37° (1024 cross-track, 1472 along track)
Focal plane temperature: <115 K
Spectral range: 0.95 – 3.95 μm (1.0 – 3.8 μm required)
Spectral resolution: variable, < 10 nm
Frame rate: 0.38 sec to 21 sec (for 512 × 1472 pixels)
Minimum frame rate (10 × 1472 pixels): 20 msec

active region) are used to clear any residual charge in the serial readout before the optically active pixels are read out. The cross-track dimension of 5000 elements allows images to be taken that meet the spatial resolution requirements for all encounters, including the Patroclus-Menoetius binary if the two objects are displaced from each other in the cross-track direction, including pointing uncertainties. All MVIC pixels are 13 × 13 μm^2 which, given the optical properties above, corresponds to a single pixel IFOV of 28.8 × 28.8 μrad^2.

The filter, which was provided by VIAVI Solutions, Santa Rosa, CA, is mounted about 500 microns above the surface of the array. It has six segments, five with the passbands described in Table 2 placed directly over the five CCDs forming the color segment of MVIC. The remaining CCD is overlain by a clear filter so that the focus position is the same for all six arrays. The scan mirror is used to scan the image of each segment of the surface across the focal plane in a pushbroom fashion. As Fig. 5 illustrates, the clocking rate of each 5024-pixel row is determined by the scan mirror rotation rate so that a row charge transfer occurs each time the image has moved a pixel width. Each CCD has two outputs, one on each side of the cross-track serial readout, which allows faster readouts, and thus

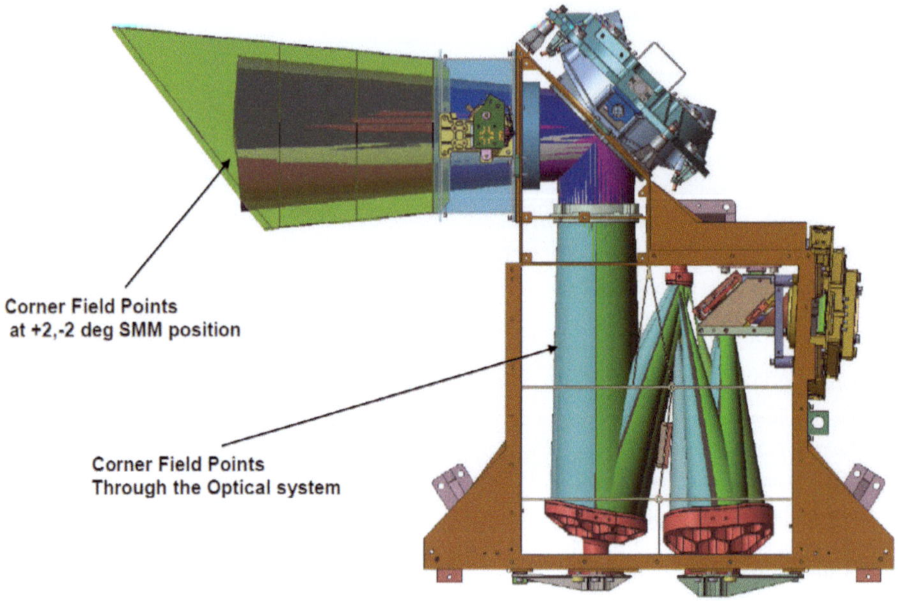

Fig. 3 Interior of the L'Ralph TDA showing the light path at the +2 degree and −2 degree positions of the scan mirror (+4 and −4 degrees of image motion)

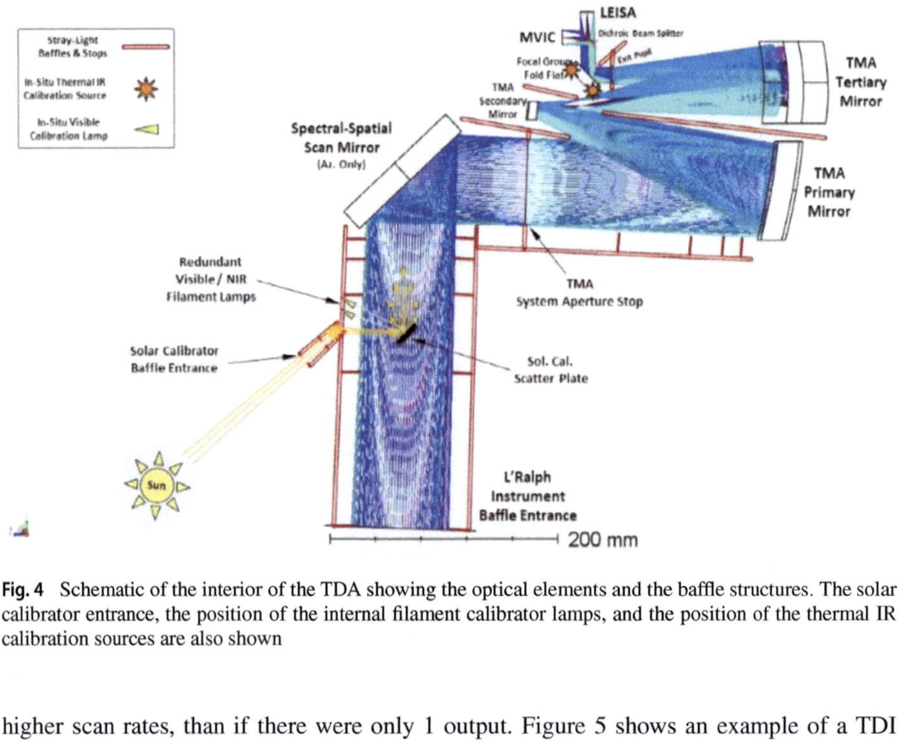

Fig. 4 Schematic of the interior of the TDA showing the optical elements and the baffle structures. The solar calibrator entrance, the position of the internal filament calibrator lamps, and the position of the thermal IR calibration sources are also shown

higher scan rates, than if there were only 1 output. Figure 5 shows an example of a TDI using 32 along-track rows. In L'Ralph, each CCD may be independently commanded to

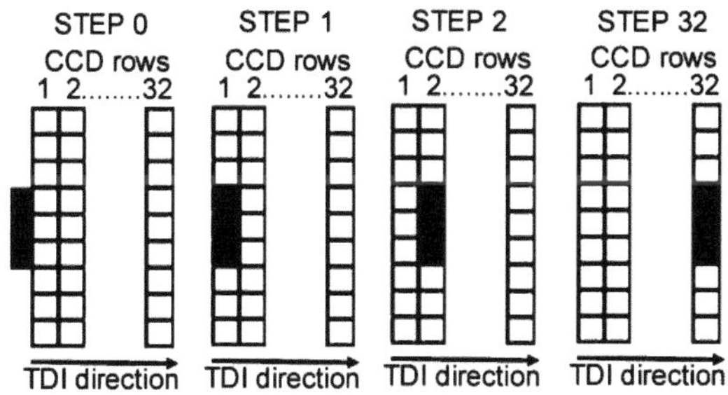

Fig. 5 Illustration of TDI in operation using 32 rows as an example. In step 0, a target is about to enter the field of view of a TDI array. In step 1, the image of the target has moved a single pixel width into the first row of the array. In step 2, the charge is transferred to the second row of the array and integration continues. A new target has moved into row 1. The next charge transfer occurs when the image of the target has moved another row. In step 32, the target has moved 32 rows, with the charge being transferred each time the image moves another row. At the next charge transfer, the charge in row 32 is transferred to a serial column for pixel readout. In L'Ralph,, each CCD may be independently commanded to use either 64, 32, 16, 8 or 4 along-track rows for TDI

use either 64, 32, 16, 8 or 4 along-track rows but, since the last row is shielded from light, there are 63, 31, 15, 7 and 3 optically active rows, allowing each channel to have a different integration time at a given scan rate. This allows data to be taken in all 6 channels in a single scan while maintaining good signal to noise ratios in the channels with lower illumination levels but not saturating in the channels with higher illumination levels. The minimum time between row transfers is 7.2 millisec, which gives a maximum scan rate of 28.8/0.0072 $\sim$ 4000 μrad/sec which, for a 4 row TDI, corresponds to an integration time of 0.0216 sec and 0.4536 sec for a 64 row TDI. This scan rate will be used for observation taken at lower solar phase angles. For most of the encounters the Level 1 data is obtained at relatively high phase angles requiring longer integration times and scan rates of about 1000–3000 μrad/sec will be used.

3.2 The LEISA Focal Plane

The LEISA focal plane assembly was provided by Newton Engineering and Product Development, Riverdale, MD. It consists of an HgCdTe H2RG detector array mounted to a heat sink with an LVF (also called a wedged filter or a linear etalon) directly over the array. The detector, which has a long wave cutoff wavelength of $\sim$3.85 μm, was supplied by Teledyne Imaging Sensors (TIS) of Camarillo CA. It is a 2048 $\times$ 2048-pixel array and each pixel is 18 $\times$ 18 μm^2 in area. The detector is read out using up to 32 outputs, each of which corresponds to a 64-pixel (along-track) $\times$ 2048-pixel (cross-track) section of the array. The HgCdTe was grown on a CdTe substrate using Molecular Beam Epitaxy (MBE) to provide good lattice matching and low dark currents. The detector was bump bonded to a standard multiplexer and the resulting hybrid was mounted to molybdenum pad. This process reduces mechanical stress induced during cooling to operational temperature. It is estimated that the assembly can safely undergo at least 1000 thermal cycles. The LEISA array is back illuminated, but, as is a standard practice, the substrate has been thinned so that the photo-active active area of the array is close to the surface (see, e.g. Beletic et al. 2008).

The filter, supplied by VIAVI Solutions. of Santa Rosa CA, consists of three segments covering 0.95–1.644 μm, 1.53–2.64 μm, and 2.55–3.99 μm respectively, which cover slightly more spectral range than the 1–3.8 μm requirement. In all segments, the transmitted wavelength varies close to linearly with position in the along-track direction. The three segments were bonded together to form a single filter element. This filter was, in turn, bonded into a holder and mounted such that the filter surface is less than 100 μm above the surface of the array. To meet the Lucy spectral and spatial coverage requirements, only 23 detector outputs are needed, with 1024 cross-track pixels in each output. The filter holder covers most of the remaining pixels in the array to reduce possible crosstalk and scattered light. The wedged filter effectively makes the pixels in each cross-track column of the array responsive to only a narrow wavelength (<10 nm) band, so that in concept, when 23 outputs are used, the focal plane may be considered as consisting of 1472 (23 × 64) adjacent narrow band filters. As with MVIC, the image is scanned over the LEISA focal plane by the scan mirror motion with the frame rate synchronized to the scan rate. The refractive index of the array is approximately 2.7 so that the total optical path between the filter and photo-active area of the array is about 100 μm. When the focus position is optimized between the array and filter surface, the f/6 beam spreads about 0.5 pixel.

LEISA may be operated in two modes: correlated double sample (CDS) mode or raw mode. In CDS mode, the reset voltage level of each pixel is subtracted from the voltage read out at the end of the integration time, and that difference is stored in instrument memory. This is the typical mode used for science scans. In raw mode, both the reset voltage level and the voltage read out at the end of the integration time are stored for each pixel. This mode is used for some calibration collects. In addition, when data is stored in spacecraft memory, it can either be on a pixel-by-pixel basis or in the form of 2-pixel × 2-pixel "superpixel" sums. Superpixeling reduces spacecraft data volume by a factor of 4 and it is the standard process used for LEISA scans. The effective IFOV in this format is 80 μrad × 80 μrad and scans are planned accordingly. Note, however, that the data stored in instrument memory is still at the single pixel 40 μrad × 40 μrad level. For the planned Trojan scan, the required FOV of LEISA covers 512 (cross-track) × 1472 (along-track) pixels and the minimum integration time when reading this out is ~0.38 seconds. However, shorter integration times may be obtained when reading out fewer than 512 cross-track pixels.

3.3 Scan Mirror System

The Scan Mirror System (SMS) is comprised of the redundant Mechanism Control Electronics (MCE) cards in the MEB, the Scan Mirror Mechanism (SMM) located on the Telescope Detector Assembly, and the redundant Differential Position Sensor System (DPSS) sensors (located on the SMM) and electronics (located on the exterior of the MEB). The SMM is a single axis flexure- based device capable of positioning the 140 mm × 100 mm open aperture scanning mirror through a ±2° range of physical motion to accommodate the ±4° line-of-sight scan requirements. The SMM employs e-beam welded flexural pivots to support the scan mirror with respect to the fixed base structure. Voice coil actuators provide electromotive force that rotates the mirror. Supported by three kinematic mounts (Ti-6-4 blade flexures), the SMM is oriented in a 2-2-2 configuration (three tangential bipod flexures spaced 120 degrees apart about a circle) centered on the central ray of the aperture. Fixed sides of the flexures are supported by a housing providing a mechanical interface to the instrument optical housing. The housing also acts as a protective environment for handling, transport, and test set ups.

The DPSS is an eddy current differential position sensor system provided by Blue Line Engineering which provides angular position knowledge of the SMM. This subsystem is comprised of a pair of redundant sensor heads and redundant Differential Position Sensor Electronics (DPSE) boxes mounted externally to the MEB. The sensor heads provide inductive single axis sensing provided by pair of sensor heads on the SMM. The DPSE provides the interface between the MCE and sensor heads for sensor head power and the return analog position signal back to the MCE.

The redundant Field-Programmable Gate Array (FPGA) based MCE cards housed in the MEB provide the communication, control, telemetry, and power connectivity between the SMS and the rest of the L'Ralph instrument. Functionally, the MCE performs closed-loop control of the SMM with filtering and compensation parameters that can be uploaded to fine tune the system, necessary for achieving the highest performance. A Digital-to-Analog Converter (DAC) provides an analog signal to a linear power amplifier to drive the mirror actuators. The MCE is capable of commanding various scan rates from various start and stop scan positions including staring at specified position locations for instrument alignment. The MCE controls the SMM through a DAC and power amplifier/current driver. Scans are specified by three parameters: the scan rate, end-of-scan location, and a duration to dwell at end of scan before declaring the scan to be complete.

The SMS can scan at a maximum rate of 70 mrad/s, however that rate is only used for repositioning the mirror. The scan rates used for imaging typically range from 2 – 4000 μrad/sec with the rate controlled to better than 10 μrad/sec and the position known to better than 5 μrad. For MVIC the range of scan rates that correspond to the allowed integration times is 40 μrad/sec to 4000 μrad/sec. For LEISA the range of scan rates that correspond to the allowed integration times for 512 crosstrack pixels is 2 μrad/sec to 210 μrad/sec.

3.4 Inflight Calibration Sources: The Filament Calibrator, the Miniaturized Blackbody Calibrator, and the Solar Port

L'Ralph has three onboard calibration sources to provide a means to monitor the radiometric stability of the instrument, one is a solar port while the other two, a filament calibrator and a miniaturized blackbody calibrator, are electronically activated (see Fig. 4). These sources were not expected to provide absolute calibration. In flight, the absolute calibration of both MVIC and LEISA will be monitored by observing stellar sources. The solar port uses an input aperture, which is a tube mounted to the L'Ralph input baffle pointing about 43 degrees off axis from the primary aperture to obtain the calibration signal. When pointed at the sun, it allows solar radiation to strike a spectralon scatter plate that is placed in front of the scan mirror. The plate scatters light over a wide enough range of angles that it fills the entrance pupil (the System Aperture Stop (see Fig. 4)) and provides illumination over the entire FOV of both the MVIC and LEISA focal planes. Because it is solar illumination, it provides signal over the entire spectral range of both MVIC and LEISA when appropriate integration times are used. In ground testing, the solar calibrator was illuminated with a laser with similar power and angular spread as the sun at 5.2 AU to verify the signal obtained under those conditions was sufficient to provide adequate signal-to-noise ratio for calibration (SNR>50). The solar calibrator was not further calibrated during thermal vacuum testing because it was not being used for absolute calibration. The optical properties of spectralon can be affected by exposure to solar UV radiation, however, the scatter plate is well shielded from UV exposure, and only sees solar radiation during solar calibration. Also, since it is known that the changes in spectralon properties are associated with contamination (e.g. Leland and Arecchi 1995), significant effort was made to avoid contamination. The scatter plate is also shielded from cosmic radiation as well. Since use of the solar calibrator involves pointing the spacecraft, calibration operations that employ it will be done on an occasional basis. However, the filament and blackbody calibrators will be used to monitor the calibration on a much more frequent cadence. Thus, during encounters, the radiometric stability of the instrument will be determined by frequent filament and blackbody measurements, complimented by less frequent solar calibrations and verified by even less frequent stellar calibrations.

The filament calibration source is a set of four small tungsten lamps (Chicago miniature, CM715) contained within an enclosure that is also mounted to the input baffle which illuminates the same scatter plate as the solar port. The four bulbs are divided into two pairs of two bulbs each to provide a redundant system. The filaments operate at a temperature of $\sim$ 2500 K at their maximum power level, providing usable signal over the entire LEISA spectral range and to the panchromatic and three longer wavelength color bands of MVIC. The miniaturized blackbody calibrator consists of a set of 4 elements, each of which contains 4 small silicon rectangles that can be heated to about 800 K. This is warm enough to provide usable radiance to LEISA at wavelengths longer than about 2.5 μm. Because this is not warm enough to produce usable signal for MVIC, the sources are located after all optical elements, very near the cold shield for the LEISA focal plane. As with the filaments, the 4 sources are powered as two redundant pairs of two sources each, so there are 16 total miniaturized blackbodies, divided into two sets of 8. The blackbody and filament calibrators were radiometrically calibrated with NIST traceable sources at several power settings during thermal vacuum testing. The filament and blackbody calibrations sources are the same as used on OVIRS, though in a different location, and have shown better than 2% stability during the first two years of that mission (Simon et al. 2018) and continue to be stable at that level 4 years later. The stability of these electronic sources will be monitored using the solar calibrations and the stellar calibrations.

3.5 Electronics

The L'Ralph MEB consists of a chassis containing ten boards all connected to a backplane: one IR focal plane electronics (FPE) card containing two redundant sides (the FPR-IR card); one CCD FPE card containing two redundant sides (the FPE-vis card); two redundant command and data handling (C&DH) cards; two redundant low voltage power supply (LVPS) cards; two redundant multi-mission mass memory (M4) cards; and two redundant mechanism control electronics (MCE) cards. All boards are contained within the MEB mounted directly to the IPP below the TDA. Two differential position sensor electronics (DPSE) boxes are mounted to the outside of the chassis. The MEB does not have temperature control electronics but does have survival heaters and temperature sensors.

The L'Ralph FPE boards in the MEB provide clocking and biases to the CCD and the H2RG, read out the analog image data from the arrays, amplify the signals from the arrays, and perform the 16 bit A/D conversion of the data using System Image, Digitizing, Enhancing, Controlling, And Retrieving (SIDECAR) Applications Specific Integrated Circuits (ASICs) (see, e.g. Loose et al. 2005). The L'Ralph FPE-IR design and SIDECAR operating process are near copies of what was used for the H1RG array on OVIRS which has been operating flawlessly since launch. The difference between the OVIRS operation and the L'Ralph operations is that on OVIRS eight 512-pixel long, 64-pixel wide, data channels were read out while on L'Ralph 23 1024-pixel long, 64-pixel wide, data channels were read out. The SIDECAR can support 32 2048-pixel long, 64-pixel wide, channels. The L'Ralph FPE-vis design is also based on the OVIRS application. The FPEs were provided by Jackson & Tull. The C&DH board interprets the commands, performs the A/D conversion of the low-speed engineering data, and provides both the high-speed imaging data interface and the low-speed housekeeping (HK) data interface. It also performs the programmable pixel summing and correlated double sampling functions. The LVPS converts the 28 V spacecraft power to the various voltages required by L'Ralph, including those supplied to the settable calibration sources. The MCE controls the scan rate of the mirror and senses its position. Each of the redundant M4 cards contains 256 Gbits of nonvolatile memory in total. The MCE was provided by GSFC while the rest of the MEB boards were provided by SwRI. SwRI also provided the MEB enclosure. Flight Software (FSW) controlling the C&DH was provided by Big Head Endian (BHE) LLC. The FSW that controls the SIDECARs in the FPEs was provided by Markury Scientific.

In a long duration mission such as Lucy, reliability of the electronics is of paramount importance, particularly for a core instrument that addresses many Level 1 objectives. As is standard practice, the dosage levels of various radiation elements (e.g. Total Ionizing Dose (TID) in Silicon, Solar Proton Fluence etc.) are modeled over the life of the mission by the project and are used to generate requirements for the mission subsystems, including instruments (for example, the TID is 14.9 (krad-Si) at the 95% confidence level for 100 mils of spherical aluminum shielding). This includes possible Destructive Single-Event Effects. For Lucy, these requirements are contained within the Lucy Environmental Requirements Document (ERD). Electronic components and shielding are then defined in terms of meeting those requirements with margin. Some Single Event Effects will require power cycling, which is included in the mission design. Parity bits are included in the data transmission to the and spacecraft which enables a level of Error Detection and Correction (EDAC) in the image data. During encounters, data is transmitted to the spacecraft memory as soon as possible, typically within hours to a few days. It remains stored on L'Ralph as well. Data transmission to the ground depends on data rate and DSN availability and can take up to several months.

In addition, to ensure that L'Ralph is robust, almost all of the electronics are redundant. As illustrated in Fig. 6, L'Ralph can operate on two separate sides (side 1 or 2), the only common elements are the detector arrays themselves. For MVIC, the potential single point failure mode of the array is somewhat mitigated by having readouts on both sides of the serial output row in each CCD, so that if one 2500 pixel-wide side fails, the other might still be available. Although LEISA's use of 23 outputs corresponding to the twenty-three 64 row wide segments of the array is done to allow an acceptable frame rate, it also provides some redundancy. If any output should fail (or more, depending on the distribution), reasonably complete spectra may still be acquired. However, the likelihood that either MVIC or LEISA will have detector failures is extremely low.

MVIC always produces image data in CDS mode. LEISA "raw" mode is used to monitor the LEISA A/D converter offset level for each segment of the array. The same offset is used in both CDS and "raw" modes. The selectable offset compensates for voltage drifts in the analog signal train over the life of the mission, maintaining dynamic range without sacrificing signal resolution (increasing quantization noise).

For both MVIC and LEISA, the dominant noise source at low light levels is the system electronics noise, including array read noise. For MVIC this is about 20 e- ($\sim$ 200 μV) and for LEISA it is about 25 e- ($\sim$ 100 μV). The overall average gain for MVIC is about 28 e-/DN (Digital Number, or least significant bit), while for LEISA it is approximately 20 e-/DN.

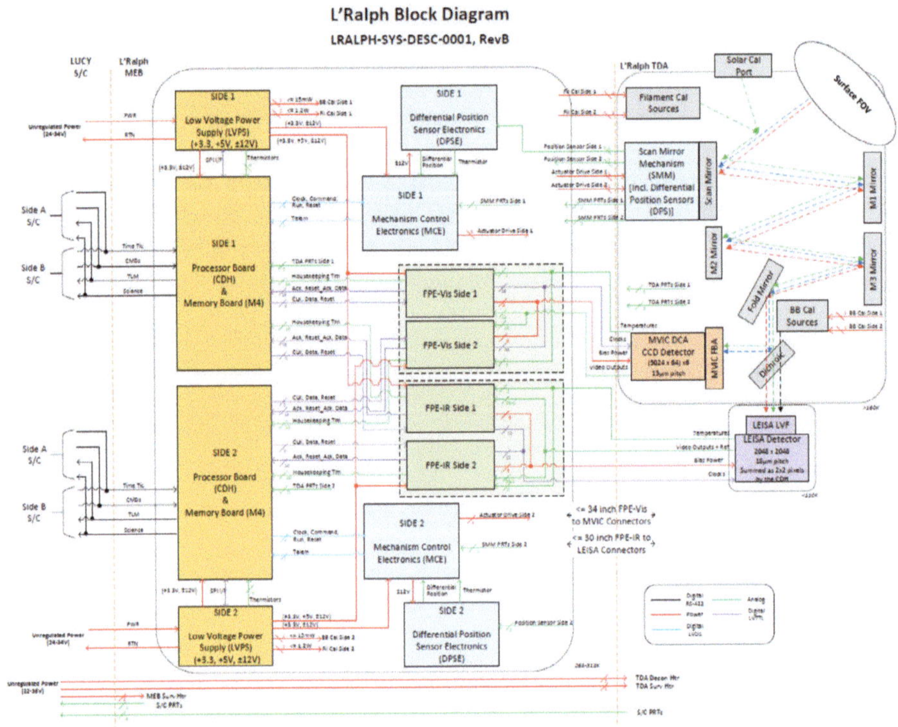

Fig. 6 Schematic of the L'Ralph electronics in the MEB and optical assembly in the TDA. Note that the system is fully redundant except for the MVIC and LEISA detectors, and the optical elements in the TDA. Both the MEB Side 1 and Side 2 electronics can be run off either the spacecraft Side A or Side B power interfaces

3.6 Pre-Launch Instrument Characterization

An extensive pre-launch program of performance verification measurements was carried out for L'Ralph at both the component level and the full instrument level. The component level characterization included measurements of the wavelength dependent quantum efficiency for the MVIC and LEISA array/filter assemblies and measurements of the wavelength dependence of the other optical elements (i.e., reflectance of the mirrors, transmission of the filters, and throughput of the beamsplitter). It also included measurements of the SMS scan rates and pointing accuracy as a function of the scan mirror sensor outputs, which allowed the development of command algorithms for accurate control of image motion scans. The primary performance characteristics determined in the instrument level tests were absolute radiometric calibration of the full system, spectral resolution, spatial resolution, signal to noise level, saturation level, calibration of the on-board calibration systems, optical pointing, and scattering. The stability of all these parameters was determined for both instrument sides, and under the thermal variations (LEISA detector temperatures of 90–115 K, MVIC detector temperatures of 155–200 K, TDA temperatures of 153–195 K) and voltage ranges (26–34 V) expected to be seen in flight. The directional characteristics of the solar calibration input were also measured, but only under ambient conditions.

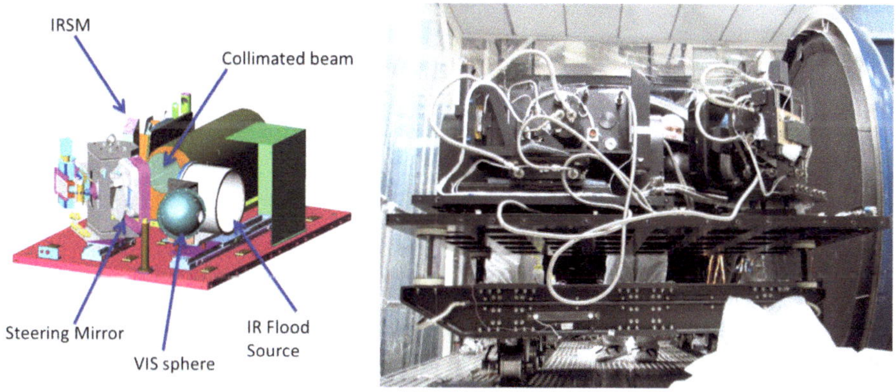

Fig. 7 (Left) Schematic of the in-chamber calibration system used to characterize L'Ralph performance. (Right) The calibration system being readied to enter the chamber

The instrument-level tests employed a system originally developed for calibrating the Thermal Infrared Sensor (TIRS) instrument on Landsat 8 and which has since been used for both the OSIRIS-REx OVIRS instrument and TIRS-2 on Landsat 9. This system consists of an infrared source module (IRSM) which illuminates collimating optics through a variety of apertures to simulate targets far removed from the output aperture. The output from the collimation source goes to a steering mirror that projects the beam onto the input of the instrument at a wide range of angles. This allows accurate location of the instrument's boresight. It also allows the off-axis scattering characteristics to be investigated. In addition to the IRSM, there are two NIST-traceable calibration sources in the calibration system, a visible integrating sphere providing visible and near IR radiation and a flood source providing light from the short-wave IR (SWIR) to the Far-IR. The entire calibration system, including the IRSM, the collimation system, the mirrors, the integrating sphere, and the flood source, goes into the chamber with the instrument. During testing, the components are cooled to near LN_2 temperatures so that unwanted thermal signal is minimized. In the chamber, the system may be configured such that light from an outside source can take the place of the IRSM and illuminate the collimating optics. This configuration is used, for example, to introduce the output of a grating monochromator into the system for spectral characterization. Figure 7 shows a diagram of the calibration system and a picture of it as it is about to enter the vacuum chamber.

3.7 Spectral Characterization

LEISA: The component-level spectral characterization of the LEISA focal plane assembly was performed at the Detector Characterization Laboratory at GSFC. This involved both determining the central wavelength of the pixels and measuring the spectral response function. Measurements were done in a temperature controlled dewar by illuminating the focal plane assembly with light from the output slit of a grating. For any wavelength, the halfwidth of the grating signal was always a factor of 4 (or more) less than the expected LEISA halfwidth. The grating output was collimated so that the entire unshielded detector array/filter assembly was lit, and multiple grating orders could be measured at each grating position. The central wavelength of each pixel was determined by stepping the lowest grating order about 10 pixels per step and fitting the cross-track weighted average maximum to a quadratic function at

Fig. 8 Image showing the dependence of pixel central wavelength as a function of position on the LEISA focal plane. All 3 filter segments are shown, and the dark regions between them are the inter-filter bonds made with absorptive bonding material. The quadratic dependence in the cross-track direction is as expected from the filter deposition process

each step, and the along-track wavelength change to a linear function in each filter segment over all steps. The spectral response was determined every 40 to 60 rows in the along-track direction by stepping the grating such that about 8 samples were obtained over the FWHM, and 20 samples were obtained overall, As shown in Fig. 8, the central wavelengths cover more than the required 1 – 3.8 μm range over the 23 detector outputs that are being used for LEISA. The FWHM obtained at the along-track sampled regions are given in Table 3 and meet the ≤ 10 nm requirement for 1 – 3.4 μm. The spectral results were checked at several locations during instrument level TVac testing. Figure 9 shows an example of a spectral response curve measured during TVac at about 1.8 μm. As expected from the filter models developed by VIAVI, the spectral response near the central wavelength drops at close to a Gaussian rate as a function of wavelength displacement from the peak. These high spectral resolution measurements are not appropriate to determine the full out-of-band (OOB) performance because the signal generated by a narrow spectral band source far from the central wavelength would not generate sufficient signal. To determine the full out-of-band spectral performance, the set of broader filters used to measure the QE of the unfiltered detector were used to measure the spectral response of the filtered detector at wavelengths more than 10 HWHM (i.e. < 50 nm) from the central wavelength. The summed OOB from the measurements was always < 3%.

MVIC: As with LEISA, the initial spectral characterization of the MVIC components was performed by the component providers; STA Inc measured the spectral response of

Table 3 LEISA spectral resolution (FWHM) as a function of wavelength for a row near the center of the array. Note that, as shown in Fig. 8, there is some wavelength curvature in the columns perpendicular to this row, the resolution does not vary significantly

Center Wavelength (nm)	Pixel	Bandwidth FWHM (nm)
1008.8	1692	6.3
1111.3	1625	6.3
1213.5	1560	6.3
1314.4	1496	6.3
1415.9	1432	6.5
1516.9	1368	6.7
1616.5	1305	7.1
1612.3	1241	5.8
1710.1	1202	6.2
1812.5	1161	6.6
1910.9	1122	6.9
2014.3	1081	7.3
2113.1	1042	7.7
2211.7	1003	8.0
2310.8	964	8.5
2414.9	923	9.0
2513.9	884	9.1
2611.3	845	9.2
2611.3	799	8.8
2714.1	757	9.4
2814.1	716	9.7
2914.6	675	9.7
3012.3	635	9.7
3112.5	594	9.9
3210.8	553	10.0
3313.9	512	9.9
3414.7	471	9.8

the CCDs every 10 nm over the 300 to 1000 nm range and VIAVI Solutions measured the filer response every 0.5 nm over the same range. These were used to model the expected performance of the focal plane. The modeled responses were verified at the focal plane assembly level at SwRI and at the instrument level during TVac testing at Goddard. Figure 10 shows the instrument level spectral response curves for the MVIC bands.

3.8 Image Quality and Enclosed Energy

As noted in Table 1, the image quality requirements for both LEISA and MVIC are defined in terms of enclosed energy as a function of distance from the center of a pixel. This is a measure of how much signal is contributed by an area outside the IFOV of the pixel. For example, in the MVIC requirements shown in Table 1, a 91% enclosed energy at 168 μrad means that 9% of the pixel's signal comes from the region more than that 84 μrad from

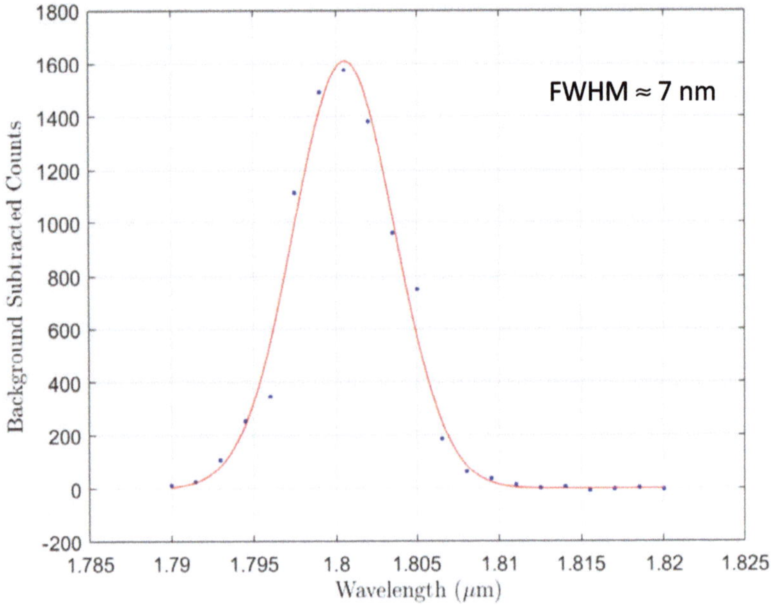

Fig. 9 Plot of the spectral band shape for a pixel near the cross-track center of the LEISA array. These data were obtained during Tvac testing at the instrumnet level and agree well with the results shown in Table 3. This is only displaying data obtained for measuring the FWHM

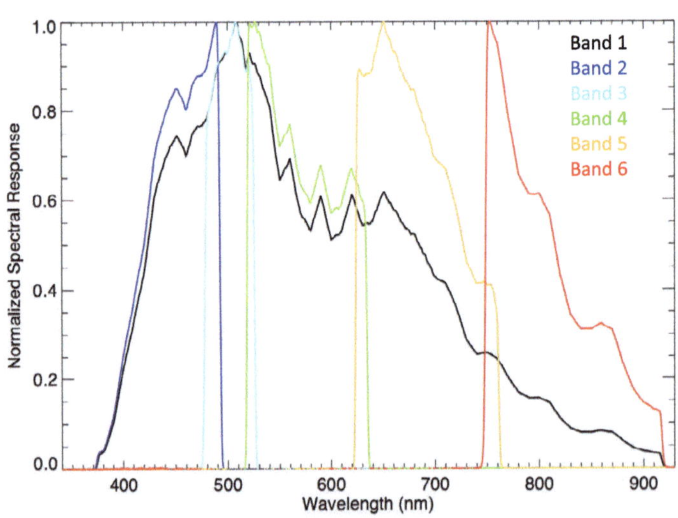

Fig. 10 Image showing the normalized spectral response for each of the MVIC channels. These were originally modeld as the product of measured spectrally dependent CCD QE, filter transmittance, mirror reflectances and dichroic filter transmittaance. The model was verified at the focal plane assembly level at SwRI and at the instrument level during TVac testing

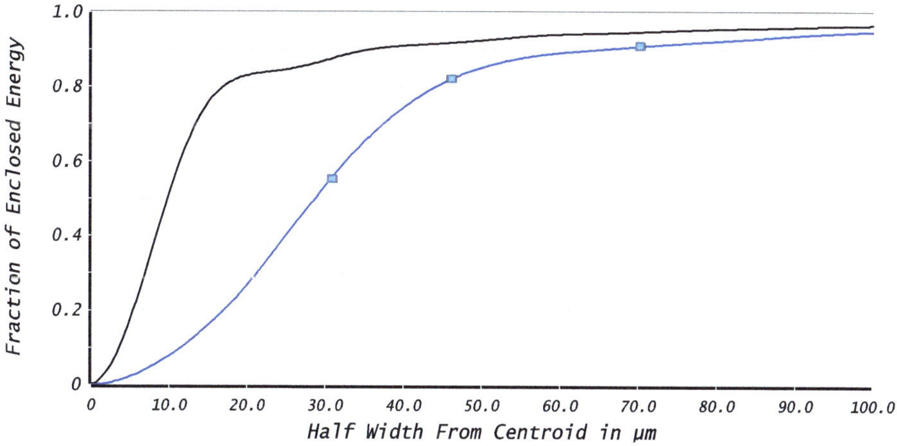

Fig. 11 (Top line) Graph of the enclosed energy fraction for a diffraction limited model of the L'Ralph optical system on the LEISA focal plane at a wavelength of 3.4 μm. The X-axis is in terms of distance from the center of a pixel in the center of the 3.4 μm spectral column. It may be converted to angular distance by multiplying by (80 μrad/36 μm). (Bottom line) Graph of the model degraded by introducing a wavefront error (WFE) term so that the performance as a function of distance meets the requirements as shown by the blue squares. The WFE was 840 nm RMS. For MVIC, the WFE error used to model the requirements was 245 nm

the center of the 28.8 μrad × 28.8 μrad MVIC IFOV. The enclosed energy measurements were done at the instrument level using collimated targets of various angular sizes which illuminated the entrance aperture and formed images on the focal plane. The change in the measured signal at the central pixel as a function of target size was compared to a model of the optical system that degraded the diffraction limited enclosed energy performance to the level of the required enclosed energy performance by including a wavefront error component in the model. In other words, if the instrument performance behaved as the model does, it would meet the requirements. The model process is illustrated in Fig. 11, which shows a comparison of the diffraction limited performance to the somewhat degraded required performance for pixels at the long-wavelength end of the LEISA spectral range. Using the model means that ratios of the observations made at smaller angular sizes to the observation at the maximum angular size may be used to verify requirements, and the maximum angular size does not have to have an enclosed energy of 100%. Note that the modeling also allows us to interpolate the required performance to measured target sizes that are not specified in the requirements. The actual measured performance is being compared to the requirements. This methodology was approved by several independent review teams.

LEISA: The enclosed energy measurements for the LEISA focal plane were made during TVac using the IRSM using targets of 50, 70, 141, 188, 282, and 800 μrad. For targets larger than 800, the intensity variation over the exposed area is affected by the spectral change over the more than 20 pixels, as well as the change arising from the optical system performance. In other words, for targets larger than 800 μrad, the shape of the image is affected by the spectral distribution of the source illuminating the target and thus is not uniform. Therefore, the largest target used in verifying the spatial response by the ratios of the central pixel intensity was 800 μrad. Note that the requirements for the LEISA enclosed energy performance are specified for the observations that are done to meet Level 1 requirements, which are made with the 2 × 2 superpixels, where the IFOV is 80 μrad.

Table 4 Ratio of enclosed energy as a function of target size to the 800 μrad enclosed energy for the measured and modeled LEISA focal plane

Size (μrad)	Measured Fraction	Modelled Fraction	Ratio
50	0.107	0.051	2.1
70	0.289	0.099	2.9
141	0.618	0.399	1.5
188	0.867	0.596	1.5
282	0.957	0.788	1.2

Table 5 Ratio of enclosed energy as a function of target size to the 435 μrad enclosed energy for the measured and modeled MVIC focal plane. The modeled performance is adjusted to the MVIC required level by including a 245 nm RMS WFE

Size (μrad)	Measured Fraction	Modelled Fraction	Ratio
27	0.176	0.153	1.2
43	0.324	0.328	1.0
65	0.650	0.548	1.2
218	0.974	0.922	1.1

The results for the ratio of the enclosed energy as a function of distance to the enclosed energy at 800 μrad is shown in Table 4. As may be seen, LEISA in 2 × 2 mode easily meets its enclosed energy requirements, as the driver for the requirement is the MVIC focal plane.

MVIC: The enclosed energy measurements for the 28.8 μrad IFOV MVIC focal plane were made using targets of 27, 43, 65, 218, and 435 μrad. Since MVIC is the driver for the enclosed energy performance, it was decided that a source with a larger focal length than the 1-meter focal length of the TVac calibration system would be used to ensure the light was fully collimated, hence the measurements were made under ambient conditions using a source with a 2.8-meter focal length. The measurements were made in the panchromatic channel because the highest spatial resolution measurements will be made with that channel. Since MVIC only operates in TDI mode it was necessary to use the scan mirror to scan the target image along the focal plane at the same rate as the charge transfer occurred during the CCD readout. Hence the measured enclosed energy includes the contribution of errors in the scan rate and is very representative of what will be seen in actual usage. The results obtained at ambient were verified in Tvac by carrying out focus scans which gave the same shape as was seen at ambient.

The results for the ratio of the enclosed energy as a function of distance to the enclosed energy at 435 μrad is shown in Table 5. As may be seen, MVIC meets or exceeds its enclosed energy requirements at the measured target sizes. These measurements were made at a TDA temperature of ∼175, near the midpoint of the expected TDA temperature range. Focal point measurements made at the upper and lower TDA temperature limits indicated that the focal point moved by ∼ 30 microns in the focus direction over the full field of view, which, for an f/6 beam corresponds to a broadening of ∼ 5 microns, which, when convolved with the 25 micron diameter of the FWHM, produces a negligible change of enclosed energy with temperature over the full field of view. Note that, since the diameter of the LEISA 2 × 2 superpixel FWHM is about 60 microns, the effect on LEISA is even smaller.

3.9 Radiometric Calibration and Signal-to-Noise Ratio (SNR)

As noted above, the pre-launch absolute radiometric calibration of the MVIC and LEISA components of L'Ralph was carried out using the visible-NIR integrating sphere and the IR flood source that are contained within the TVac calibration system. Both sources are NIST calibrated to better than 2% accuracy. Various settings of the temperature of the IR flood source and the aperture of the integrating sphere are used to get a range of radiances that are representative of the signal that will be seen from the Trojans over the range of solar distances, phase angles, and inherent reflectance values of the objects. These data were used to determine calibration coefficients for the detector elements. The calibration coefficients were found to be stable over the voltage and temperature ranges defined above. The resulting SNRs for LEISA and MVIC for the driving case of Polymele are described below.

LEISA: For all the Trojans the assumed reflectance range is 0.04 to 0.4 for all LEISA wavelengths. Under the assumption that the reflectance changes by 0.04 magnitudes/degree as a function of phase angle, this corresponds to an I/F range of 0.0019 to 0.019 for the Level 1 observations at Polymele which has the highest phase angle (82°) and, at 5.7 AU, is also the Trojan furthest from the sun that Lucy will visit. To obtain radiances consistent with these parameters the integrating sphere with varying apertures was used for wavelengths < 2.2 μm and the flood source at several temperatures (345 K, 315 K, 275 K, and 225 K) was used for wavelengths > 2.2 μm. Encounter design has limited the time available to do this observation to 800 sec, so for the 1 to 3.4 μm spectral range, this allows an integration time of 1.2 seconds to complete a scan of the 500 superpixels across the target. Figure 12 shows the SNR observed for the I/F of 0.0019. Note that this shows both superpixeled SNRs and non superpixeled SNRs. Since this is for the required 70 nm Full Width tenth Maximum (FWTM) spectral resolution, pixels are summed in the spectral range. The spectral sampling is always 5 nm or less, so at least 14 pixels are summed in the spectral direction. As can be seen, LEISA easily meets the required SNR over the 1 to 3.4 μm spectral range even without superpixeling. With superpixeling, it meets the requirements over essentially the full spectral range covered by LEISA. Under these conditions there is no saturation for an I/F of 0.019. Because scans are done using the SMM and it takes less than 22 seconds to transition from 1 scan to the next, scans covering the full wavelength range will typically be done in two segments. The first will cover from 0.95 to about 2.6 microns and the second will cover from 2.6 to 3.9 microns. Since the intensity is greater at the shorter wavelengths, the second segment can be done at a rate about 3 times slower than the first (while still maintaining the overall time of the scan). This will improve the quality of the observation.

For Level 2 observations near close approach at near 0 phase angle LEISA will saturate at wavelengths < 2.7 micron for an I/F of 0.4 at the shortest integration time for reading out 512 cross-track pixels (0.38 seconds). There are two simple ways around this, though: 1) if only 122 cross-track pixels are read out, the integration time can be reduced to 0.1 sec and no wavelengths saturate or 2) only data at wavelengths > 2.7 microns is taken. Which solution is taken will depend on in-flight data and the science goal of the observation.

MVIC: For all the Trojans the assumed reflectance range is 0.03 to 0.4 for all MVIC channels. Under the assumptions above this corresponds to an I/F range of 0.0014 to 0.019. To obtain radiances consistent with these parameters the integrating sphere with varying apertures was used for all MVIC channels, since the flood source cannot be set to a high enough temperature to produce sufficient radiance in any channel. The encounter requirements are that one must be able to obtain an SNR > 25 when scanning a 130 mrad target in the pan band in < 45 seconds (a scan rate of 2900 μrad/sec or greater) and an SNR > 25 in all the color channels when scanning a 130 mrad target in < 180 seconds (a scan

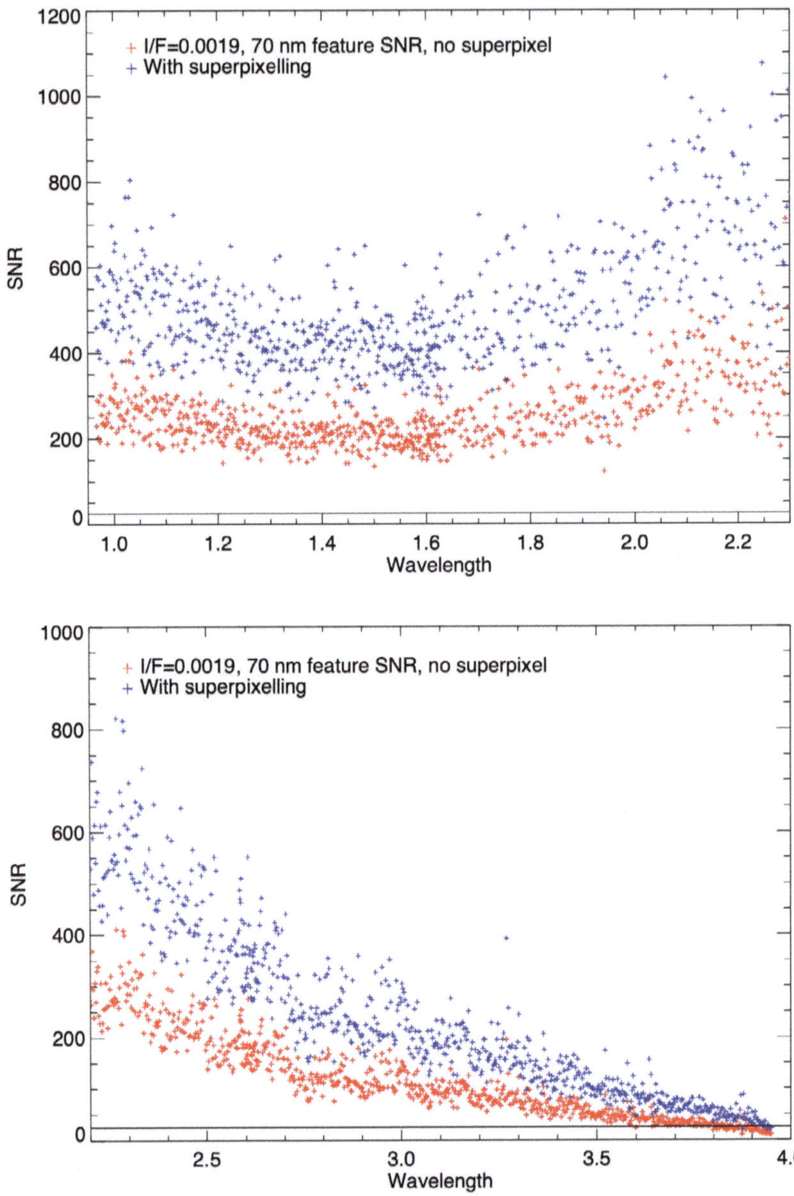

Fig. 12 SNR ratio for an I/F of 0.0019 at a heliocentric distance of 5.7 AU for a FWTM of 0.07 nm and a 1.2 second integration time. The requirement is that it must be above 25 for all wavelengths from 1 to 3.4 μm

rate of 800 μrad/sec or greater when taking the FOV between the MVIC channels into account). Table 6 shows the average SNR observed for an I/F of 0.0014 for a pan scan rate of 2900 μrad/sec using 32 TDI rows and a color channel scan rate of 800 μrad/sec also with 32 TDI rows in each channel. As may be seen from the table, the SNR requirement is easily met for all channels. In fact, the same results would be obtained for a 1600 μrad/sec color scan using 64 TDI rows. The actual scan operations remain to be determined from in-flight

Table 6 SNR for MVIC channels at 5.7 AU for I/F = 0.0014. The scan rate for pan is 2900 µrad/sec. The scan rate for the color channels is 800 µrad/sec. All cases use 32 TDI rows

BAND	AVERAGE SNR
pan	78.6
Violet	80.6
Green	60.2
Orange	133.4
Phyllosilicate	86.4
Near IR	32.9

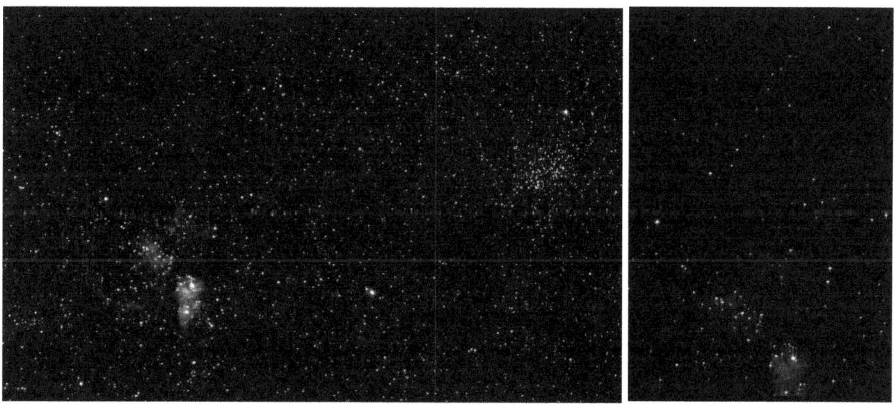

Fig. 13 Left; MVIC pan image showing the Wishing Well star cluster and a nebula. Right; a colorized version of the nebula obtained using data from the MVIC Green, Orange, and Phyllosilicate color channels during the same scan

observations. Note that, for the color scans shown in Table 6, there would be saturation in channel 4 for I/F = 0.019. This can easily be eliminated while still easily meeting the SNR > 25 requirement by reducing the number of TDI rows for this channel to 16.

For Level 2 observations near closest approach the color scans can be done using the fastest scan rate of 4000 µrad/sec and 4-8 TDI rows (depending on the band) to avoid saturation for I/F = 0.4 (even at 5.2 AU), however, the pan band would saturate for I/F > 0.15, even at TDI 4. There is no strong driver for obtaining pan scans at closest approach.

3.10 In-Flight Calibration Data

The Lucy Mission launched on October 16, 2021. Since its launch, L'Ralph has obtained data during a functional test done at approximately Launch + 20 days (L+20 days), calibration measurements done at approximately L+4 months and L+7 months, and a series of observations of the Earth and the Moon done during the Earth Gravity Assist (EGA-1) approximately 12 months after launch. The L+20 days L'Ralph test only verified that the operational parameters (voltages, currents, timing, etc.) were as expected. No images were obtained. The L+4 and L+7 month data collections included MVIC scans of M6 (the Butterfly Cluster, MGC 6405), M7 (NGC 6475) and the Wishing Well (NGC 3532). Figure 13 shows an example of a data obtained from the Wishing Well. The M6/M7 and Wishing Well scans were done at five different cross track positions to determine if there was a dependence of the radiometry on the placement of the image. The measurements did not indicate

Table 7 FWHMs in terms of number of pixels derived from stellar scans for all MVIC channels. The top row shows the results of fitting the two-dimensional PSF to a Gaussian function with a single variable (σ). The bottom row shows the results of fitting the same PSFs to a Gaussian function with two variables (σx and σy). The FWHM given in the bottom row is the geometric average $(\sigma x * \sigma y)^{1/2}$. To obtain the FWHM in terms of µrad, the numbers given above are multiplied by the IFOV (28.8 µrad)

	Pan	Violet	Green	Orange	Phyllo	NIR
# stars analyzed	**325**	**171**	**106**	**257**	**305**	**44**
FWHM: single variable	1.70 +/- 0.21	1.66 +/- 0.30	1.62 +/- 0.29	1.60 +/- 0.31	1.64 +/- 0.38	1.64 +/- 0.38
FWHM: two independent variables	1.66 +/- 0.24	1.62 +/- 0.29	1.59 +/- 0.27	1.56 +/- 0.29	1.61 +/- 0.35	1.61 +/- 0.35

there was any significant variation with the cross-track position and that the radiances obtained from the stars were consistent with pre-launch calibration. An analysis of the Point Spread Function (PSF) of the stellar observations also showed consistency of the FWHM measured in flight with the pre-launch value. Table 7 shows the results of fitting the measured 2-dimensional PSF to both a Gaussian with a single variable (σ) and a Gaussian with two variables (σx and σy) for numerous stars in the 6 MVIC channels. As may be seen from the table the FWHM obtained from the single variable fittings agrees well with the geometric average $(\sigma x^* \sigma y)^{1/2}$ of the 2 variable fittings. That, and the fact that in the 2-D fits $\sigma x \sim= \sigma y$ indicates that the PSF is primarily driven by defocus at off-nominal temperatures and not by higher order asymmetric effects. (The nominal temperature ranges given in Sect. 3.6 are: LEISA detector temperatures of 90–115 K, MVIC detector temperatures of 155–200 K, TDA temperatures of 153–195 K. At times, during some in-flight observations, the TDA was at 215 K and/or MVIC was at 212 K.) The inflight results of a FWHM of ~ 1.7 pixels (~50 µrad) agree well with the pre-launch MVIC enclosed energy results discussed above and presented in Table 5 and are consistent with the requirements. MVIC measurements made during L+7 also verified the pointing of the solar calibrator and provided the baseline intensity levels for comparison with future observations. MVIC measurements of the on-board filament calibration source intensities were consistent with pre-launch values. We note that, as expected for heliocentric distances less than about 3 AU, the temperature of the LEISA detector was too high to allow any stellar observations to be made without dark current saturation. LEISA solar and on-board calibration source measurements were also not possible for the same reason.

During EGA-1, MVIC scans of the Earth were done at 5 different cross-track positions. As expected, these observations were mostly saturated and did not provide any radiometric or extended source optical performance information. However, they were extremely useful in providing measurements of the scattering characteristics of L'Ralph. They have verified that, for the angular range of the observations to be made at the Trojans, scattering is significantly less than 1% and therefore is not an issue for the reflectance range specified in the requirements. MVIC color scans of the Moon did provide unsaturated images over most of the surface in the Violet, Green and NIR channels. The Phyllosilicate and Orange images are not saturated in the area near the terminator. Since the lunar scans were repeated at 5 cross-track positions, averages in the unsaturated regions allowed further verification that any dependence of signal level on cross-track location is accounted for in the pre-launch calibration. Figure 14 shows the calibrated radiance obtained in each of the 5 scans for the

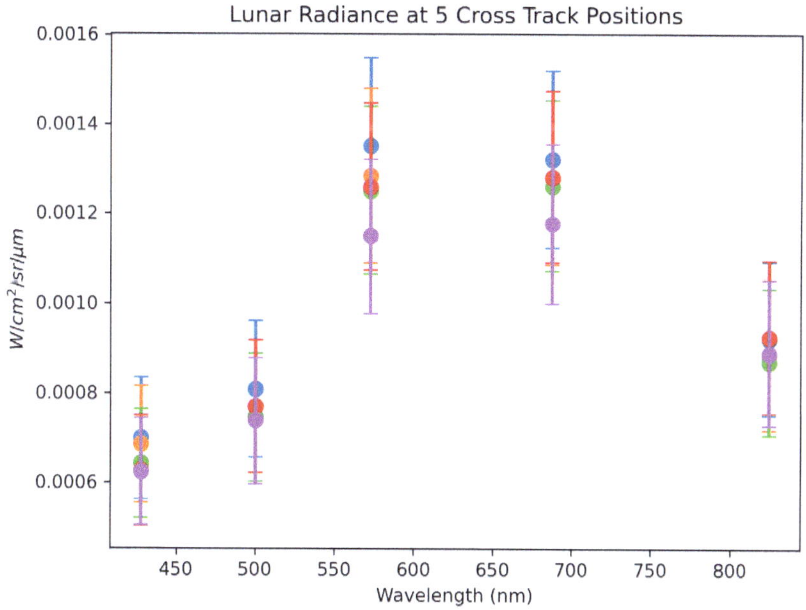

Fig. 14 Plot of the mean radiances obtained in an area near the terminator for each of the 5 MVIC color channels in each of the 5 lunar scans. The error bars represent the standard deviation over the area

5 MVIC color channels at an area near the terminator where none of the channels are saturated. As may be seen from the figure, in each channel the data at all 5 scan positions fall within the standard deviations, and there is no apparent dependence on what part of the array is used to obtain the data. The MVIC pan channel was still saturated in this region. Figure 15 shows a comparison between MVIC I/F data (based on pre-launch calibration) obtained in the Lunar Mare and Highlands regions, where none of the color channels were saturated, with the Moon Mineralogy Mapper (M3) reflectance data in those areas. The MVIC data is corrected to the incidence angle (i) = 30°, emissivity (e) = 0°, and phase (α) = 30° standard used by M3 (Green et al. 2011; Besse et al. 2013). As may be seen, the MVIC and M3 data agree well in these areas (the M3 data does not extend into the wavelengths of the MVIC Violet and Green channels). Figure 15 also shows a comparison between the MVIC data and the ROLO standard modeled data (Kiefer and Stone 2005) over the entire lunar surface (note that saturated areas are excluded from the MVIC data, which primarily affects the Phyllosilicate and Orange channels). The ROLO model does cover all MVIC channels. Once again, there is good agreement between the MVIC data and the standard.

Finally, during EGA-1, some unsaturated LEISA spectral observations of the Earth were made. This was made possible by three non-standard conditions. First, for several hours before the scans were done, the IPP could be pointed in a direction that minimized heating of the TDA and the focal plane due to radiation from the solar cells. While the scans could not be carried out in this orientation, this lowered the detector temperature to about 150 K before the scans were done. This is still about 40 K higher than the normal operating temperature at the Trojans, where heating from the solar cells will not be problematic. Second, the cross-track readout size of the LEISA detector was reduced from 512 pixels to 20 pixels, which allowed the minimum integration time to be reduced from 0.38 seconds to about 0.026 seconds. Of course, this significantly reduced the image size, but it meant that,

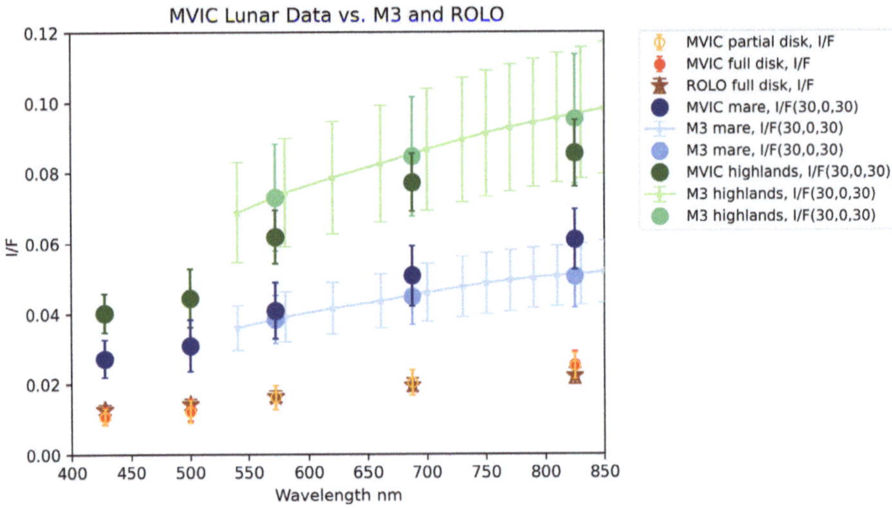

Fig. 15 An I/F comparison between MVIC data taken on the moon with appropriately spectrally sampled 1) ROLO global model data and 2) Moon Minerology Mapper (M3) reflectance data over the lunar Mare and lunar Highland areas where MVIC was not saturated in any channel. For the latter cases the MVIC data were corrected to the incidence angle (i) = 30°, emissivity (e) = 0°, and phase (α) = 30° standard used by M3. The ranges given in the comparison between the M3 and MVIC data represent the range of I/F observed in the Mare and Highland areas. For the Rolo comparisons the MVIC data are not corrected for (i.e., α) and the MVIC uncertainties are the standard deviations over the area observed. The ROLO uncertainties are those produced by the model. Note that the M3 data do not extend to the wavelengths of the two shortest wavelength MVIC channels, but the ROLO model does

even though the detector dark current was several orders of magnitude higher than it will be at the Trojans, data not saturated by dark current could be obtained. Finally, the much higher signal produced by the Earth than will be produced by the Trojans was sufficient to produce a measurable signal in the much shorter integration time. Figure 16 shows an example of a spectrum obtained in a region near Australia. As may be seen in this figure, the LEISA spectrum does show the expected structure for Earth. Although, because of the non-standard condition under which these data were obtained, they cannot be used to verify radiance levels etc., they do indicate that the LEISA hyperspectral imager is working.

3.11 Conclusion

This paper describes the design, operation, and performance of L'Ralph, a highly capable, very flexible, Vis/NIR multispectral imager/SWIR hyperspectral imager flying on the Lucy mission to the Trojan asteroids in Jupiter's orbit. L'Ralph consists of a telescope feeding two focal planes, the visible/NIR MVIC imager and the LEISA. MVIC will provide very sensitive, high fidelity, panchromatic and multi-spectral maps of the Trojans at a spatial resolution of 1.5 km/FWHM or higher over most of the surface and 0.6 km/FWHM over a good fraction of it. LEISA will provide full hemispheric, SWIR, spectral maps with spatial resolutions of 1-5 km/FWHM or better. These will be used to accurately characterize the surface composition, and variations associated with fresh surface regions in particular. At near closest approach, MVIC can obtain images with spatial resolutions on the order of fifty to one hundred meters, while LEISA can measure spectra at the 100 to 200 m spatial scale. L'Ralph has been extensively tested at the component and full instrument level and its in-flight operation has verified that MVIC meets all its performance requirements with margin.

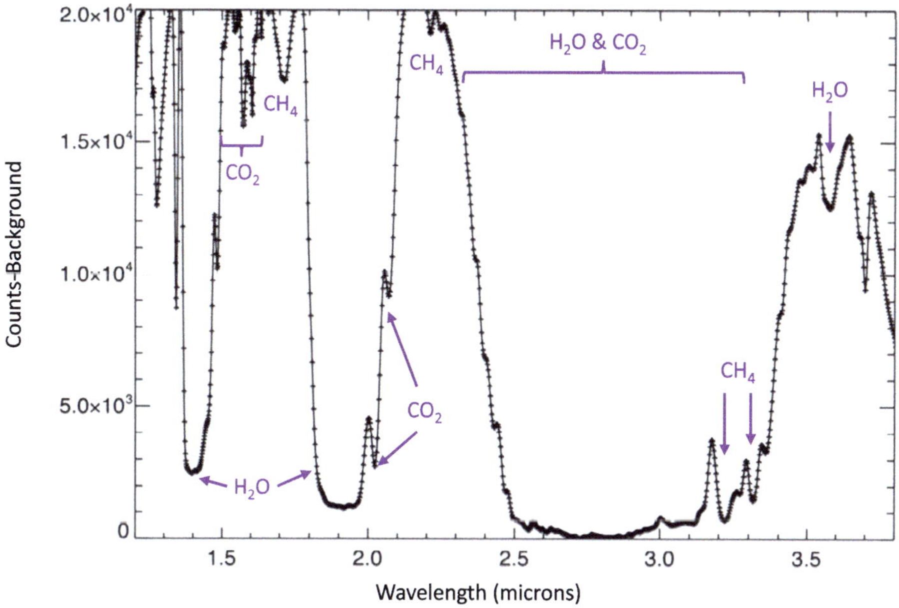

Fig. 16 Spectrum of the Earth obtained by LEISA near the northeast coast of Australia. Expected spectral features are evident

(LEISA requirements have not been verified in-flight because, as expected, for heliocentric distances of 3 AU or less its focal plane has not cooled sufficiently to allow operation at the conditions that will exist at the Trojans. However, observations of Earth taken at warmer-than-optimal operating temperatures indicate LEISA is operating nominally.) L'Ralph will provide a wealth of information on the geology, composition, morphology and thermal characteristics of the Trojan objects. The data it produces during its flybys will shed new light on the formation and evolution of our solar system.

Acknowledgements The authors would like to thank the entire L'Ralph support teams at GSFC and SwRI (San Antonio) for their untiring efforts in making L'Ralph a reality. The contributions of VIAVI Solutions, STA Inc, TIS, NuTek, Coherent/Tinsly, Jackson and Tull, Newton Engineering and Product Development, Big Head Endian LLC, Markury Scientific and Blue Line Engineering are also gratefully acknowledged. The authors would also like to thank the PI office at the Southwest Research Institute, the Project Office at GSFC, the Program Office at MSFC and NASA HQ for their support.

Declarations

Competing Interests The authors declare that they have no competing interests.

References

Beletic JW et al (2008) Teledyne Imaging Sensors: infrared imaging technologies for astronomy and civil space. In: Proc. SPIE 7021, High energy, optical, and infrared detectors for astronomy III, vol 70210H

Besse B, Sunshine J, Staid M, Boardman J, Pieters C, Guasqui P, Malaret E, Mclaughlin S (2013) A visible and near-infrared photometric correction for Moon Mineralogy Mapper (M3). Icarus 222:229–242

Brown ME (2016) The 3-4 μm spectra of Jupiter Trojan asteroids. Astron J 152:159

Buie M, Keeny B, Levison H, Olkin C, Lucy Occultations Team (2022) Shape and duplicity of Lucy Mission target Polymele from occultation observations. AAS DPS 54, London, Ontario CA

Campins H, Hargrove K, Pinilla-Alonso N, Howell ES, Kelley MS, Licandro J, Mothe-Diniz T, Fernandez Y, Ziffer J (2010) Water ice and organics on the surface of the asteroid 24 Themis. Nature 464:1320–1321

Cruikshank DP et al (2008) Hydrocarbons on Saturn's satellites iapetus and phoebe. Icarus 193:334–343

Dalle Ore CM, Cruikshank DP, Clark RN (2012) Infrared spectroscopic characterization of the low-albedo materials on Iapetus. Icarus 221:735–743

Green RO, Pieters C, Mouroulis P, Eastwood M, Boardman J, Glavich T, Isaacson P, Annadurai M, Besse S, Barr D, Buratti B, Cate D, Chatterjee A, Clark R, Cheek L, Combe J, Dhingra D, Essandoh V, Geier S, Goswami JN, Green R, Haemmerle V, Head J, Hovland L, Hyman S, Klima R, Koch T, Kramer G, Kumar ASK, Lee K, Lundeen S, Malaret E, McCord T, McLaughlin S, Mustard J, Nettles J, Petro N, Plourde K, Racho C, Rodriquez J, Runyon C, Sellar G, Smith C, Sobel H, Staid M, Sunshine J, Taylor L, Thaisen K, Tompkins S, Tseng H, Vane G, Varanasi P, White M, Wilson D (2011) The Moon Mineralogy Mapper (M^3) imaging spectrometer for lunar science: instrument description, calibration, on-orbit measurements, science data calibration and on-orbit validation. J Geophys Res, Planets 116:E10

Kiefer HH, Stone TC (2005) The spectral irradiance of the Moon. Astron J 129:2887–2901

Leland JE, Arecchi AV (1995) Analysis of spectralon® material for use in on-board calibration systems for the medium resolution imaging spectrometer (MERIS). SPIE, vol 2475

Levison HF, Olkin CB, Noll KS, Marchi S, Bell JF III, Bierhaus E, Binzel R, Bottke W, Britt D, Brown M, Buie M, Christensen P, Emery J, Grundy W, Hamilton VE, Howett C, Mottola S, Pätzold M, Reuter D, Spencer J, Statler TS, Stern SA, Sunshine J, Weaver H, Wong I (2021) Lucy mission to the Trojan asteroids: science goals. Planet Sci J 2:171

Levison HF et al (2024) Lucy science and mission overview. Space Sci Rev 220

Loose M, Beletic J, Blackwell J, Garnett J, Wong S, Hall D, Jacobson S, Rieke M, Winters G (2005) The SIDECAR ASIC: focal plane electronics on a single chip. In: Proc. SPIE 5904, Cryogenic optical systems and instruments XI, 59040V

Noll KS, Brown ME, Weaver HA, Grundy WM, Porter SB, Buie MW, Levison HF, Olkin C, Spencer JR, Marchi S (2020) Detection of a Satellite of the Trojan asteroid (3548) Eurybates—a Lucy mission target. Planet Sci J 1:44

Olkin CB, Levison HF, Vincent M, Noll KS, Andrews J, Gray S, Good P, Marchi S, Christensen P, Reuter D, Weaver H, Pätzold M, Bell JF III, Hamilton VE, Dello Russo N, Simon A, Beasley M, Grundy W, Howett C, Spencer J, Ravine M, Caplinger M (2021) Lucy mission to the Trojan asteroids: instrumentation and encounter concept of operations. Planet Sci J 2:172

Olkin C B et al (2024) Mission design and concept of operations for the Lucy mission. Space Sci Rev 220

Reuter DC, Stern SA, Scherrer J, Jennings DE, Baer J, Hanley J, Hardaway L, Lunsford A, McMuldroch S, Moore J, Olkin C, Parizek R, Reitsma H, Sabatke D, Spencer J, Stone J, Throop H, Van Cleve J, Weigle GE, Young LA (2008) Ralph: a visible/infrared imager for the new horizons Pluto/Kuiper belt mission. Space Sci Rev 140:129–154

Reuter DC, Simon AA, Hair J, Lunsford A, Manthripragada S, Bly V, Bos B, Brambora C, Caldwell E, Casto G, Dolch Z, Finneran P, Jennings D, Jhabvala M, Matson E, McLelland M, Roher W, Sullivan T, Weigle E, Wen Y, Wilson D, Lauretta DS (2018) The OSIRIS-REx Visible and InfraRed Spectrometer: spectral maps of the asteroid Bennu. Space Sci Rev 214:54

Rivkin AS, Emory JP (2010) Detection of ice and organics on an asteroidal surface. Nature 464:1322–1323

Rosenberg KP, Hendrix KD, Jennings DE, Reuter DC, Jhabvala MD, La AT (1994) Logarithmically variable infrared etalon filters. In: SPIE proceedings, Optical thin films IV: new developments, San Diego, vol 2262, pp 25–27

Simon AA, Reuter DC, Gorius N, Lunsford A, Wind G, Cosentino RG, Lauretta D, The OSIRIS-REx Team (2018) In-flight calibration and performance of the OSIRIS-REx visible and IR spectrometer (OVIRS). Remote Sens 10:1486

Takir D, Emery JP (2012) Outer main Belt asteroids: identification and distribution of four 3- μm spectral groups. Icarus 219:641–654

Takir D, Emery JP, McSween HY, Hibbitts CA, Clark RN, Pearson N, Wang A (2013) Nature and degree of aqueous alteration in CM and CI carbonaceous chondrites. Meteorit Planet Sci 48:1618–1637

Space Science Reviews (2023) 220:1
https://doi.org/10.1007/s11214-023-01029-y

The Lucy Thermal Emission Spectrometer (L'TES) Instrument

P.R. Christensen[1] · V.E. Hamilton[2] · G.L. Mehall[1] · S. Anwar[1] · H. Bowles[1] · S. Chase[1] · Z. Farkas[1] · T. Fisher[1] · A. Holmes[1] · I. Kubik[1] · I. Lazbin[3] · W. O'Donnell[1] · C. Ortiz[1] · D. Pelham[1] · S. Rogers[1] · K. Shamordola[1] · T. Tourville[1] · R. Woodward[1]

Received: 12 July 2023 / Accepted: 20 November 2023 / Published online: 19 December 2023

Abstract
The Lucy Thermal Emission Spectrometer (L'TES) will provide remote measurements of the thermophysical properties of the Trojan asteroids studied by the Lucy mission. L'TES is build-to-print hardware copy of the OTES instrument flown on OSIRIS-REx. It is a Fourier Transform spectrometer covering the spectral range 5.71–100 μm (1750–100 cm^{-1}) with spectral sampling intervals of 8.64, 17.3, and 34.6 cm^{-1} and a 7.3-mrad field of view. The L'TES telescope is a 15.2-cm diameter Cassegrain telescope that feeds a flat-plate Michelson moving mirror mounted on a linear voice-coil motor assembly to a single uncooled deuterated L-alanine doped triglycine sulfate (DLATGS) pyroelectric detector. A significant firmware change from OTES is the ability to acquire interferograms of different length and spectral resolution with acquisition times of 0.5, 1, and 2 seconds. A single ~0.851 μm laser diode is used in a metrology interferometer to provide precise moving mirror control and IR sampling at 772 Hz. The beamsplitter is a 38-mm diameter, 1-mm thick chemical vapor deposited diamond with an antireflection microstructure to minimize surface reflection. An internal calibration cone blackbody target, together with observations of space, provides radiometric calibration. The radiometric precision in a single spectrum is $\leq 2.2 \times 10^{-8}$ W cm^{-2} sr^{-1} /cm^{-1} between 300 and 1350 cm^{-1}. The absolute temperature error is <2 K for scene temperatures >75 K. The overall L'TES envelope size is 37.6 × 29.0 × 30.4 cm, and the mass is 6.47 kg. The power consumption is 12.6 W average. L'TES was developed by Arizona State University with AZ Space Technologies developing the electronics. L'TES was integrated, tested, and radiometrically calibrated on the Arizona State University campus in Tempe, AZ. Initial data from space have verified the instrument's radiometric and spatial performance.

Keywords Thermal emission spectrometer · Asteroid · Trojan · Lucy

✉ P.R. Christensen
phil.christensen@asu.edu

1 School of Earth and Space Exploration, Arizona State University, Tempe, AZ, USA

2 Southwest Research Institute, Boulder, CO, USA

3 AZ Space Technologies, Gilbert, AZ, USA

Abbreviations
ADC analog to digital converter
ARM antireflection microstructure
ASU Arizona State University
BCU Bench Checkout Units
CDR Critical Design Review
CPT Comprehensive Performance Test
CVD chemical vapor deposited
DFT discrete Fourier transform
DLATGS deuterated L-alanine doped triglycine sulfate
FET Field Effect Transistor
FPGA Field Programmable Gate Array
FPU Field Programmable Unit
FWHM full-width, half-maximum
GSE Ground Support Equipment
IFOV instantaneous field of view
IFT Instrument Functional Test
IMU inertial measure unit
IPP Instrument Pointing Platform
IRF instrument response function
LMC Lockheed Martin Corporation
L'TES Lucy Thermal Emission Spectrometer
MER Mars Exploration Rover
MGS Mars Global Surveyor
Mini-TES Miniature TES
MLI Multi-Layer Insulation
MO Mars Observer
NE$\Delta\varepsilon$ Noise Equivalent Delta Emissivity
NESR noise equivalent spectral radiance
OTES OSIRIS-REx Thermal Emission Spectrometer
PDR Preliminary Design Review
S/C spacecraft
SRR System Requirement Review
TES Thermal Emission Spectrometer
THEMIS Thermal Emission Imaging System
ZPD zero path difference

1 Introduction

The Lucy Thermal Emission Spectrometer (L'TES) instrument will aid in the characterization of the Trojan asteroids observed by the Lucy mission (Levison et al. 2021; Olkin et al. 2021) through the determination of their thermophysical properties. The primary science objective of the L'TES investigation is to determine the regolith physical properties (primarily grain size) using daytime and nighttime temperature measurements. This objective is addressed using thermal infrared spectral observations between 1750 and 100 cm^{-1} (5.7 to 100 μm). The L'TES spectral range, spectral sampling, radiometric accuracy, and spatial resolution provide high-quality, mid-infrared spectra at sufficient ground resolution to investigate the physical properties of each of the Trojan asteroids that Lucy will visit – Eurybates,

Leucus, Orus, Patroclus and Menoetius, which orbit each other, and Polymele. The L'TES calibrated radiance spectra will be converted to surface temperature(s) by fitting mixtures of Planck blackbody functions, with specified emissivity maxima (typically 0.97-1.0). Thermal inertia will be determined by fitting thermal models to the diurnal variation of the derived surface temperature(s) (e.g. Christensen et al. 2001; Kieffer 2013; Emery et al. 2014; Rozitis et al. 2020)

In this paper we present a description of the L'TES as-built instrument, calibration methods, and system performance. We also describe the basic instrument operational strategy, data processing methodology, and the plans for archiving the data through the Planetary Data System.

2 Instrument Description

2.1 Instrument Overview

The L'TES instrument continues the progression of infrared spectrometers and multispectral imagers developed by Arizona State University (ASU) for use on an array of planetary missions, including the Mars Observer (MO) and Mars Global Surveyor (MGS) Thermal Emission Spectrometer (TES) instruments (Christensen et al. 1992, 2001), the two Mars Exploration Rover (MER) Miniature TES (Mini-TES) instruments (Christensen et al. 2003), the Mars Odyssey Thermal Emission Imaging System (THEMIS) instrument (Christensen et al. 2004), the OSIRIS-REx OTES (Christensen et al. 2018), and the Emirates Mars Mission Infrared Spectrometer (EMIRS) (Edwards et al. 2021). A key feature of these instruments is that all have used uncooled bolometric detectors, which has allowed them to extend to longer wavelengths and be less complex with lower mass and cost than instruments using cooled photon detectors. This long-wave performance is essential for the measuring the Trojan asteroid temperatures, which are expected to be in the 60-120 K range.

L'TES is a Fourier-transform interferometer that collects hyperspectral thermal infrared data over the spectral range from 1750 to 100 cm^{-1} (5.7 to >100 μm) in a single detector. Unlike its predecessor, OTES, L'TES has selectable spectral sampling at 8.64 cm^{-1}, 17.3 cm^{-1}, and 34.6 cm^{-1}. The three spectral samplings were chosen to allow short (0.5 sec and 1 sec), lower-spectral-resolution scans to be completed on the asteroid night sides where spatial sampling is more important than spectral resolution at the very low expected temperatures. L'TES has a 7.3-mrad full-width, half-maximum (FWHM) instantaneous field of view (IFOV) that allows nightside and dayside data to be collected during each flyby. L'TES is a build-to-print mechanical copy of OTES but has redesigned electronics from a different vendor than OTES.

The primary measurement requirement is to obtain spectra with sufficient absolute radiometric accuracy to allow the thermal inertia to be determined with sufficient accuracy to characterize the surface materials. L'TES will acquire full-disk integrated spectra during the approach mission phase (Levison et al. 2021; Olkin et al. 2021) and localized spatial information that will vary in resolution from $\sim$3 to 9 km at closest approach during the different flybys (Olkin et al. 2021). Because no single surface will be observed at different times of day, the L'TES diurnal observations taken of different locations will be combined to provide an estimate of the average asteroid thermal inertia.

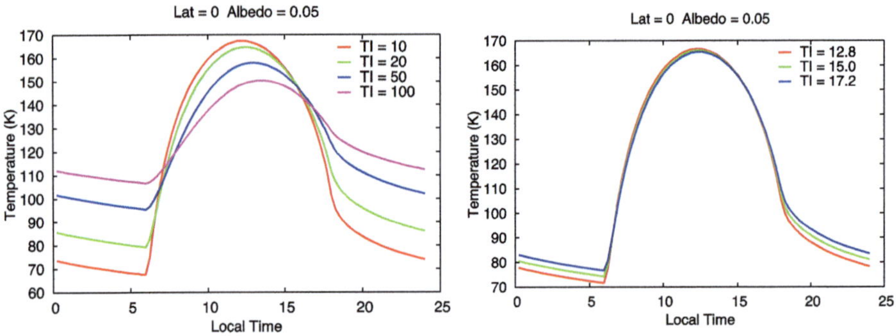

(a) (b)

Fig. 1 Modeled Trojan asteroid surface temperatures at the equator. **(a)** A representative range of expected thermal inertia values with a typical albedo value of 0.05. **(b)** Surface temperatures for a nominal thermal inertia of 15 SI and values of thermal inertia 15% higher (17.2 SI) and 15% lower (12.8 SI). These models were used to determine the L'TES absolute temperature requirement (2 K) necessary to determine the thermal inertia to ±15%

2.2 L'TES Measurement Requirements

Thermal Inertia The L'TES thermal inertia requirement is to determine the surface thermal inertia to ±15%. Figure 1a shows modeled diurnal temperature variations for expected Trojan thermal inertias of 10 to 100 $\mathrm{J\ m^{-2}\ K^{-1}\ sec^{-1/2}}$ (hereafter referred to as SI) at the equator with an albedo of 0.05 using the krc thermal model (Kieffer 2013). Figure 1b gives the modeled diurnal temperatures for a surface with an inertia of 15 SI, together with models for 15% higher (17.2 SI) and 15% lower (12.8 SI) thermal inertias. Nighttime temperature differences between these models are ±2 K with the coldest temperature of ~75 K occurring just prior to sunrise. These models result in an absolute temperature measurement accuracy requirement of 2 K for temperatures ≥75 K.

2.3 L'TES Design

L'TES is a mechanical built-to-print copy of the OTES instrument (Christensen et al. 2018). The interferometer, detector, telescope, internal calibration assembly, and housing are identical to OTES. The only mechanical difference is a reduction in the length of the stray light baffle ("sunshade) that is appropriate for the geometric orientation of observations acquired at Jupiter's orbit (Fig. 2). The only new development was the incorporation of electronics developed by a different vendor (AZ Space Technologies) than developed the OTES electronics. The L'TES electronics are based on those developed by AZ Space Technologies for the EMIRS instrument (Edwards et al. 2021) and incorporate enhancements to the digital servomechanism for the interferometer moving mirror that improved its performance, provided selectable moving mirror travel, and accommodated minor electronic parts obsolescence. The L'TES functional block diagram is shown in Fig. 3 and the instrument properties are summarized in Table 1.

2.3.1 Opto/Mechanical

The L'TES optical system uses a compact Cassegrain telescope design with a 15.2-cm diameter f/3.91 Ritchey-Chretien telescope that directs the incoming energy through a collimating off-axis parabolic mirror to create an optical beam with an afocal ratio of eight.

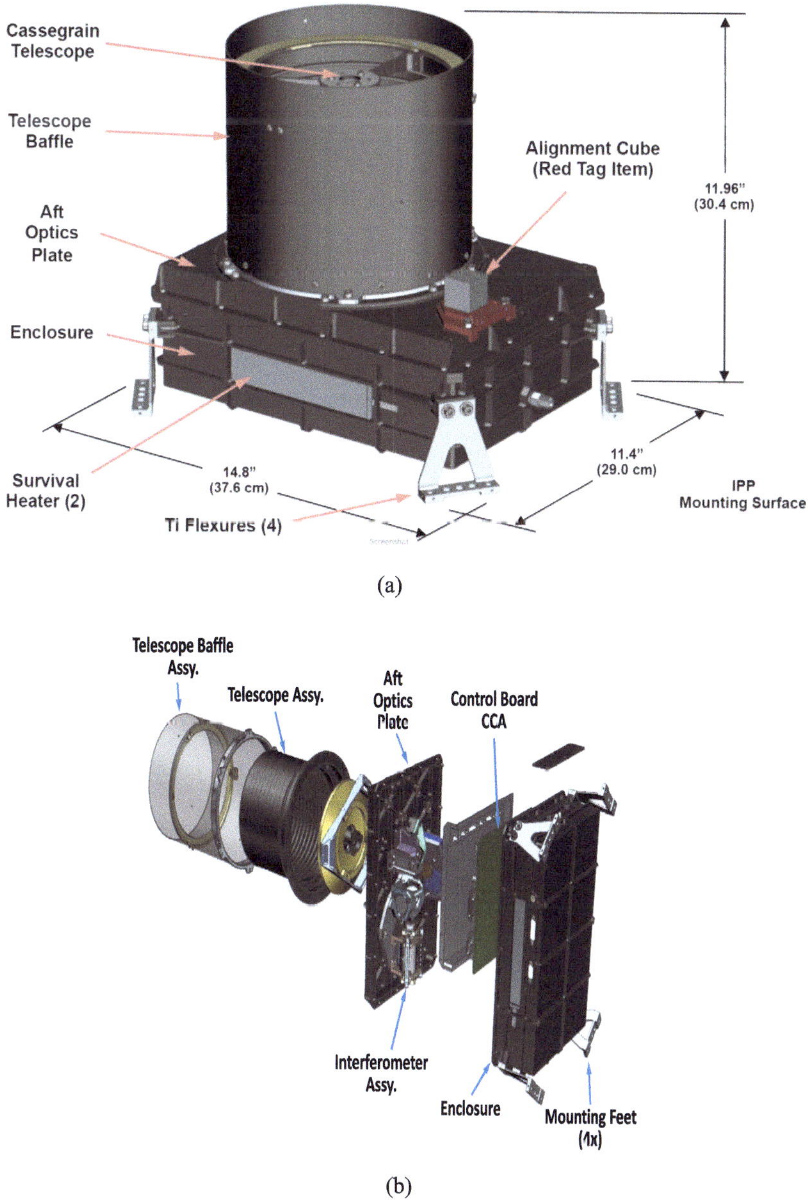

Fig. 2 The L'TES CAD model. **(a)** The assembled instrument with dimensions. **(b)** L'TES, in an exploded view, showing the major elements of the L'TES modular design

This beam passes to the flat-plate interferometer moving mirror mounted on a voice-coil motor assembly (Fig. 4). The beam radiance passes through a CVD diamond beamsplitter installed in a radial three-point mount. Approximately half of the radiance is reflected to a fixed mirror, the other half is transmitted to the moving mirror, which travels ± 0.281 mm to achieve the nominal 8.64 cm^{-1} spectral sampling. Selectable travel distances produce the additional spectral samplings of 17.3 cm^{-1} and 34.6 cm^{-1}. The two optical beams recom-

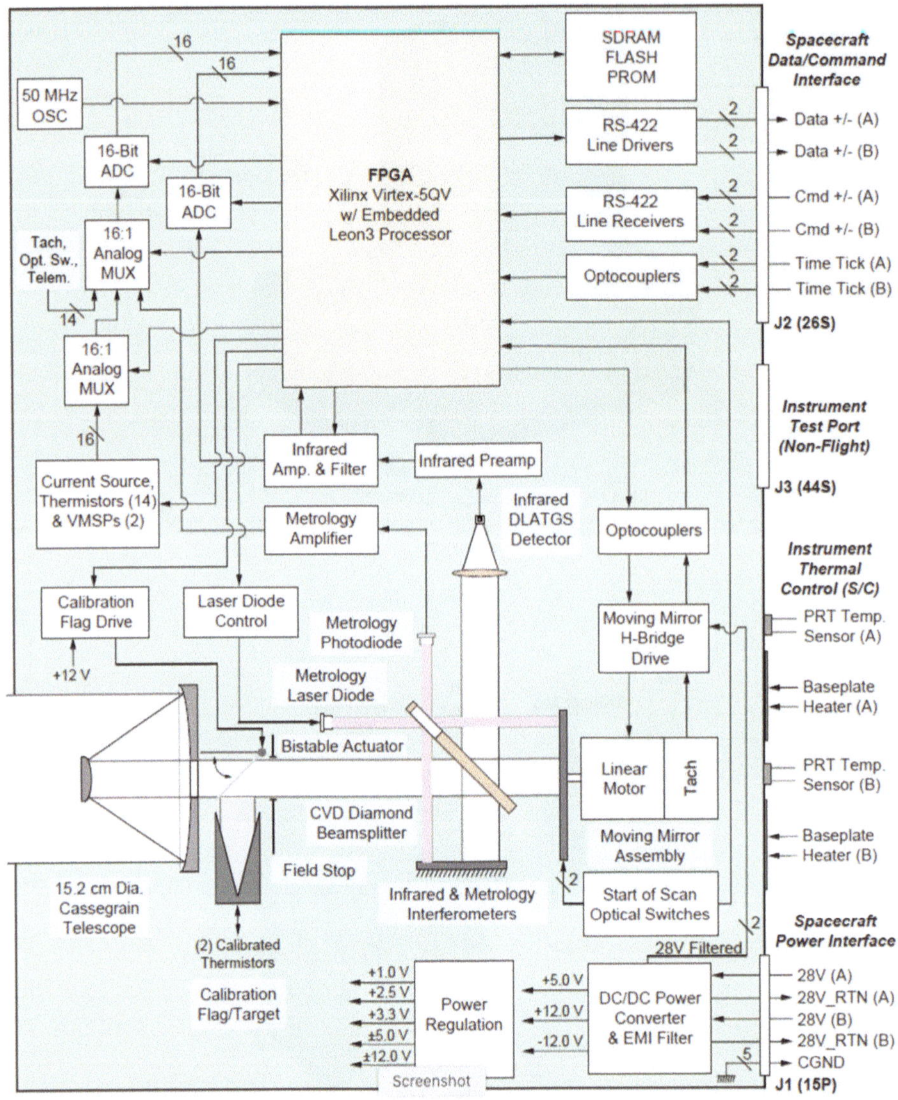

Fig. 3 The L'TES block diagram, showing the key functional elements and the instrument-to-spacecraft interfaces

bine at the beamsplitter, producing the interference that determines the spectral distribution of the scene radiance, and travel to a parabolic mirror that images the beam onto a single on-axis detector element with a CVD diamond lens. A single monochromatic ULM VCSEL laser diode with a wavelength of 0.851 μm feeds a fringe counting metrology interferometer that uses the same moving and fixed mirrors and beamsplitter as used in the IR signal chain. The metrology interferometer is used to precisely control the velocity, track the position of the moving mirror, and trigger the sampling of the IR signal. The laser diode provides four times oversampling of the shortest scene bandpass wavelength. All mirror surfaces are diamond-turned and gold-coated, with reflectances of >0.99 across the L'TES spectral

Table 1 L'TES as-built instrument parameters

Parameter	Value
Spectral range	1750 to 100 cm^{-1} (5.71–100 μm)
Spectral sampling	8.64 cm^{-1}, 17.28 cm^{-1}, and 34.56 cm^{-1}
Telescope aperture	15.2 cm
f#	3.91
Field of view (FWHM)	7.31 mrad in azimuth and 7.32 mrad in elevation
Detector	Uncooled deuterated L-alanine doped triglycine sulfate (DLATGS) pyroelectric
Detector D*	1.2 10^9 cm $Hz^{1/2}$ W^{-1} at 10 Hz, 22 °C
NESR (2-sec scan)	1.6×10^{-8} W cm^{-2} str^{-1}/cm^{-1} from 300 to 1300 cm^{-1}
Cycle times per measurement	2, 1, 0.5 s, including scan reversal time
Metrology laser self-apodized wavelength (25 °C)	0.851 μm
Michelson mirror travel	±0.281 mm; ±0.141 mm; ±0.070 mm
Michelson mirror velocity	0.319 mm/s
Sampling frequency	772 Hz
Number of bits per sample	16
Number of samples per interferogram—Filled	1360, 680, 340
Nominal data volume per 2 s scan	data: 21,760 bits; telemetry: 200 bits
In-flight calibration	Two-point; internal calibration blackbody and space
Cal blackbody emissivity	0.98 ±0.005
Thermal requirements	Operation in-spec: −10 °C to 30 °C
	Flight allowable operational range: −10 °C to +40 °C
	Non-operational protoflight survival range: −30 °C to +55 °C
Solar protection	Cal mirror in stowed position
Mass	6.47 kg
Power	12.6 W average; 18.0 W peak
Dimensions	37.6 × 29.0 × 30.4 cm

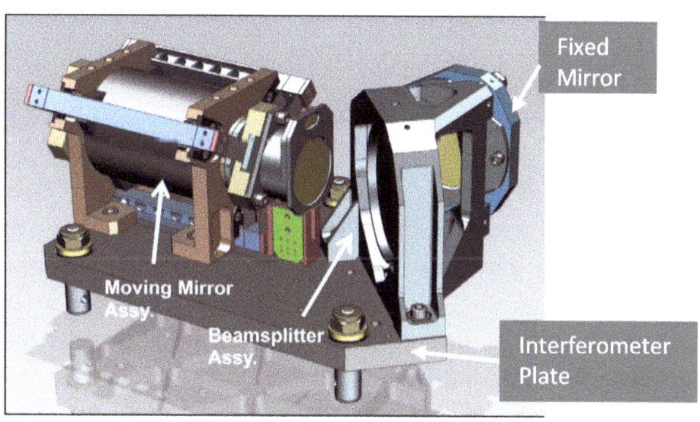

Fig. 4 The L'TES CAD model showing the interferometer layout on the aft optics plate

range. All of the L'TES machined components are aluminum for light weight and strength while meeting the launch load and pyro shock requirements. The use of baffles around the telescope housing and secondary mirror, and the use of diffuse black Z306 paint around the optics and within the cavity, minimize stray light effects.

L'TES utilizes a 38-mm-diameter, 1-mm-thick CVD diamond beamsplitter substrate. This diamond was fabricated by Diamond Materials and is installed in a radial three-point mount that provides precise alignment over the −10 °C to +30 °C operation-in-specification range, maintains mechanical integrity over the −30 °C to +55 °C nonoperational protoflight survival range, and accommodates the low thermal expansion coefficient and high heat conductivity of diamond. An antireflection microstructure (ARM) developed by TelAztec increases the transmission of raw diamond by $\sim$25% from 7 to >25 μm. Because light passes through the beamsplitter twice, this improvement represents an 50% increase in beamsplitter throughput. The low dispersion properties of diamond allow a single substrate to be used without the need for a compensator component, which would have increased the number of surface reflections and reduced the instrument throughput. The IR beamsplitting function is provided by a Ge beam-dividing coating. A parabolic mirror reimages the optical pupil onto an on-axis thermal IR detector that has a small (4.4 mm) CVD diamond lens also fabricated by Diamond Materials.

The L'TES internal calibration blackbody is a 1-cm-diameter, 15° half-angle cone that has two precision thermistors installed in its walls. These thermistors were calibrated to an absolute accuracy of 0.1 K, and the interior surface was painted with PT-401 paint resulting in a derived emissivity of the internal cal target of >0.99. The energy from this target reflects off a gold-plated mirror toward the interferometer. This cal mirror rotates 60° from its stowed position to reflect the radiance from the blackbody to the interferometer. The cal mirror is not instrumented directly, but it is stowed in the cal mirror cavity, whose temperature is determined from redundant, precision thermistors. (Christensen et al. 2018).

2.3.2 Detector

L'TES uses an uncooled deuterated L-alanine doped triglycine sulfate (DLATGS) pyroelectric detector fabricated by Leonardo Electronics (formerly Selex-Galileo). The measured D*, including the diamond lens, is 1.2×10^9 cm·Hz$^{1/2}$/W at 10 Hz, 22 °C. The detector has a bias voltage applied to the Field Effect Transistor (FET) to ensure it remains properly poled. An identical detector was used on the OTES instrument. Preamplification and front-end filtering are performed on the preamplifier circuit board to amplify the signal and to AC couple the detector output to block high-frequency oscillations.

2.3.3 Electronics

A ±12-volt regulator supplies power to the detector and preamplifier electronics. The spectrometer circuit board performs the bulk of the analog electronics processing. L'TES command, control, and data flow tasks are controlled by logic in the command-and-control field programmable gate array (FPGA). The interface electronics parse out the instrument commands that control L'TES hardware functions. The flow of the digital interferometer data is controlled by additional logic in the command-and-control board FPGA. After each interferometer scan, the 16-bit interferogram data and 16-bit telemetry data are moved from the analog to digital converter (ADC) to the input memory buffer on a 16-bit parallel data bus. The 16-bit data are then serialized for transfer to the spacecraft (S/C) via RS-422 interface. The analog multiplexer, digital serializer, and data formatting logic are included in the command-and-control FPGA. The DC power converter accepts 24 to 34 volts unregulated input voltage and supplies +3.3, ±5 and ±15 volts regulated output voltage.

The L'TES timing sequencing electronics are implemented in the command-and-control board FPGA. These electronics generate the timing waveforms necessary to control and synchronize instrument operation and provide the control and synchronization of the amplification, track/hold, multiplexing, and analog to digital conversion of the detector and telemetry signals. They also control and synchronize the interferometer servo electronics with the data acquisitions. The timing sequencing electronics include the fringe delay electronics that are used to correct the sampling error due to the phase delays between the fringe and IR analog channels. All clocks in the timing sequencer are generated from the master clock crystal oscillator that operates at a frequency of 20 MHz and is synced to the spacecraft clock. The L'TES interferometer servo electronics were completely redesigned from previous instruments to use a digital servo drive running in the FPGA. This digital servo receives scan timing clocks from the timing sequencer electronics and the fringe clock from the fringe detection electronics. The motor control logic uses these clocks to synchronize the mirror movement with the spectrometer data acquisitions. The moving mirror uses a direct drive BEI Kimco linear motor with tachometer feedback. The moving mirror tachometer signal is returned to the interferometer control electronics to allow active feedback control of the actuator.

The digital servo control loop is implemented using a custom processor core with a double precision Field Programmable Unit (FPU) that runs inside an FPGA mounted on the command control board. The control loop uses digitized metrology fringe, tachometer, and end-of-travel optical switch as inputs; performs fringe data processing, feedback compensation, sample timing, and command/telemetry functions; and outputs the motor command to the D/A converter controlling the voice-coil driver circuit for the interferometer moving mirror.

High sample frequency of the fringe signal (20 kHz) enables the control loop to estimate the servo velocity at a higher frequency than would be achievable by only monitoring the metrology interferometer zero crossings. A rate estimation algorithm inside the servo controller uses the sinusoid model of the metrology fringe to derive a servo rate estimate at a 2-kHz rate. This model is also used to predict the appropriate times to sample the interferometer detector. The detector samples are based on estimated fringe signal zero crossings and are delayed to compensate for the phase delay of the Bessel filter in the detector electronics. An optical switch is used near the end of travel to measure the absolute position and control the servo motion so that it remains centered about the zero-path difference (ZPD) position. The control algorithm includes adjustable optical switch thresholds so that the scan pattern can be centered without physically moving the optical switch assembly.

There is a great deal of flexibility in the design and adjustment of the digital servo. Controller gains and limits, as well as some other configuration options, such as parameters controlling scan timing, can be changed via commands while the controller is executing out of SRAM. Also, because the entire algorithm is stored in nonvolatile memory, it can be modified without changes to hardware.

The flow of the digital interferometer data is controlled by additional logic in the command-and-control board FPGA. After each interferometer scan, the 16-bit interferogram data and 16-bit telemetry data are moved from the A/D to the input memory buffer on a 16-bit parallel data bus.

2.3.4 Thermal Design

L'TES is a thermal instrument that requires thermal stability of <0.1 °C per minute in order to maintain calibration between periodic calibration observations. This stability is achieved by conductively isolating L'TES from the spacecraft to prevent rapid thermal transients, while radiatively coupling L'TES to the spacecraft to provide a long-term thermal sink. Replacement heaters and thermal blanketing are used to maintain the instrument within its flight allowable temperature limits. L'TES will operate within specification over a temperature range of −10 to +30 °C, operate from −10 to +40 °C, and survive form −30 to +55 °C.

2.3.5 S/C Interface

The L'TES spectrometer provides data to the spacecraft across two synchronous RS-422 serial lines at 57.6 Kbps during the 200-msec interferometer scan reversal period whenever L'TES is powered. Instrument commanding, including operation of the internal cal mirror and the selection of redundant components, is provided across two synchronous RS-422 serial lines at 57.6 Kbps. Synchronization of the interferometer is performed by L'TES using the spacecraft clock. The L'TES power interface uses redundant unregulated power lines to the input of the L'TES DC/DC converter. The unregulated input power can be from +24 to +34 VDC, with a nominal value of 28 VDC.

2.3.6 L'TES Radiation and Contamination Mitigation

L'TES is based on the OSIRIS-REx OTES and MGS TES instruments that were designed to survive 20 krad environments. It is estimated that the MGS TES exceeded this radiation dose in its 10-year life, with no detectable radiation damage effects. OTES has operated for >5 years in a radiation environment similar to that of Lucy with no detectable degradation in performance. L'TES was built following the Goddard Space Flight Center and Lucy Project assembly and test procedures to minimize any organic contamination carried on the instrument. Contamination on the instrument would have to be extreme (>10 s of μm) to degrade the performance based on experience with Mini-TES in the dusty surface environment of Mars (Ruff et al. 2011). Analysis and monitoring of witness samples through the L'TES development at ASU demonstrate that L'TES was delivered with a cleanliness level D per NASA CR 4740.

2.3.7 L'TES Operational Modes

L'TES begins collecting telemetry data when power is applied. Upon receipt of the data collection command, L'TES collects and transmits interferograms to the spacecraft continuously when operational. L'TES will begin acquiring data of each Trojan several hours prior to closest approach when the asteroid fills ∼1% of the L'TES field of view and continue for several hours after closest approach. Spatially resolved observations will be collected during the flyby through two-axis scanning of the instrument pointing platform (IPP) to point L'TES at the desired locations. L'TES has two operational modes – data collection and standby. In data collection mode L'TES collects one interferogram every 0.5, 1, or 2 seconds; in standby mode no interferogram data are collected, but telemetry data are collected once every two seconds. Instrument commands are used to insert the internal cal flag mirror into the optical path, to select the scan period, and to switch between the redundant start-of-scan optical switches.

3 Calibration Methodology

3.1 Calibrated Radiance

The L'TES calibration approach is identical to that of OTES and is described in detail in Christensen et al. (2018). In summary, the L'TES interferometer measures the difference in the modulated spectral radiance (I; in W cm^{-2} sr^{-1} /cm^{-1}) coming alternatively from the external scene and from the detector itself as the interferometer moving mirror translates. The external radiance is the sum of the radiance from the scene that enters the aperture, as limited by the field stop (I_{scene}), the radiance from the optics themselves (I_{optics}), as well as any radiance from the field stop and the interior of the instrument that is modulated through the interferometer ($I_{instrument}$). The radiance from the interior of the instrument that does not pass through the interferometer is not modulated and is a DC term in the interferogram that will be absent from the transformed spectrum. The difference between the modulated external radiance and that from the detector ($I_{detector}$) is the signal measured by the detector.

The optics radiance (I_{optics}) can be divided into the emitted radiance from the fore optics (I_{fore}; i.e. the telescope) and the aft optics (I_{aft}; collimating, transfer, and detector focusing optics). The aft optics radiance does not pass through the interferometer and is not modulated, so only the fore optics radiance becomes a component of the interferogram signal. The radiance from the instrument ($I_{instrument}$) includes the radiance emitted from the field stop, which is modulated through the interferometer. In practice, however, this modulated radiance is indistinguishable from the detector radiance; this radiance will be included in the detector radiance in the subsequent discussion.

Using these definitions, the electrical signal (V) produced by the interferometer is given by:

$$V = \left(I_{scene} + I_{fore} - I_{detector}\right) * IRF \tag{1}$$

where IRF, in units of Volt/(W cm^{-2} /cm^{-1} sr^{-1}), is the instrument response function that converts the optical input to electrical output. Observations of two calibration targets allow the I and IRF terms to be determined and used to determine the scene radiance.

The radiance from the scene is given by:

$$I_{scene} = \varepsilon_{scene} B_{scene} \tau_{optics} \tag{2}$$

where:

ϵ_{scene} = emissivity of the scene
B_{scene} = Planck emission from the scene
τ_{optics} = throughput of the full optics

The determination of the desired scene radiance (I_{scene}) from Eq. (1) requires knowledge of the two unknown terms ($I_{fore} - I_{detector}$) and the IRF, which in turn requires the observation of two calibration targets of known radiance. As with OTES, the L'TES internal calibration blackbody is located behind the telescope and the IRF cannot be determined simply from observations of the two calibration targets because the same optical path is not viewed in the scene, space, and cal blackbody observations, and the I_{fore} term does not cancel as it does in the two-target, full-optics method. In this case the fore optics throughput and emission must be estimated. The uncertainties in these estimates result in errors in the scene radiance, the magnitude of which can be investigated by modeling the scene radiance as the optical throughput and emission are varied over their uncertainties (see Sects. 3.2 and 6.7).

As derived in Christensen et al. (2018), the scene radiance ($\epsilon_{\text{scene}}\text{B}_{\text{scene}}$) is given by:

$$
\begin{aligned}
&\varepsilon_{scene} B_{scene} \\
&\quad = \left(\frac{V_{scene} - V_{space}}{V_{cal} - V_{space}}\right) \\
&\quad \times \left(\frac{\varepsilon_{cal} B_{cal} \rho_{flag} + \varepsilon_{flag} B_{flag} - \left(\varepsilon_{primary} B_{primary} \rho_{secondary} + \varepsilon_{secondary} B_{secondary}\right)}{\tau_{fore}}\right. \\
&\quad \left. - \varepsilon_{space} B_{space}\right) + \varepsilon_{space} B_{space} \qquad (3)
\end{aligned}
$$

where *cal* is the internal calibration target, *flag* is the gold coated flag that is inserted into the beam path to observe the cal target, *space* refers to the space calibration observation, and *primary* and *secondary* refer to the two Cassegrain telescope mirrors.

3.2 Absolute Calibration

The absolute accuracy of the scene radiance is a function of the uncertainty in the knowledge of each of the terms in Eq. (3). The uncertainty of the signal terms (V) is determined by the instrument noise of each measurement, whereas the uncertainty in the emissivity, reflectance, and temperature of the calibration target and mirrors is determined by the accuracy to which each of these terms can be measured. Averaging multiple observations can reduce the uncertainty due to instrument noise; however, the uncertainties in the remaining terms cannot be reduced and are the driving terms in the absolute accuracy. The uncertainties due to the emissivity, reflectance, and temperature of the cal target and fore optics were modeled by assessing the effects of the uncertainties in each of these parameters using Eq. (3). This equation has been applied in a Monte Carlo model to assess the absolute error in the scene radiance using: (1) the measured random noise errors in the signals (V_{scene}, V_{space}, and V_{cal}); (2) the uncertainty in the radiance from the cal and space targets (B_{cal}, and B_{space}) and the primary and secondary mirrors ($B_{primary}$, $B_{secondary}$) that are due to temperature errors in the measurement of each of these elements; and (3) the uncertainty in the emissivity of the calibration blackbody and the reflectivity of the primary and secondary mirrors and the calibration flag.

The worst-case expected absolute error in the cal target temperature is 0.5 C and its emissivity is 0.005 (Christensen et al. 2018). These estimated errors were included in a Monte Carlo model that incorporated the measured L'TES flight instrument response function and instrument noise (noise equivalent spectral radiance; NESR) (see Sect. 6). The simulated L'TES spectra were converted into scene radiance that was then converted to brightness temperature at each wavenumber, assuming an emissivity of unity. The surface kinetic temperature, which is the desired parameter for temperature mapping, was determined by averaging the brightness temperature over a specified wavenumber range. The full spectral range was used for warm targets, whereas narrower low-wavenumber ranges were used at lower surface temperatures. As a result, the NESR begins to affect the accuracy of the derived surface kinetic temperature at temperatures below ∼125 K because fewer spectral channels are available to average the brightness temperature. This process is roughly equivalent to fitting a blackbody radiance curve to the measured radiance to provide a best-fit determination of the surface temperature. For the very low temperatures expected for the Trojan asteroids, the measurement noise becomes a significant contributor and dominates the temperature determination for scene temperatures below ∼120 K. Figure 5 demonstrates the effect of

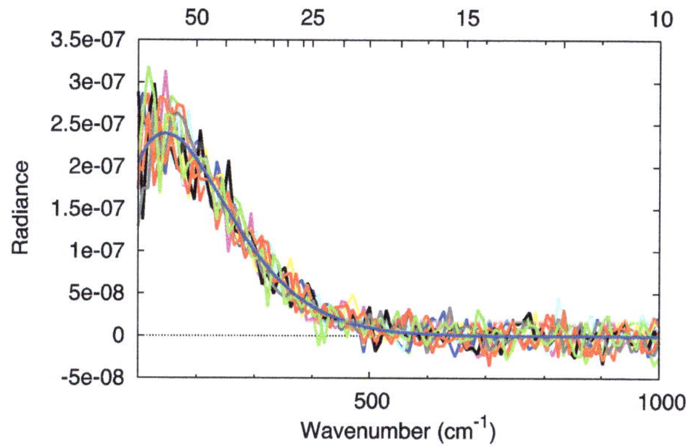

(a)

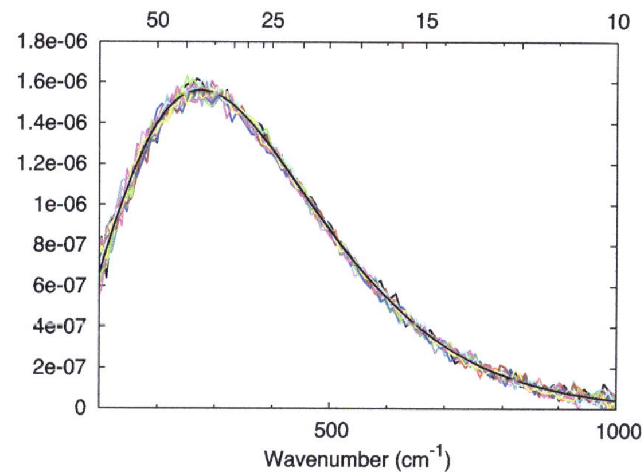

(b)

Fig. 5 Modeled L'TES radiance spectra compute by applying random noise equivalent to the measured NESR versus wavenumber to a Planck function assuming an emissivity of unity. **(a)** Radiance from a surface at 75 K. **(b)** Radiance from a surface at 140 K. Upper scale is micrometers

instrument noise at low temperatures for modeled spectra at scene temperatures of 75 and 140 K using the measured L'TES noise performance (Sect. 6.6).

The model was run for 10,000 random cases for surface temperatures from 70 to 400 K at 10 K intervals, and the standard deviation of the derived surface kinetic temperature was computed for each surface temperature (Fig. 6). These values represent the best estimate of the L'TES absolute temperature accuracy errors. As seen in Fig. 6 L'TES meets the absolute temperature requirement of ± 2 K for surface temperatures ≥ 75 K. The measured absolute temperature performance using the results from the pre-launch test data are given in Sect. 6.

4 Experimental Setup

4.1 Test Equipment and Facilities

The L'TES instrument was assembled, tested, and calibrated on the Arizona State University Tempe campus in an ISO Class 7 (class 10,000) cleanroom in building ISTB4, with Class 6 (1,000) flow benches for component assembly (Christensen et al. 2018). L'TES calibration was done in both ambient and thermal vacuum conditions using test equipment that was calibrated to NIST standards on a routine basis. Bench-level testing of the L'TES instrument

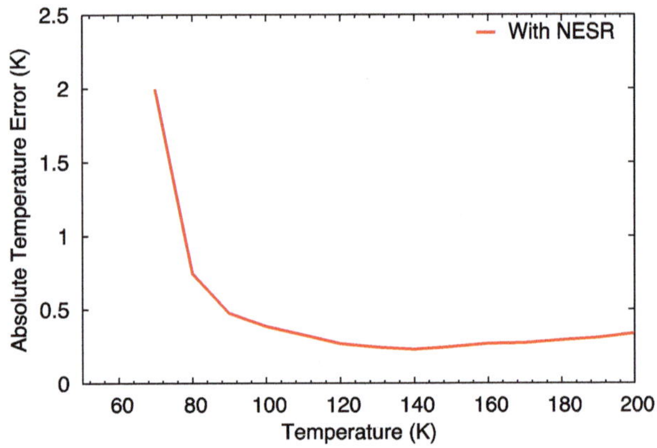

Fig. 6 Modeled error in the derived surface temperature. The model incorporates a Monte Carlo analysis of the expected uncertainties in each of the calibration parameters (Eq. (3)) together with the measured NESR to compute a series of radiance spectra. The kinetic temperature was derived from these model spectra and the 1-sigma variation was determined and plotted versus scene temperature

consisted of piece-part and system-level testing of each sub-assembly under ambient conditions, followed by system performance testing that was conducted before and after each environmental test and throughout thermal-vacuum testing. Ambient tests determined the instrument functional performance, the field-of-view definition and alignment, the out-of-field response (encircled energy), the spectrometer spectral range and sample position, and verified the command and data links. The thermal vacuum calibration testing determined the instrument response function in vacuum over the expected range of instrument temperature, the signal gain value, the emissivity and temperature of the internal calibration blackbody, and the radiometric precision and accuracy.

Throughout the integration and test phase an Instrument Functional Test (IFT) was performed using the L'TES ambient target and/or two calibration reference blackbody standards (Bench Checkout Units; BCU) (Christensen et al. 2018) to provide trending of performance throughout the test program. A Comprehensive Performance Test (CPT) was performed using the ambient and/or BCU targets and provided the baseline radiometric performance assessment for the instrument. This test was also performed before and after environmental tests. The L'TES optical field of view was characterized using an 8-inch-diameter f/7.5 off-axis collimator that was scanned in elevation and azimuth using computer-controlled actuators.

The L'TES thermal vacuum radiometric calibration was performed using the two BCU calibration targets that were placed inside the thermal vacuum chamber. The absolute temperature of these targets was determined using precision thermistors that were NIST-calibrated to ±0.1 °C. Their emissivities were determined to be 0.99897 ±0.0002 based on the spectral properties of the PT-401 paint and the analysis of a cone blackbody (Bedford 1988; Prokhorov et al. 2009). The temperature of the primary and secondary mirrors was determined using the NIST calibration of the flight thermistors mounted to these elements, and the reflectivity was determined using laboratory measurements of witness samples of the gold coatings that were applied to the optical surfaces.

L'TES was radiometrically calibrated in vacuum at instrument temperatures of −10, 0, 10, and 25 °C. Calibration tests were performed by viewing the two BCU targets, one set to a temperature of 85 K to simulate space observations and the second set to temperatures of 100, 150, 180, 210, 240, 270, 300, and 330 K to span the expected range of target surface temperatures. L'TES was installed in a LACO thermal vacuum chamber developed for the L'TES project in the Class 8 (100,000) cleanroom in ISTB4 on the ASU campus. L'TES was mounted on a metallic L-plate with thermostatically controlled strip heaters that simulated the instrument pointing platform deck for thermal balance and was used to control the instrument temperature during thermal vacuum testing and thermal cycling. A liquid-nitrogen-cooled plate was attached to the back of the mounting plate to provide a cooling source for the bracket/instrument. L'TES was configured in the chamber to allow its aperture to point between the two external BCU blackbody calibration targets using a vacuum-compatible rotary stage attached to the underside of the L-plate (Christensen et al. 2018). The instrument and mounting bracket were fully blanketed with flight-like Multi-Layer Insulation (MLI) thermal blankets to simulate the IPP thermal configuration. Ground Support Equipment (GSE) controlled the L'TES, L-plate, targets, and rotary stage via cabling through the chamber feedthroughs.

4.2 Data Processing Software/Scripts

The L'TES data were processed using the open-source vector math software package called *davinci* that was developed at ASU for instrument data processing and has been used for calibration and processing of the TES, Mini-TES, THEMIS, OTES, and EMIRS data. The L'TES performance varies slightly between the forward and reverse moving mirror scan directions, so all processing is done separately for the two scan directions. The discrete Fourier transform (DFT) is performed on the interferograms, bad spectra resulting from chamber- or spacecraft-induced noise are identified and removed, and the internal and external calibration targets and scene target are identified using telemetry. Groups of calibration spectra are averaged and linearly interpolated over time and the radiometric calibration and data analysis are performed.

5 L'TES Development

5.1 Flight Instrument Development

The L'TES instrument began development in January 2017 with the selection of the Lucy mission for Phase B. The System Requirement Review (SRR) was held on Oct. 24, 2017. The Preliminary Design Review (PDR) was held on April 13, 2018, and the Critical Design Review (CDR) on March 19, 2019. L'TES was delivered to the spacecraft on December 13, 2020, with a total development time from SRR to delivery of 38 months. The most challenging aspect of the L'TES development program was the assembly, test, and calibration of the instrument during the height of the COVID-19 pandemic between January and December 2020. The L'TES team undertook stringent measures to reduce the risk of infection, including stringent use of mask and glove protocols, limiting the number of personnel in the assembly cleanroom at any time to three, separating assembly and inspection shifts into morning and afternoon operations, and holding technical interchange discussions via video conferencing. The most challenging technical issue was the alignment of the interferometer. The interferometer fixed and moving mirrors and the beamsplitter are mounted on a vibration isolated plate (Fig. 4). The fixed mirror is mounted to the beamsplitter housing on a three-point mount and aligned using nuts on the three threaded studs. These nuts are injection bonded following final alignment to prevent movement during launch. Following the initial L'TES thermal cycling, the interferometer alignment was observed to have degraded, as evidenced by a decrease in the IRF at short wavelengths. The spectral shape of this reduction was accurately modeled as a tilt in the fixed mirror relative to the beamsplitter housing. Subsequent analysis revealed that minor oxidation on the three studs and minor roughness on the nuts holding the fixed mirror resulted in a 13 arcsec tilt of the fixed mirror

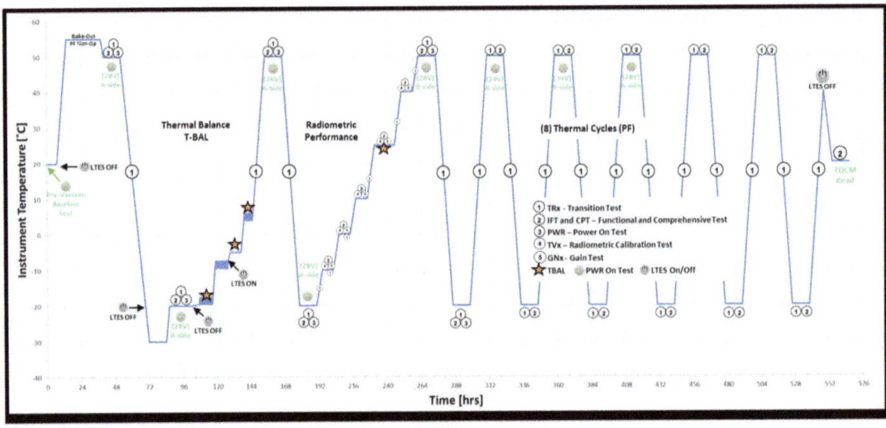

Fig. 7 The L'TES thermal balance and thermal vacuum test timeline

following initial heating. Once this first-time "settling" occurred, the fixed mirror exhibited the expected 2-3 arcsec tilt over the full operational temperature range from −10 to +40 °C. Through test it was determined that torquing, releasing, and re-torquing the fixed mirror mounting nuts significantly reduced the first-time shift. The original flight beamsplitter assembly, whose nuts had been bonded, was replaced with the spare assembly and testing continued. A second, related issue was a change in the alignment of the metrology laser over temperature that resulted in a variable laser signal with temperature. If the laser signal increased beyond the analog-to-digital saturation limit, then the servo control algorithm would not function properly. The tilt over temperature issue was corrected as discussed above, but as a precaution the metrology laser lens was redesigned to reduce its signal sensitivity to interferometer tilt.

L'TES underwent its final thermal vacuum testing from October 30 through November 23, 2020, with the radiometer calibration occurring between November 9 and 15th. These tests included radiometric calibration at multiple instrument temperatures, as well as eight thermal cycles from −30 to +50 °C (Fig. 7).

5.2 Post-Delivery Testing

Following the delivery of the flight instrument to the spacecraft developer and flight system integrator, Lockheed Martin Corporation (LMC; Olkin et al. 2021), a noise source was observed during system-level testing. This noise was associated with the spacecraft inertial measurement units (IMU), which produce discrete vibrations in the 500–700 Hz range that have a magnitude of ∼5–10 milli-G at the L'TES spacecraft mounting surface. In its original configuration, the L'TES IR interferometer collected samples at 656 Hz. The servo control uses the signal from the metrology laser signal to predict when the laser signal will go to zero and when the IR signal should be sampled. The IMU-induced vibrations resulted in small, periodic jitter in the moving mirror position, and the zero-position of the mirror had slight (∼1% of the sampling distance) offsets. The magnitudes of these periodic offsets are small (0.01 × 1/2 × 0.849 µm or 0.004 µm), but they result in artificial sinusoids at specific frequencies that correspond to the difference in frequency between the L'TES 656 Hz sampling frequency and the IMU frequencies. Two of these beat frequencies occurred at ∼40 and 100 Hz, which is within the L'TES information band from 10 to 120 Hz, and resulted in spikes in the L'TES IR spectrum. A similar problem was observed on OTES, and the same solution used for OTES was implemented on L'TES. This solution involved increasing the servo velocity to 0.321 mm/s, giving a sampling frequency of 772 Hz, which moves the IMU frequency disturbances outside the L'TES information band. In-flight data from OTES confirm that this solution works very well. The values given in Table 1 correspond to the higher servo velocity that will be used in flight.

6 Pre-Launch Calibration Results

6.1 Gain

The L'TES gains were set to nominal values of 1×, and 2× using individual precision resistors. Gain 2× was set for the nominal instrument performance and expected space and target temperatures. A gain of 1× was included as a safety factor to accommodate unexpected instrument behaviors or asteroid characteristics. The actual gain values were determined using data acquired during the thermal vacuum testing at five instrument (detector) temperatures of −5.8, 3.4, 14.1, 27.5, and 51.8 °C. The method used the interferogram peak-to-peak (P-P) value, averaged for 90 spectra in each of the forward and reverse scan directions at each of the two gain states for each instrument temperature. The peak-to-peak values correspond to the ZPD position in the interferogram and have the highest signal-to-noise ratio, providing the best indicator of the true gain values. For each test the target temperature was ∼27 °C. The computed 2× gain was 1.973 ±0.00489 with no systematic temperature variation in the gain values. In normal operations all data within an observing sequence will be acquired using the high gain, so no correction for gain is required. If comparing data acquired with different gain settings, or if different gain settings are used (off-nominal case), then a gain correction of 1.973 should be applied.

6.2 Field of View

The L'TES field of view was determined in azimuth and elevation using a collimator viewing a glowbar target through 1-mrad-wide vertical or horizontal slits placed at the collimator focus (Christensen et al. 2018). The collimator was oriented parallel to the L'TES optical axis (0° azimuth, 0° elevation). A thermally stable shutter was used that was alternatively opened to view the glowbar/slit or closed to view the shutter to remove any thermal drift in the system. Twenty spectra were collected with the shutter open and 20 with the shutter closed at each azimuth or elevation step. The slit was moved at 0.5-mrad steps, starting and ending seven degrees outside of the nominal field of view. The analysis averaged the value of the peak-to-peak of the ZPD position of the interferogram to maximize the signal-to-noise ratio. The thermal drift was removed and the data were normalized to the minimum and maximum P-P values. The field of view results following the environmental testing and final pre-ship alignments are given in Fig. 8. The field of view, defined at the full-width, half-maximum (FWHM) point in the normalized data, is 7.31 mrad in azimuth and 7.32 mrad in elevation. The L'TES optical axis was measured relative to the L'TES alignment cube, and a transformation matrix for the optical axis relative to the spacecraft-mounting interface was provided to LMC for the L'TES spacecraft alignment and pointing determination.

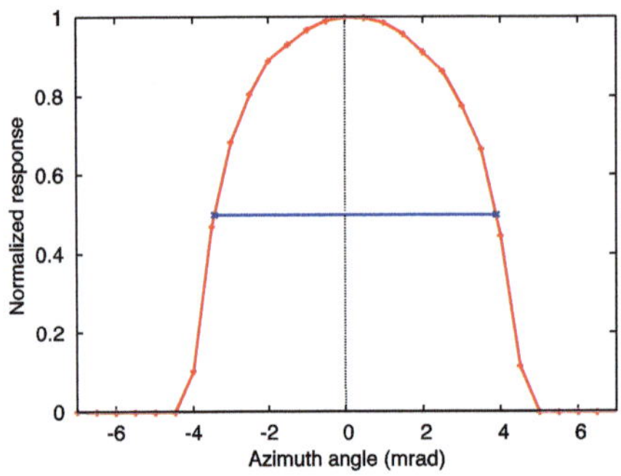

(a)

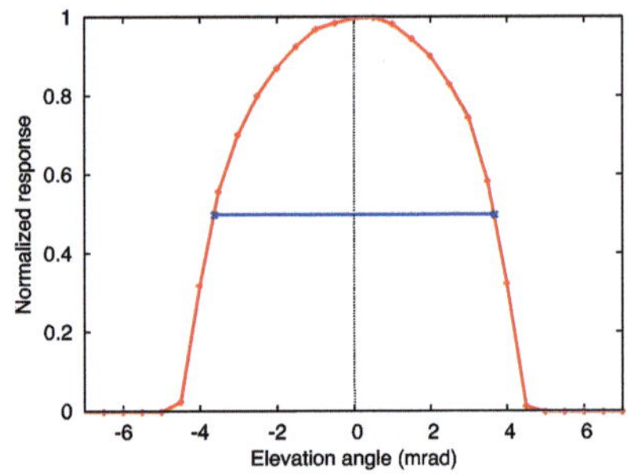

(b)

Fig. 8 The L'TES field of view measured following environmental and thermal vacuum testing, and immediately prior to delivery to the spacecraft. **(a)** Azimuth. **(b)** Elevation

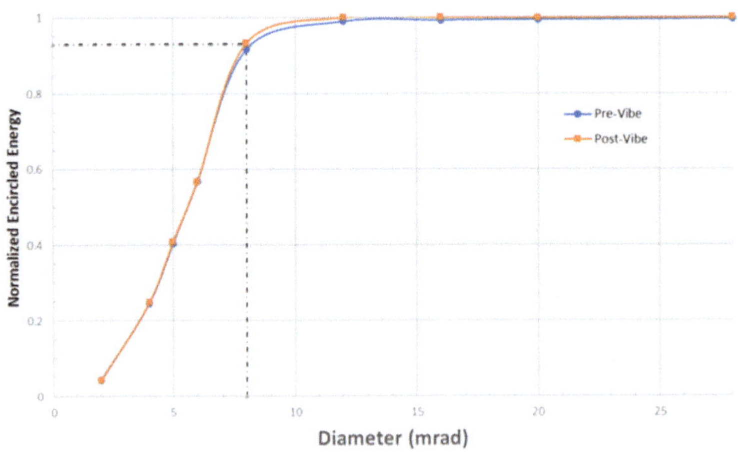

Fig. 9 The L'TES encircled energy measured pre- and post-vibration testing and immediately prior to delivery to the spacecraft. The purple data point shows the requirement of $\geq$85% of the energy in an 8-mrad field

6.3 Out-of-Field Response

The L'TES out-of-field response was determined by measuring the encircled energy performance using the collimator with a set of fixed apertures with diameters of 2, 4, 6, 8, 12, 16, 20, 28, and 36 mrad that were placed at the collimator focus. The encircled energy test collected 30 spectra with the shutter open and 30 with the shutter closed at each azimuth or elevation step. The analysis used the averaged values of the peak-to-peak of the interferogram to maximize the signal-to-noise ratio. The thermal drift was removed, and the data were normalized to the minimum and maximum P-P values. The results are shown in Fig. 9 and demonstrate that L'TES easily exceeds its requirement for 85% of the measured energy coming from one geometric scene footprint of 8 mrad.

6.4 Instrument Response Function

The L'TES Instrument Response Function (IRF) provides the transfer from photons coming into the instrument to the output signal (voltage) from the instrument as an interferogram. The IRF varies with wavenumber due to the wavelength variations of the detector response, the diamond beamsplitter and lens transmission, and all the beamsplitter and mirror finishes and coatings. The IRF was determined as a function of instrument temperature through a two-point calibration using one of the BCU calibration blackbody standards together with the internal calibration blackbody during the thermal vacuum radiometric calibration. Figure 10a gives an example of the IRF for an instrument temperature of 10 °C, at the original servo velocity and IR sampling frequency of 656 Hz. The variation of IRF with instrument temperature for a representative set of wavenumbers is given in Fig. 10b. The primary cause of variation in response function is the variation in detector performance with temperature, and to a lesser extent, changes in the interferometer alignment with temperature. Over the L'TES in-specification operational temperature range of −10 to 30 °C and the performance spectral range of 300 to 1350 cm^{-1}, the L'TES IRF varies by less than 15%.

The data in Fig. 10 were collected during pre-delivery thermal vacuum testing at ASU with the original IR sampling frequency of 656 Hz. Unfortunately, no precision radiometric data could be collected with the flight unit after the IR sampling frequency was increased to 772 Hz (see Sect. 5.1). The sampling frequency of the OTES instrument was also changed after thermal vacuum calibration (Christensen et al. 2018), and in-flight experience with OTES confirms that the IRF spectral performance is essentially identical between the original 656 Hz and final 772 Hz sampling frequencies. Therefore, we conclude that the IRF data presented in Fig. 10 is the performance that is expected for L'TES in flight. It is important to note that the IRF will be directly measured during all flight operations using observations of space and the internal calibration blackbody, and the radiometric calibration does not rely on the pre-launch IRF data.

6.5 Spectral Sample Position and Spectral Range

The L'TES spectral sample position is a function of the total Michelson mirror displacement, which is determined by the metrology laser wavelength and the number of samples that were collected in each interferogram. During ambient testing atmospheric CO_2 was observed and the L'TES spectrum was fit to a line-by-line modeled CO_2 spectrum. The results from this test found the best-fit value for the laser wavelength to be 0.851 µm. Over the full expected L'TES operational temperature, the laser wavelength only varies by $< \pm 0.17\%$, which results in the L'TES IR sample spacing varying by $< \pm 0.015$ cm^{-1}. Based on this small error the laser wavelength will be assumed to be constant over temperature.

6.6 Precision: Noise Equivalent Spectral Radiance

The L'TES precision, or noise equivalent spectral radiance (NESR), was determined by calculating a scene radiance using the two precision BCU calibration targets as outlined in Eq. (3). The NESR is defined as the standard deviation of the scene radiance and was determined during thermal vacuum testing operational instrument temperatures of −10, 0, 10, and 25 °C. In addition to the intrinsic noise in the system, any changes in the temperature of the target or the instrument during the radiometric test will result in changes in the derived scene radiance, so that the measured NESR is the upper limit on the true system performance. To minimize these additional noise sources, the L'TES instrument and the BCU

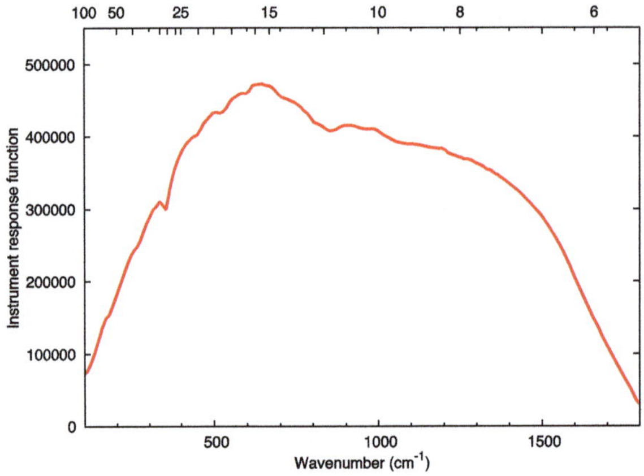

(a)

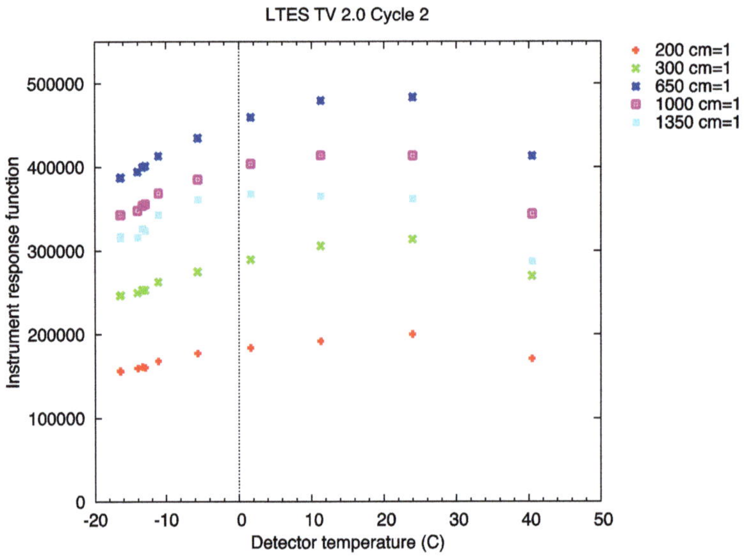

(b)

Fig. 10 (a) The L'TES Instrument Response Function (IRF in units of Volt/(W cm^{-2} /cm^{-1} sr^{-1}) for an instrument temperature of 10 °C. Upper scale is micrometers. **(b)** The variation in the IRF (in units of Volt/(W cm^{-2} /cm^{-1} sr^{-1}) with temperature at five representative wavenumbers. These data were collected at the different instrument temperatures and target temperatures during thermal vacuum radiometric testing

targets were maintained at as stable a temperature as possible and the observations of each target were limited to 200 s.

The thermal vacuum NESR results are given in Fig. 11. These data were collected during pre-delivery thermal vacuum testing at ASU with the original moving mirror velocity and an IR sampling frequency of 656 Hz.

At the original sampling frequency L'TES meets the performance requirement over the full range from 300 to 1350 cm^{-1} (Fig. 11). However, as discussed in Sect. 5.1, the moving mirror velocity and IR sampling frequency were increased to move the observed mi-

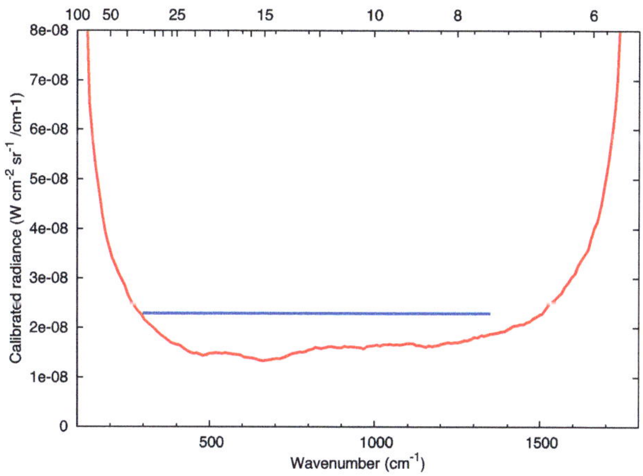

Fig. 11 The L'TES noise equivalent delta radiance (NESR in units of W cm^{-2} /cm^{-1} sr^{-1}), which represents the 1-sigma variation in calibrated radiance. These data were collected in thermal vacuum radiometric testing at the original moving mirror sampling frequency of 656 Hz

crophonic interference from the IMU out of the L'TES information band. Unfortunately, a complete and systematic set of data could not be collected at LMC to determine the L'TES NESR after the sampling velocity was increased. The L'TES instrument performance model predicts a 20% increase in the NESR at the higher servo velocity, but half of this increase will be recovered due to the larger number of spectral bands. Because the primary science objective is to determine the asteroid temperature by fitting a set of blackbody curves to the measured spectrum, the slightly higher NESR will have an insignificant effect on the primary L'TES science objective.

6.7 Absolute Radiometric Calibration

During the L'TES thermal vacuum testing the calibrated radiance was computed using Eq. (3) through observations of one of the NIST-calibrated BCU targets, whose temperature was set to 85 K to simulate space observations, together with observations of the internal cal target. The second BCU was treated as the "scene", and its temperature was varied from 100 K to 330 K to simulate the full range of expected scene temperatures that will be observed by L'TES in flight. In deriving the calibrated radiance, it was assumed the emissivities of both BCU targets and the internal cal target were exactly 1.0, and that the temperatures of the targets were accurately given by their thermistor measurements. The results are shown in Fig. 12, which gives both the L'TES-derived calibrated radiance and the computed Planck emission. As can be seen the L'TES absolute calibration is excellent.

The L'TES absolute temperature requirement is 2 K for target temperature ≥75 K (Sect. 2.2). The absolute accuracy is controlled by the knowledge of each of the parameters in Eq. (3), with the primary error sources being the calibration blackbody temperature and emissivity (Christensen et al. 2018). The absolute temperature performance was verified by using the calibrated radiance (Fig. 12) to compute a brightness temperature at each wavenumber assuming unit emissivity/ The resulting brightness temperature spectrum was filtered to reduce noise effects and the maximum of the brightness temperature was then assumed to be the kinetic temperature of the BCU target. Figure 13 shows the results for each of the eight scene temperatures acquired at each of the four instrument temperatures (Sect. 4.1) and confirms that L'TES readily meets its 2 K absolute temperature requirement.

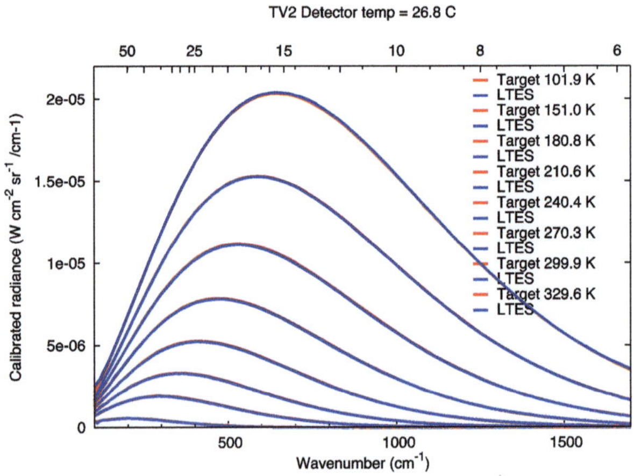

Fig. 12 The calibrated radiance computed by L'TES during thermal vacuum testing. The data were collected at an instrument (detector) temperature of 26.8 °C viewing the BCU target whose temperature was varied from ~100 to ~300 K. The blue curved are the L'TES-derived radiances; the red curves are the computed Planck emission

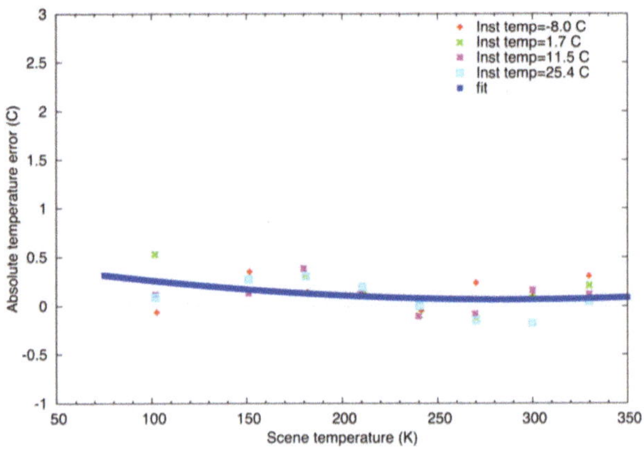

Fig. 13 The measured absolute temperature error. The calibrated radiance was computed for each target temperature at each instrument temperature and the kinetic temperature was derived. The absolute temperature error is the difference between this temperature and the measured temperature of the BCU target (see text for details)

7 L'TES Mission Operations

7.1 In-Flight Operation

L'TES is operated using spacecraft-commanded sequences that consist of power on, a warm-up and thermal stabilization period, a series of alternating observations of the external scene and the internal calibration blackbody, and power off. Upon receipt of power the L'TES moving mirror begins operating in the default 2-sec scan mode and L'TES begins collecting and outputting telemetry information. In this "stand-by" mode, no interferogram data are output. An L'TES command from the spacecraft puts the instrument into its interferogram output "science" mode. In the science mode L'TES collects groups of interferograms viewing the external scene with the calibration flag open, and groups of interferograms viewing the internal calibration blackbody with the calibration flag closed. During each asteroid encounter L'TES will begin viewing the asteroid as an unresolved point source. Periodic slews of the IPP away from the asteroid will provide L'TES with space observations, and periodic observations will be obtained of the internal cal target. The timing of these observations will be determined from in-flight experience, with the goal of acquiring observations at the appropriate intervals to adequately remove the L'TES temperature variations while minimizing movement of the IPP and interruptions of the asteroid observations. During closest

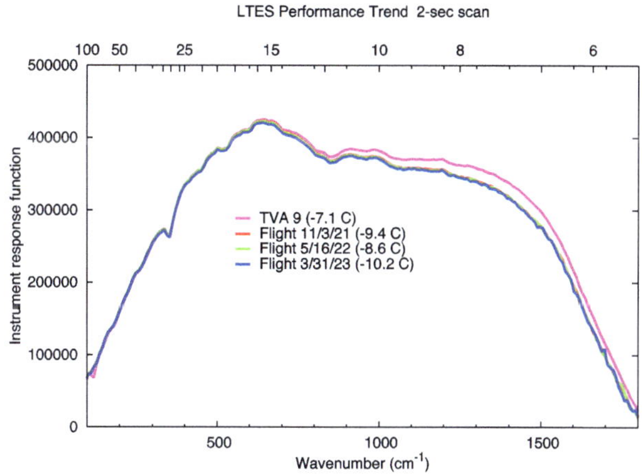

Fig. 14 The L'TES Instrument Response Function. All of the data were collected at an instrument temperature of approximately -10 °C and confirm that the L'TES performance remains constant in space and is very close to the performance measured in thermal vacuum (TVA 9)

approach the observations of the internal cal target will be minimized to maximize the asteroid science data collection. If no space views are available, then a previously determined Instrument Response Function will be used.

7.2 In-Flight Calibration and Observations

The L'TES instrument data will be calibrated in flight using Eq. (3), together with periodic views of space and the internal calibration blackbody. Typical operation will collect 10 interferograms of the internal calibration blackbody and ~10 interferograms of space when possible. All interferograms are converted to spectra through a discrete Fourier transform that is performed on the ground. Following this transform, the spectra in each group of space and cal observations will be averaged and then interpolated between successive space or calibration observations. If only a single group of space or cal observations are collected, these will be used for the entire sequence. If science spectra are collected before or after the first or last space or cal observation, these bounding space or cal observations will be extended forward or backward in time as appropriate. If only calibration blackbody or space observations are acquired in a sequence, then the pre-launch IRF, at the appropriate instrument temperature, will be used with the available calibration observation determine the scene radiance.

In-flight experience with the TES and OTES instruments demonstrated that the instrument response function varied by less than 5% over the life of those missions (Christensen et al. 2001), and a similar stability is expected for the L'TES instrument. The calibration requires knowledge of the instrument radiance (I_{detector}), which is determined from observations of space and the cal target. Because the instrument temperature will vary slightly over time, periodic observations of space and the cal target are required, and the instrument radiance used for each scene spectrum is assumed to be a linear interpolation of the values derived from each bounding set of calibration observations.

As of May 2023 L'TES had been turned on three times in space for instrument checkout and once to collect Earth-Moon observations. The three checkouts were acquired at instrument temperatures close to -10 °C and the derived IRFs are essentially identical and close to the IRF measured prior to instrument delivery at a temperature of -7 °C (Fig. 14). The slight change in IRF post-launch is due to the expected slight change in the interferometer alignment once it is operating in a zero-G environment. L'TES observations of the Earth

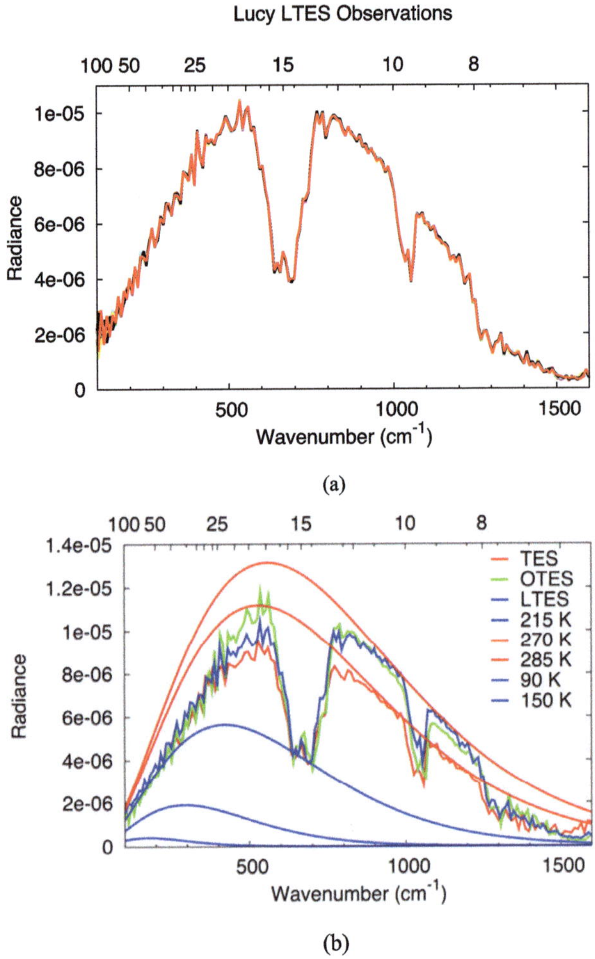

Fig. 15 L'TES observations of the Earth. a) Ten consecutive TES spectral collected at 10 cm^{-1} spectral resolution during the Lucy Earth gravity assist flyby in Oct. 2022. The spectral signatures are due to gases in the Earth's atmosphere and their repeatability between spectra confirms the excellent L'TES noise performance

were acquired during the Lucy Earth gravity assist flyby on Oct. 13, 2022. Ten consecutive spectra acquired at 10 cm^{-1} resolution are shown in Fig. 15a. The spectral features observed are due to water vapor, CO_2, and methane and the consistency of these spectra is an excellent indicator of the low spectral noise that is achieved with L'TES. Figure 15b shows a comparison of similar observations acquired in 1996 by the Mars Global Surveyor TES instrument (Christensen and Pearl 1997) and in 2017 by the OSIRIS-REx OTES instrument (Christensen et al. 2018). The radiance differences in the 8-14 μm atmospheric windows are due to the differences in ocean and land surface temperatures.

8 Data Processing and Archiving

The L'TES data processing pipeline is straightforward and based on the flight-proven OTES and methodology and software (Christensen et al. 2018). Each L'TES sequence, from power on to power off, will be processed independently, and the interferograms are returned and processed separately on the ground in the forward and reverse scan directions. Observations of space will be identified using the spacecraft pointing data. Each interferogram will be transformed to a spectrum and processed to a calibrated radiance spectrum. Bad or noisy spectra will be identified and removed prior to averaging groups of space or cal target data. The temperatures of the internal blackbody and optical elements will be determined using the redundant thermistor data, with testing done to eliminate erroneous data. These data provide the inputs to Eq. (3) that are used to compute the calibrated radiance for each scene spectrum.

The L'TES standard data product is calibrated spectral radiance. These data, along with all of data used in the calibration, will be delivered to the Planetary Data System on the schedule agreed to by the Lucy Project. For each spectrum the calibrated radiance is converted to brightness temperature at each wavenumber, assuming unity emissivity. The processing continues to compute surface temperature in a full emissivity-temperature separation process. Temperature uncertainties are computed for use in the thermal inertia error analyses. These uncertainties are computed using an initial estimate of the surface temperature from the brightness temperature to constrain the temperature-dependent wavenumber range over which to average the brightness temperature, computing a new surface temperature, and repeating until the results converge.

Acknowledgements We would like to thank John Andrews and Michael Vincent and their team at the Southwest Research Institute for their excellent support in the Lucy Payload Office, Sam Pellicori for his engineering support, Tara Fisher, Ashley Toland, and MarKetta Roemke for priceless administrative support, and Leonardo Electronics, Diamond Materials, General Dynamics, Avior, BEI, TelAztec, and L&M Machining for their support in the L'TES fabrication. And we thank Christopher Edwards for his review that improved the final version.

Funding This material is based upon work supported by the National Aeronautics and Space Administration under contract GR35682 issued through the NASA Discovery Program.

Declarations

Competing Interests The authors have no competing interests to declare that are relevant to the content of this article.

References

Bedford RE (1988) Calculation of effective emissivities of cavity sources of thermal radiation. In: DeWitt DP, Nutter GD (eds) Theory and practice of radiation thermometry. Wiley, New York, pp 653–772

Christensen PR, Pearl J (1997) Initial data from the Mars global surveyor thermal emission spectrometer experiment: observations of the Earth. J Geophys Res 102:10,875–810,880

Christensen PR, Anderson DL, Chase SC, Clark RN, Kieffer HH, Malin MC, Pearl JC, Carpenter J, Bandiera N, Brown FG (1992) Thermal emission spectrometer experiment: Mars observer mission. J Geophys Res, Planets 97(E5):7719–7734

Christensen PR et al (2001) The Mars global surveyor thermal emission spectrometer experiment: investigation description and surface science results. J Geophys Res 106:23,823–823,871

Christensen PR, Mehall GL, Silverman SH, Anwar S, Cannon G, Gorelick N, Kheen R, Tourville T, Bates D, Ferry S (2003) Miniature thermal emission spectrometer for the Mars Exploration rovers. J Geophys Res 108(E12)

Christensen PR et al (2004) The Thermal Emission Imaging System (THEMIS) for the Mars 2001 Odyssey mission. Space Sci Rev 110:85–130

Christensen PR et al (2018) The OSIRIS-REx Thermal Emission Spectrometer (OTES) instrument. Space Sci Rev 214:87

Edwards CS, Christensen PR, Mehall GL, Anwar S, Tunaiji EA, Badri K, Bowles H, Chase S, Farkas Z, Fisher T (2021) The Emirates Mars Mission (EMM) Emirates Mars InfraRed Spectrometer (EMIRS) instrument. Space Sci Rev 217:1–50

Emery JP, Fernández YR, Kelley MSP, Warden KT, Hergenrother C, Lauretta DS, Drake MJ, Campins H, Ziffere J (2014) Thermal infrared observations and thermophysical characterization of OSIRIS-REx target asteroid (101955) Bennu. Icarus 234:17–35

Kieffer HH (2013) Thermal model for analysis of Mars infrared mapping. J Geophys Res 116:451–470

Levison HF, Olkin CB, Noll KS, Marchi S, Bell JF III, Bierhaus E, Binzel R, Bottke W, Britt D, Brown M (2021) Lucy mission to the Trojan asteroids: science goals. Planet Sci J 2(5):171

Olkin CB, Levison HF, Vincent M, Noll KS, Andrews J, Gray S, Good P, Marchi S, Christensen P, Reuter D (2021) Lucy mission to the Trojan asteroids: instrumentation and encounter concept of operations. Planet Sci J 2(5):172

Prokhorov AV, Hanssen LM, Mekhontsev SN (2009) Calculation of radiation characteristics of blackbody radiation sources. In: Zhang Z, Tsai B, Machin G (eds) Radiometric temperature measurements, I. Fundamentals. Experimental Methods in the Physical Sciences, vol 42. Academic Press, New York, pp 181–240

Rozitis B, Ryan A, Emery J, Christensen P, Hamilton V, Simon A, Reuter D, Al Asad M, Ballouz R-L, Bandfield J (2020) Asteroid (101955) Bennu's weak boulders and thermally anomalous equator. Sci Adv 6(41):eabc3699

Ruff SW, Farmer JD, Calvin WM, Herkenhoff KE, Johnson JR, Morris RV, Rice MS, Arvidson RE, Bell JF III, Christensen PR (2011) Characteristics, distribution, origin, and significance of opaline silica observed by the Spirit rover in Gusev crater, Mars. J Geophys Res 116:E00F23

Space Science Reviews (2024) 220:3
https://doi.org/10.1007/s11214-023-01034-1

The First Lucy Earth Flyby (EGA1)

John R. Spencer[1] · James F. Bell[2] · Phillip R. Christensen[2] · Neil Dello Russo[3] · Hannah H. Kaplan[4] · Dennis C. Reuter[4] · Amy A. Simon[4] · Michael A. Vincent[1] · Harold A. Weaver[3]

Received: 13 July 2023 / Accepted: 29 November 2023 / Published online: 8 January 2024

Abstract
The Lucy spacecraft successfully performed the first of two Earth Gravity Assist maneuvers on October 16th 2022, flying 360 km above the Earth's surface at 11:04 UT. The flyby was essential for the Lucy mission design, but also provided a wealth of data for scientific, calibration, and public engagement purposes. The Earth and Moon provided excellent calibration targets, being large, bright, and well-characterized, though instrument saturation was sometimes an issue, as the instruments are designed for operation 5 AU from the sun. Calibration data of the Earth and/or Moon were taken with all Lucy instruments, improving knowledge of instrument alignment, stray light characteristics, and sensitivity to resolved targets. In addition, Lucy obtained scientifically valuable thermal emission spectra of the Moon, and extensive images of the DART mission impact into the Didymos system, from a unique geometry, 20 days before the Earth flyby.

Keywords Lucy mission · Spacecraft · Moon · DART mission · Flyby

1 Introduction

1.1 Purpose of the Earth Gravity Assist

On October 16th 2022, exactly 1 year after launch, the Lucy spacecraft performed the first of three Earth flybys in its nominal mission. This flyby was designated Earth Gravity Assist (EGA) 1. EGA1 was designed to raise the aphelion of Lucy's orbit from 1.16 AU to 2.26 AU. The second flyby, EGA2 on Dec. 13 2024, will further raise aphelion to 5.71 AU, sufficient to reach the Trojan asteroids that are the mission's primary science targets. Had the spacecraft missed its primary October 2021 launch opportunity, it could have been launched directly into its post-EGA1 trajectory during a backup launch window in October 2022,

✉ J.R. Spencer
john.spencer@swri.org

1 Southwest Research Institute, 1050 Walnut St., Suite 300, Boulder, CO, USA

2 School of Earth & Space Exploration, Arizona State University, Tempe, AZ, USA

3 Johns Hopkins University Applied Physics Laboratory, Laurel, MD, USA

4 NASA Goddard Space Flight Center, Greenbelt, MD, USA

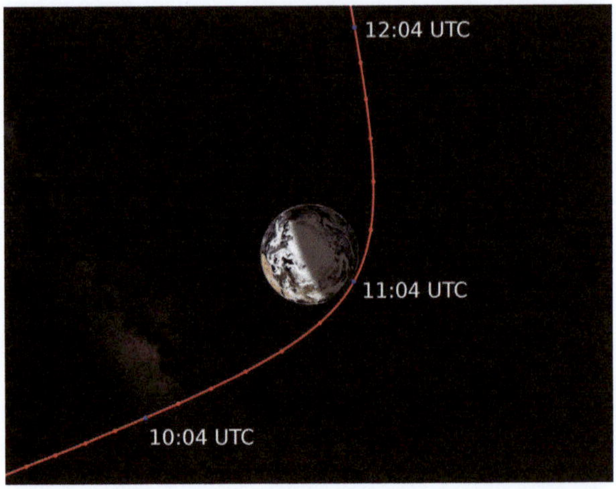

Fig. 1 Lucy Earth flyby trajectory for EGA1, looking down from the north. UTC times are for October 16th 2022. Tick marks are shown every 10 minutes along the trajectory

thus allowing the spacecraft to execute an identical science mission regardless of the launch window used. EGA1 offered an early opportunity to test and demonstrate the performance of Lucy's science instruments and systems, and a series of observations were developed to take advantage of this event.

1.2 EGA1 Geometry

The flyby geometry for EGA1 is shown in Fig. 1. The spacecraft approached the Earth from close to the direction of the sun, viewing both Earth and Moon at low phase angles. Closest approach was at 11:04:26 UT, shortly after Earth shadow ingress at 11:02 UT, at an altitude of 359.7 km off north-west Australia. The flyby was originally designed for a lower altitude of 300 km, but concern over Lucy's aerodynamic stability in the Earth's upper atmosphere, given the partially-deployed state of the one of its solar arrays, resulted in the decision to raise the flyby altitude to reduce aerodynamic forces, at the cost of a modest expenditure of fuel. Shadow egress occurred at 11:26 UT, over the Eastern Pacific. The spacecraft passed the Moon at a minimum range of 56,100 km at 5:37 UT on October 17th, though lunar observations were made much earlier, a greater range (Table 1). For more details on the EGA trajectory and execution, see Olkin et al. (2024).

1.3 Observation Priorities

EGA1's highest priority was a successful and accurate gravity assist. Remote sensing observations of the Earth and Moon were lower priority, and were limited to three short observing windows at roughly -3, -1.3, and +0.3 days from closest approach (Table 1). Within those windows, most observations were designed for instrument calibration, taking advantage of the Earth and Moon as well-characterized targets with large angular diameters. One observation, however, was designed primarily for public engagement.

The only observations of the Earth or Moon taken primarily for scientific purposes were two thermal emission spectroscopic observations of the Moon, taking advantage of the L'TES instrument's ability to obtain high SNR lunar spectra at wavelengths inaccessible from Earth. In addition, the Double Asteroid Redirection Test (DART) spacecraft's impact with Dimorphos, the moon of the near-earth asteroid Didymos, fortuitously occurred just 20

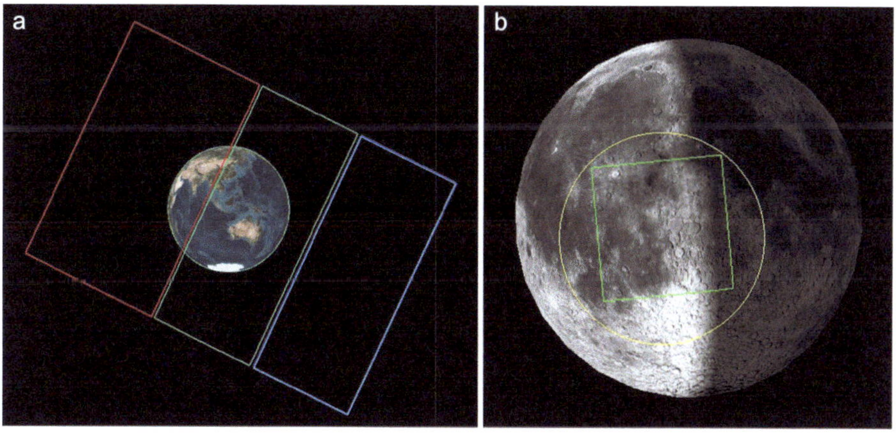

Fig. 2 Examples of observations made during EGA1. a: A L'Ralph/LEISA scan of the Earth (observation E3, Table 1, and Fig. 4). The red, green, and blue rectangles show different wavelength channels of the LEISA instrument, which were swept across the Earth using the L'Ralph instrument scan mirror. b: A L'LORRI observation of the lunar terminator (observation M4, and Fig. 3c). The green rectangle shows the L'LORRI field of view. The yellow circle is the L'TES field of view, though L'TES data were not obtained during this particular observation

days before the Earth flyby, and science observations of the DART impact were included in the EGA1 mission plan.

Observations taken by Lucy's science instruments during the EGA1 period are listed in Table 1. Observations are identified with informal designations in the first column, which are referred to in the text.

2 Calibration Observations

All of Lucy's remote sensing instruments (L'LORRI, L'Ralph (MVIC and LEISA), TTCAM, and L'TES) obtained calibration data during EGA1. Data saturation was an issue for L'LORRI, MVIC, and LEISA, which were not designed to operate at 1 AU, but observations were designed to minimize this saturation, and all instruments obtained useful calibration data.

2.1 L'LORRI

The Lucy Long-Range Reconnaissance Imager (L'LORRI), a panchromatic long focal length camera, is the highest-resolution imager on Lucy (Weaver et al. 2023a,b,c). Both the Earth and Moon were so bright on EGA approach that L'LORRI images were saturated even at the fastest possible exposure time (1 millisecond). However, the brightness of the Earth made it a good target to characterize off-axis scattered light within the instrument, important for instance for assessing L'LORRI's ability to detect faint objects, such as potential small satellites, near its primary targets. To this end, L'LORRI obtained a series of images pointed up to 3.4 degrees from the Earth's limb, to determine the brightness of scattered Earth light, and its distribution across the detector (observations E1, E6, Table 1).

The most important L'LORRI calibration activity during EGA was to use images of the lunar terminator to determine L'LORRI's effective angular resolution under flight conditions

Table 1 Summary of Lucy EGA1 Observations

Observation[1]	Activity	Instruments	Primary Purpose	Start Time Relative to EGA, days	Start Date/Time (UT)	Duration, hours	Primary target	Range, km	Solar Phase, Degrees	Details
D1	DART impact support	L'LORRI	Science	-20.00	9/26/22 11:15	36.00	Didymos	18,500,000	32	High time resolution imaging before and after impact
EM1	PR imaging of the Earth and Moon	TTCAM	Public Engagement	-3.00	10/13/22 11:08	0.05	Earth / Moon midpoint	1,440,000	1	Geometry is for the Earth
E1	L'LORRI Earth scattered light	L'LORRI	Calibration	-2.99	10/13/22 11:16	0.33	Earth	1,440,000	1	Multiple pointings near the Earth's limb.
E2	Earth spectral calibration	L'TES, TTCAM	Calibration	-2.97	10/13/22 11:44	0.41	Earth	1,420,000	1	
M1	L'LORRI/L'TES alignment, L'TES FOV mapping	L'TES, L'LORRI, TTCAM	Calibration	-2.96	10/13/22 12:12	0.75	Moon	1,720,000	9	Raster scan
E3	LEISA Earth radiometry	LEISA	Calibration	-1.30	10/15/22 3:54	0.07	Earth	643,000	2	
E4	MVIC Earth radiometry	MVIC	Calibration	-1.30	10/15/22 4:03	0.15	Earth	640,000	2	All 5 color filters
E5	MVIC scattered light	MVIC	Calibration	-1.29	10/15/22 4:14	0.47	Earth	636,000	2	Scan, 1 - 5 degrees from Earth, 7 positions. Panchromatic filter only
E6	L'LORRI Earth scattered light, TTCAM radiometry	L'LORRI, TTCAM	Calibration	-1.26	10/15/22 4:52	0.42	Earth	624,000	2	Multiple L'LORRI pointings near the Earth's limb, Earth in TTCAM field of view

Table 1 (*Continued*)

Observation[1]	Activity	Instruments	Primary Purpose	Start Time Relative to EGA, days	Start Date/Time (UT)	Duration, hours	Primary target	Range, km	Solar Phase, Degrees	Details
M2	Moon thermal IR spectroscopy (disk-integrated)	L'TES, TTCAM	Science	-1.24	10/15/22 5:29	0.62	Moon	865,000	23	Raster scan
M3	MVIC radiometry	MVIC	Calibration	0.26	10/16/22 17:15	0.15	Moon	261,000	81	All 5 color filters
M4	Moon crater imaging test, L'LORRI absolute calibration	L'LORRI	Calibration	0.27	10/16/22 17:31	0.41	Moon	256,000	82	Inertial Stares
S1	L'LORRI point spread function	L'LORRI	Calibration	0.28	10/16/22 17:55	0.02	Star Field	N/A	N/A	
M5	Moon trigger mode test	L'LORRI	Calibration	0.25	10/16/22 18:04	0.01	Moon	244,000	82	Scan
M6	TTCAM extended source calibration	TTCAM, L'LORRI	Calibration	0.30	10/16/22 18:14	0.05	Moon	241,000	82	Stare
M7	Moon thermal IR spectroscopy (resolved)	L'TES, L'LORRI, TTCAM	Science	0.30	10/16/22 18:21	0.85	Moon	232,000	83	Raster scan

[1]Observation names are informal, and their use is restricted to this article.

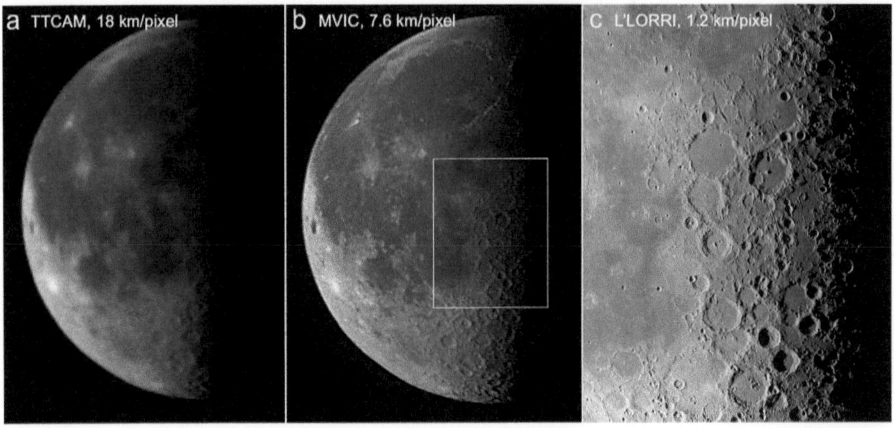

Fig. 3 Images of the lunar near side taken October 16th 2022 from similar range by Lucy's three visible-wavelength cameras, providing comparison of imaging capabilities. The TTCAM image (a) was taken at 18:14 UT from a range of 241,000 km (observation M6, Table 1). The MVIC image (b), which uses MVIC's violet, green, and near-infrared channels to create a color composite, was taken at 17:15 UT from a range of 261,000 km (observation M3, Table 1). The white box shows the location of the L'LORRI image. The L'LORRI image (c) of the lunar terminator was taken at 17:37 UT from a range of 254,000 km (observation M4, Table 1). The resolution has been improved by Lucy-Richardson deconvolution

(observation M4, Table 1). This information will be used to optimize L'LORRI imaging of its primary Trojan science targets, ensuring that spatial resolution is sufficient to meet Lucy's primary science goals. L'LORRI pointed at an inertially fixed location while the Moon's terminator drifted across the field of view, obtaining 400 images with alternating 2 msec and 5 msec exposure times. These images covered the center of the Moon's Earth-facing hemisphere at a scale of 1.3 km/pixel (Fig. 3c). Additional images covered the northern and southern parts of the terminator. Images of a nearby starfield (observation S1, Table 1) characterized the point spread function for the lunar images. Comparison to higher-resolution images from other spacecraft determined that impact craters can be identified in raw images with 90% reliability down to diameters of $\sim$18 pixels (Robbins et al. 2023a, 2023b). The same study showed that image subsampling and stacking, followed by deconvolution using the point spread function obtained from the nearby starfield images, improved the 90% crater detection limit significantly to $\sim$13 pixels, though the specific numbers vary considerably with details such as solar incidence angle and the crater measurement technique.

Because the radiometric properties of the Moon as a function of wavelength, surface location, and geometry are well known (e.g. Keifer and Stone 2005, Sato et al. 2014), the lunar data will also be valuable for refinement of the absolute radiometric calibration of L'LORRI for resolved sources.

One other calibration activity involved using the moon to test L'LORRI trigger mode (observation M5, Table 1). Trigger mode, which saves images only when a target is autonomously detected in the camera field of view, was developed for the original New Horizons LORRI instrument. However, it has never been used on science targets on New Horizons, because it proved impossible to validate trigger mode in flight on a resolved target in advance of attempting to use it on a high-priority science target. In the Lucy LORRI test, performed by scanning L'LORRI across the Moon, trigger mode performed perfectly, saving a series of images when the Moon appeared in the L'LORRI field of view. The mode can now be used with confidence if needed for future science observations.

Fig. 4 A LEISA false-color composite spectral image of the Earth at 1.95 (blue), 2.0 (green) and 3.5 (red) μm wavelengths, from observation E3 (Table 1). To enable short exposures to limit saturation, only narrow imaging swaths could be obtained, in 3 separate scans. Despite some image artifacts, many details are visible, including bright clouds in blue, the coastlines of the Arabian Peninsula and Iran on the far left in green, and Australia in yellow on the right

2.2 L'Ralph MVIC

The Multicolor Visible Imaging Camera, MVIC, is part of the L'Ralph instrument (Reuter et al. 2023), and shares optics with the LEISA instrument described below. As with L'LORRI, Ralph color images of the Earth (observations E4, Table 1) were partially or entirely saturated, as expected. However, panchromatic images taken near the Earth (observation E5, Table 1) were used, as with L'LORRI, to characterize off-axis scattered light within MVIC, verifying that scattered light levels were low enough to meet requirements at the Trojans. In addition, images of the Moon (observation M3, Table 1) were obtained in all MVIC color filters. While parts of the Moon were saturated in several filters, all contained useful unsaturated data near the terminator, and three filters were almost entirely unsaturated, yielding a three-color image (Fig. 3b). As with L'LORRI, these data are valuable for refinement of the absolute radiometric calibration of MVIC for resolved sources, by comparison with the known wavelength-dependent photometric characteristics of the Moon (e.g. Keifer and Stone 2005). As reported in Reuter et al. (2023), there was good agreement between the absolutely calibrated radiances derived from the MVIC pipeline, and published radiance data at comparable geometries.

2.3 L'Ralph LEISA

The Linear Etalon Imaging Spectral Array (LEISA) is the other component of L'Ralph, providing infrared spectral mapping from 1.0 to 3.8 μm (Reuter et al. 2023). However, LEISA is designed to operate at the Trojan asteroids, > 5 AU from the sun, and cannot be cooled to normal operating temperature at 1 AU. Useful LEISA scans of the Earth (observation E3, Table 1) were nonetheless obtained during EGA1, by pre-cooling the detector before the observations, and reading out narrow swaths of pixels to minimize integration time. Due to the warm instrument temperature, some of the data were saturated, though unsaturated data were obtained at several wavelengths and Earth locations, including areas where nearly complete spectra were observed (see Fig. 4, and Reuter et al. 2023 for an example). Scans of the Moon were not attempted due to unfavorable geometry, which resulted in instrument temperatures even further from nominal than was possible for the Earth observations.

2.4 L'TES

The Lucy Thermal Emission Spectrometer (L'TES) obtains thermal IR spectra from 6 to 75 microns, with a single detector covering a 7.3 mrad FWHM field of view (Christensen et al. 2024). A raster scan of the Moon during EGA1 provided an invaluable opportunity to

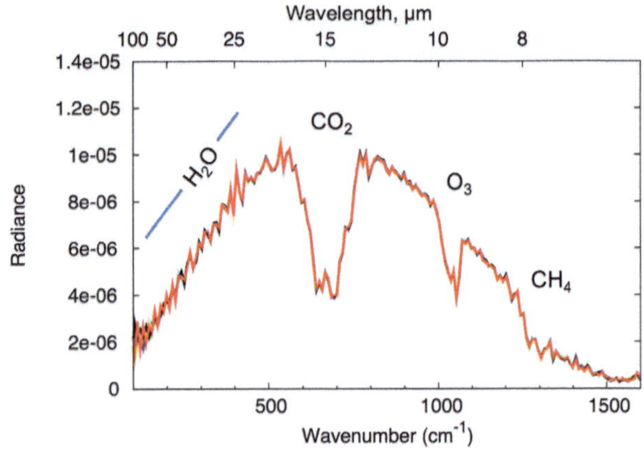

Fig. 5 Ten consecutive L'TES spectrum of the Earth, plotted in different colors, obtained during EGA1 (observation E2, Table 1), with prominent spectral features identified. The spectra overlap almost perfectly

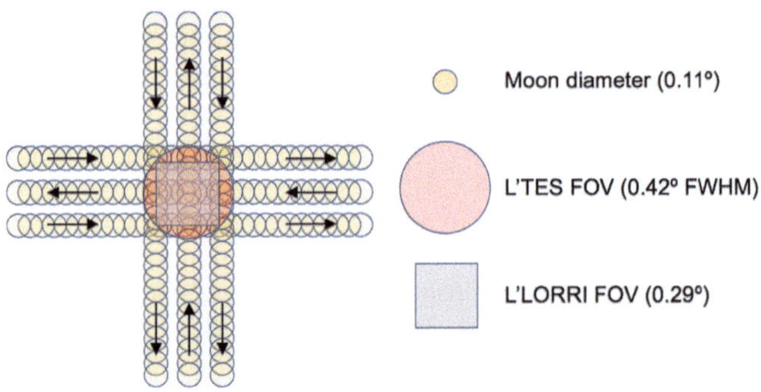

Fig. 6 Cartoon illustrating the use of the Moon to map the shape and location of the L'TES field of view (FOV) relative to the L'LORRI field of view, 2.96 days before Earth close approach (observation M1, Table 1). The observation is shown in instrument-fixed coordinates, with the moon moving three times across the field of view in each of two orthogonal directions. Imaging by L'LORRI during the scans, showing background stars and/or the Moon itself, established the location of the Moon relative to L'LORRI as a function of time, which could then be compared to the signal seen by L'TES to determine the L'LORRI/L'TES alignment. Simultaneous TTCAM images were also used, but are not shown (the TTCAM field of view is larger than the figure)

characterize the spatial response across the L'TES field of view (observation M1, Table 1, and Fig. 5), because the Moon, seen at low phase on approach, provided a very bright, relatively uniform, source that was smaller than the L'TES field of view (Fig. 6). Simultaneous L'LORRI and TTCAM imaging also provided the location of the L'TES aperture relative to the boresights of the L'LORRI and TTCAM apertures- essential information for accurately targeting L'TES science observations at the Trojan asteroids.

In addition, spectra of the Earth and Moon provided checks on the wavelength calibration and absolute sensitivity of L'TES (Observations E2, M1, M2, M7, Table 1, and Fig. 5) (Christensen et al. 2024).

2.5 TTCAM

Lucy's Terminal Tracking Camera (TTCAM) is primarily used for automated tracking of Lucy's Trojan asteroid targets during flybys, ensuring that the primary science instruments are pointed at the targets (Bell et al. 2023; Zhao et al. 2023). However, the cameras also address Level 1 Lucy science goals, in particular by providing high-cadence global imaging of the Trojan targets in the minutes surrounding closest approach. As with L'LORRI and MVIC, EGA1 imaging of the Earth and Moon with a wide range of exposure times (observations E6, M6, Table 1, and Figs. 3a, 8) provided an opportunity to calibrate instrument radiometric performance and assess image quality on extended targets. Derived radiance values for lunar images were within 15% of expected values (Zhao et al. 2023). TTCAM also provide support imaging for L'TES scans of the Earth and Moon (observations E2, M1, M2, M7, Table 1, and Fig. 6).

3 Science Observations

3.1 DART Impact

The DART spacecraft's impact with Dimorphos, moon of near-Earth asteroid Didymos, occurred at 23:14 UT on September 26th 2022, as a test of the effectiveness of spacecraft impacts in changing the trajectory of potentially hazardous asteroids (Cheng et al. 2023; Daly et al. 2023). The impact was observed directly by telescopes on or near the Earth, including the Hubble Space Telescope (Li et al. 2023), and multiple ground-based facilities (Graykowski et al. 2023). L'LORRI also obtained extensive observations of the impact from the vantage point of the Lucy spacecraft (observation D1, Table 1). A total of 1549 images were obtained, using L'LORRI's 4x4 pixel binning mode to minimize data volume. The L'LORRI observations provided both backup for ground-based observations of the impact, in case of bad weather at terrestrial observatories on the right side of the Earth to observe the impact, and a unique viewing geometry. Range (0.126 AU) was larger than for Earth observations (0.0757 AU), but solar phase angle (31.5 degrees) was significantly lower than the view from Earth (53.2 degrees). Images showed the dramatic brightening of Didymos due to the impact ejecta, and spatially resolved some detail of the impact cloud (Weaver et al. 2022, 2023b, 2023c). The lower phase angle of the Lucy observations provided unique information on the scattering properties of the ejecta cloud.

3.2 Lunar Thermal Spectroscopy

In addition to its calibration observations of the Moon, obtained when the Moon was much smaller than the L'TES field of view, L'TES observed the moon twice more for scientific purposes, first later during the approach, when it filled more of the instrument field of view (observation M2, Table 1), and second after the Earth flyby, when the moon was slightly larger than the field of view (observation M7, Table 1, and Fig. 2). These data provided higher SNR lunar observations, with some modest spatial resolution in the case of post-EGA images. The resulting spectra are scientifically valuable, because there are very few available well-calibrated thermal IR spectra of the moon, in particular at wavelengths outside the 10 and 20 μm atmospheric windows observable from the Earth's surface (Salisbury et al. 1995; Greenhagen et al. 2010).

4 Public Engagement Observations

While many of the observations described above produced scenic images that were the subject of public releases after the flyby, only one observation (EM1, Table 1) was made purely

Fig. 7 The Earth and Moon seen by TTCAM, taken at 11:08 UT on October 13th 2022, with an exposure of 0.1 msec, during observation EM1 (Table 1). Contrary to the usual practice with similar images, the Moon (far left) has not been brightened relative to the Earth, preserving the true relative albedos of the two bodies

Fig. 8 Lucy's best image of the Earth from EGA1, a 0.1 msec exposure taken with TTCAM at 04:54 UT on October 15th 2022 from a range of 624,000 km, during observation E6 (Table 1). Image scale is 46 km/pixel. The image is centered on the Indian Ocean and includes, near the left limb, the Ethiopian discovery site of the Australopithecus fossil "Lucy" that gave its name to the mission

for public engagement purposes. This was an approach view of the Earth and Moon in a single TTCAM frame (Fig. 7), effectively showing the relative sizes, separation, and albedos of the two bodies. Unlike LORRI and MVIC, TTCAM was able to obtain unsaturated images of the Earth, due to its ability to use exposure times shorter than 1 msec. Figure 8 shows TTCAM's best full-frame image of the Earth, taken during the TTCAM Earth radiometric calibration observations (E6, Table 1).

5 Summary

EGA1, Lucy's first planetary encounter, was a success from an engineering standpoint, delivering Lucy into the next phase of its orbital journey to the Trojans. But the flyby also provided valuable calibration data for all of Lucy's instruments, including radiometric, spectral, and geometric calibrations using both the Earth and the Moon. Valuable science data, and images for public engagement, were also obtained during the flyby. Because EGA1 met all the desired calibration and other goals, and because of other demands on the flight team, EGA2, in December 2024, will likely include a much smaller set of observations of the Earth/Moon system. The EGA1 data provide a small appetizer for the wealth of science results anticipated when Lucy reaches its primary Trojan targets, starting in late 2027.

Declarations

Competing Interests The authors declare they have no financial or non-financial interests.

References

Bell JF et al (2023) The Terminal Tracking Camera system on the NASA Lucy Trojan asteroid Discovery mission. Space Sci Rev 219. https://doi.org/10.1007/s11214-023-01030-5

Cheng AF et al (2023) Momentum transfer from the DART mission kinetic impact on asteroid Dimorphos. Nature 616:457–460. https://doi.org/10.1038/s41586-023-05878-z

Christensen PR et al (2024) The Lucy Thermal Emission Spectrometer (L'TES) instrument. Space Sci Rev 219. https://doi.org/10.1007/s11214-023-01029-y

Daly RT et al (2023) Successful kinetic impact into an asteroid for planetary defence. Nature 616:443–447. https://doi.org/10.1038/s41586-023-05810-5

Graykowski A et al (2023) Light curves and colours of the ejecta from Dimorphos after the DART impact. Nature 616:461–464. https://doi.org/10.1038/s41586-023-05852-9

Greenhagen BT, Lucey PG, Wyatt MB, Glotch TD, Allen CC, Arnold JA, Bandfield JL, Bowles NE, Hanna KLD, Hayne PO (2010) Global silicate mineralogy of the Moon from the Diviner Lunar Radiometer. Science 329(5998):1507–1509. https://doi.org/10.1126/science.1192196

Keifer HH, Stone TC (2005) The spectral irradiance of the moon. Astron J 129:2887–2901. https://doi.org/10.1086/430185

Li J-Y et al (2023) Ejecta from the DART-produced active asteroid Dimorphos. Nature 616:452–456. https://doi.org/10.1038/s41586-023-05811-4

Olkin C et al (2024) Mission design and concept of operations for the Lucy mission. Space Sci Rev 220

Reuter DC et al (2023) L'Ralph: A visible/infrared spectral imager for the Lucy mission to the Trojans. Space Sci Rev 219:69. https://doi.org/10.1007/s11214-023-01009-2

Robbins SJ, Bierhaus EB, Barnouin OS, Lauer TR, Spencer JR, Marchi S, Weaver HA, Levison HF, Dello Russo N (2023b) Imaging Lunar Craters with the Lucy Long Range Reconnaissance Imager (L'LORRI): a Resolution Test for NASA's Lucy Mission. Planet Sci J 4:234. https://doi.org/10.3847/PSJ/ad01b4

Robbins SJ, Bierhaus E, Spencer JR, Lauer TR, Weaver H, Marchi S, Barnouin OS, Dello Russo N, Levison HF (2023a) Testing the ability for deconvolution and Nyquist-sampling to allow detection of small impact craters: Lunar proof-of-concept with Lucy's L'LORRI camera. In: 2023 Lunar and Planetary Science Conference

Salisbury JW, Murcray DG, Williams DW, Blatherwick R (1995) Thermal infrared spectra of the Moon. Icarus 115(1):181–190. https://doi.org/10.1006/icar.1995.1087

Sato H, Robinson M, Hapke B, Denevi BW, Boyd AK (2014) Resolved Hapke parameter maps of the Moon. J Geophys Res, Planets 119:1775–1805. https://doi.org/10.1002/2013JE004580

Weaver HA et al (2022) Investigating the DART impact event with the Lucy Long Range Reconnaissance Imager. In: 2022 AGU fall meeting, Chicago, IL. https://ui.adsabs.harvard.edu/abs/2022AGUFM.P54C.01W

Weaver HA et al (2023a) The Lucy LOng Range Reconnaissance Imager (L'LORRI). Space Sci Rev 219:82. https://doi.org/10.1007/s11214-023-01028-z

Weaver HA et al (2023b) Investigation of the DART impact event from NASA's Lucy spacecraft. In: Abstract, Asteroids, Comets, Meteors 2023 meeting, Flagstaff, Arizona

Weaver HA et al (2023c) Lucy Observations of the DART Impact Event. Planet Sci J (in press)

Zhao Y, Bell JF et al (2023) Pre-flight and in-flight calibration and performance of the Terminal Tracking Cameras. (TTCams) on the NASA Lucy Mission. To be submitted to Earth and Space Science

Space Science Reviews (2025) 221:28
https://doi.org/10.1007/s11214-024-01123-9

Technical STEM Workforce Development Re-Imagined: The Lucy Student Pipeline and Competency Enabler (L'SPACE)

Sheri Klug Boonstra[1] · Katherine Kretke[2] · Dann Garcia[1] · Heather Fischer[3] · Martin Storksdieck[4]

Received: 18 June 2023 / Accepted: 1 November 2024 / Published online: 12 March 2025

Abstract
NASA's Lucy Student Pipeline and Competency Enabler (L'SPACE) is a novel student collaboration program designed and implemented for the NASA Lucy Mission. This technical STEM workforce development program has been designed to provide an authentic, team-based learning environment to support STEM majors as they prepare to enter the space exploration ecosystem. L'SPACE recruits and serves from colleges and universities all over the United States of America and is designed at every level to foster a STEM identity in the students, working to create a more diverse workforce in the future. From the program's inception in Fall of 2018 through the Fall of 2023, over 8967 unique students have been accepted and participated in the program, with 8070 having completed the L'SPACE academies. Through active mentoring and support, the completion rate for the L'SPACE Program is 90%. L'SPACE participants come from all 50 states, the District of Columbia, and Puerto Rico and represent 915 US colleges and universities, including 328 community colleges. The level of effort of the participants across all the elements of L'SPACE represents over 750,000 hours of workforce development training delivered and completed. The L'SPACE Team's staff, internal and external evaluators actively evaluate all elements of the program to make sure that the program is meeting the participants' and industry's needs. L'SPACE uses a fast track and iterative approach, being informed by the data to make needed adjustments to the program as often as needed to reflect participant-centered and/or industry changes in the rapidly evolving workforce development environment.

Keywords STEM Workforce Development · Virtual Workforce Training · STEM Job Readiness · STEM Retention · STEM Mentoring

✉ S.K. Boonstra
sklug@asu.edu

1 Mars Space Flight Facility, 201 E. Orange Mall, School of Earth and Space Exploration, Arizona State University, Tempe, USA

2 Southwest Research Institute Boulder, Boulder, USA

3 Oregon State University, Corvallis, USA

4 Seaview Downs, Seaview Downs, Australia

1 Introduction

The space exploration industry is rapidly growing and evolving, demanding a Science, Technology, Engineering and Math (STEM) workforce that is willing to innovate and push boundaries. To keep up with the demand, methodologies used to train the emerging STEM workforce must be agile enough to keep pace with these ever-changing needs. To accommodate this growing need for highly qualified workers in STEM, a new model of workforce training at scale is necessary, one which can attract talented students who might otherwise get missed. Under the current model of STEM workforce training (e.g. internship opportunities), the space exploration industry is losing out on a significant amount of qualifiable talent, as well as experiencing a narrowing of perspectives, which can only hinder the creative thinking and innovative ideas that the industry thrives on. Current and long-used STEM workforce models need to be examined not only through their success in preparing the emerging workforce for current needs, but also be vetted to address evolving critical and complicated challenges. Preparation of this workforce cannot rely simply on past ways if the STEM ecosystem is to progress and be successful in achieving NASA's exploration goals.

The current model of STEM preparation has been firmly rooted in academic preparation. Students are steeped in their discipline knowledge through the coursework of their degree program. These courses are for the most part following traditional framework foundations of knowledge that have been designated appropriate for meeting long-standing academic standards needed to confer a particular discipline degree. Academic programs are, in many cases, locked into this traditional framework of knowledge and changes in content are not easy to implement without many complicated steps within the institution. As a result, updates to higher education STEM-based curriculum to match the pace of a changing world and industry is minimal. Many established tenured professors have not had industry-related experience, and adjunct professors- of practice[1] are still relatively rare (Benjamin et al. 2006). Building a solid foundation of knowledge and theory is an important step in learning. However, with shrinking academic budgets, many courses are diminishing their hands-on and applied projects. Students rarely see the current application of their coursework knowledge embedded in their classes. The result is that students who experience academic preparation alone do not, in most cases, have the actual practice that will be needed to be prepared for jobs as they move from school to the STEM workforce.

Partnering with industry to expand the number of hands-on opportunities (such as internships) seems like a good solution but becomes more of a geographic problem, especially for rural area students. Traditionally, place-based opportunities rely on regional assets within industry to provide STEM workforce training. As the economy changes within sectors of the country, many areas are deprived of such workforce development opportunities which tends to limit the economic horizon of many students. Many of these areas' institutions, such as smaller colleges and community colleges already suffer from lack of funding which impacts basic resources such as labs and commonly have a deficiency of staff with current, relevant experience to teach evolving disciplines (e.g., space exploration topics). Industry often focuses on and recruits from higher-profile institutions for their workforce training

[1] One avenue to bring the current state of industry into the higher education instruction is to include industry experts as professors of practice (many of which have less than Ph.D. degrees). These professors of practice have usually built their credentials and expertise through non-academic careers but can bring current knowledge of industry standards, methodologies, practices, protocols, and procedures that will help students to become more job ready. The number of such professors of practice are not currently quantified, however, the numbers across higher education institutions are probably low.

and job opportunities, which leaves many geographically distributed students uncompetitive and lacking a pathway for applied STEM opportunities.

To help address STEM workforce training challenges and gaps, the Lucy Team in collaboration with Arizona State University's School of Earth and Space Exploration developed the Lucy Student Pipeline and Competency Enabler (L'SPACE), a novel NASA mission student collaboration program designed and implemented for the NASA Lucy Mission. The Lucy Mission will explore a record-breaking number of asteroids, flying by three asteroids in the solar system's main asteroid belt, and by eight Trojan asteroids that share an orbit around the Sun with Jupiter. The Lucy spacecraft launched from Cape Canaveral, FL on October 16, 2021.

The L'SPACE Program is designed to provide team-based workforce development training in a student-friendly manner *at scale*. To train the next generation of mission-savvy scientists, engineers, and other mission team members (e.g., programmers, mission planners, etc.), the L'SPACE Team implemented a direct engagement strategy that has inspired, engaged, and retained the interest of thousands of undergraduate STEM students by providing access and practice to workforce relevant skills, knowledge, and highly-skilled NASA-based academy faculty to provide instruction for the academy participants.

Prior to the design and implementation of the L'SPACE Program, an informal needs assessment was conducted by the L'SPACE Program leadership across major stakeholders within the exploration ecosystem to determine current and future STEM workforce needs. These stakeholders included NASA leadership, NASA mission teams, and industry partners. In person, conversational discussions were held with people who were working within the NASA ecosystem - mostly principal investigators, project managers, mission team members, and aerospace company managers and leadership. Focus areas of the needs assessment included stakeholders' perspective about attributes and traits most valued in the emerging workforce candidates, gaps they are seeing in essential skills, and forecasted future needs for their workforce. The results of the needs assessment showed the dominant attribute needed for their ideal candidate was the experience and competency of successfully working on interdisciplinary teams to optimize a project. Additional workforce skills of value included competencies in communication and collaboration. Basically, the most requested attributes were soft skills that are not widely taught and practiced in academia. Other major areas of stakeholder needs included the ability to recruit and retain a stable STEM workforce and the transfer of knowledge.

The L'SPACE Program has three overarching goals:

1. Provide an authentic and agile subject-matter expert (SME)-led NASA workforce development training with rigorous NASA-focused team projects at a national scale with opportunities for undergraduate STEM majors, graduate students, post-doctoral scientists and engineers, and early career professionals that will lead to a growth mindset and job readiness.
2. Provide learning environment where all STEM students can gain experience in critical industry-standard skills and practices. The desired skills and practices (taken from the conversational data from the needs assessment) include working in multi-disciplinary teams, communicating within projects where understanding is built and valued across disciplines and cultures, collaboration to optimize a project, problem-solving to be successful in difficult environments, delivering the project on time and within budget, and embedding critical thinking throughout.
3. Provide a holistic technical workforce development environment where students can become a part of a national STEM community to build their STEM identity and self-confidence to see that they *do* belong in the STEM field. This type of environment not

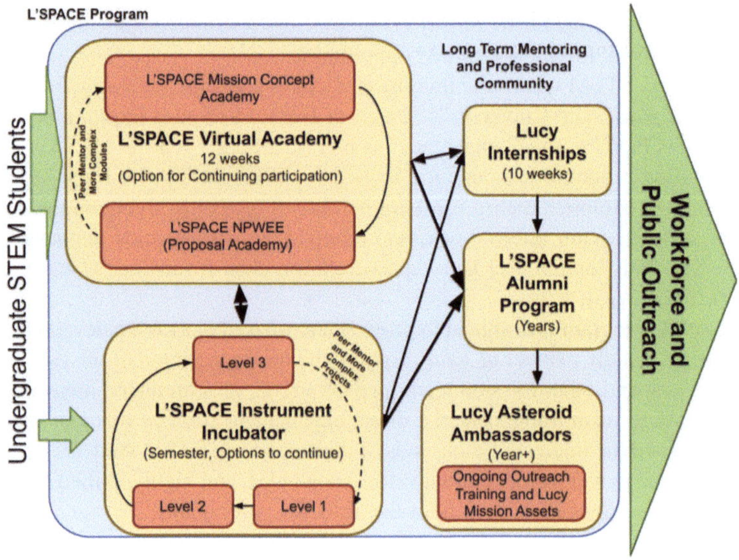

Fig. 1 A diagram indicating the elements of the L'SPACE Program and how students move through those elements. Undergraduate STEM students enter the program through either the virtual academies or the Instrument Incubator. As alumni, they can then participate in the Lucy Internships, and the Lucy Ambassador Program. All alumni of the program remain part of the L'SPACE Community as they proceed into the workforce

only encourages the necessary knowledge transfer that is important to NASA competencies, but also facilitates the building of their professional networks, including both professional SMEs and other students who will become part of their professional community.

The L'SPACE program has two core elements for different student populations: the free, on-line, team-based L'SPACE Virtual Academies available to students at all U.S. colleges and universities, and the in-person, for-credit Lucy Instrument Incubator/Space Works courses at ASU. The virtual academies, described in detail in Sect. 2, serve the largest number of students and are the main focus of this article (the other elements of the L'SPACE program provide complementary, targeted training to a smaller group of students and an interested reader may jump ahead to Sect. 6 to read more about those elements, with the Lucy Instrument Incubator/Space Works courses described in Sect. 6.1). The L'SPACE Program model is designed to be significantly agile (unlike traditional academia) and expandable to be able to add other types of relevant job readiness training. The students flow through the program and can select which program aspects are most useful for their own career development including paid internships (Sect. 6.2), and training for volunteer opportunities in science outreach (the Lucy Asteroid Ambassadors, Sect. 6.3), as well as an active alumni community (Sect. 6.4). Figure 1 shows a visual representation of these program elements, as well as possible student trajectories through the program.

In Sect. 3, we describe how our student-centered approach has influenced the L'SPACE Program design, including statistics about the population this program has served. In Sect. 4, we describe two key elements of the L'SPACE model, teamwork, and mentoring. In Sect. 5, we look more deeply at evaluation. In Sect. 6, we describe the other components of the

L'SPACE program. In Sect. 7, we describe how the L'SPACE Program has changed students' lives. In Sect. 8, we summarize the outcomes and future plans.

2 The L'SPACE Virtual Academies

The core element of the L'SPACE program that serves the largest number of students are the virtual academies. The technical training is delivered within the framework of 12-week academies which are held three times per year (spring, summer, and fall). Each academy's live plenary sessions are 90-minutes long and meet once a week, with the expectation that students will spend a total of approximately 6-10 hours per week on the course, mostly in independent and group work outside of the sessions. The meeting times for the academies are designed to not interfere with most academic schedules which allows individuals who are students to participate in the L'SPACE academies while taking classes. The L'SPACE program currently has two academies:

1) The **Mission Concept Academy** (MCA) is a NASA Science Mission Directorate (SMD) focused, team-based small spacecraft solar system mission concept formulation and design project. In MCA, students learn the basics of NASA protocols, processes, and procedures through designing a unique NASA mission to fit within the designated requirements and destination given. The teams then mature their concept into an industry- standard professional Preliminary Design Review (PDR) level technical document for NASA review. 12 different SMEs provide students with real-world applications from their experience. The MCA team's PDR documents average 100 - 150+ pages of technical writing in length. Professional feedback is provided to the teams on the aspects of their design and PDR technicalities, and teams are allowed to resubmit the updated content to professionalize their work. This updating and resubmission format allows for team knowledge growth, as the first drafts of PDR sections may lack critical elements. Discussions with their mentors to identify industry-standard shortfalls help the participants understand the rigor it takes to meet those industry standards. The participants use their final PDR deliverable as evidence of their collaborative skills and knowledge during interviews for internships and jobs, walking their interviewers through their particular areas of responsibility within the project.

   The **NASA Proposal Writing and Evaluation Experience Academy** (NPWEE), co-funded with support from the NASA Space Technology Mission Directorate (STMD), is an authentic proposal writing academy led by the NASA Marshall Space Flight Center Chief Technologist and their office to help participants learn how to write a selectable proposal. Even NASA has recognized the unique training NPWEE provides and asked for early career NASA employees and contractors to be able to join, offering NASA professional development credit for their participation in their SATERN system. STEM graduate students and postdoctoral students are also permitted to join NPWEE. The participants are then divided up into teams by the L'SPACE staff and the NASA employees, contractors, and graduate students are mixed in with the undergraduate students to form the teams.

   Participants receive training by the Chief Technologist and NASA L'SPACE Leadership Team on the current list of items needed for NASA exploration. The teams, in response to a solicitation created for use in the NPWEE Academy by NASA, respond with ideas that would be useful to NASA's exploration efforts agency wide. The team's proposal is submitted according to the NASA solicitation guidelines and coupled with an authentic NASA review panel experience in which they are trained by the Chief Technologist's staff to serve as NASA reviewers. As such, they score the quality of the academy proposals according to the review matrix provided by NASA. By seeing the proposal and review process earlier in their careers, participants can become better prepared to propose ideas of value to NASA and learn how to be an innovator. NASA STMD awards $10,000 for the highest scoring proposal as start-up funding for the teams to mature the idea. As of Fall 2023, NASA STMD had awarded funding to 22 NPWEE teams. Through NPWEE, winning teams have formed companies, filed for patents, and worked with NASA and industry to further their proposed ideas. NPWEE teams that have received the initial $10K of funding are reporting they have gone on to receive additional funding awards and support for testing from NASA centers and leveraged industry partnerships to continue their work.

Learning within the L'SPACE online community is designed to be very interactive and uses the Zoom platform. L'SPACE is designed to help open pathways for any pre-STEM and STEM students through early career who are looking for ways to gain career-enabling experiences that can provide rigorous workforce development. Our program seeks to provide support to all students to help them become competent in their fields and acquire the skills needed to be competitive and hirable into the STEM workforce. To enable all students, L'SPACE provides 8 asynchronous skill modules that help to fill gaps for students who might not have had access to training or information related to NASA missions or mission practices available within their institutions. These skill modules contain relevant, foundational skill information and students take an end of module assessment to verify the knowledge transfer.

Through Fall 2023, the L'SPACE faculty roster lists 53 subject matter experts (SMEs) which include NASA, academic, and industry practitioners whose presentations focus on important elements needed for exploration, the current and future state of exploration, and lessons-learned from past missions – thus enabling knowledge transfer. Lucy Team members have comprised 21% of the L'SPACE faculty and have contributed excellent Lucy mission-related narratives and examples of how exploration highlighting the Lucy mission with mission updates as the mission continues its journey. During this period of performance, the L'SPACE Program has delivered 33 12-week academies. Many L'SPACE faculty members continue to share their knowledge and expertise each session, as they are seeing the impact this program has in workforce preparation and job readiness and enjoy participating in the virtual format to reach a large, diverse audience with whom their experience is shared.

The L'SPACE Program's focus on working in teams is both an authentic representation of space exploration—all space exploration is a collaborative effort—and an important component of making sure that students can achieve practice and skills in the areas most valued by their future employers. To deliver technical workforce development at scale, the virtual method was the best option. Traditionally, many online courses have extremely high attrition levels (e.g., massive online courses (MOOCs) have an <10% completion rate) [Khalil and Ebner 2014]; However, the L'SPACE Program has been able to achieve an online program design for technical training that has allowed them to achieve a 90% completion rate – even through the disruptive time of COVID-19. This completion rate is quite a significant achievement since, in the L'SPACE model, the students receive no academic credit or stipend for their participation. As discussed previously, to be cost effective to reach a population of students at scale, the model was designed to be complementary to the students' academic and work schedules, as to allow them to gain valuable workforce training experience at no cost within the confines of their regular life and schedule, meaning no travel or schedule disruption was necessary. The participants' view the training they receive through the L'SPACE academies as valuable, relevant workforce development training (as our data shows from the program evaluation surveys) and is additive to their career pathways which, in many cases, is not available to them in most of their traditional academic settings.

For the period of performance considered in this paper, the L'SPACE participants hail from 915 US colleges and universities. The logistics of getting individual institutional and departmental permissions to have this training added to the student's individual degree program as an element towards their graduation would have been cost-prohibitive to the program, though the L'SPACE Program does supports students who do decide on their own to pursue the credit option with their institution/department. To date, this request is a rare occurrence. Instead, the students who fulfil all the requirements of the L'SPACE academies do receive a certificate of completion which includes badges for any of the L'SPACE skill modules they completed. This document, along with the evidence of their L'SPACE projects (e.g., their team project - preliminary design reviews (PDRs) and descriptions of their proposed innovation for the NASA NPWEE academy are often used by the students during interviews for jobs or internships and deemed valuable by both the students and the interviewers.

L'SPACE intentionally builds workforce development skills aligned to NASA's SMD's mission focus and the evolving needs of the STEM workforce. Participants learn, through working on their unique team-based projects, foundational practices, protocols, and procedures used on NASA projects. L'SPACE emulates the real world by having their participants work in multi- disciplinary teams to produce industry-standard mission designs to a Preliminary Design Review level. While the L'SPACE participants are not physically building hardware as part of this experience, the level of technicality that the students are gaining and practicing has translated to valuable skills based on industry's level of interest and hiring of L'SPACE participants.

One of the indicators as to the value that L'SPACE is providing to the emerging workforce is that L'SPACE has become a known and respected program within NASA and industry for talented prospective hires. These employers have seen the value-added knowledge in skills and process that L'SPACE students represent and are seeking more of the same types of candidates.

The L'SPACE Program is designed to supplement undergraduate STEM students' academic training by providing a more equitable and inclusive model for workforce development and delivering it at scale. The L'SPACE model changes the approach to STEM workforce development and moves it from the realm of higher education institutions by empowering individual students to choose to participate in the workforce training. The L'SPACE Program offers any student who attends a US college and university the opportunity to learn industry-level skills that are highly aligned to NASA practices, protocols, and procedures through unique team- based projects.

Unlike academic degree programs, the L'SPACE Program is agile and serially iterative informed by program evaluation and industry input. The L'SPACE Program is designed to pivot or enhance the curriculum as industry changes to stay relevant to the evolving STEM workforce needs. Examples of this type of change include helping students gain a "growth mindset" in terms of their future and current work. With the advent of artificial intelligence and machine learning and other changes happening within the current workforce, the influence of these tools is dramatically shifting the landscape of what the workplace and work in general will look like. The emerging workforce will need to expect changes to be taking place within the span of their career and embrace the upskilling that will be necessary to stay in step with industry's changing needs, which embodies the "growth mindset". The L'SPACE Program can help inform and manage the expectations of the emerging workforce with the reality of the skills and job readiness needed for the STEM workforce.

All L'SPACE activities are strategically aligned with NASA processes, protocols, and procedures. In doing so, L'SPACE participants are better prepared for the technical environment. NASA professionals who have worked with or hired L'SPACE students have remarked that they need little or no onboarding. In some cases, NASA managers have observed that the L'SPACE students are so prepared, they are ready to immediately put on a technical project upon hiring. Other companies within the NASA exploration ecosystem have discovered how value-added the L'SPACE students are and are requesting access to resumes or for L'SPACE to post their job openings within the L'SPACE alumni network.

3 L'SPACE Program through a Student-Centered Lens

The L'SPACE Academy model has worked to uncover and address many challenges and barriers to entering STEM career pathways. Through our team's research and evaluator interviews, the L'SPACE Program has documented many examples of challenges encountered by students pursuing STEM (and higher education in general). These examples include: lack of workforce development opportunities for students who attend a college or university part-time, students who are caregivers (to children or have adult care responsibilities) as many of these students cannot travel away from home to take internships to gain the needed and relevant experience that would enhance their chances of getting hired, students experiencing homelessness and, as a result, lack the necessary traditional avenues for communication such as having a permanent address or phone and find themselves at a distinct disadvantage, students with a lower than 3.0 GPA (the minimum GPA for NASA and most industry internships, students who must depend on a recommendation from someone in authority at their institution (e.g. professors or deans) who may not see them as *their* ideal candidate or someone they would want to represent *their institution and /or class* often are not given recommendations to move their application forward, students with food insecurities, and students who are experiencing PTSD or domestic violence are also more at risk of disappearing from the STEM pool.

From our experience, economics is also an opportunity limiter. Many students find that to stay in school and achieve their STEM undergraduate degrees, they must work to be able to support themselves and, in many cases, their families as well. The traditional NASA and industry internship model offers paid internships which seem on the surface to be helpful. However, interns must use the internship funding to cover their housing, food, and sometimes even the transportation to get to and from their internship. All these expenses are in addition to their regular bills which continue as well. With cost of living rising faster than the internship stipends, most internships end up being break-even prospects or even worse, cost the participants out-of-pocket to be able to gain the experience. The economics of the traditional internship models inhibits many students from being able to afford participating in workforce development experiences, thus putting them at a disadvantage when competing for STEM jobs. Once the barriers and challenges were identified, the L'SPACE Program looked for innovations in program design that would mitigate barriers in workforce preparation opportunities and open the pathway in ways that would allow for wider participation and access.

In addressing the economics barrier, the L'SPACE Academies are both offered as free programs. With the academies being a virtual opportunity, no travel is needed, providing cost savings and the convenience of being able to not interrupt care giving and other commitments. The Zoom platform can accept up to 500 students per academy and by being a browser-based platform, the academy can accommodate students with minimal technology capabilities, such as joining using a phone and a hotspot. Access to the academy is not regulated by the college or university. L'SPACE has democratized the capability of receiving workforce development training for all by not requiring letters of recommendations. The students themselves can opt into this program and it is through their commitment and dedication, not through having to ask institutional gatekeepers' permission, that they are able to raise their skill levels and gain more job readiness. As with the ASU Lucy internships, there are no GPA requirements and students can be attending a college or university for any percentage of time.

The welcoming culture of L'SPACE has been one of the most important aspects of the program. From the onset, a consistent and strong message to all participants has been "You belong here!" and "We will see you to the finish line!" Many students, particularly those who have been economically disadvantaged, are tentative to feel they can survive or qualify for STEM career fields. The lack the lack of relatable STEM role models, the lack of career advisors at their institutions, the complicated lives that accompany students with fewer family resources are formidable obstacles for at-risk students to complete a STEM (or any) degree program. As a countermeasure to these barriers, L'SPACE intentionally offers and provides avenues of support with ongoing opportunities for students to ask any questions and seek career and workforce development guidance. We welcome and encourage their participation in the L'SPACE community – or L'Fam as the students call themselves as a long-term engagement. The term L'Fam emerged at the very first L'SPACE Academy session in the Fall of 2018. Students coined this term to refer to the community that L'SPACE created, playing off the Lucy convention for naming the spacecraft's instruments. The term has stuck as it resonated with the students and became the indicator that the community that L'SPACE was building was the result of the many intentional aspects of the program that were designed to promote the feeling of belonging and sense of family – albeit their STEM family. Through these family ties, students can have access to L'SPACE staff for letters of recommendation for internships and graduate school, help with job interview readiness, and other opportunities long after their completion of the academies.

The evolving curation of the social and emotional support of the L'SPACE students has emerged as one of the most impactful elements of program success. L'SPACE has offered empowerment and upward mobility to students who might not otherwise easily find paths forward. The fact that the L'SPACE Program is able to attract and retain an average of over 2000 students per year, with a program completion rate of 90%, makes L'SPACE one of NASA's most cost-effective workforce training programs. The success of the program has been ever more notable when realizing that those percentages have been maintained consistently across the entire period of performance through national issues such as the historic COVID-19 pandemic and associated economic challenges. The L'SPACE model is helping NASA to reach workforce-training goals that have been elusive in the past.

3.1 The Students in L'Fam

The L'SPACE Program is designed to serve a multidisciplinary student body. During the period of performance described in this paper (Fall 2018 through Fall 2023), 8967 unique students have participated in L'SPACE and completed the programs, including students from 915 US colleges and universities, 328 community colleges. Students in the program attend colleges and universities in all 50 states, the District of Columbia, and the United States' Territories of Puerto Rico and Guam. Using the student surveys, we have also identified that 44% of the students are eligible for Pell Grants and 45% need to work to pay for college. Additionally, 18% of students are first generation college students. The program has leveraged a

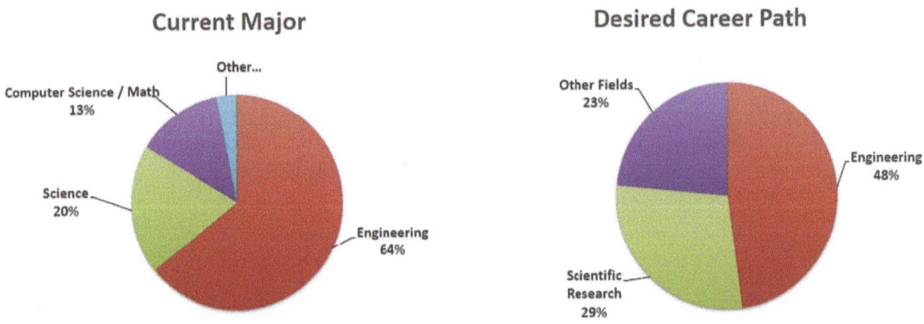

Fig. 2 The distribution of the current majors of the L'SPACE students (left), and their desired career path (right). Roughly two-thirds of the L'SPACE students are engineering majors, one-fifth are science majors, and the remaining are computer science, math, or other majors. Roughly half of the students intend to enter into an engineering career while the remaining students are split between interest in pursuing scientific research or another field

strong connection to the NASA Community College Aerospace Scholars (NCAS) program. Additionally, the team has worked to boost peer to peer recruitment as well as using NASA distribution networks to let people know of this workforce development opportunity. All these numbers roughly align with population averages, indicating we are reaching a wide and representative swath of the undergraduate student population.

Additionally, the evaluation confirmed that the program is reaching its target audience. The student survey found that most of the students have decided to participate in the L'SPACE Program to benefit their future career (91%) and because they are interested in NASA Science (93%). Students are admitted to the program in any stage of their undergraduate careers, so students range from Freshmen through Seniors. The students in the program have a mixture of majors (Fig. 2). The students are fairly evenly split between students who want to attend graduate school or stay in academia, students who want to work in the government, and students who are interested in working in the private sector. Roughly half of the students hope to pursue careers in engineering, while the remainder of the students are fairly evenly split between desiring to pursue scientific research and other STEM fields.

Once the participants were accepted into the program, L'SPACE model was able to maintain interest and grow the program throughout the period of performance discussed in this article – Fall of 2018 through Fall of 2023, which included the COVID-19 pandemic. The program growth showed that participants' interest in attaining technical readiness preparation did not wane and the emerging workforce could still be trained and skilled up in important ways, even during times of adversity that would help to meet STEM workforce needs, and especially NASA's STEM workforce needs.

4 Teamwork and Mentoring

Teamwork is a foundational skill of value and was identified by employers as the number one desirable trait for ideal candidates in our initial L'SPACE needs assessment. As such, teamwork is at the core of our program design and is a common thread throughout all the L'SPACE activities. Student success in the L'SPACE Program is aided by the intentional formulation of the teams. The teams are individually created and balanced to be a mixture of grade-levels, backgrounds, and experience with all the elements balanced for project success. The L'SPACE Program promotes the concept that each person has an important role to

Fig. 3 The students were asked to list three words that are important for good teamwork. Communication, trust, collaboration, cooperation were some of the most common words chosen to describe the most important element of teamwork

contribute to the project and working together, teams can achieve their goals and objectives. This approach helps the students grow in their identities as scientists and engineers, as well as helps retention in the program as it shows that with the team, challenges can be overcome.

Foundational understanding of healthy team relationships, contributing to the state of the profession, and the honoring of differences within the team is an ongoing theme throughout the L'SPACE Academies. A session on teaming is scheduled within the first two weeks of each Mission Concept Academy to give participants a good roadmap as to how to work productively in teams. At the final session of each 12-week academy, we ask the participants to list three words using a word cloud app that best describes what they know to be good teamwork, based on their academy experiences. Figure 3 shows an example of what attributes are being called out.

To improve teamwork and make sure that all students have a positive team experience, we piloted a peer mentoring program for MCA teams in Fall 2020. The peer mentors were chosen from applicants who had successfully completed the L'SPACE Mission Concept Academy, had adequate discipline knowledge, demonstrated interpersonal skills, technical writing skills, project management and teaming skills, and had a strong desire to help others succeed. This program was well received by both the student participants and the mentors, so it has been a part of the MCA program since that time.

Peer mentors are trained in areas such as conflict resolution, servant leadership, providing constructive technical feedback, and understanding the International Traffic in Arms Regulation (ITAR). This training is focused on helping them gain leadership experience and practice in helping teams address mission and team challenges, develop, and strengthen their transfer of knowledge skills, and use best practices for enhancing their communication and collaboration skill sets. Each mentor supported on average 3 teams, spending a substantial amount of time, around 50 hours, with each team. The mentors are a demographically diverse population, 49% women and 39% underrepresented students of color. Since the implementation of the mentorship program, we have had fewer students note issues in team dynamics as the mentors can help make sure each team is a welcoming and inclusive environment and can address any issues early. The student surveys indicated that these efforts have successfully helped the students gain confidence working in a team (17%

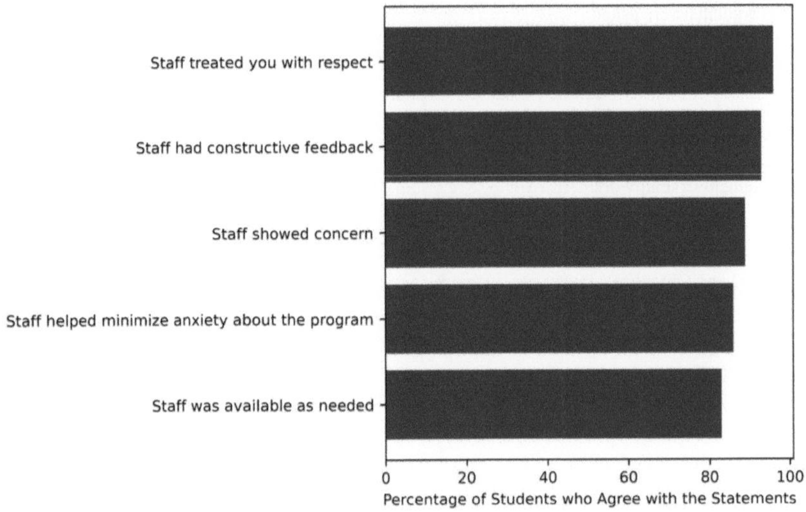

Fig. 4 Survey results about how the L'SPACE staff/leadership impacted the student experience. The 80-90% of the students found that the staff created a positive experience for the students by being available, giving constructive and supportive feedback, and treating all students with respect and care

fewer students felt insecure when working on a team in the post-course survey compared to the pre-survey).

In addition to the peer mentors, mentoring by the L'SPACE staff have also been central to the program's success. The responsiveness and flexibility of the staff and leadership have allowed the program to thrive during the COVID-19 pandemic. The staff and program leadership have also supported the cultivation of the L'Fam, which in the surveys is referred to by many students. In the alumni survey, we asked how the staff/leadership of the program was impactful. The results showed that an overwhelming majority of students found the staff helpful and responsive to their needs (Fig. 4).

5 The Role of Evaluation and Iteration to Improve the L'SPACE Program

The L'SPACE Program works with external evaluators to conduct student surveys and provide feedback to the L'SPACE team. This evaluation aims to help the L'SPACE Program optimize strategies and activities that contribute to the project's success and provide initial evidence for program quality and success. The evaluation focuses on assessing how the program works toward achieving its goals. It was designed to focus on these student outcomes and provide useful feedback to the L'SPACE Team as they continue to improve the program.

Our evaluation questions focus on assessing whether, how, and to what degree the L'SPACE Program reached its goals and focused on whether the program was beneficial to participating students. The overall evaluation goal is to *Understand how the L'SPACE Program supports the acquisition and transfer of skills to the workforce for all students.*

To achieve this evaluation goal, we addressed the following questions and objectives:

1. To what degree does the program emphasize working in teams, communication, problem-solving, and critical thinking?
2. In what ways and to what degree do students who participate in the program benefit?

3. To what degree did the students feel the program was valuable overall, and what elements of the program contributed most to the student's perceived value?
4. What challenges does the program face in reaching its goals?
5. How did the program respond to early feedback, and did changes lead to improved perceptions of value by students?

The assessment of the program was informed by the experiential learning program literature, specifically on programs that seek to facilitate skill development and transfer into the workplace. Experiential learning is an applied learning approach in which students engage in professional, productive learning activities, thereby constructing knowledge and practicing skills (CAS 2009; Cantor 1997). Experiential learning programs include (but are not limited to) internships, externships, practicums, student teaching, cooperative education programs, research experiences for undergraduate students (REU), and work-integrated learning programs. Universities, or facilities at universities, often offer these programs. These programs can also be associated with a national lab, a government agency, a corporation (as part of workplace learning), or a nonprofit organization.

Many of these programs aim to facilitate skill transfer, which is the ability of students to apply cognitive gains from one learning situation to another (Dori and Sasson 2013; National Academies of Sciences and Medicine 2018). Specifically, these programs aim to facilitate student skills transfer gained in an experiential learning setting to the workforce, referred to as graduate transfer (Leberman and McDonald 2016). There are examples of programs that facilitate long-term impact on the skill transfer of their participants. For instance, Bradberry and de Maio (2019) found that students in an experiential program reported being able to transfer the skills they learned to their current jobs. Mayorga (2019) found that students in an experiential learning program focused on business incubation could speak to their experience with specific essential skills by referring back to the program, which made them more competitive in the job market. Hart (2019) reviewed the education literature to assess how different types of project-based learning programs facilitate skill transfer and found that interdisciplinary programs were significantly associated with perceived gains in interdisciplinary effectiveness with communication and teamwork skills.

The transfer of skill is not entirely dependent on the program in which the student acquires the skill. It is also impacted by the learner's characteristics and the characteristics of the new context within which they apply the skills. Jackson's 2016 model of skill transfer postulates that various learner, program, and workplace characteristics impact skill transfer. The learner's personality, motivation, and related experience are key characteristics that influence the transfer (Chiaburu and Marinova 2005; Jackson 2016). The learner's workplace is also a factor. The culture of the workplace, peer support, supervisory support, and capacity or opportunity for transfers all impact degrees of transfer (Chiaburu and Marinova 2005; Jackson 2016).

The evaluation process employed Jackson's 2016 model of graduate transfers to inform how to characterize the L'SPACE Program and the program's characteristics and how it facilitates skills acquisition and transfer. L'SPACE employs many of Jackson's program characteristics that encourage skill transfer but also fosters the development of learning characteristics and incorporates workplace characteristics.

Specifically, the L'SPACE Program was designed to have the following characteristics:

1. The similarity between learning and application contexts
2. Learner appreciation for skill relevance
3. Clearly defined learning goals and objectives for target skills
4. Collaborative engaged learning

5. Reflecting on target skills and strategies for development

Additionally, the program fosters learner characteristics, including motivation to learn and transfer abilities. The program also incorporates workplace characteristics of the potential employers of the students (NASA, aerospace companies, etc.), including access to role models and coaching, a collaborative environment with peer support, group norms, and a culture that is open to innovation and change.

To test the degree that the program successfully achieved these design goals, the evaluation team implemented a pre-post-survey design in the early parts of the program (and continued with a retrospective post-survey design in later phases), reviewed peer evaluation assessments, administered a survey of program alums, and conducted select course observations. The evaluation team also routinely held meetings with the L'SPACE Program Team to discuss preliminary evaluation findings and optimize the program.

The pre-post and L'SPACE alumni survey provided the majority of evaluation data. In 2018-2020 the evaluators implemented pre-post surveys. Pre-surveys were sent to all students in the program's first week, and post-surveys were sent in the last week. These pre-post surveys were used to compare students' confidence in their STEM skills and career path plans from the beginning of the program to the end. Starting in 2020, the evaluators employed retrospective post-surveys to address potential survey fatigue. In Fall of 2021, the evaluators also sent an alumni survey to all former students.

6 Other L'SPACE Elements

While the virtual academies are the largest component of the L'SPACE program, serving the greatest number of students, the other components of the L'SPACE program provide an opportunity for students to remain engaged in the program and community and receive more specialized trainings. These elements, which were shown in Fig. 1, are described in the subsequent subsections.

6.1 Lucy Instrument Incubator/Space Works Courses

The School of Earth and Space Exploration at Arizona State University's Space Works courses preceded and served as effectively a pilot for the L'SPACE program, as the first of these courses served as the inspiration for the MCA virtual academy. The courses were incorporated into the L'SPACE program as the Lucy Instrument Incubator program and, by 2023, have been expanded into set of 4 courses within the ASU. These courses form a second core to the L'SPACE program and alternative entry point available to students at Arizona State University.

These courses bring together students from science and engineering to *apply* their coursework using team and project-based coursework. The Instrument Incubator students have access to a large selection of laboratory equipment where they can not only design, but also prototype and test their ideas in a hands-on environment. Through the Lucy Instrument Incubator/Space Works Program, students gain traction in the important areas of working on multidisciplinary teams to optimize a project. These students move through foundational learning that is common to companies that are part of the exploration ecosystem (e.g., schedules, design, NASA processes, protocols, and procedures) to more industry-standard complex skills (e.g., design, manufacturing, working with electronics, coding, and environmental testing). These skills are deeply aligned to the job readiness practices that will help the students to be more effective in their discipline and within interdisciplinary teams and become "added value" to companies who are looking for new hires. The ASU Fulton College of Engineering also recognized the value of this applied class and granted their engineering students cross-listed credit for these courses towards their ABET-based engineering degrees. The Lucy Instrument Incubator/Space Works students also have access to one-on-one mentoring with the instructional team.

While the Lucy Instrument Incubator/Space Works program supports a smaller number of students than the Virtual academies (319 unique students who take multiple levels of Space Works from Fall 2018 to the Fall of 2023), the L'SPACE team has been able to take lessons learned from this smaller, more accessible group of Space Works students to improve the student experience in the virtual L'SPACE Academies. The L'SPACE team also found that ideas emerged from the virtual academies which were applied to the Space Works courses. Another unique aspect of these two programs is that there have also been times where The Lucy Instrument Incubator/Space Works students and L'SPACE students have worked together on projects or field trips and many of the Lucy Instrument Incubator students at ASU also participate in the virtual L'SPACE Academies as the programs have different focus areas that build their industry-standard skill sets.

Unlike the Lucy Instrument Incubator/Space Works courses, the L'SPACE program does not offer college credit. At the beginning of the formulation of L'SPACE, this option was considered. However, to offer an at-scale program that crosses so many higher educational institutions, the mechanics of trying to align the L'SPACE program with all the different colleges and universities' degree requirements proved to be unattainable. Instead, at the end of L'SPACE, the participants have evidence of their learned skills and can use their Preliminary Design Review at internship and job interviews to highlight their industry-standard skill sets.

6.2 Lucy Internships

The L'SPACE Program offers a Lucy Mission summer internship program. There are two types of internships available to students. The first is a more traditional, 10-week internship where students are supervised by mentors on the Lucy Team over the summer. The applications for these internships are only available to participants who have completed at least one L'SPACE Academy. A searchable compilation of the intern applications is then provided to the internship coordinator at the Lucy Co-Investigator (Co-I) institutions where they choose who they would like to extend the intern offer to. The selected interns are then mentored by individuals at the Co-I institutions: Southwest Research Institute, NASA Goddard Space Flight Center, Lockheed Martin, KinetX, and Johns Hopkins Applied Physics Laboratory. Most internships are on-site. However, when circumstances dictate, some (or all) of these internships can be offered as virtual internships, as was the case during the COVID-19 pandemic. This traditional type of internship is modeled after the NASA internship model where interns participate in their host organization's projects, usually associated with the Lucy mission. They participate in activities such as team meetings that help build up networks with professionals and their time spent within the projects helps them to visualize their place more closely within the STEM workforce. The interns get to see the application of their discipline coursework in the real world rather than only seeing it theoretically as a school project. Internships also can expand their working knowledge of process, protocols, and procedures (how things really work) within the world of exploration. As an additional outcome, we have seen students from these traditional internships being offered follow-on, company sponsored internships and permanent positions as a result of their initial Lucy internship.

A second type of Lucy internship was created in response to the need of finding a more equitable pathway for underrepresented students to participate in experiential workforce development. The traditional model for internships (10 or 15 weeks, stipend based) assumes that 1) students can relocate and spend 10 to 15-weeks living and working at another location; and 2) the stipend will be adequate to cover their expenses. Our experience has shown that when all the factors that affect many underrepresented students such as complex family situations, attending school part- time because of economic challenges, and having to care for family members (as mentioned in Chap. 2) are considered, underrepresented students find fewer pathways to take part in workforce development opportunities.

When internships are looked at through a purely economic lens, the stipends which are given to cover housing and food costs during the internship period fail to account for students who are already paying rent/leases for housing where they live. Even though the student is relocating and must rent another space to live, they are, in the majority of cases, paying two rents. In addition, their monthly bills also need to be paid. For many students, the summer months (the most common times students apply for internships) are also very often dedicated to working full-time to earn money to stay in school during the fall and spring. These students often face a choice of upskilling to try to become more competitive and hirable or being able to stay in school. Even under the best of circumstances, given the rising cost of housing and food and the lack of stipends being increased to keep pace, traditional internships are becoming a break-even at best or even worse, an out-of-pocket, unaffordable experience. The L'SPACE Team researched the costs that all the Lucy Interns (the traditional and ASU interns) would encounter, factored in inflation, and determined a stipend that would adequately meet the costs for their internship. Air travel to and from the internship sites and costs for two bags up to 50 lbs. and the cost of a taxi or rideshare from and to the airport to the also was also covered. For the interns that decided to drive to their internship site instead of using the air transportation, they were reimbursed up to the amount of the ASU travel agent quoted roundtrip coach fare.

The L'SPACE Program has introduced an alternative type of internship opportunity at one of the Co-I sites, Arizona State University (ASU). This internship is aimed at providing skill-building professional development for students L'SPACE students unable to participate in more traditional internships. This new Lucy internship format is shorter (5-weeks instead of 10-weeks). The length of time for this internship was determined by interviews with students who had the desired characteristics and were already in the L'SPACE Academies. We determined that the 5-week timeframe was achievable for them to be able to leave their jobs or family situations and still be able to return without major ramifications.

The ASU Lucy internship is designed to be an accelerated, STEM workforce development experience that intentionally focuses more on exposure and practice in foundational skill acquisition such as working and communicating on a multi-disciplinary team, optimizing a project through the lifecycle of an instrument, and meeting milestones and deadlines. In selecting candidates for this internship, the selection team looks for students who are mostly from community colleges, or 4-year colleges that don't have exploration-focused courses or maker space facilities available to students. We target freshman and sophomore college students to help motivate them to continue their schooling to achieve their degree. The earlier level students also are not the prime selection candidates within most of the traditional internship pools as well. However, we have found that the ASU Lucy intern students, once upskilled, are seen as more selectable candidates because of their ASU experience in future internship application rounds.

The ASU interns spend 5-weeks at ASU's 3000 sq ft Instrument Incubator/Space Works maker laboratory getting an intensive hands-on, project-based training similar to what is offered during the semester Instrument Incubator/Space Works courses. The interns work in

Fig. 5 ASU Lucy interns learn to work in teams to conceive, design, build, and test their projects. Courtesy of S.K. Boonstra, Arizona State University

teams, are taught and practice proper cleanroom and electrostatic discharge (ESD) protocols, learn and apply an industry-level NX CAD in their project design, use additive manufacturing rapid prototyping to validate their design and then fabricate their final design in nylon carbon fiber. Once these steps are completed, they assemble the electronics, add the necessary code, and test their projects under vacuum conditions to capture their required science. Figure 5 shows some of the students working on their projects.

The ASU Lucy interns also experience an immersive science and engineering field experience - a planetary comparison tour as part of their ASU Lucy internship. This field experience leverages the planetary analog assets that are present in the northern Arizona region and includes Meteor Crater, the Grand Canyon, Sunset Volcanics, Lowell Observatory, and numerous field sites on the Colorado Plateau. This valuable experience allows the science and engineering students to gain insight into how "ground-truthing" of mission concepts are done through analog sites and testing.

For the 2018–2021 period of performance (with the Lucy internships starting in the summer of 2019), a total of 35 L'SPACE students participated in traditional Lucy Co-Investigator internships while 85 students participated in the ASU Lucy internship 5 week program for a total of 118 Lucy Interns for the period of performance. Students in both types of internships rated their experiences positively.

6.3 Lucy Mission Ambassadors

The Lucy Asteroid Ambassador Program is an opportunity for L'SPACE alumni to volunteer to teach their communities about the Lucy Mission and the L'SPACE Program. All graduates of the L'SPACE Virtual Academy are eligible and invited to apply to this program, which is modeled after NASA's highly regarded Solar System Ambassador Program.

The students selected as ambassadors receive specific training on the Lucy Mission and on best practices in science public outreach. The ambassadors are provided a continually updated, power-point presentation about the mission as well as resources to help them with all aspects of science communication, from identifying and reaching out to possible audiences, to gaining confidence as a science communicator, and becoming more effective at sharing the Lucy story and their part of it. The ambassadors also receive Lucy mission pins and patches and a Lucy name badge to identify them as Lucy Ambassadors, as well as a small assortment of mission stickers or other promotional items.

From the first cohort of students in the summer 2019 through December 2023, the Lucy Asteroid Ambassador Program trained 254 volunteer ambassadors who went on to carry out

Fig. 6 A graphic made by ambassador David Dezell Turner who used popular culture references to help make Lucy accessible to a broader audience. He used this graphic in social media posts putting the Lucy Mission in context with other NASA asteroid missions. Courtesy of D.D. Turner

outreach to their communities. These students directly interacted with an estimated 6500 people in their communities as well as thousands of more people online through social media activities. We leverage the fact that the L'SPACE alumni come from diverse backgrounds all over the United States to speak to audiences that might not normally have an opportunity to learn about NASA missions. Students are most likely to present to middle and high school classrooms in their areas, as well as student organizations at their universities. Ambassadors are also commonly present at local community centers.

While most ambassadors carried out their outreach by presenting the supplied Lucy Mission Presentation, some ambassadors chose to use their varied talents to create new content to reach out to their communities in creative ways. Some students have made short videos or graphics that were shared online, such as the social media post shown in Fig. 6. One student wrote an article in Turkish about the mission for their ambassador social media post. The ambassadors have found fun ways to speak about the mission to a younger, more diverse, generation.

While the ambassadors are a relatively small subset of students compared to the large L'SPACE Program, they are able to extend the impact of L'SPACE to an even wider audience. As compelling, talented students, they serve as an inspiration to their communities and help expand the public's view of what it means to be part of the NASA community.

6.4 L'SPACE Alumni Group

After completing a L'SPACE virtual academy or an Instrument Incubator/Space Works course, we continue to stay connected to the students though the alumni program. At this point, this is most active through a L'SPACE group on the social media platform LinkedIn. Students are encouraged to join this group to have access to announcements of opportunities, and 30% of the students who complete the academy join the LinkedIn Alumni group. By using the overall LinkedIn platform, we are also able to longitudinally track the students after they have completed the program and see how their careers develop. The group is active to maintain interest, both serving as a location for students to share their own success stories, as well as a place for the broader space exploration community to advertise opportunities. Approximately 85% of our students follow L'SPACE on LinkedIn.

7 L'SPACE - Transforming Students' Lives

One of the most important elements that emerged from the L'SPACE Program was how powerful the aspect of being part of a STEM community and having people who believe in your capability of being able to reach for and achieve your STEM career goals led to the long-term commitment of the participants to staying within the L'Fam community and on a pathway to achieving the goals. So many of the L'SPACE student participants had no clear path forward to gain STEM workforce experience. L'SPACE became the pathway they had been seeking in terms of feeling welcomed into STEM and being supported to reach their goals. The students who came to L'SPACE were fully capable of learning, but because of limitations such as coming from schools that were not endowed with state-of-the-art laboratories and research faculty or attending an institution that had a lack of hands-on project funding to involve students in work. Students who were experiencing challenges such as having to support their families (or sometimes extended families), students who were experiencing homelessness, food insecurity, or domestic violence were common themes from our data from our evaluators who conducted interviews.

The L'SPACE community is one of support, mentoring, and helping students go beyond their circumstances. The most powerful driver behind the L'SPACE Program is not teaching the students the technical skills that will be required in their careers (although by providing training in programs such as geospatial information systems (GIS) and computer aided design (CAD) the program does allow the students to add specific, marketable technical skills to their resumes). The most powerful driver is helping the students see themselves as scientists and engineers capable of contributing to the space and exploration economy. Therefore, it is key to promote a student focused environment that is safe to really achieve success for *all* students. Additionally, the team makes sure that the students hear from a diverse population of speakers. Allowing the students to hear from experts of different ages, gender, ethnicities, races, career fields and stages makes it easier for the students to see themselves as someday being one of those experts.

As L'SPACE alumni graduate and move into the workforce, more data will be available to determine the trajectory of the L'SPACE participants. In our last L'SPACE alumni survey conducted at the end of 2021, of the alumni who had graduated with a 4-year degree and are working full-time, 74% are on the career path they want to be on. And while they are satisfied with their current position, they are also looking to advance in the next few years. Preliminary data on L'SPACE alumni who have gone on to have full time jobs indicates that students valued their participation in the L'SPACE Program and believe that it has helped them in their careers. In Fig. 7, we show the survey results for how these students rated the L'SPACE Program in helping them further their career. As a near universal outcome, the students felt that the program helped empower them to believe that they can have a career in STEM.

Students elaborated on these beliefs in the open-ended survey questions. One L'SPACE student wrote, "No other program has changed my life to such a degree. Everyone I have met has been just truly wonderful and a blessing to be involved with. I really cannot stress how much this program has saved me in light of the pandemic. I would be very lost without my L'SPACE experiences and nowhere near the trajectory I am currently on. It prepared me for the workforce, it introduced me to other wonderful peers, it networked me with career professionals, and most importantly it gave me confidence that I can actually be involved and make a difference in the effort of space exploration. So yes, if my classmate is interested in any of the above (or would even care to listen!) I would happily recommend L'SPACE to them."

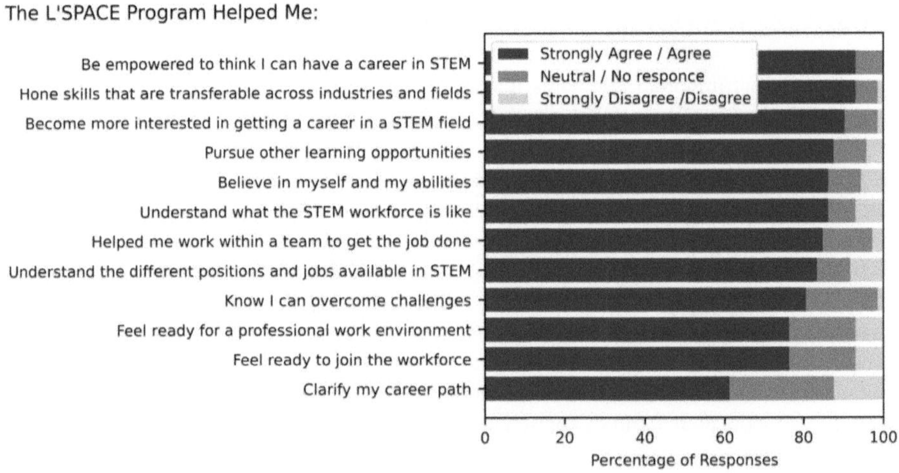

Fig. 7 Survey results from L'SPACE alumni who are now working full time. The majority of students agree or strongly agree that the L'SPACE Program has helped reinforce their STEM identity and aided them in the career pathways ($n = 73$, Represents alumni who are working full time. Total response to the alumni survey was 437, which includes students who are still in their undergraduate program, pursuing graduate school, or are looking for work)

Outside of the formal evaluation, L'SPACE participants would often send emails outlining their gratitude and progress. The L'SPACE Team was so moved by these submissions the team started to post L'SPACE testimonials to the L'SPACE website (https://lspace.asu.edu). Here is an example of how L'SPACE is transformational:

Kori Eliaz, shown in Fig. 8, is one of the students who was able to experience the L'SPACE Program's transformational capabilities. Kori had started the L'SPACE Program in 2019. At 19, Kori Eliaz dropped out of community college in California to help care for a loved one who suffered a stroke. She became a round-the-clock caregiver and found herself homeless and sleeping on a friend's floor for six months. Because of the unique way L'SPACE was designed, she was able to still participate in a NASA workforce development project despite her circumstances. Kori Eliaz participated in both L'SPACE Academies and excelled in all her projects. She was subsequently chosen as a Lucy Intern in 2019 by Lucy Co-Investigator Lockheed Martin. She was able to move to Colorado and find more stable circumstances as a result. In 2021, Kori was honored as an Astronaut Scholar, one of 68 awards in the US. She transferred schools to continue working with Lockheed Martin, being hired as a full-time systems engineer in 2022.

8 Outcomes and Future Plans

The L'SPACE program is a novel model for workforce development, equipping the next generation of students with the skills they need to succeed in their careers in an evidence-based manner. In the five years of the program, a population of 8967 students from across the country including U.S. territories participated in the program. Altogether, L'SPACE produced over 750,000 hours of workforce development training delivered by subject matter experts and completed by the participants.

Surveys of the L'SPACE alumni have focused on gathering evidence around a key claim of the program, namely that by design, it allows students to transfer course skills into the

"During my time as a Lucy intern, I was constantly exposed to industry standards and best practices and asked to uphold those standards. Working on a deep space project at Lockheed Martin means following the routines and processes that have labeled it as the most successful aerospace corporation in the world. I was put on the spot and expected to follow the rigorous testing and evaluation processes that veterans of the corporation used. During those ten weeks, I was an engineer on Lucy with a goal and a purpose and a timeline. Being a Lucy intern propelled my career, not just because of the amazing resume boost it gave me, but because of the confidence it gave me and myself, the confidence it gave me to say I'm writing code that will protect the spacecraft. I know exactly how it works. I know how to test it, and I know how to take these ideas out of my head and make them into a reality."

Fig. 8 Kori Eliaz, L'SPACE Program participant, NPWEE awarded project winner, and L'SPACE intern. Courtesy of K. Eliaz.

workplace or the next educational experience. Our evidence suggests that L'SPACE was highly successful in that regard: 88% of alumni respondents agreed that the experience they had working on a project team from L'SPACE was very helpful or even essential for their current position. Additional plans to harvest the data within the L'SPACE LinkedIn community and future surveys to look longitudinally at how the L'SPACE participants have been hired. These findings will be published as part of our future work.

With the success of the peer mentors, plans are underway to add an application-based, competitively selected L'SPACE Leadership Internship which will offer stipends to the peer mentors in the spring of 2024. This new internship will replace the existing volunteer peer mentor program, have the same requirements of proven success in the L'SPACE Mission Concept Academy. As an internship, this type of workforce training will add more legitimacy to their training for their resume, as the L'SPACE Leadership Internship will be an enhanced way for students to show evidence-based growth and practice in important leadership skills as they mentor their L'SPACE teams. This experience will offer deep practice in STEM workforce leadership skills (especially in mentoring practices) that industry needs and values but currently finds this type of training difficult to implement within their existing workforce. L'SPACE will be helping to fill an industry gap with this specialized internship.

In conclusion, the L'SPACE Program will continue to prepare the emerging STEM workforce using an evolving academy model that strives to stay in step with the changing exploration ecosystem's needs. The authenticity and excitement of being partnered with of an active NASA mission offers the rare, shared experience of having access to first-person accounts of how exploration works from the Lucy team members. In addition, moving through the actual timeline of the mission allows participants to share the operational side of a mission: mission operations, science capture, science return, and science data analysis, giving participants the opportunity to reflect and learn from the broader aspects of the mission lifecycle. Lucy mission highlights and mission discoveries have been and will continue to be integrated into the academies in real time to showcase the excitement and challenges of space exploration.

The L'SPACE Program has received funding to continue for at least five years into the 12-year Phase E – the operational phase of the Lucy Mission. Program evaluation will remain to be a focus for the L'SPACE Program which will allow for continued evidence-based improvements to the design and application aspects of the program, and to assess how and to what degree the project has achieved its goals of workforce preparedness, mainly in areas of knowledge acquisition and transfer of essential skills to the workforce. The program will continually actively evaluate itself, as the research and evolution of this project is helping to define best practices in this type of at-scale technical training. With 8967 L'SPACE alumni, the L'SPACE leadership team looks forward to maintaining L'Fam and continuing to follow and support these students as they continue in their careers.

Funding This work was supported by NASA through the Lucy Mission PI Institution - Southwest Research Institute as the Student Collaboration of the Lucy Mission [Contract# NNM16AA08C and Subcontract# L9905IRG] and NASA NPWEE funding for the student team innovation awards [Grant Agreement 80NSSC19M0186].

Declarations

Competing Interests The authors have no competing interests to declare that are relevant to the content of this article.

References

Benjamin E, Hollinger DA, Knight J (2006) Professors of practice. American Association of University Professors. https://www.aaup.org/report/professors-practice

Bradberry LA, de Maio J (2019) Learning by doing: the long-term impact of experiential learning programs on student success. J Polit Sci Educ 15(1):94–111. https://doi.org/10.1080/15512169.2018.1485571

Cantor, JA (1997) Experiential learning in higher education: linking classroom and community. ASHE-ERIC Higher Education Report No 7, George Washington University, Washington DC. https://eric.ed.gov/?id=ED404949

CAS (2009) Civic engangement and service-learning programs standards. Council for the Advancement of Standards in Higher Education. Fort Lauderdale, FL

Chiaburu DS, Marinova SV (2005) What predicts skill transfer? An exploratory study of goal orientation, training self-efficacy and organizational supports. Int J Train Dev 9(2):110–123. https://doi.org/10.1111/J.1468-2419.2005.00225.X

Dori YJ, Sasson I (2013) A three-attribute transfer skills framework–part I: establishing the model and its relation to chemical education. Chem Educ Res Pract 14(4):363–375

Hart JL (2019) Interdisciplinary project-based learning as a means of developing employability skills in undergraduate science degree programs. J Teach Learn Grad Empl 10(2):50–66. https://doi.org/10.21153/jtlge2019vol10no2art827

Jackson D (2016) Modeling graduate skill transfer from university to the workplace. J Educ Work 29(2):199–231

Khalil H, Ebner M (2014) MOOCs completion rates and possible methods to improve retention - a literature review. In: Proceedings of world conference on educational multimedia, hypermedia and telecommunications 2014, pp 1236–1244. AACE, Chesapeake

Leberman S, McDonald L (2016) The transfer of learning: Participants' perspectives of adult education and training. Routledge, London. https://doi.org/10.4324/9781315552118

Mayorga LK (2019) HEIs and workforce development: helping undergraduates acquire career-readiness attributes. Ind High Educ 33(6):370–380. https://doi.org/10.1177/0950422219875083

National Academies of Sciences and Medicine (2018) How people learn II: learners, contexts, and cultures. The National Academies Press, Washington DC. https://doi.org/10.17226/24783

Publisher's Note Springer Nature remains neutral with regard to jurisdictional claims in published maps and institutional affiliations.

MIX
Papier aus verantwortungsvollen Quellen
Paper from responsible sources
FSC® C105338

If you have any concerns about our products,
you can contact us on
ProductSafety@springernature.com

In case Publisher is established outside the EU,
the EU authorized representative is:
Springer Nature Customer Service Center GmbH
Europaplatz 3, 69115 Heidelberg, Germany

Printed by Libri Plureos GmbH
in Hamburg, Germany